T0076247

Oxidative Stress in Applied Basic Research and Clinical Practice

Editor-in-Chief
Donald Armstrong

More information about this series at http://www.springer.com/series/8145

Note from the Editor-in-Chief

All books in this series illustrate point-of-care testing and critically evaluate the potential of antioxidant supplementation in various medical disorders associated with oxidative stress. Future volumes will be updated as warranted by emerging new technology, or from studies reporting clinical trials.

Donald Armstrong
Editor-in-Chief

Mark A. Babizhayev • David Wan-Cheng Li
Anne Kasus-Jacobi • Lepša Žorić • Jorge L. Alió
Editors

Studies on the Cornea and Lens

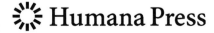 Humana Press

Editors
Mark A. Babizhayev
Innovative Vision Products, Inc.
New Castle, DE, USA

Innovative Vision Products, Inc.
Moscow, Russia

Anne Kasus-Jacobi
Department of Pharmaceutical Sciences
and Oklahoma Center for Neuroscience
University of Oklahoma
Health Sciences Center
Oklahoma City, OK, USA

Jorge L. Alió
Ophthalmology
Miguel Hernandez University
Alicante, Spain

David Wan-Cheng Li
Department of Ophthalmology
and Visual Sciences
Truhlsen Eye Institute, College of Medicine
University of Nebraska Medical Center
Omaha, NE, USA

Key Laboratory of Protein Chemistry and
Developmental Biology of Education
Ministry of China
College of Life Sciences
Hunan Normal University
Changsha, Hunan, China

Lepša Žorić
Faculty of Medicine
Department of Ophthalmology
University of Pristina
Kosovska Mitrovica, Serbia

ISSN 2197-7224   ISSN 2197-7232 (electronic)
ISBN 978-1-4939-1934-5   ISBN 978-1-4939-1935-2 (eBook)
DOI 10.1007/978-1-4939-1935-2
Springer New York Heidelberg Dordrecht London

Library of Congress Control Number: 2014955055

Contents

Part III Clinical Practice

Contributors

Kumarasamy Anbarasu Department of Marine Biotechnology, Bharathidasan University, Tiruchirappalli, Tamil Nadu, India

J.M. Asara Division of Signal Transduction and Mass Spectrometry Core, Beth Israel Deaconess Medical Center and the Department of Medicine, Harvard Medical School, Boston, MA, USA

Felipe Ávila Laboratory of Cellular Biology of Aging, UR4-IFR83, Pierre-and-Marie-Curie University, Paris, France

Interdisciplinary Research Program of Excellence in Healthy Aging (PIEI-ES), Universidad de Talca, Talca, Chile

Vibudhuta Awasthi Department of Pharmaceutical Sciences, University of Oklahoma Health Sciences Center, Oklahoma City, OK, USA

Mark A. Babizhayev Innovative Vision Products Inc., New Castle, DE, USA

Innovative Vision Products Inc., Moscow, Russia

Richard Bone Department of Physics, Florida International University, Miami, FL, USA

Daniel J.J. Carr Departments of Ophthalmology and Microbiology and Immunology, University of Oklahoma Health Sciences Center, Oklahoma City, OK, USA

Ana J. Chucair-Elliott Department of Ophthalmology, University of Oklahoma Health Sciences Center, Oklahoma City, OK, USA

Alex Cohen Department of Ophthalmology, University of Oklahoma Health Sciences Center, Oklahoma City, OK, USA

Alex W. Cohen Dean McGee Eye Institute, University of Oklahoma, Oklahoma City, OK, USA

Iñigo Corcóstegui Crespo Surgical Clinical Institute of Ophthalmology (ICQO), Bilbao, Spain

Michael H. Elliott Department of Ophthalmology, University of Oklahoma Health Sciences Center, Oklahoma City, OK, USA

Nuran Ercal Department of Chemistry, Missouri University of Science and Technology, Rolla, MO, USA

Clare L. Fraser Save Sight Institute, University of Sydney, Sydney, NSW, Australia

Sydney Eye Hospital, Sydney, NSW, Australia

Bertrand Friguet Laboratory of Cellular Biology of Aging, UR4-IFR83, Pierre-and-Marie-Curie University, Paris, France

Adaptation Biologique et Vieillissement, Institute of Biology Paris-Seine, INSERM U1164, UPMC University, Paris, France

Avinash Gurbaxani Save Sight Institute, University of Sydney, Sydney, NSW, Australia

Sydney Eye Hospital, Sydney, NSW, Australia

Rijo Hayashi Department of Ophthalmology, Koshigaya Hospital, Dokkyo Medical University, Koshigaya, Saitama, Japan

Wen-Feng Hu Department of Ophthalmology and Visual Sciences, Truhlsen Eye Institute, College of Medicine, University of Nebraska Medical Center, Omaha, NE, USA

Key Laboratory of Protein Chemistry and Developmental Biology of Education Ministry of China, College of Life Sciences, Hunan Normal University, Changsha, Hunan, China

Xiao-Hui Hu Department of Ophthalmology and Visual Sciences, Truhlsen Eye Institute, College of Medicine, University of Nebraska Medical Center, Omaha, NE, USA

Key Laboratory of Protein Chemistry and Developmental Biology of Education Ministry of China, College of Life Sciences, Hunan Normal University, Changsha, Hunan, China

Zhao-Xia Huang Department of Ophthalmology and Visual Sciences, Truhlsen Eye Institute, College of Medicine, University of Nebraska Medical Center, Omaha, NE, USA

Key Laboratory of Protein Chemistry and Developmental Biology of Education Ministry of China, College of Life Sciences, Hunan Normal University, Changsha, Hunan, China

Weike Ji Department of Ophthalmology and Visual Sciences, Truhlsen Eye Institute, College of Medicine, University of Nebraska Medical Center, Omaha, NE, USA

Rama Kannan Departments of Ophthalmology and Biochemistry, University of Missouri–Columbia School of Medicine, Columbia, MO, USA

Humeyra Karacal Department of Ophthalmology and Visual Sciences, Washington University, St. Louis, MO, USA

D. Karamichos Department of Ophthalmology, University of Oklahoma Health Sciences Center, Oklahoma City, OK, USA

Anne Kasus-Jacobi Department of Pharmaceutical Sciences and Oklahoma Center for Neuroscience, University of Oklahoma Health Sciences Center, Oklahoma City, OK, USA

Eunbi Kim Department of Ophthalmology, Hallym University College of Medicine, Seoul, Korea

Bojana Kisic Institute of Biochemistry, University of Pristina, Kosovska Mitrovica, Serbia

James K. Kubilus Department of Integrated Physiology and Pathobiology, Tufts University Medical School, Boston, MA, USA

Rasiah Pratheepa Kumari Department of Marine Biotechnology, Bharathidasan University, Tiruchirappalli, Tamil Nadu, India

Marlyn P. Langford Department of Ophthalmology, Louisiana State University Health Sciences Center, Shreveport, LA, USA

David Wan-Cheng Li State Key Laboratory of Ophthalmology, Zhongshan Ophthalmic Center, Sun Yat-Sen University, Guangzhou, Guangdong, China

Department of Ophthalmology and Visual Sciences, Truhlsen Eye Institute, College of Medicine, University of Nebraska Medical Center, Omaha, NE, USA

Key Laboratory of Protein Chemistry and Developmental Biology of Education Ministry of China, College of Life Sciences, Hunan Normal University, Changsha, Hunan, China

Thomas F. Linsenmayer Department of Integrated Physiology and Pathobiology, Tufts University Medical School, Boston, MA, USA

Fang-Yuan Liu State Key Laboratory of Ophthalmology, Zhongshan Ophthalmic Center, Sun Yat-Sen University, Guangzhou, Guangdong, China

Yi-Zhi Liu State Key Laboratory of Ophthalmology, Zhongshan Ophthalmic Center, Sun Yat-Sen University, Guangzhou, Guangdong, China

Zhong-Wen Luo State Key Laboratory of Ophthalmology, Zhongshan Ophthalmic Center, Sun Yat-Sen University, Guangzhou, Guangdong, China

Peter McCluskey Save Sight Institute, University of Sydney, Sydney, NSW, Australia

Sydney Eye Hospital, Sydney, NSW, Australia

Dijana Miric Institute of Biochemistry, University of Pristina, Kosovska Mitrovica, Serbia

Okhil K. Nag Department of Pharmaceutical Sciences, University of Oklahoma Health Sciences Center, Oklahoma City, OK, USA

Quan Dong Nguyen Department of Ophthalmology and Visual Sciences, Truhlsen Eye Institute, College of Medicine, University of Nebraska Medical Center, Omaha, NE, USA

H. Anne Pereira Departments of Pharmaceutical Sciences, Pathology, and Cell Biology and Oklahoma Center for Neuroscience, University of Oklahoma Health Sciences Center, Oklahoma City, OK, USA

Con Petsoglou Save Sight Institute, University of Sydney, Sydney, NSW, Australia

Sydney Eye Hospital, Sydney, NSW, Australia

Srinivasagan Ramkumar Department of Marine Biotechnology, Bharathidasan University, Tiruchirappalli, Tamil Nadu, India

Thomas B. Redens Department of Ophthalmology, Louisiana State University Health Sciences Center, Shreveport, LA, USA

Mahsa Rezaei Chicago Medical School, Rosalind Franklin University of Medicine and Science, Chicago, IL, USA

Stuart Richer Ocular Preventive Medicine, Captain James A. Lovell Federal Health Care Center, Chicago, IL, USA

Anna M. Roszkowska Ophthalmology Unit, Cornea and Refractive Surgery Section, University Hospital of Messina, Messina, Italy

Puttur Santhoshkumar Department of Ophthalmology, University of Missouri–Columbia School of Medicine, Columbia, MO, USA

Bill Sardi Longevinex® Associates, Las Vegas, NV, USA

Mehul Shah Drashti Netralaya, Dahod, Gujarat, India

Shreya Shah Drashti Netralaya, Dahod, Gujarat, India

K. Krishna Sharma Departments of Ophthalmology and Biochemistry, University of Missouri–Columbia School of Medicine, Columbia, MO, USA

Department of Ophthalmology, University of Missouri School of Medicine, Columbia, MO, USA

Young Joo Shin Department of Ophthalmology, Hallym University College of Medicine, Seoul, Korea

Eduardo Silva Faculty of Chemistry, Department of Physical Chemistry, Pontifical Catholic University of Chile, Santiago, Chile

Simon E. Skalicky Save Sight Institute, University of Sydney, Sydney, NSW, Australia

Sydney Eye Hospital, Sydney, NSW, Australia

William Stiles Ophthalmology, Captain James A. Lovell Federal Health Care Center, Chicago, IL, USA

Christopher Talbot Department of Integrated Physiology and Pathobiology, Tufts University Medical School, Boston, MA, USA

Xiang-Cheng Tang State Key Laboratory of Ophthalmology, Zhongshan Ophthalmic Center, Sun Yat-Sen University, Guangzhou, Guangdong, China

Department of Ophthalmology and Visual Sciences, Truhlsen Eye Institute, College of Medicine, University of Nebraska Medical Center, Omaha, NE, USA

Donald E. Texada Department of Ophthalmology, Louisiana State University Health Sciences Center, Shreveport, LA, USA

Bency Thankappan Department of Marine Biotechnology, Bharathidasan University, Tiruchirappalli, Tamil Nadu, India

Shakila Tobwala Department of Chemistry, Missouri University of Science and Technology, Rolla, MO, USA

Elío Díez-Feijóo Varela Surgical Clinical Institute of Ophthalmology (ICQO), Bilbao, Spain

Theresa Vo Optometry, Captain James A. Lovell Federal Health Care Center, Chicago, IL, USA

Ling Wang Department of Ophthalmology and Visual Sciences, Truhlsen Eye Institute, College of Medicine, University of Nebraska Medical Center, Omaha, NE, USA

Key Laboratory of Protein Chemistry and Developmental Biology of Education Ministry of China, College of Life Sciences, Hunan Normal University, Changsha, Hunan, China

Zhengfeng Wang Department of Ophthalmology and Visual Sciences, Truhlsen Eye Institute, College of Medicine, University of Nebraska Medical Center, Omaha, NE, USA

Zachary Woodward Department of Ophthalmology and Visual Sciences, Truhlsen Eye Institute, College of Medicine, University of Nebraska Medical Center, Omaha, NE, USA

Edward Wylegala Ophthalmology Clinic, Railway Hospital, Medical University of Silesia, Katowice, Poland

Sam Young Yoon Department of Ophthalmology, Hallym University College of Medicine, Seoul, Korea

Pejman Zargar Chicago Medical School, Rosalind Franklin University of Medicine and Science, Chicago, IL, USA

J.D. Zieske Schepens Eye Research Institute/Massachusetts Eye and Ear and the Department of Ophthalmology, Harvard Medical School, Boston, MA, USA

Lepša Žorić Department of Ophthalmology, University of Pristina, Kosovska Mitrovica, Serbia

About the Editors

Dr. Mark A. Babizhayev is a biophysicist and holds a Ph.D. in Biophysics and Pathophysiology from the Moscow Helmholtz Research Institute of Eye Diseases, Moscow, Russia. He has dedicated many years to scientific research and has written more than 102 published articles and has 15 patents to his name. The recipient of numerous awards, Dr. Babizhayev is currently Executive Director of Innovative Vision Products, Inc., Delaware, USA, Senior Research Investigator at Moscow Helmholtz Research Institute of Eye Diseases, and Scientific Consultant at EXSYMOL SAM, Monte Carlo, Monaco, and Bruschettini SRL, Genoa, Italy. He is also one of the principal researchers involved in the development of a breakthrough for the treatment and prevention of senile cataract.

Dr. David Wan-Cheng Li is an Associate Professor of Ophthalmology and Visual Sciences in the Truhlsen Eye Institute, University of Nebraska Medical Center, an elected Lotus Scholar Professor of Cell and Developmental Biology of Hunan Normal University by Hunan Province Government, and a Visiting Professor of Ophthalmology and Visual Sciences in the State Key Laboratory in the Zhongshan Ophthalmic Center of Sun Yat-sen University Medical Center. He received his Ph.D. from University of Washington, Seattle, USA, and completed his postdoctoral training at Columbia University, New York, USA. His scientific contributions include the major discoveries that stress-induced apoptosis acts as a common cellular mechanism for non-congenital cataract formation, that human hTERT is compatible with the RNA template from bovine and rabbit, and that different SUMOs have distinct functions in regulating cell differentiation. He received the outstanding cataract research award from National Foundation for Eye Research and the Lotus Scholar Contribution Award from Hunan Province Government. He is currently the Editor-in-Chief for *Current Molecular Medicine*.

Dr. Anne Kasus-Jacobi is Assistant Professor of Pharmaceutical Sciences and adjunct faculty in the Departments of Neuroscience and Physiology at the University of Oklahoma Health Sciences Center, USA. She was previously Assistant Professor of Ophthalmology at the Dean McGee Eye Institute in the University of Oklahoma

Health Sciences Center. She received her Ph.D. in Endocrinology from the University of Paris, France, and continued her postdoctoral training in the laboratory of Drs. Michael Brown and Joseph Goldstein at the University of Texas Southwestern Medical Center, USA. Dr. Kasus-Jacobi's current interests include the development of innovative peptide treatments for corneal wounds and antibiotic-resistant bacterial infections. She is a member of the Association for Research in Vision and Ophthalmology and serves as an editorial board member of the *Journal of Ocular Diseases and Therapeutics*.

Dr. Lepša Žorić works in clinical ophthalmology on the Faculty of Medicine of University of Pristina, Serbia. She completed her Ph.D. at the University of Pristina with a thesis on antioxidation status in aqueous humor, lens, and serum of patients with age-related cataracts. A member of several international and national medical and neuro-ophthalmology societies, Dr. Zoric has presented or published more than 100 articles in both domestic and international journals.

Dr. Jorge L. Alió M.D., Ph.D. is professor and chairman of Ophthalmology at the Miguel Hernández University, Alicante, Spain, and formerly chairman of Ophthalmology at the University of Alicante, Spain. He has been appointed with several visiting professorships at universities in the United States and Europe. He attended the Complutense University of Madrid, Spain, where he received a medical degree and a Ph.D. in biological thermodynamics and its application in the study of ophthalmic pathology. Recipient of the 2012 Lifetime Achievement Award from the International Society of Refractive Surgery, Dr. Alió's main research interests include refractive, lens and corneal surgery, ocular inflammation, and preventative ophthalmology. His main contributions have been in the area of excimer laser refractive surgery, microincisional lens surgery, and multifocal accommodative and premium IOLs.

Part I
Basic Science in Cornea

Chapter 1
Oxidative Stress in Cornea

Eunbi Kim, Sam Young Yoon, and Young Joo Shin

1.1 Oxidative Stress

Oxidative stress has been implicated in many pathophysiological events [1]. A disturbance in the pro-oxidant and antioxidant balance leads to potential damage [2]. The oxidative stress caused by reactive oxygen species (ROS) plays a pivotal role. ROS are not only considered to be the damaging factors in various pathologies, but they also participate in a wide variety of physiological processes such as signaling transduction [3, 4].

1.2 Oxidative Stress in Cornea

1.2.1 Antioxidant Defenses of the Tears, Aqueous Humor, and Cornea

The tear film contains superoxide dismutase-1 (SOD-1) activity, but no catalase or glutathione peroxidase (GPx) activity [5]. Low molecular-weight nonenzymatic antioxidants or antioxidant precursors including ascorbic acid, glutathione, uric acid, cysteine, and tyrosine were found in tear content [6]. High concentrations of lactoferrin in tears effectively chelate redox-active iron, thereby obviating·OH formation by iron-assisted Fenton chemistry [7].The aqueous contains significant

E. Kim • S.Y. Yoon • Y.J. Shin, M.D. (✉)
Department of Ophthalmology, Hallym University College of Medicine,
948-1 Daerim1-dong, Youngdeungpo-gu, Seoul 150-950, South Korea
e-mail: schinn@hanmail.net; schinn7@gmail.com

© Springer Science+Business Media New York 2015
M.A. Babizhayev et al. (eds.), *Studies on the Cornea and Lens*,
Oxidative Stress in Applied Basic Research and Clinical Practice,
DOI 10.1007/978-1-4939-1935-2_1

amounts of ascorbic acid, glutathione, uric acid, cysteine, and tyrosine and little SOD activity [7]. Ascorbic acid within the aqueous Humor might modulate lipoxygenase activity within the cornea and afford extracellular protection against ROS released by infiltrating leukocytes during ocular inflammation [8]. The cornea, on account of its exposure to UV light and high oxygen tension, requires robust antioxidant defenses. The human cornea is rich in SOD activity [7].

1.2.2 Oxidative Stress as a Toxin

1.2.2.1 Oxidative Damage

Imbalance between the production of ROS and the defence mechanisms against the reactive intermediates or the repair system from the oxidative damage produces oxidative stress [2]. Human corneal endothelial cells (HCECs) play an essential role in maintaining corneal transparency and thickness using adenosine-5′-triphosphate (ATP) [8–10]. HCECs confront the situations including oxidative stress because ROS are generated as by-products of ATP production [11, 12]. A lot of diseases and surgeries including ultrasound phacoemulsification, senescence, bullous keratopathy and inflammation, which can damage HCECs, are associated with oxidative stress [13]. Oxidative stress has been reported as a mediator of apoptosis [14]. ROS has been implicated in the cellular damage [14].

Ultrasound (US) phacoemulsification is currently the most widely used means by which cataract surgery in the developed world is carried out [15, 16]. Phacoemulsification is associated with the potentially injurious effects on HCECs [16, 17]. HCECs' damage during phacoemulsification can lead to permanent corneal injury requiring corneal allotransplantation as well as interfere with early visual rehabilitation after surgery [16]. Thus, we reported the effect of phacoemulsification energy on the redox state of cultured HCECs [16]. Redox fluorometry was used for measure redox ratio [18]. Redox fluorometry is at present an in vitro, inexpensive, rapid, and reproducible means to quantitatively assess HCECs injury following US phacoemulsification [18]. The redox ratio has been reported to be well correlated with oxidative stress [16, 19] because it reflects changes in the relative amounts of reduced nicotinamide nucleotides and oxidized flavins [16, 20]. HCECs exposed to increasing phacoemulsification times and increasing ultrasonic energy levels displayed dose-dependent decreases in measured redox ratios. Lower redox ratios in response to phacoemulsification did not associate with decreases in cell size or altered patterns of mitochondrial localization [16]. One mechanism by which phacoemulsification energy can reduce the redox ratio in cells is by generation of extracellular ROS, such as hydroxyl radical and other free radical species. These are known to be formed during phacoemulsification and to come into contact with the HCECs [21–24]. Such radicals generated near cellular membranes may result in protein oxidation as well as lipid oxidation [23]. These free radicals can act as external oxidative stressors in a manner similar to exposure to a chemical oxidant such as tert-butylhydroperoxide (tBHP) [16]. Secondly, phacoemulsification energy can reduce the redox ratio in cells through phacoemulsification-induced intracellular apoptosis-mediated ROS [16]. Apoptotic processes can decrease intracellular redox ratios due to oxygen consumption and oxidative stress in the presence of complex I or complex II substrates that are consumed during the apoptotic process [25, 26]. In the our study, redox ratios were significantly reduced in a dose-dependent manner in response to increasing phacoemulsification times as well as increasing phacoemulsification power, each reflecting an indicator of increasing total energy transfer to the HCEC. Predictably, the percentage of nonviable, stained cells increased with both longer phacoemulsification times and higher phacoemulsification powers. Redox ratios were well correlated with the percentage of nonviable, stained cells. However, cell size was not significantly altered by either greater phaco-emulsification time or increased ultrasonic power. Moreover, phacoemulsification did not have a significant effect on mitochondrial distribution patterns. While US phacoemulsification of HCECs results in shifts in the redox ratio that implicate ROS, the injury pattern is distinct from that seen following extracellular exposure of HCECs to the chemical oxidant tBHP. Concentrations of tBHP sufficient to cause both injury and apoptosis do result in similar redox ratio shifts but are also associated with characteristic changes in cell morphology such as decreased cell size (area) and a perinuclear pattern of mitochondrial distribution. Therefore, the redox fluorometry data presented herein support a role for oxidative injury as a component of phacoemulsification injury as well as implicate additional modes of injury in this complex HCEC stressor. Redox fluorometry in combination with HCEC morphometric measurements has potential to serve as an indicator of HCEC injury resulting secondary to US phacoemulsification. Differences in injury response between chemical oxidant injury and US phacoemulsification injury may provide fundamental insights into the mechanisms of injury involved.

Radiotherapy has been widely used as the treatment of ocular diseases including orbital tumor, conjunctival lymphoma, and intraocular malignant tumors [27]. Radiation causes oxidative damage on the ocular tissue including cornea, lens, and lacrimal glands [27–30]. It has been reported that 20 Gy radiation induces the transient damage in the corneal epithelium and the delayed and irreversible changes in the corneal endothelium [3, 29, 31]. Sicca syndrome due to lacrimal gland dysfunction has been one of the well-known complications after high doses of radiation therapy [27, 32–34]. Low-dose radiation has been known to induce cataract because radiation induces ROS as an oxidative stress [35].

1.2.2.2 Cell Arrest

Aging is characterized by a progressive deterioration in physiological functions and metabolic processes, leading to an increase in morbidity and mortality. The free radical theory of aging originally stated that free radicals generated endogenously cause oxidative modification of cellular components; the accumulation of oxidative damage with aging results in cellular dysfunction and eventually cell death [36, 37].

Lipid peroxidation and the oxidative damage of proteins and DNA increase with age. The oxidative stress theory of aging proposes that ROS, which primarily results from normal mitochondrial metabolism, cause progressive damage resulting in the functional decline that defines aging. Oxidative stress-induced apoptosis involves a very early degradation of mitochondrial polynucleotides (both DNA and RNA), apparently by a calcium-activated mitochondrial nuclease. Oxidative damage to mitochondrial enzymes, and to the mitochondrial genome itself, plays major roles in various age-related degenerative processes.

We reported several drugs to prevent the HCECs from oxidative stress- induced death [13, 38]. Clusterin has been described to attribute to many cellular physiologic functions, including cell–cell interactions, lipid transportation, cell survival, and apoptosis [39]. Clusterin induced under cytotoxic conditions to protect cells from cytotoxic stress [40]. Tert-butyl hydroperoxide (tBHP) is a ROS generating agent that causes lipid oxidation, stress-induced premature senescence [40]. tBHP damages DNA, resulting in cell death [17]. Clusterin reduced tBHP-induced intracellular ROS formation, increased cell viability against tBHP-induced oxidative stress, and decreased cell apoptosis induced by oxidative stress. Thus, clusterin showed the protective effect on oxidative stress-induced cell death of HCECs [41].

Rapamycin also has been suggested to be a protective agent against oxidative stress [13]. Rapamycin (sirolimus) is a macrolide antibiotic with potent immuno-suppressive properties. It has been used as an immunosuppressive agent in organ transplantation. Rapamycin has been described to induce autophagy through inhibition of mTOR1, 2 and to reduce ROS in cells [42, 43]. HCECs play an essential role in maintaining corneal transparency and thickness using ATP [9, 10, 44]. These cells confront the situations including oxidative stress because ROS are generated as by-products of ATP production [11, 12]. Rapamycin reduced 2′,7′-dihydrodichlorofluorescein oxidation and increased GSH in HCECs. Rapamycin significantly inhibited tBHP-induced ROS production. Cells treated with rapamycin showed higher viability compared to control at 5 mM tBHP. Rapamycin effectively protected HCECs from ROS-induced cell death through increasing intra-cellular GSH. Glutathione (GSH) has been well-known to protect against oxidative damage caused by ROS which are generated during normal metabolism [45, 46]. The recycling of GSH constituents maintains homeostasis of GSH within cells [46, 47]. GSH may be one of the major defense mechanisms in HCECs under oxidative stress relative to other antioxidative pathways. Glutathione (GSH) and GSH-associated metabolism constitute the principal line of defense for the protection of cells against oxidative and other forms of toxic stress [48]. GSH can scavenge free radicals, reduce peroxides, and can be conjugated with electrophilic compounds, thereby eliminating both ROS and their toxic by-products [49, 50]. GSH provides a critical defense system for the protection of cells against many forms of stress [48]. The mechanisms of rapamycin against oxidative stress have been reported to include antioxidative gene expression [51] and autophagy [52].

Cysteamine is a reducing aminothiol compound generated in animals [49]. Cysteamine has been previously employed for the treatment of cystinosis, since cysteamine interacts with free cystine inside the lysosome, leading to the formation of cysteine or the disulfide cysteine–cysteamine [50, 53]. Cysteamine can deplete cells of 90 % of their cystine contents [54]. Topical cysteamine has also proven useful in the treatment of corneal cystine crystals [55–57]. It has been recently reported that cysteamine appears to be a primary source of taurine with membrane-stabilizing and antioxidative properties [58–60] and has also been suggested as a novel approach to enhancing cellular glutathione levels [59]. Cysteamine reduced 2′,7′-dihydrodichlorofluorescein oxidation and increased glutathione. Cysteamine significantly inhibited tBHP-induced ROS production. Cysteamine-treated cells evidenced higher viability relative to the controls at 5 mM tBHP, and cysteamine also effectively protected HCECs against ROS-induced cell death via an increase in intracellular glutathione. Our data indicate that low-dose cysteamine can protect HCEC against oxidative injury-mediated cell death via the inhibition of ROS production, although high-dose cysteamine was found to be toxic to normal HCECs. Cysteamine may prove useful as a protective agent in patients under oxidative stress conditions. Thus, cysteamine may be helpful as a pre- or post-medication after cataract surgery for protection against bullous keratopathy. Further study to correlate cysteamine concentration in anterior chamber is necessary for clinical application.

1.2.3 Oxidative Stress as Signal Molecules

Cellular oxidation/reduction (redox) states regulate various aspects of cellular function and maintain homeostasis [61]. Moderate levels of reactive oxygen species/reactive nitrogen species (ROS/RNS) function as signals to promote cell proliferation, regulation, and survival [62], although increased levels of ROS/RNS can induce cell death [61, 62].

1.2.3.1 Inflammation

Allograft corneal transplantation is widely performed to treat corneal diseases, including bullous keratopathy, Fuchs' endothelial dystrophy, and corneal opacity [63]. Rejection is a major complication after allograft corneal transplantation [63–65]. The cornea has been described as an immune-privileged site because cornea has no blood or lymphatic vessels [66]. However, corneal endothelial rejection can cause persistent graft edema and failure [63, 67] while epithelial rejection is usually self limited. Suppression of corneal endothelial rejection is important to maintain endothelial function and transparency of grafts [68]. Many drugs have been developed to suppress the inflammation associated with organ transplantation including corneal endothelial rejection after corneal transplant [69–71]. The mainstay is the use of immunosuppressive agents including rapamycin, steroid, and cyclosporine A [72]. Recently, ROS has been described to induce the inflammation [73, 74]. It has been reported that antioxidants play an essential role in inhibiting inflammation [74]. However, there has been no study about the role of antioxidants in corneal endothelial rejection after corneal transplantation. Cysteamine (CYS) is a simple aminothiol that participates in various physiologic processes, including amino acid transport, disulfide reduction, protein synthesis, and protection against oxidative damage [59, 75–77]. CYS has been described as an antioxidant [78]. CYS enhances glutathione (GSH) synthesis under various forms of toxic stress [59]; furthermore, it exerts a potent and prolonged anti-inflammatory effect [79, 80]. We reported that CYS effectively suppresses human peripheral blood mononuclear cells, HCEC reaction via ROS reduction [81]. Endothelial rejection involves inflammatory signals, including keratic precipitates and inflammation in the anterior chamber [63, 67]. Leukocyte counts in the graft bed are inversely correlated with actuarial graft survival [82]. Corneal endothelial rejection has been known as the most severe form of rejections [83, 84]. Two forms of endothelial rejection have been described [84]. Khodadoust line in endothelial rejection usually originates at a vascularized area of the cornea [84], or at the site of an anterior synechiae [83]. Within a few days, the line extends across the donor cornea, destroying endothelial cells, depositing keratic precipitates. The other form is diffuse endotheliitis although diffuse keratic precipitates are not visible. Endothelial rejection is one of the inflammatory responses. Endothelial rejection is one of inflammatory responses although the corneal endothelial cell rejection has been known to be mediated by a chronic adaptive immunity [85] and it is not well established that simply inflammation plays a major role in this process. Thus, the interaction between PBMC and HCEC is important for graft survival. We showed that CYS suppresses PBMC proliferation and decreases intracellular ROS levels in a dose-dependent manner [65]. CYS is an antioxidant that enhances intracellular GSH levels. Moreover, CYS may act as a scavenger of superoxide free radicals and hydrogen peroxide [78]. ROS have been described to modulate inflammation and tissue repair [86]. Although ROS causes cell death at high level, a low level of ROS has been described to mediate cell-signaling events in inflammation as a second messenger [86]. Reduction of intracellular ROS can reduce the cell proliferation and inflammatory response. Nuclear factor-erythroid 2-related factor 2 (Nrf2), a redox-sensitive transcription factor, has been implicated in cellular responses to oxidative stress; Nrf2 inhibits nuclear factor kappa B (Nf-ƙB) [87, 88], which plays a key role in immune response and inflammation [89]. Thus, CYS may exert an anti-inflammatory effect by reducing intracellular ROS formation. In addition, CYS suppressed the levels of IL-6 and TGF-ß1 [65], which are the cytokines secreted by PBMCs [90, 91]. Nf-ƙB has been reported to induce the activation of IL-6 gene expression [92, 93] and to be regulated by TGF-ß1 [94, 95]. IL-6 is a proinflammatory cytokine that plays multiple roles during injuries and inflammation [88]. IL-6 levels in the human aqueous humor have been reported to rise during corneal endothelial immune reactions [96, 97]. It has been reported that the specific release of proinflammatory cytokines from alloreactive infiltrating cells, results in apoptosis in the corneal endothelium [98]. TGF- ß 1 is a multifunctional cytokine [99]. Although the role of TGF- ß 1 in corneal endothelial rejection is yet unclear, TGF-ß1 has been reported to be induced by tissue injuries, participate in tissue fibrosis [96], and regulate inflammation and fibrosis [100]. TGF-ß1 has been reported to modulate Nf-ƙB [94, 95], which plays a key role in inflammatory response [89].

Thus, TGF-ß1 might regulate the immune response during corneal endothelial rejection. This study showed that CYS decreased PBMC proliferation, IL-6 and TGF-ß1 levels via ROS formation. Our results suggest that CYS could suppress inflammation associated with PBMCs to corneal endothelial cells [81]. However, further in vivo studies are required.

In addition we reported combined treatment with antioxidants and immuno-suppressants on cytokine release by human peripheral blood mononuclear cells - chemically injured keratocyte reaction [101]. Corneal chemical burn can induce a devastating and permanent damage to ocular surface resulting in corneal blindness [102]. Corneal chemical burn injuries can induce a large extent of cell death [102]. Especially, exposure to alkali agent may cause extensive damage to ocular tissues because alkali can progress rapidly and penetrate into deep tissues [103]. Although there have been many studies about treatment of chronic ocular damages including amniotic membrane transplantation, oral mucosal transplantation and limbal trans-plantation [103, 104], suppression of acute and chronic inflammation induced by chemical burn still has been challenging. A variety of medical therapies including topical and systemic drugs have been investigated to control inflammation and promote ocular surface healing [105, 106]. After ROS has been reported to be able to induce inflammation [86, 107], there have been many studies to report the effect of antioxidants on inflammation [86, 102, 107–109]. Corneal chemical burn injury can induce inflammation [104]. Alkali agents saponify the cellular membrane and induce cell death [103]. Cellular membrane damage is the cause of inflammation in necrosis [105]. Necrosis induces inflammation while apoptosis doesn't [110]. Thus, chemically injured keratocyte may play an important role in inducing inflammation in corneal chemical burn. Interactions between chemical-burned keratocyte and PBMC may play an essential role in chemical burn because keratocytes are the major cellular components of cornea [108]. We found that antioxidants as well as immunosuppressants suppress the PBMC proliferation [101]. The combination of immunosuppressants and NAC was more effective in suppression of PBMC prolif-eration. It was reported that low concentrations of ROS can stimulate cellular proliferation as a second messenger [109, 111]. ROS are implicated in the regulation of several cellular processes depending on their intracellular levels. Although high levels of ROS are toxic to the cells [2, 105], low levels of ROS have various physi-ologic roles for appropriate signal transduction, kinase activation, and biologic responses associated with receptor signaling [63]. In addition, ROS has been described to modulate the inflammation [64] and promote inflammatory cytokines [106]. Thus, antioxidants can have a potential to suppress inflammation. On the other hand, the combination of immunosuppressants and ALA increased PBMC proliferation except for the combination of dexamethasone and ALA [101]. Although it has been reported that ALA alone inhibits airway inflammation [112], we found the combina-tion of immunosuppressants and ALA may increase inflammation on the contrary. Further molecular-based studies are needed to clarify these results [101].

MMP-9 is a collagenase IV produced by PBMCs [113, 114]. The production of MMPs by PBMC are triggered by chemokines and cytokines produced by other inflammatory cells at the site of inflammation [115]. MMP-9 is involved in degradation of extracellular matrix [116]. Corneal stroma can be thinned progressively by MMP-9 upregulation in corneal chemical burn [117]. Suppression of MMP-9 is important to prevent cornea from thinning. MMP-9 levels were lower in ALA group and NAC group [101]. The combination of immunosuppressants and antioxidants was more effective in suppression of MMP-9 production except for the combination of MPA and ALA [101].

IL-6 is a proinflammatory cytokine produced by PBMC [90]. IL-6 levels decreased in treatment with dexamethasone and antioxidants [101]. The combination of immunosuppressants and antioxidants are more effective to suppress the production of IL-6 [101]. MIF levels were not significantly different in treatment with each drug [101]. MIF levels increased in MLKR treated with the combination of immunosuppressants and antioxidants except for the combination of rapamycin and ALA [101]. MIF has been described to be an innate immunity molecule leading to induction of proinflammatory activities [118]. MIF was originally described as a regulator of macrophage responses [118]. It directly or indirectly promotes expression of a variety of proinflammatory cytokines including IL-6, TNF-alpha, IL-8[124–126], MMP-9 [127], and TGF-β1 [119, 120]. MIF levels were elevated in the groups which showed lower PBMC proliferation rates and lower IL-6 levels in this study. MIF might be elevated as a master regulator of inflammation [124] in response to suppression of inflammation. TGF-β1 is a multifunctional cytokine that participates in a wide range of biologic events, including inflammation and wound repair [119]. TGF-β1 induces synthesis and accumulation of extracellular matrix protein and has been implicated as the potent and key mediator of fibrogenesis [120]. TGF-β1 has been reported to be induced by MIF in IgA nephropathy [120] while reduced TGF-β1 and increased MIF has been described in severe malaria [121]. TGF-β1 plays an essential role in corneal chemical burn via inducing inflammation and repairing wound [122]. TGF- β1 inducing transdifferentiation from keratocyte to myofibroblast contribute these processes [122]. Therefore, antioxidants as well as immunosuppressants suppressed the PBMC proliferation induced by chemically injured keratocyte [101]. The combination of immunosuppressants and antioxidants had a synergic effect on MLKR [101]. Cytokine production was different depending on the combination of drugs. Our results suggest that the different drugs should be selected for treatment according to the phases of corneal chemical burn [101].

Furthermore, previous studies reported antioxidative capacity of rapamycin [13]. We revealed the effect of rapamycin on corneal chemical burn [123]. Corneal chemical injuries may produce extensive tissue damage resulting in permanent visual impairment [2]. Although it is essential to control inflammation in the acute phase, the development of corneal opacity and neovascularization (NV) in the later phase has been reported to be the main cause of permanent visual loss. The effectiveness of several drugs for the treatment of chemical burn-induced corneal NV and opacity has been investigated [101, 128, 129]. The role of vascular endothelial growth factors (VEGF) in corneal chemical burn has been investigated in previous studies [131, 132]. Rapamycin has the suppressive effect on corneal NV and opacity as well as an antioxidative capacity [13, 123]. Chemical burn has been reported to cause a chronic inflammatory state. Inflammation of the cornea results in corneal NV [129]. Thus, suppression of the inflammatory cytokine can reduce corneal inflammation [132] and angiogenesis [133]. Furthermore, NV often causes fibrosis [134], which results in corneal opacity [135]. TGF-b1, which can be modulated by IL-6 [136], has been reported to be involved in the initiation of the fibrotic response in vivo [137]. Thus, the drugs that suppress inflammation and reduce fibrosis can be the most effective in treatment of corneal chemical burn. Rapamycin is an immunosuppressant used in several cancer treatments [138] and in allograft rejection treatment [139]. Rapamycin acts through inhibition of mammalian target of rapamycin (mTOR) [140]. Multiple upstream signals, including those triggered by growth factors and nutrients, are integrated into the mTOR signaling pathway. Inhibition of mTOR leads to cell cycle arrest, inhibition of cell proliferation, immunosuppression, and autophagy induction [140–142]. Recently, rapamycin has been reported to have an effect on pulmonary fibrosis and on intraperitoneal fibrosis [143]. Corneal neovascularization and corneal opacity scores measured 4 weeks after the chemical burn corneal injury were lower in the rapamycin-treated group than in the control group [123]. Two weeks after the chemical burn injury, a significant elevation in the corneal IL-6 levels of the positive control group was observed, compared to the levels in the negative control group or the rapamycin-treated group ($P<0.05$) [123]. Corneal TGF-b1 levels were lower in the rapamycin-treated group than in the control group at 4 weeks after chemical burn injury ($P<0.05$) [123].

Moreover, rapamycin inhibited TGF-b1-induced a-SMA expression and augmented ERK 1/2 phosphorylation. Corneal chemical burn may induce corneal blindness by increasing the corneal opacity and corneal NV [101, 103]. Rapamycin has been reported to inhibit pulmonary and peritoneal fibrosis [141, 142, 144] as well as allograft rejection [140]. In this study, the corneal opacity and NV scores were lower in the rapamycin group than those in the positive control group at 4 weeks after the chemical injuries. Rapamycin treatment reduced IL-6 levels in the rapamycin group compared to those in the positive group 2 weeks after the chemical injuries. IL-6 is a proinflammatory cytokine produced in corneal chemical burn tissue [132] and is known to induce angiogenetic factors [133]. IL-6 levels in the rapamycin group decreased at 2 weeks and then did not differ at 4 weeks after chemical injury. The TGF-b1 levels in chemically burned corneas increased 4 weeks after chemical injury infliction. In vivo studies revealed that corneal TGF-b1 levels were higher in the positive control group than in the rapamycin group and the negative control group. TGF-b1 has been reported to stimulate transdifferentiation of corneal stromal cells to myofibroblasts, which then increase the corneal opacity [135]. In vitro studies showed that the expression of a-SMA, a marker of myofibroblast [135], increased with the TGF-b1 treatment. Furthermore, TGF-b1 treatment stimulated the corneal stromal cells to transdifferentiate into myofibroblasts, whereas rapamycin inhibited TGF-b1-induced myofibroblast transdifferentiation. Rapamycin treatment also elevated phospho-ERK 1/2 levels. The ERK 1/2 signaling cascade is a central pathway in the regulation of cellular processes such as proliferation, differentiation, and survival [145]. The ERK 1/2 cascade is activated by many extracellular stimuli and internal processes [146], and the ERKs also phosphorylate a large number of substrates, including Elk1, c-fos, and smad-linker [146–148]. Crosstalk between ERKs and SMAD signaling pathways has been reported [149]. The activation of SMAD pathway plays a central part in the development of fibrosis [150]. Rapamycin targets the mTOR protein [151], which is a serine/threonine kinase robustly regulated by a diverse array of upstream signals [152]. ERK has been reported to involve regulation of the mTOR pathway [153]. The mTOR-mediated protein synthesis requires coincident and mutually dependent activity in the PI3K and ERK pathways [153]. ERKs phosphorylate distinct sites on tuberous sclerosis complex 2 (TSC2), leading to greater repression of its GTPase-activating protein (GAP) activity and, consequently, a magnified stimulation of mTOR complex 1 (mTORC1) signaling [154]. mTOR is a major effector of cell growth and protein synthesis via the direct functional control of its downstream targets, ribosomal protein S6 kinase (S6k), and eukaryotic initiation factor 4E-binding protein (4EBP)-1 [151]. IL-6 expression was not detected in an in vitro study, while IL-6 levels of cornea in vivo increased 2 weeks after chemical injuries. In vivo, chemical corneal injuries induce acute and chronic inflammation of ocular surface, 9 where a variety of inflammatory cells are involved [132]. Rapamycin, which is an anti-inflammatory agent [140], reduced IL-6 level of cornea in vivo study. Even though IL-6 has been described as being secreted by many different cell types including T cells, macrophages and vascular endothelial cells [155], it is secreted by corneal stromal cells in response to exogenous stimuli including ultraviolet irradiation [156], infection [157], and immunological challenge [156]. However, cultured corneal stromal cells may produce extremely low levels of IL-6 if inflammatory stimuli are absent [158]. In an in vitro study, TGF-b1 was used for induction of myofibroblast transdifferentiation, and it did not cause IL-6 secretion as an inflammatory stimuli. TGF-b1 expression in cultured corneal stromal cells was not investigated in the in vitro study because the cells were treated with TGF-b1 exogenously.

1.2.3.2 Proliferation

Park et al. reported that cellular redox state predicts in vitro corneal endothelial cell proliferation capacity [159]. Mitochondria are considered to be major organelles in the generation of ROS and play an essential role in the redox balance, with NAD(P) H and FAD as cofactors [20, 160]. As such, mitochondria have been considered valuable targets for therapy. Therapeutic agents acting on redox metabolism at mitochondrial inner membranes have been recently developed [161, 162]. Redox fluorometry is a noninvasive method that measures the cellular autofluorescence emitted from reduced NAD(P)H and oxidized flavoproteins in living cells. Reduced NAD(P)H and oxidized flavoprotein signals originate from the mitochondrial matrix space [18, 163, 164], and the measurement of the ratio of the two fluorescence intensities equates the reduction–oxidation ratio of reduced pyridine nucleotide to oxidized flavoprotein [18, 163]. The redox ratio evaluated through fluorometry has been used in the past to evaluate the redox state, and the inverse of the redox ratio corresponds to the cellular metabolic activity [163]. In addition, mitochondrial autofluorescence provides high-resolution mitochondrial images with accurate structural detail, as documented in live HCECs [164], since most of NAD(P)H autofluorescence originates from intracellular mitochondria [19, 20, 163, 165]. Redox ratio computation by using redox fluorometry is a useful predictor of cellular proliferation [18, 164].

Mitochondria play essential roles in energy production, ROS generation, and apoptosis. Mitochondria are considered to be the major source of cellular ROS [166, 167]. Most cancer cells have increased metabolic activity and produce large amounts of ROS [168]. Altered mitochondrial respiratory metabolism increases the production of ROS in cells [169]. Although an appropriate level of intracellular ROS plays an important role in maintaining redox balance and signaling cellular proliferation [170–172], higher levels of ROS could lead to cellular damage. Physiologic fluctuations of ROS can change the cellular redox state as positive and negative effectors in cellular proliferation [173]. Although redox signaling may be related to cell density, it has been reported that steady-state levels of intracellular ROS were significantly lower in confluent cells than in sparse cultures [174]. However, in the present study, cells cultured on collagen-coated dishes showed higher intracellular ROS (lower redox ratio) by redox fluorometry, associated with a higher proliferation rate [18]. Thus, a lower redox ratio might be due to a higher proliferation rate, before the cells reach confluence. Additionally, dividing cells require high energy metabolism, and this creates an electron flux from NADH to the mitochondrial electron-transport chain to generate more adenosine triphosphate (ATP), and consequently, the NADþ/NADH couple is oxidized [173]. In addition, the relation between redox and proliferation rate has been investigated in several cultured cell types, including HCECs and mesenchymal stem cells [159, 165]. Redox fluorometry provides quantitative information on the redox status of living cells because it represents the changes of reduced NAD(P)H and oxidized flavoproteins [18, 165]. Redox fluorometry provides a method of estimating the redox state in individual cells. It can be measured by autofluorescence of cellular NAD(P)H and flavoproteins. NAD(P)H exists as a cofactor in the mitochondria and cytoplasm [18]. However, most NAD(P)H exists in mitochondria [175]. Flavoproteins exist mostly as electron-receptor equivalents to the fluctuations of NADþ in cells [175]. They are cofactors for enzymes in oxidation–reduction reactions, and it has been found that only lipoamide dehydrogenase and electron-transfer flavoprotein in the mitochondrial matrix contribute significantly to cellular flavoprotein fluorescence [176].

1.2.3.3 Differentiation

It has been reported that ROS plays an essential role in differentiation [177]. However, further study is essential to investigate the role of ROS in limbal stem cell and corneal endothelial stem cells. Thus, the balance of ROS is important for cell functions. The concentration, context, and location of ROS may be important to decide the role of ROS.

1.2.4 Other Corneal Diseases

Pterygium has been well-known to be associated with oxidative stress induced by UV. It is an inflammatory, proliferative, and vascularized connective tissue derived from the conjunctiva [178]. UV irradiation has an oxidative damage on conjunctiva tissue either directly by UV phototoxic effects or indirectly by formation of radical oxygen species (ROS) [178]. Oxidative stress plays an essential role in p53 gene mutation in pterygium. 8-hydroxydeoxyguanosine (8-OHdG), which is an oxidative DNA damage and a ubiquitous marker of oxidative stress, expressed at high levels in pterygium [178]. Oxidative stress-mediated induction of the protein, survivin has been correlated with DNA oxidation and downregulation of the cell-growth inhibitor p53, leading to a hyperproliferative state [179].

Oxidative stress has recently been described to be associated with dry eye syndrome [180, 181]. It has been linked to corneal, conjunctival, and lacrimal gland injury [181]. Corneal epithelial alterations in a blink-suppressed dry eye are accompanied by increases of oxidative stress markers, and ROS production in corneal epithelia, suggesting a relationship between accumulation of oxidative stress and the etiology of corneal surface disorder in blink-suppressed dry eye [180]. Oxidative marker including 4-HNE and 8-OHdG expressed strongly in the lacrimal glands of aged Sod1 knock-out mice [181]. The association of Sjøgren's syndrome-associated oxidative stress and mitochondrial dysfunction has been reported [182]. Oxidative stress has been reported to be connected with inflammation.

Diabetic keratopathy has been often accompanied in diabetic patients. Hyperglycemia has been reported to cause the oxidative stress [183]. It resulted from the generation of ROS and impairment of intracellular antioxidant defense system, in the cornea. The accumulation of advanced glycation end products (AGEs), which is associated with oxidative stress [183], might cause the decrease of corneal endothelial cells with aging.

References

1. Buddi R, Lin B, Atilano SR, Zorapapel NC, Kenney MC, Brown DJ. Evidence of oxidative stress in human corneal diseases. J Histochem Cytochem. 2002;50:341–51.
2. Freeman BA, Crapo JD. Biology of disease: free radicals and tissue injury. Lab Invest. 1982;47:412–26.
3. Afanas'ev I. ROS and RNS signaling in heart disorders: could antioxidant treatment be successful? Oxid Med Cell Longev. 2011;2011:293769.
4. Bisbal C, Lambert K, Avignon A. Antioxidants and glucose metabolism disorders. Curr Opin Clin Nutr Metab Care. 2010;13(4):439–46.
5. Crouch RK, Goletz P, Snyder A, Coles WH. Antioxidant enzymes in human tears. J Ocul Pharmacol. 1991;7(3):253–8.
6. Gogia R, Richer SP, Rose RC. Tear fluid content of electrochemically active components including water soluble antioxidants. Curr Eye Res. 1998;17(3):257–63.
7. Shoham A, Hadziahmetovic M, Dunaief JL, Mydlarski MB, Schipper HM. Oxidative stress in diseases of the human cornea. Free Radic Biol Med. 2008;45(8):1047–55.

8. Williams RN, Paterson CA. The influence of topical corticosteroid therapy upon polymorphonuclear leukocyte distribution, vascular integrity and ascorbate levels in endotoxin-induced inflammation of the rabbit eye. Exp Eye Res. 1987;44(2):191–8.
9. Maurice DM. The structure and transparency of the cornea. J Physiol (Lond). 1957;136:263–86.
10. Dikstein S, Maurice DM. The active control of corneal hydration. Isr J Med Sci. 1972;8:1523–8.
11. Brown DI, Griendling KK. Nox proteins in signal transduction. Free Radic Biol Med. 2009;47:1239–53.
12. Gillespie MN, Pastukh V, Ruchko MV. Oxidative DNA modifications in hypoxic signaling. Ann N Y Acad Sci. 2009;1177:140–50.
13. Shin YJ, Cho DY, Chung TY, Han SB, Hyon JY, Wee WR. Rapamycin reduces reactive oxygen species in cultured human corneal endothelial cells. Curr Eye Res. 2011;36(12):1116–22.
14. Loh KP, Huang SH, De Silva R, Tan BK, Zhu YZ. Oxidative stress: apoptosis in neuronal injury. Curr Alzheimer Res. 2006;3:327–37.
15. Nishi Y, Engler C, Na DR, Kashiwabuchi RT, Shin YJ, Cano M, Jun AS, Chuck RS. Evaluation of phacoemulsification-induced oxidative stress and damage of cultured human corneal endothelial cells in different solutions using redox fluorometry microscopy. Acta Ophthalmol. 2010;88(8):e323–7.
16. Shin YJ, Nishi Y, Engler C, Kang J, Hashmi S, Jun AS, Gehlbach PL, Chuck RS. The effect of phacoemulsification energy on the redox state of cultured human corneal endothelial cells. Arch Ophthalmol. 2009;127:435–41.
17. Augustin AJ, Dick HB. Oxidative tissue damage after phacoemulsification: influence of ophthalmic viscosurgical devices. J Cataract Refract Surg. 2004;30:424–7.
18. Shin YJ, Tata DB, Waynant RE, Gehlbach PL, Chuck RS. Fluorometric determination of the redox state and distribution of mitochondria in human malignant glioblastoma cells grown on different culturing substrates. Photomed Laser Surg. 2010;28 Suppl 1:S105–10.
19. del V Cano M, Reyes JM, Park CY, et al. Demonstration by redox fluorometry that sulforaphane protects retinal pigment epithelial cells against oxidative stress. Invest Ophthalmol Vis Sci. 2008;49(6):2606–12.
20. Huang S, Heikal AA, Webb WW. Two-photon fluorescence spectroscopy and microscopy of NAD(P)H and flavoprotein. Biophys J. 2002;82(5):2811–25.
21. Shimmura S, Tsubota K, Oguchi Y, Fukumura D, Suematsu M, Tsuchiya M. Oxiradical-dependent photoemission induced by a phacoemulsification probe. Invest Ophthalmol Vis Sci. 1992;33(10):2904–7.
22. Holst A, Rolfsen W, Svensson B, Ollinger K, Lundgren B. Formation of free radicals during phacoemulsification. Curr Eye Res. 1993;12(4):359–65.
23. Cameron MD, Poyer JF, Aust SD. Identification of free radicals produced during phacoemulsification. J Cataract Refract Surg. 2001;27(3):463–70.
24. Takahashi H, Sakamoto A, Takahashi R, Ohmura T, Shimmura S, Ohara K. Free radicals in phacoemulsification and aspiration procedures. Arch Ophthalmol. 2002;120(10):1348–52.
25. Ricci JE, Gottlieb RA, Green DR. Caspase-mediated loss of mitochondrial function and generation of reactive oxygen species during apoptosis. J Cell Biol. 2003;160(1):65–75.
26. Fleury C, Mignotte B, Vayssière JL. Mitochondrial reactive oxygen species in cell death signaling. Biochimie. 2002;84(2–3):131–41.
27. Finger PT. Radiation therapy for orbital tumors: concepts, current use, and ophthalmic radiation side effects. Surv Ophthalmol. 2009;54:545–68.
28. Kiuchi T, Tatsuzaki H, Wakabayashi T, Okamoto F, Kaji Y, Oshika T. Long-term changes in rabbit cornea after ionizing radiation. Cornea. 2004;23:S87–90.
29. Kiuchi T, Tatsuzaki H, Hommura S, Oshika T. Specular microscopic study of X-ray-irradiated rabbit cornea. Eye (Lond). 2004;18:929–34.
30. Blodi FC. The effects of experimental x-radiation on the cornea. Arch Ophthalmol. 1960;63:20–9.

31. Jooyandeh F, Moore JS, Morgan RE, Phillips GO. Chemical effects of gamma-irradiation of aqueous solutions of heparin and keratan sulphate. Radiat Res. 1971;45:455–61.
32. Beutel J, Schroder C, von Hof K, Kosmehl H, Wedel T, Sieg P, et al. Pharmacological prevention of radiation-induced dry eye-an experimental study in a rabbit model. Graefes Arch Clin Exp Ophthalmol. 2007;245:1347–55.
33. Fard-Esfahani A, Mirshekarpour H, Fallahi B, Eftekhari M, Ekhari M, Saghari M, et al. The effect of high-dose radioiodine treatment on lacrimal gland function in patients with differentiated thyroid carcinoma. Clin Nucl Med. 2007;32:696–9.
34. Solans R, Bosch JA, Galofré P, Porta F, Roselló J, Selva-O'Callagan A, et al. Salivary and lacrimal gland dysfunction (sicca syndrome) after radioiodine therapy. J Nucl Med. 2001;42:738–43.
35. Bigsby RM, Valluri S, Lopez J, Mendonca MS, Caperell-Grant A, DesRosiers C, et al. Ovarian hormone modulation of radiation-induced cataractogenesis: dose-response studies. Invest Ophthalmol Vis Sci. 2009;50:3304–10.
36. Beckman KB, Ames BN. The free radical theory of aging matures. Physiol Rev. 1998;78:547–81.
37. Sohal RS. Role of oxidative stress and protein oxidation in the aging process. Free Radic Biol Med. 2002;33:37–44.
38. Shin YJ, Seo JM, Chung TY, Hyon JY, Wee WR. Effect of cysteamine on oxidative stress-induced cell death of human corneal endothelial cells. Curr Eye Res. 2011;36:910–7.
39. Rosenberg ME, Silkensen J. Clusterin: physiologic and pathophysiologic considerations. Int J Biochem Cell Biol. 1995;27:633–45.
40. Dumont P, Chainiaux F, Eliaers F, Petropoulou C, Remacle J, Koch-Brandt C, Gonos ES, Toussaint O. Overexpression of apolipoprotein J in human fibroblasts protects against cytotoxicity and premature senescence induced by ethanol and tert-butylhydroperoxide. Cell Stress Chaperones. 2002;7(1):23–35.
41. Shin YJ, Kim JH, Seo JM, Lee SM, Hyon JY, Yu YS, Wee WR. Protective effect of clusterin on oxidative stress-induced cell death of human corneal endothelial cells. Mol Vis. 2009;15:2789–95.
42. Tunon MJ, Sanchez-Campos S, Gutierrez B, Culebras JM, Gonzalez-Gallego J. Effects of FK506 and rapamycin on generation of reactive oxygen species, nitric oxide production and nuclear factor kappa B activation in rat hepatocytes. Biochem Pharmacol. 2003;66:439–45.
43. Suzuki M, Endo M, Shinohara F, Echigo S, Rikiishi H. Rapamycin suppresses ROS-dependent apoptosis caused by selenomethionine in A549 lung carcinoma cells. Cancer Chemother Pharmacol. 2011;67(5):1129–36.
44. Huang B, Blanco G, Mercer RW, Fleming T, Pepose JS. Human corneal endothelial cell expression of Na+, K+-adenosine triphosphatase isoforms. Arch Ophthalmol. 2003;121:840–5.
45. Ng LE, Halliwell B, Wong KP. Nephrotoxic cell death by diclofenac and meloxicam. Biochem Biophys Res Commun. 2008;369:873–7.
46. Schulz JB, Lindenau J, Seyfried J, Dichgans J. Glutathione, oxidative stress and neurodegeneration. Eur J Biochem. 2000;267:4904–11.
47. Thiele K, Bierhaus A, Autschbach F, et al. Cell specific effects of glucocorticoid treatment on the NF-kappaBp65/IkappaBalpha system in patients with Crohn's disease. Gut. 1999;45:693–704.
48. Maher P. The effects of stress and aging on glutathione metabolism. Ageing Res Rev. 2005;4:288–314.
49. Dickinson DA, Forman HJ. Glutathione in defense and signaling. Ann N Y Acad Sci. 2002;973:488–504.
50. Hayes JD, McLellan LI. Glutathione and glutathione-dependent enzymes represent a co-ordinately regulated defence against oxidative stress. Free Radic Res. 1999;31:273–300.
51. Kofman AE, McGraw MR, Payne CJ. Rapamycin increases oxidative stress response gene expression in adult stem cells. Aging (Albany NY). 2012;4(4):279–89.

52. Dutta D, Xu J, Kim JS, Dunn Jr WA, Leeuwenburgh C. Upregulated autophagy protects cardiomyocytes from oxidative stress-induced toxicity. Autophagy. 2013;9(3):328–44.
53. Pisoni RL, Park GY, Velilla VQ, Thoene JG. Detection and characterization of a transport system mediating cysteamine entry into human fibroblast lysosomes. Specificity for aminoethylthiol and aminoethylsulfide derivatives. J Biol Chem. 1995;270:1179–84.
54. De Biaggi CP, Barros PS, Silva VV, Brooks DE, Barros SB. Ascorbic acid levels of aqueous humor of dogs after experimental phacoemulsification. Vet Ophthalmol. 2006;9:299–302.
55. Gahl WA, Thoene JG, Schneider JA. Cystinosis: a disorder of lysosomal membrane transport. In: Scriver CR, Beaudet al, Sly WS, et al., editors. The metabolic & molecular bases of inherited disease, vol. 3. 8th ed. New York: McGraw Hill; 2001. p. 5085–108.
56. Jones NP, Postlethwaite RJ, Noble JL. Clearance of corneal crystals in nephropathic cystinosis by topical cysteamine 0–5%. Br J Ophthalmol. 1991;75:311–2.
57. Tavares R, Coelho D, Macario MC, Torres A, Quadrado MJ, Murta J. Evaluation of treatment with cysteamine eyedrops for cystinosis with confocal microscopy. Cornea. 2009;28:938–40.
58. Kessler A, Biasibetti M, da Silva Melo DA, Wajner M, Dutra-Filho CS, de Souza Wyse AT, Wannmacher CM. Antioxidant effect of cysteamine in brain cortex of young rats. Neurochem Res. 2008;33:737–44.
59. Maher P, Lewerenz J, Lozano C, Torres JL. A novel approach to enhancing cellular glutathione levels. J Neurochem. 2008;107:690–700.
60. Dominy Jr JE, Simmons CR, Hirschberger LL, Hwang J, Coloso RM, Stipanuk MH. Discovery and characterization of a second mammalian thiol dioxygenase, cysteamine dioxygenase. J Biol Chem. 2007;282:25189–98.
61. Kamata H, Hirata H. Redox regulation of cellular signalling. Cell Signal. 1999;11:1–14.
62. Cross AR, Jones OT. Enzymic mechanisms of superoxide production. Biochim Biophys Acta. 1991;1057:281–98.
63. Panda A, Vanathi M, Kumar A, Dash Y, Priya S. Corneal graft rejection. Surv Ophthalmol. 2007;52:375–96.
64. Des Marchais B, Bazin R, Boisjoly HM, Laughrea PA, Dubé I, Lille S, Roy R. Role of presensitization and donor-recipient crossmatching in corneal graft outcome. Cornea. 1998;17:141–5.
65. Jonas JB, Rank RM, Budde WM. Immunologic graft reactions after allogenic penetrating keratoplasty. Am J Ophthalmol. 2002;133:437–43.
66. Streilein JW. Anterior chamber associated immune deviation: the privilege of immunity in the eye. Surv Ophthalmol. 1990;35:67–73.
67. Levenson JE, Brightbill FS. Endothelial rejection in human transplants. Arch Ophthalmol. 1973;89:489–92.
68. Coster DJ, Williams KA. The impact of corneal allograft rejection on the long-term outcome of corneal transplantation. Am J Ophthalmol. 2005;140:1112–22.
69. Morris RE, Wu J, Shorthouse R. A study of the contrasting effects of cyclosporine, FK 506, and rapamycin on the suppression of allograft rejection. Transplant Proc. 1990;22:1638–41.
70. McMahon G, Weir MR, Li XC, Mandelbrot DA. The evolving role of mTOR inhibition in transplantation tolerance. J Am Soc Nephrol. 2011;22:408–15.
71. Chon WJ, Josephson MA. Leflunomide in renal transplantation. Expert Rev Clin Immunol. 2011;7:273–81.
72. Collier SJ. Immunosuppressive drugs. Curr Opin Immunol. 1989–1990;2:854–8.
73. Li YJ, Takizawa H, Kawada T. Role of oxidative stresses induced by diesel exhaust particles in airway inflammation, allergy and asthma: their potential as a target of chemoprevention. Inflamm Allergy Drug Targets. 2010;9:300–5.
74. Eberlein M, Scheibner KA, Black KE, Collins SL, Chan-Li Y, Powell JD, Horton MR. Antioxidant inhibition of hyaluronan fragment-induced inflammatory gene expression. J Inflamm (Lond). 2008;5:20.
75. Szabo S, Reichlin S. Somatostatin in rat tissues is depleted by cysteamine administration. Endocrinology. 1981;109:2255–7.

76. Deleuze S, Goudet G. Cysteamine supplementation of in vitro maturation media: a review. Reprod Domest Anim. 2010;45:e476–82.
77. Anand T, Kumar D, Chauhan MS, Manik RS, Palta P. Cysteamine supplementation of in vitro maturation medium, in vitro culture medium or both media promotes in vitro development of buffalo (Bubalus bubalis) embryos. Reprod Fertil Dev. 2008;20:253–7.
78. Kessler A, Biasibetti M, Feksa LR, Rech VC, Melo DA, Wajner M, Dutra-Filho CS, Wyse AT, Wannmacher CM. Effects of cysteamine on oxidative status in cerebral cortex of rats. Metab Brain Dis. 2008;23:81–93.
79. Salam OM. Modulation of inflammatory paw oedema by cysteamine in the rat. Pharmacol Res. 2002;45:275–84.
80. El-Shenawy SM, Abdel-Salam OM, Baiuomy AR, El-Batran S, Arbid MS. Studies on the anti-inflammatory and antinociceptive effects of melatonin in the rat. Pharmacol Res. 2002;46:235–43.
81. Shin YJ, Hyon JY, Kim S, Koh JW, Kwon SI, Wee WR. Cysteamine suppresses human peripheral blood mononuclear cells–human corneal endothelial cell reaction via reactive oxygen species reduction. Mol Vis. 2011;17:3371–8.
82. Williams KA, White MA, Ash JK, Coster DJ. Leukocytes in the graft bed associated with corneal graft failure: analysis by immunohistology and actuarial graft survival. Ophthalmology. 1989;96:38–44.
83. Smolin G, Biswell R. Corneal graft rejection associated with anterior iris adhesion: case report. Ann Ophthalmol. 1978;10:1603–4.
84. Khodadoust AA, Silverstein AM. Transplantation and rejection of individual cell layers of the cornea. Invest Ophthalmol. 1969;8:180–95.
85. Maenz M, Morcos M, Ritter T. A comprehensive flowcytometric analysis of graft infiltrating lymphocytes, draining lymph nodes and serum during the rejection phase in a fully allogeneic rat cornea transplant model. Mol Vis. 2011;17:420–9.
86. Khodr B, Khalil Z. Modulation of inflammation by reactive oxygen species: implications for aging and tissue repair. Free Radic Biol Med. 2001;30:1–8.
87. George L, Lokhandwala MF, Asghar M. Exercise activates redox-sensitive transcription factors and restores renal D1 receptor function in old rats. Am J Physiol Renal Physiol. 2009;297:F1174–80.
88. Vallières L, Rivest S. Interleukin-6 is a needed proinflammatory cytokine in the prolonged neural activity and transcriptional activation of corticotropin-releasing factor during endotoxemia. Endocrinology. 1999;140:3890–903.
89. Tak PP, Firestein GS. NF-κB: a key role in inflammatory diseases. J Clin Invest. 2001;107:7–11.
90. Schindler R, Mancilla J, Endres S, Ghorbani R, Clark SC, Dinarello CA. Correlations and interactions in the production of interleukin-6 (IL-6), IL-1, and tumor necrosis factor (TNF) in human blood mononuclear cells: IL-6 suppresses IL-1 and TNF. Blood. 1990;75:40–7.
91. Korpinen E, Groop PH, Fagerudd JA, Teppo AM, Akerblom HK, Vaarala O. Increased secretion of TGF-beta1 by peripheral blood mononuclear cells from patients with Type 1 diabetes mellitus with diabetic nephropathy. Diabet Med. 2001;18:121–5.
92. Libermann TA, Baltimore D. Activation of interleukin-6 gene expression through the NF-kappa B transcription factor. Mol Cell Biol. 1990;10:2327–34.
93. Bhat-Nakshatri P, Newton TR, Goulet Jr R, Nakshatri HNF. κB activation and interleukin 6 production in fibroblasts by estrogen receptor-negative breast cancer cell-derived interleukin 1α. Proc Natl Acad Sci U S A. 1998;95:6971–6.
94. Mou HB, Lin MF, Huang H, Cai Z. Transforming growth factor-β1 modulates lipopolysaccharide-induced cytokine/chemokine production and inhibits nuclear factor-κB, extracellular signal-regulated kinases and p38 activation in dendritic cells in mice. Transplant Proc. 2011;43:2049–52.
95. Murillo MM, Carmona-Cuenca I, Del Castillo G, Ortiz C, Roncero C, Sánchez A, Fernández M, Fabregat I. Activation of NADPH oxidase by transforming growth factor-beta in hepatocytes mediates up-regulation of epidermal growth factor receptor ligands through a nuclear factor-kappaBdependent mechanism. Biochem J. 2007;405:251–9.

96. Flynn TH, Mitchison NA, Ono SJ, Larkin DF. Aqueous humor alloreactive cell phenotypes, cytokines and chemokines in corneal allograft rejection. Am J Transplant. 2008;8:1537–43.
97. Funding M, Vorum H, Nexo E, Moestrup SK, Ehlers N, Moller HJ. Soluble CD163 and interleukin-6 are increased in aqueous humour from patients with endothelial rejection of corneal grafts. Acta Ophthalmol Scand. 2005;83:234–9.
98. Sagoo P, Chan G, Larkin DF, George AJ. Inflammatory cytokines induce apoptosis of corneal endothelium through nitric oxide. Invest Ophthalmol Vis Sci. 2004;45:3964–73.
99. Klass BR, Grobbelaar AO, Rolfe KJ. Transforming growth factor β1 signalling, wound healing and repair: a multifunctional cytokine with clinical implications for wound repair, a delicate balance. Postgrad Med J. 2009;85:9–14.
100. Ma FY, Tesch GH, Ozols E, Xie M, Schneider MD, Nikolic-Paterson DJ. TGF-β1-activated kinase-1 regulates inflammation and fibrosis in the obstructed kidney. Am J Physiol Renal Physiol. 2011;300:F1410–21.
101. Yi K, Chung TY, Hyon JY, Koh JW, Wee WR, Shin YJ. Combined treatment with antioxidants and immunosuppressants on cytokine release by human peripheral blood mononuclear cells—chemically injured keratocyte reaction. Mol Vis. 2011;17:2665–71.
102. Kubota M, Shimmura S, Kubota S, Miyashita H, Kato N, Noda K, Oawa Y, Usui T, Ishida S, Umezawa K, Kurihara T, Tsubota K. Hydrogen and N-acetyl-L-cysteine rescue oxidative stress-induced angiogenesis in a mouse corneal alkali-burn model. Invest Ophthalmol Vis Sci. 2011;52:427–33.
103. Wagoner MD. Chemical injuries of the eye: current concepts in pathophysiology and therapy. Surv Ophthalmol. 1997;41:275–313.
104. Nuijts RM. Autologous limbal transplantation in unilateral chemical burns. Doc Ophthalmol. 1999;98:257–66.
105. Valko M, Leibfritz D, Moncol J, Cronin MT, Mazur M, Telser J. Free radicals and antioxidants in normal physiological functions and human disease. Int J Biochem Cell Biol. 2007;39:44–84.
106. Dröge W. Free radicals in the physiological control of cell function. Physiol Rev. 2002;82:47–95.
107. Bulua AC, Simon A, Maddipati R, Pelletier M, Park H, Kim KY, Sack MN, Kastner DL, Siegel RM. Mitochondrial reactive oxygen species promote production of proinflammatory cytokines and are elevated in TNFR1-associated periodic syndrome (TRAPS). J Exp Med. 2011;208:519–33.
108. Zhou Q, Wang Y, Yang L, Wang Y, Chen P, Wang Y, Chen P, Wang Y, Dong X, Xie L. Histone deacetylase inhibitors blocked activation and caused senescence of corneal stromal cells. Mol Vis. 2008;14:2556–65.
109. Burdon RH, Gill V, Rice-Evans C. Oxidative stress and tumor cell proliferation. Free Radic Res Commun. 1990;11:65–76.
110. DiBartolomeis SM, Moné JP. Apoptosis: a four-week laboratory investigation for advanced molecular and cellular biology students. Cell Biol Educ. 2003;2:275–95.
111. Burdon RH, Rice-Evans C. Free radicals and the regulation of mammalian cell proliferation. Free Radic Res Commun. 1989;6:345–58.
112. Cho YS, Lee J, Lee TH, Lee EY, Lee KU, Park JY, Moon HB. Alpha-lipoic acid inhibits airway inflammation and hyperresponsiveness in a mouse model of asthma. J Allergy Clin Immunol. 2004;114:429–35.
113. Zhang WJ, Wei H, Hagen T, Frei B. Alpha-lipoic acid attenuates LPS-induced inflammatory responses by activating the phosphoinositide 3-kinase/Akt signaling pathway. Proc Natl Acad Sci U S A. 2007;104:4077–82.
114. Saja K, Babu MS, Karunagaran D, Sudhakaran PR. Antiinflammatory effect of curcumin involves downregulation of MMP-9 in blood mononuclear cells. Int Immunopharmacol. 2007;7:1659–67.
115. Coussens LM, Tinkle CL, Hanahan D, Werb Z. MMP-9 supplied by bone marrow-derived cells contributes to skin carcinogenesis. Cell. 2000;103:481–90.
116. Egeblad M, Werb Z. New functions for the matrix metalloproteinases in cancer progression. Nat Rev Cancer. 2002;2:161–74.

117. Matsubara M, Zieske JD, Fini ME. Mechanism of basement membrane dissolution preceding corneal ulceration. Invest Ophthalmol Vis Sci. 1991;32:3221–37.
118. Calandra T, Roger T. Macrophage migration inhibitory factor: a regulator of innate immunity. Nat Rev Immunol. 2003;3:791–800.
119. Wahl SM. Transforming growth factor beta (TGF-beta) in inflammation: a cause and a cure. J Clin Immunol. 1992;12:61–74.
120. Leung JC, Chan LY, Tsang AW, Liu EW, Lam MF, Tang SC, Lai KN. Anti-macrophage migration inhibitory factor reduces transforming growth factor-beta 1 expression in experimental IgA nephropathy. Nephrol Dial Transplant. 2004;19:1976–85.
121. Chaiyaroj SC, Rutta AS, Muenthaisong K, Watkins P, Na Ubol M, Looareesuwan S. Reduced levels of transforming growth factor-β1, interleukin-12 and increased migration inhibitory factor are associated with severe malaria. Acta Trop. 2004;89:319–27.
122. Andresen JL, Ehlers N. Chemotaxis of human keratocytes is increased by platelet-derived growth factor-BB, epidermal growth factor, transforming growth factor-alpha, acidic fibroblast growth factor, insulin-like growth factor-I, and transforming growth factor-beta. Curr Eye Res. 1998;17:79–87.
123. Shin YJ, Hyon JY, Choi WS, Yi K, Chung ES, Chung TY, Wee WR. Chemical injury-induced corneal opacity and neovascularization reduced by rapamycin via TGF-β1/ERK pathways regulation. Invest Ophthalmol Vis Sci. 2013;54:4452–8.
124. Gadjeva M, Nagashima J, Zaidi T, Mitchell RA, Pier GB. Inhibition of macrophage migration inhibitory factor ameliorates ocular Pseudomonas aeruginosa-induced keratitis. PLoS Pathog. 2010;6:e1000826.
125. Flaster H, Bernhagen J, Calandra T, Bucala R. The macrophage migration inhibitory factor-glucocorticoid dyad: regulation of inflammation and immunity. Mol Endocrinol. 2007;21:1267–80.
126. Calandra T, Bernhagen J, Metz CN, Spiegel LA, Bacher M, Donnelly T, Cerami A, Bucala R. MIF as a glucocorticoid-induced modulator of cytokine production. Nature. 1995;377:68–71.
127. Yu X, Lin SG, Huang XR, Bacher M, Leng L, Bucala R, Lan HY. Macrophage migration inhibitory factor induces MMP-9 expression in macrophages via the MEK-ERK MAP kinase pathway. J Interferon Cytokine Res. 2007;27:103–9.
128. Panda A, Jain M, Vanathi M, et al. Topical autologous platelet-rich plasma eyedrops for acute corneal chemical injury. Cornea. 2012;31:989–93.
129. Dohlman CH, Cade F, Pfister R. Chemical burns to the eye: paradigm shifts in treatment. Cornea. 2011;30:613–4.
130. Kwon YS, Kim JC. Inhibition of corneal neovascularization by rapamycin. Exp Mol Med. 2006;38:173–9.
131. Kwon YS, Hong HS, Kim JC, Shin JS, Son Y. Inhibitory effect of rapamycin on corneal neovascularization in vitro and in vivo. Invest Ophthalmol Vis Sci. 2005;46:454–60.
132. Sakimoto T, Sugaya S, Ishimori A, Sawa M. Anti-inflammatory effect of IL-6 receptor blockade in corneal alkali burn. Exp Eye Res. 2012;97:98–104.
133. Tzeng HE, Tsai CH, Chang ZL, et al. Interleukin-6 induces vascular endothelial growth factor expression and promotes angiogenesis through apoptosis signal-regulating kinase 1 in human osteosarcoma. Biochem Pharmacol. 2013;85(4):531–40.
134. Ito TK, Ishii G, Chiba H, Ochiai A. The VEGF angiogenic switch of fibroblasts is regulated by MMP-7 from cancer cells. Oncogene. 2007;26:7194–203.
135. Wilson SE. Corneal myofibroblast biology and pathobiology: generation, persistence, and transparency. Exp Eye Res. 2012;99:78–88.
136. Luckett-Chastain LR, Gallucci RM. Interleukin (IL)-6 modulates transforming growth factor-betaexpression in skin and dermal fibroblasts from IL-6-deficientmice. Br J Dermatol. 2009;161:237–48.
137. Tandon A, Tovey JC, Sharma A, Gupta R, Mohan RR. Role of transforming growth factor Beta in corneal function, biology and pathology. Curr Mol Med. 2010;10:5655–78.

138. Sonpavde G, Choueiri TK. Biomarkers: the next therapeutic hurdle in metastatic renal cell carcinoma. Br J Cancer. 2012;107:1009–16.
139. Gurk-Turner C, Manitpisitkul W, Cooper M. A comprehensive review of everolimus clinical reports: a new mammalian target of rapamycin inhibitor. Transplantation. 2012;94:659–68.
140. Kahan BD. Sirolimus: a comprehensive review. Expert Opin Pharmacother. 2001;2:1903–17.
141. Nair S, Ren J. Autophagy and cardiovascular aging: lesson learned from rapamycin. Cell Cycle. 2012;11:2092–9.
142. Tulek B, Kiyan E, Toy H, et al. Anti-inflammatory and anti-fibrotic effects of sirolimus on bleomycin-induced pulmonary fibrosis in rats. Clin Invest Med. 2011;34:E341.
143. Xu T, Xie JY, Wang WM, Ren H, Chen N. Impact of rapamycin on peritoneal fibrosis and transport function. Blood Purif. 2012;34:48–57.
144. Kure T, Chang JH, Kato T, et al. Corneal neovascularization after excimer keratectomy wounds in matrilysin-deficient mice. Invest Ophthalmol Vis Sci. 2003;44:137–44.
145. Xing D, Sun X, Li J, Cui M, Tan-Allen K, Bonanno JA. Hypoxia preconditioning protects corneal stromal cells against induced apoptosis. Exp Eye Res. 2006;82:780–7.
146. Shaul YD, Seger R. The MEK/ERK cascade: from signaling specificity to diverse functions. Biochim Biophys Acta. 2007;1773:1213–26.
147. Kretzschmar M, Doody J, Timokhina I, Massague J. A mechanism of repression of TGFbeta/Smad signaling by oncogenic Ras. Genes Dev. 1999;13:804–16.
148. Alarcon C, Zaromytidou A-I, Xi Q, et al. Nuclear CDKs drive Smad transcriptional activation and turnover in BMP and TGF-b pathways. Cell. 2009;139:757–69.
149. Hayashida T, Decaestecker M, Schnaper HW. Cross-talk between ERK MAP kinase and Smad signaling pathways enhances TGF-beta-dependent responses in human mesangial cells. FASEB J. 2003;17:1576–8.
150. Massagué J, Seoane J, Wotton D. Smad transcription factors. Genes Dev. 2005;19:2783–810.
151. Hay N, Sonenberg N. Upstream and downstream of mTOR. Genes Dev. 2004;18:1926–45.
152. Corradetti MN, Guan KL. Upstream of the mammalian target of rapamycin: do all roads pass through mTOR? Oncogene. 2006;25:6347–60.
153. Tsokas P, Ma T, Iyengar R, Landau EM, Blitzer RD. Mitogen-activated protein kinase upregulates the dendritic translation machinery in long-term potentiation by controlling the mammalian target of rapamycin pathway. J Neurosci. 2007;27:5885–94.
154. Winter JN, Jefferson LS, Kimball SR. ERK and Akt signaling pathways function through parallel mechanisms to promote mTORC1 signaling. Am J Physiol Cell Physiol. 2011;300:C1172–80.
155. Kishimoto T, Akira S, Narazaki M, Taga T. Interleukin-6 family of cytokines and gp130. Blood. 1995;86:1243–54.
156. Kennedy M, Kim KH, Harten B, et al. Ultraviolet irradiation induces the production of multiple cytokines by human corneal cells. Invest Ophthalmol Vis Sci. 1997;38:2483–91.
157. Cole N, Bao S, Willcox M, Husband AJ. Expression of interleukin-6 in the cornea in response to infection with different strains of Pseudomonas aeruginosa. Infect Immun. 1999;67:2497–502.
158. Cubitt CL, Lausch RN, Oakes JE. Differences in interleukin-6 gene expression between cultured human corneal epithelial cells and keratocytes. Invest Ophthalmol Vis Sci. 1995;36:330–6.
159. Park CY, Zhu Z, Zhang C, Moon CS, Chuck RS. Cellular redox state predicts in vitro corneal endothelial cell proliferation capacity. Exp Eye Res. 2006;83:903–10.
160. Kakkar P, Singh BK. Mitochondria: a hub of redox activities and cellular distress control. Mol Cell Biochem. 2007;305:235–53.
161. Armstrong JS. Mitochondria: a target for cancer therapy. Br J Pharmacol. 2006;147:239–48.
162. Ushio-Fukai M, Nakamura Y. Reactive oxygen species and angiogenesis: NADPH oxidase as target for cancer therapy. Cancer Lett. 2008;266:37–52.
163. Chance B, Schoener B, Oshino R, Itshak F, Nakase Y. Oxidation-reduction ratio studies of mitochondria in freeze-trapped samples: NADH and flavoprotein fluorescence signals. J Biol Chem. 1979;254:4764–71.

164. Ramey NA, Park CY, Gehlbach PL, Chuck RS. Imaging mitochondria in living corneal endothelial cells using autofluorescence microscopy. Photochem Photobiol. 2007;83:1325–9.
165. Reyes JM, Fermanian S, Yang F, et al. Metabolic changes in mesenchymal stem cells in osteogenic medium measured by autofluorescence spectroscopy. Stem Cells. 2006;24:1213–7.
166. Richter C, Gogvadze V, Laffranchi R, et al. Oxidants in mitochondria: from physiology to diseases. Biochim Biophys Acta. 1995;1271:67–74.
167. Gutteridge JM, Halliwell B. Free radicals and antioxidants in the year 2000: a historical look to the future. Ann N Y Acad Sci. 2000;899:136–47.
168. Szatrowski TP, Nathan CF. Production of large amounts of hydrogen peroxide by human tumor cells. Cancer Res. 1991;51:794–8.
169. Zamzami N, Larochette N, Kroemer G. Mitochondrial permeability transition in apoptosis and necrosis. Cell Death Differ. 2005;12:1478–80.
170. Murrell GA, Francis MJ, Bromley L. Modulation of fibroblast proliferation by oxygen free radicals. Biochem J. 1990;265:659–65.
171. Nicotera TM, Privalle C, Wang TC, Oshimura M, Barrett JC. Differential proliferative responses of Syrian hamster embryo fibroblasts to paraquat-generated superoxide radicals depending on tumor suppressor gene function. Cancer Res. 1994;54:3884–8.
172. Preeta R, Nair RR. Stimulation of cardiac fibroblast proliferation by cerium: a superoxide anion-mediated response. J Mol Cell Cardiol. 1999;31:1573–80.
173. Attene-Ramos MS, Kitiphongspattana K, Ishii-Schrade K, Gaskins HR. Temporal changes of multiple redox couples from proliferation to growth arrest in IEC-6 intestinal epithelial cells. Am J Physiol Cell Physiol. 2005;289:C1220–8.
174. Pani G, Colavitti R, Bedogni B, Anzevino R, Borrello S, Galeotti T. A redox signaling mechanism for density-dependent inhibition of cell growth. J Biol Chem. 2000;275:38891–9.
175. Patterson GH, Knobel SM, Arkhammar P, Thastrup O, Piston DW. Separation of the glucose-stimulated cytoplasmic and mitochondrial NAD(P)H responses in pancreatic islet beta cells. Proc Natl Acad Sci U S A. 2000;97:5203–7.
176. Hassinen I, Chance B. Oxidation-reduction properties of the mitochondrial flavoprotein chain. Biochem Biophys Res Commun. 1968;28:895–900.
177. Muzio G, Maggiora M, Paiuzzi E, Oraldi M, Canuto RA. Aldehyde dehydrogenases and cell proliferation. Free Radic Biol Med. 2012;52(4):735–46.
178. Tsai YY, Cheng YW, Lee H, Tsai FJ, Tseng SH, Lin CL, Chang KC. Oxidative DNA damage in pterygium. Mol Vis. 2005;25(11):71–5.
179. Maxia C, Perra MT, Demurtas P, Minerba L, Murtas D, Piras F, Corbu A, Gotuzzo DC, Cabrera RG, Ribatti D, Sirigu P. Expression of survivin protein in pterygium and relationship with oxidative DNA damage. J Cell Mol Med. 2008;12(6A):2372–80.
180. Nakamura S, Shibuya M, Nakashima H, Hisamura R, Masuda N, Imagawa T, Uehara M, Tsubota K. Involvement of oxidative stress on corneal epithelial alterations in a blink-suppressed dry eye. Invest Ophthalmol Vis Sci. 2007;48(4):1552–8.
181. Kojima T, Wakamatsu TH, Dogru M, Ogawa Y, Igarashi A, Ibrahim OM, Inaba T, Shimizu T, Noda S, Obata H, Nakamura S, Wakamatsu A, Shirasawa T, Shimazaki J, Negishi K, Tsubota K. Age-related dysfunction of the lacrimal gland and oxidative stress: evidence from the Cu, Zn-superoxide dismutase-1 (Sod1) knockout mice. Am J Pathol. 2012;180:1879–96.
182. Pagano G, Castello G, Pallardó FV. Sjøgren's syndrome-associated oxidative stress and mitochondrial dysfunction: prospects for chemoprevention trials. Free Radic Res. 2013;47:71–3.
183. Lal MA, Brismar H, Eklöf AC, Aperia A. Role of oxidative stress in advanced glycation end product-induced mesangial cell activation. Kidney Int. 2002;61:2006–14.

Chapter 2
Corneal Degenerations

Anna M. Roszkowska and Edward Wylegala

2.1 Introduction

Degenerations are defined as the gradual deterioration of a tissue or an organ that was previously normal with frequent loss of functional activity. Corneal degenerations are characterized by the deposition of material, thinning of tissue, and vascularization. Contrary to corneal dystrophies, they are not hereditary and could be unilateral. Moreover, because of the continuous exposure to environmental conditions, the cornea might undergo changes related to UV stimulation, and in some cases, oxidative stress might be responsible for the onset of degenerative processes [1, 2].

Corneal degenerations are characterized by a late onset; they can occur in several local diseases with mainly unilateral involvement or result from underlying systemic disorders or age-related processes. In the latter cases, they are bilateral even if asymmetric.

Corneal degenerations can be classified with relation to corneal changes. Generally, forms with abnormal peripheral corneal thinning and those with material deposition are recognized. The thinning processes are mainly of unknown etiology even if some general coexisting disorders were reported. In degenerations characterized by material deposition, the local and systemic causes underlying the corneal involvement could usually be identified [1].

A.M. Roszkowska, MD, PhD. (✉)
Ophthalmology Unit, Cornea and Refractive Surgery Section,
University Hospital of Messina, Messina, Italy
e-mail: hanka.roszkowska@gmail.com

E. Wylegala
Ophthalmology Clinic, Railway Hospital, Medical University of Silesia, Katowice, Poland

© Springer Science+Business Media New York 2015   23
M.A. Babizhayev et al. (eds.), *Studies on the Cornea and Lens*,
Oxidative Stress in Applied Basic Research and Clinical Practice,
DOI 10.1007/978-1-4939-1935-2_2

2.2 Age-Related Corneal Degenerations

Age-related corneal degenerations represent rather common ocular conditions in elderly. They appear both as corneal thinning and material deposition and generally do not require any therapy as they could lead to visual disturbances only in sporadic cases [3].

2.2.1 Arcus Senilis

Arcus senilis is considered the most frequent corneal degeneration in elderly, with a prevalence of above 90 % in over 80-year-old subjects. It is bilateral but frequently asymmetric. Clinically, it is represented by a circular band of extracellular deposits of lipoproteins with an evidently sharp external border separated from the limbus by the clear cornea. The internal border of the arcus is poorly defined, while the central cornea is always clear. Lipids deposition starts superiorly and inferiorly involving the deep stroma in proximity to the Descemet's membrane to extend to the sub-Bowman's superficial stroma and become circumferential (Fig. 2.1).

Beyond senescence, the lipid arcus can be present in young males with hypercholesterolemia and hypertriglyceridemia. In fact, such association is actually considered a sign of cardiovascular diseases, and a special medical care with cardiovascular evaluation should be recommended in the presence of the arcus in the young [3–5].

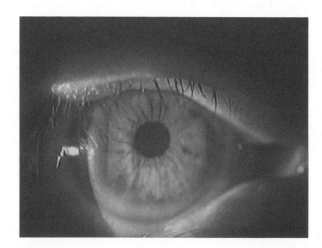

Fig. 2.1 Arcus senilis. (Archives Dr. A.M. Roszkowska)

2.2.2 Vogt Limbal Girdle

It is represented by limbal semilunar whitish lines that have an onset in the inter-palpebral rim with a prevalence on the nasal zone. Limbal girdle is characterized by stromal thinning frequently associated to subepithelial calcium deposition. This common degenerative corneal change is present in 100 % of over 80-year-old subjects [3, 6].

Vogt described two types of girdle: type I, separated from the limbus by a clear zone, includes calcium deposits and is considered an early form of band keratopa-thy, while type II consists of hyaline degeneration with hyperelastosis in the periph-eral Bowman's membrane (Fig. 2.2).

2.2.3 Crocodile Shagreen

It involves mainly the anterior corneal stroma, but in some rare cases, it can involve the posterior as well. Clinically, it presents evident polygonal cloudities separated by the clear lines that recall the crocodile shagreen (Fig. 2.3). Such clinical manifes-tation seems to be related to the particular abnormal orientation of the stromal fibers. Electron microscopy revealed the presence of vacuoles throughout the corneal stroma, with some of them containing electron-dense material. Crocodile shagreen corneal degeneration is rather uncommon and can be observed casually during a routine slit-lamp ocular examination. It is considered innocuous, but in its posterior variant, in some cases it can affect visual acuity [7, 8].

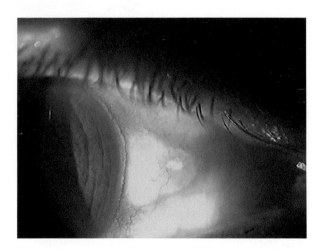

Fig. 2.2 Vogt limbal girdle. (Archives Dr. A.M. Roszkowska)

Fig. 2.3 Crocodile shagreen.
(Archives Dr.
A.M. Roszkowska)

Fig. 2.4 Hassall-Henle
bodies. (Archives Dr.
A.M. Roszkowska)

2.2.4 *Peripheral Corneal Guttae*

Age-related changes in the corneal endothelium are related above all to the progressive cell loss with reduction of the cell density [9, 10]. The peripheral guttae are also known as Hassall-Henle bodies and are formed by the localized deposition of abnormal collagen secerned by the aging endothelial cells (Fig. 2.4). They consist of sporadic excrescences on the corneal endothelium localized prevalently in the middle periphery. This differentiates them from the Fuchs' endothelial dystrophy, which starts in the center and spreads to the periphery, causing time corneal failure [11]. The number and extension of guttae are limited, so the endothelial cell density and function are not affected. In the pathogenesis of this age-related degeneration, the role of oxidative stress could be considered. In fact, oxidative stress involvement was established at the onset and progression of the Fuchs' endothelial dystrophy, where similar histopathological alterations occur [2].

2.3   Corneal Degenerations with Tissue Thinning

2.3.1   Terrien's Marginal Degeneration

Terrien described it in 1900 as a noninflammatory, painless, evolutive thinning in the peripheral cornea, and its etiology remains poorly understood. Association with autoimmune disorders was described, so the autoimmune mechanism underlying the pathogenesis was hypothesized. Recently, an association with the chronic inflammatory skin condition known as erythema elevatum diutinum was reported [12]. Terrien's marginal degeneration could be observed at any age, but prevalence in middle-aged males was observed. Generally, it is bilateral but asymmetric. Terrien's degeneration generally starts in the superior area with a local haze separated from the limbus by a clear cornea. Successively, the peripheral stromal thins with typically an intact epithelial layer that characterizes this degeneration. Then the degenerative processes extend in a circumferential manner involving progressively the whole corneal periphery, and in the advanced forms, the degeneration becomes circumferential. Progressive neovascularization could arise, as well as lipid deposits, in the central border of the thinning (Fig. 2.5). Patients are asymptomatic until visual changes occur because of the progressive increment of irregular astigmatism [13, 14]. In some sporadic cases, slight inflammation could be present, as reported by confocal findings, and it responds well to the topical steroid therapy [15, 16]. Histopathological studies showed an increased lysosomal activity together with the absence of the Bowman's membrane in the site of thinning [14]. As to the instrumental findings, the topographical examination demonstrates irregular astigmatism in advanced stages of the disease. The AC-OCT evaluation might be helpful in the evaluation of the peripheral thinning and in differential diagnosis with other peripheral corneal disorders. Recent AC-OCT findings report the stromal cavity formation in the peripheral cornea due probably to the collagen phagocytosis [17]. The differential diagnosis should consider the peripheral corneal melt, Mooren's

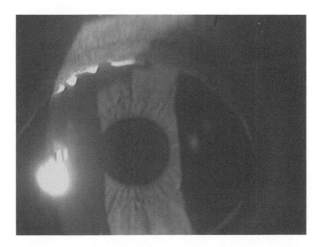

Fig. 2.5 Terrien's marginal degeneration. (Archives Dr. A.M. Roszkowska)

ulcer, pellucid marginal degeneration, and dellen. The treatment consists of optical correction of irregular astigmatism. In advanced forms, contact lenses are recommended. The most dreaded complication in severe Terrien's degeneration is related to corneal perforation. Perforation could be spontaneous or could result from any trauma, so the surgical approach should be considered when extreme thinning occurs [18]. Lamellar ring-shaped peripheral corneoscleral grafting is usually considered, but in some cases, tissue excision with appropriated suture could be performed [19, 20].

2.3.2 Dellen

Dellen is a localized area of corneal thinning in the periphery next to the limbus. It is located in proximity to the areas of tissue swelling or inflammation when correct spreading of the tear layer is highly disturbed. Frequently, it follows some surgical procedures that induce conjunctival chemosis with paralimbal elevation, which produces a localized abnormality of the tear film layer with following localized dehydration and thinning [21–24]. This relatively common condition is reversible. In fact, the restoration of the normal film layer spreading and intense lubrication constitute the appropriate therapy for this disorder. Recently, the use of a large-diameter soft contact lens was proven to be an effective therapeutic solution to treat the dellen that arose after a pterygium removal [25].

2.4 Corneal Degenerations with Material Deposits

2.4.1 Salzmann's Nodular Degeneration

Salzmann's nodular degeneration (SND) is a noninflammatory, slowly progressive disease characterized by single or multiple whitish-gray subepithelial corneal nodules. Salzmann described it in 1925 in association with keratoconjunctivitis, but with time, several different disorders coexisting with this disorder were detected. To date, its etiopathogenesis remains unknown, but it is considered to be associated with ocular surface inflammation or chronic trauma. However, several cases with no previous corneal pathology have been reported and are considered idiopathic.

Clinically, the typical clinical signs could be observed during slit-lamp examination, with either solitary or multiple nodules elevated on the corneal surface and located in the midperiphery of the cornea. In some cases, single or few nodules are present in one or more quadrants during routine ocular examination in asymptomatic patients, and sometimes, the nodules may invade the central zone. When multiple nodules are present, they have a typical circular arrangement and might be confluent in the advanced stages. A normal clear cornea between the nodules is a characteristic clinical feature (Fig. 2.6). The ocular surface disorders involved in the

Fig. 2.6 Salzmann nodular degeneration. (Archives Dr. A.M. Roszkowska)

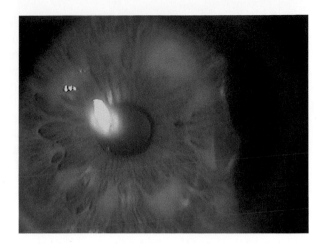

etiopathogenesis of Salzmann's degeneration are Meibomian gland dysfunction and dry eye disease followed by rosacea, phlyctenular keratitis, vernal keratoconjunctivitis, trachoma and interstitial keratitis. The chronic injuries that are most commonly associated with the development of the degenerative changes are extended contact lens use and previous corneal surgery followed by actinic exposure, recurrent erosions, and chemical or thermal injuries [26–29].

Recently, SND in patients with Crohn's disease was reported, and a hypothesis about association with systemic diseases was advanced [30, 31].

Therefore, although the etiology remains uncertain, it is hypothesized that Salzmann's nodular degeneration can be triggered by multiple events that lead to a nonspecific corneal tissue reaction based on individual predisposition. It appears to be more frequent in middle-aged women, and it is bilateral in about 60 % of cases. The main symptoms are visual disturbances and photophobia, ocular discomfort, mainly characterized by dryness, and foreign body sensation. In severe forms, characterized by very elevated nodules, painful epithelial lesions can occur. Visual impairment is progressive and due to an astigmatic defect produced by the nodules. Severe corneal irregularities in advanced stages generate high irregular astigmatism with severe visual loss. Although the diagnosis of Salzmann's degeneration is exclusively based on clinical findings, several instrumental examinations such as corneal topography, anterior segment optical coherence tomography, and confocal microscopy are useful to provide a better understanding of the corneal alterations and to quantify the amount of visual impairment. Corneal topography demonstrates the irregularities of the corneal shape produced by corneal nodules, and the induced surface alteration depends on their number and localization.

AC-OCT allows the evaluation of the nodules' dimension and depth, allowing the investigation of the stromal extension of the nodules that can help decide the surgical approach [32]. Confocal microscopy showed an increased reflectivity of the anterior stroma with abnormal and activated keratocytes and marked stromal scatter corresponding to the nodules as a result of the presence of fibrosis [33–35].

Histopathological studies have revealed that the corneal nodules are located subepithelially but may extend to one-third of the anterior stroma and are formed by a dense connective tissue with hyaline degeneration. The corneal epithelium has an irregular thickness, with extremely increased thinning over the corneal nodules. The Bowman's membrane exhibits disruption, and in the correspondence of the nodules, it is no longer appreciable and frequently replaced by fibrosis. Subepithelial fibrosis is a frequent histopathological finding, with activated fibroblasts beneath the epithelial cells. Increased expression of matrix metalloproteinase-2 (MMP-2) was recently detected in patients affected by Salzmann's degeneration, and this may be responsible for the induction of the basement membrane and Bowman's layer disruption [33, 35–37].

Therapy is related to the symptoms and severity. In early forms, treatment is aimed at reducing the presenting symptoms with conservative therapy concerning eyelid hygiene, lubricants, and anti-inflammatory eyedrops sufficiently to eliminate dry eye symptoms and foreign body sensation. Visual loss is an indication for surgical treatment, which consists of the removal of the nodules and the restoration of corneal surface regularity. In superficial nodules, manual removal with a crescent knife is sufficient and leaves a uniform smooth underlying surface. In the case of deeper ingrowth that involves the anterior stroma, after the nodule's removal, the excimer laser PTK is usually performed to smoothen the surface. In some cases with deep ingrowth, the anterior lamellar becomes necessary [38, 39]. The recurrence of Salzmann's nodules after surgical removal can occur with varying prevalence rates (18–21.9 %) and periods of time to recurrence (1 month to 6 years) [40–42].

2.4.2   Band Keratopathy

Band keratopathy consists of the deposition of calcium salts in the interpalpebral region. It starts in proximity to the limbus in the corneal periphery nasally and temporally at the same time. Successively, it spreads gradually to the center, forming a band-like whitish-gray plaque. Clear cornea separates the peripheral border of the band from the limbus. Visual acuity decreases with the advancement of deposits. In fact, in the advanced stages, the density of calcium deposition might create a severe visual impairment (Fig. 2.7). Moreover, in severe lesions, the deposits become elevated and produce considerable discomfort because of the ocular surface alterations.

Band keratopathy can affect a single eye when it arises after local disorders such as chronic anterior uveitis, interstitial keratitis, severe injury, chronic corneal inflammations, and phthisis bulbi. It is bilateral when generated by a general disorder. Systemic conditions underlying the band keratopathy onset are chronic renal failure, juvenile rheumatoid arthritis, and hyperuricemia.

Histology shows calcium salt deposits localized in the Bowman's layer, basement membrane of epithelium, and anterior stroma [43].

Fig. 2.7 Band Keratopathy.
(Archives Prof. E. Wylegala)

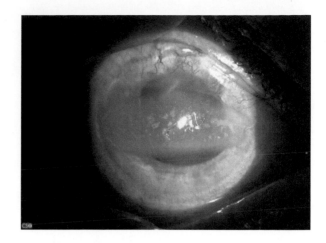

As for the treatment, the cause of band keratopathy should be addressed. The coexistence of an underlying systemic or local disorder must be recognized and treated first. Successively, calcium deposits are treated with a chelation procedure, which has been proven to be effective and easy to perform. Mechanical debridement was demonstrated to be effective as well in early stages. In case of dense and thick deposits, the manual scraping is necessary prior to chelation. Chelation is performed using a 1.5–3.0 % solution of ethylenediaminetetraacetic acid (EDTA) for 15–20 min. If required, the corneal surface might be further smoothened by excimer laser phototherapeutic keratectomy. Epithelial layer restoration could be enhanced by amniotic membrane application, which may help optimize the clinical outcome. Only in very advanced cases with severe visual impairment could the lamellar keratoplasty be considered a therapeutic option [43–45].

2.4.3 Lipid Keratopathy

Lipid keratopathy can occur spontaneously or may be a consequence of a previous ocular injury or inflammation. The first form is a primary one, rather infrequent, because of stromal deposits of cholesterol, fats, and phospholipids. It is not associated with vascularization.

The second form occurs as a secondary disorder associated to previous ocular diseases or injury. The secondary lipid keratopathy is much more common, and it is typically associated with corneal vascularization (Fig. 2.8). The most frequent corneal pathologies underlying the secondary lipid keratopathy are herpetic infections, both simplex and zoster. Treatment should focus on the underlying disease first. Successively, abnormal vascularization should be eliminated by argon laser photocoagulation or needle point cautery to induce the absorption of the lipids through the destruction of the feeder vessels.

Fig. 2.8 Lipid keratopathy.
(Archives Dr.
A.M. Roszkowska)

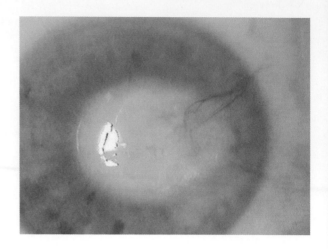

Corneal grafting remains the last treatment option, but it needs a relative quiescence, and the clinical outcome is rather poor because of the corneal thinning, hypesthesia, and persistent vascularization [46, 47].

2.4.4 Spheroidal Keratopathy

Spheroidal corneal degeneration is also known as corneal elastosis, Labrador keratopathy, climatic droplet keratopathy, Bietti's nodular dystrophy, proteinaceous corneal degeneration, elastotic degeneration, Fisherman's keratopathy, and Eskimo's corneal degeneration. It was described by Bietti in 1955 and is characterized by oil-like, amber-colored, spheroidal deposits at the limbus and in the peripheral interpalpebral cornea.

It arises typically in outdoor workers, and the main causing factor is ultraviolet exposure. It is more frequent in populations living in zones with high climatic UV exposure, and the severity is correlated positively with the time of exposure (Fig. 2.9).

With time, the corneal opacification increases, and deposits spread toward the center of the cornea. The advanced lesions are associated with localized haze. Sometimes the corneal changes coexist with the same conjunctival deposits. Spheroidal keratopathy is more frequent in males; it is bilateral and relatively innocuous, although in some cases, visual impairment might occur.

Histologic examination showed extracellular protein deposits in the anterior corneal stroma with elastotic degeneration. They are formed by UV-degraded protein materials that originated from the plasma through the limbal vessels. Progressively, such deposits slowly replace the Bowman's membrane [47–50]. In patients with spheroidal degeneration, protection against ultraviolet damage with sunglasses is mandatory. In advanced forms with central involvement, the various grades of visual impairment might occur, requiring appropriate therapeutic options. In such cases, superficial keratectomy and PTK are recommended, and good clinical and visual outcome could be expected.

Fig. 2.9 Spheroidal keratopathy, (Archives Prof. E. Wylegala)

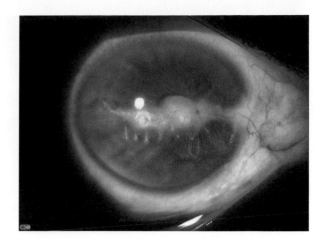

2.4.5   Vortex Keratopathy

This corneal degeneration is also known as cornea verticillata. It is bilateral and characterized by the whitish-gray or golden-brown deposits that form a whorl-like pattern in the inferior hemicornea. Its typical clinical appearance makes it easily recognizable (Fig. 2.10).

Vortex keratopathy is a form of drug-induced lipidosis because it results from chronic prolonged systemic therapies with amiodarone, chloroquine, hydroxychloroquine, indomethacin, and phenothiazine with deposition in the cornea of phospholipids not metabolized by lysosomal phospholipases. The vortex keratopathy is also typical for a lipid storage disorder called Fabry disease caused by genetically determined enzymatic deficiency. Corneal lipid depositions are localized in the basal epithelial cells [51–53]. Recent confocal investigation showed deeper stromal deposits in advanced forms, suggesting a higher toxicity than was thought until now in long-term treated patients [54]. Generally, no visual alterations are present. However, in the extended drug-induced keratopathy, some visual disturbance might occur, indicating a high toxicity. In such case, the modification of systemic therapy could be considered. Vortex keratopathy is reversible and decreases when systemic therapy is switched or reduced.

2.4.6   Pigment Deposition Keratopathies

These corneal degenerations consist of deposition of different pigmented substances in corneal layers. The most frequent are due to the epithelial iron deposits. Beyond the Hudson–Stahli line observed sporadically in normal corneas of aged individuals, the iron deposits are typically present in particular corneal conditions when the regular corneal profile is modified. This can be observed in the cases of pterygium,

Fig. 2.10 Vortex keratopathy.
(Archives Prof. E. Wylegala)

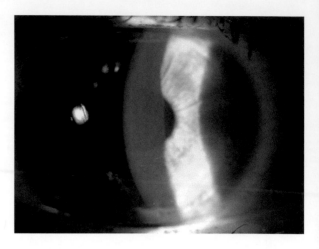

keratoconus, previous corneal refractive surgery, corneal scars, and filtering blebs [55, 56]. In these conditions, the common finding is related to the high variation of corneal shape, and the deposits are localized typically in the zones where such changes occur. Corneal profile variations generate alteration of the physiological lacrimal film distribution that probably underlies this keratopathy. Histologically, the deposits are formed by intraepithelial hemosiderin in the basal cells.

2.4.6.1 Pterygium

In the advanced forms, the iron deposits separate the advancing head of pterygium exactly where the corneal surface undergoes variations. In this case, the deposits form the so-called Stocker-Hali line [57].

2.4.6.2 Keratoconus

In keratoconus, circular iron deposits surround the base of the cone where extreme variations of the corneal curvature could be detected, and they are commonly known as Fleischer ring [58, 59] (Fig. 2.11).

2.4.6.3 Refractive Surgery

Central corneal iron deposits in corneal epithelium are a common side effect of corneal refractive surgery, both incisional and excimer laser (Fig. 2.12). Corneal refractive procedures change corneal curvature by inducing central flattening or steepening. As previously mentioned, the important changes in corneal profile, induce alteration in tear film dynamics that promote epithelial deposition of iron. So

Fig. 2.11 Fleischer's ring in keratoconus. (Archives Prof. E. Wylegala)

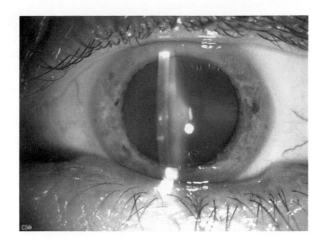

Fig. 2.12 Central corneal iron deposits in eye with previous radial keratotomy . (Archives Dr. A.M. Roszkowska)

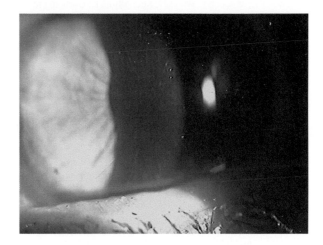

in eyes that have undergone hyperopic refractive corneal procedures with laser-induced excessive central corneal steepening pseudo-Fleischer ring, could be observed. In myopic correction, the iron lines could be appreciated in the central cornea where the surgically induced flattening occurs [60–62].

2.5 Conclusions

In this chapter, the main corneal degenerations were considered. Such corneal disorders are not hereditary, have a late onset, and might be related to local or systemic diseases or can arise with aging. In the first case, degenerations occur only in the affected eye, while in the last, the bilateral involvement is present. Some forms of degeneration might induce severe visual impairment, and appropriate therapy is necessary to recovery and maintenance of visual acuity. Therapy is aimed to resolve the underlying disorder first and only successively to treat the corneal changes. The age-related degenerations are bilateral, usually innocuous for visual capacity. Correct diagnosis is mandatory for therapeutic choice and the efficacy of clinical outcomes.

References

1. Friedlaender MH, Smolin G. Corneal degenerations. Ann Ophthalmol. 1979;11(10): 1485–95.
2. Saccà SC, Roszkowska AM, Izzotti A. Environmental light and endogenous antioxidants as the main determinants of non-cancer ocular diseases. Mutat Res. 2013;752(2):153–71.
3. Kotulak JC, Brungardt T. Age-related changes in the cornea. J Am Optom Assoc. 1980;51(8):761–5.
4. Chua BE, Mitchell P, Wang JJ, Rochtchina E. Corneal arcus and hyperlipidemia: findings from an older population. Am J Ophthalmol. 2004;137(2):363–5.
5. Ang M, Wong W, Park J, Wu R, Lavanya R, Zheng Y, Cajucom-Uy H, Tai ES, Wong TY. Corneal arcus is a sign of cardiovascular disease, even in low-risk persons. Am J Ophthalmol. 2011;152(5):864–71.
6. Sugar HS, Kobernick S. The white limbus girdle of Vogt. Am J Ophthalmol. 1960;50:101–7.
7. Belliveau MJ, Brownstein S, Agapitos P, Font RL. Ultrastructural features of posterior crocodile shagreen of the cornea. Surv Ophthalmol. 2009;54(5):569–75.
8. Meyer JC, Quantock AJ, Thonar EJ, Kincaid MC, Hageman GS, Assil KK. Characterization of a central corneal cloudiness sharing features of posterior crocodile shagreen and central cloud dystrophy of François. Cornea. 1996;15(4):347–54.
9. Roszkowska AM, Colosi P, D'Angelo P, Ferreri G. Age-related modifications of the corneal endothelium in adults. Int Ophthalmol. 2004;25(3):163–6.
10. Hillenaar T, van Cleynenbreugel H, Remeijer L. How normal is the transparent cornea? Effects of aging on corneal morphology. Ophthalmology. 2012;119(2):241–8.
11. Bergmanson JP, Sheldon TM, Goosey JD. Fuchs' endothelial dystrophy: a fresh look at an aging disease. Ophthalmic Physiol Opt. 1999;19(3):210–22.
12. Shimazaki J, Yang HY, Shimmura S, Tsubota K. Terrien's marginal degeneration associated with erythema elevatum diutinum. Cornea. 1998;17(3):342–4.
13. Beauchamp GR. Terrien's marginal corneal degeneration. J Pediatr Ophthalmol Strabismus. 1982;19(2):97–9.
14. Pouliquen Y, Dhermy P, Renard G, Goichot-Bonnat L, Foster G, Savoldelli M. Terrien's disease: clinical and ultrastructural studies, five case reports. Eye (Lond). 1989;3:791–802.
15. Ceresara G, Migliavacca L, Orzalesi N, Rossetti L. In vivo confocal microscopy in Terrien marginal corneal degeneration: a case report. Cornea. 2011;30(7):820–4.
16. Ferrari G, Tedesco S, Delfini E, Macaluso C. Laser scanning in vivo confocal microscopy in a case of Terrien marginal degeneration. Cornea. 2010;29(4):471–5.
17. Hattori T, Kumakura S, Mori H, Goto H. Depiction of cavity formation in Terrien marginal degeneration by anterior segment optical coherence tomography. Cornea. 2013;32(5):615–8.
18. Srinivasan S, Murphy CC, Fisher AC, Freeman LB, Kaye SB. Terrien marginal degeneration presenting with spontaneous corneal perforation. Cornea. 2006;25(8):977–80.
19. Cheng CL, Theng JT, Tan DT. Compressive C-shaped lamellar keratoplasty: a surgical alternative for the management of severe astigmatism from peripheral corneal degeneration. Ophthalmology. 2005;112(3):425–30.
20. Wang T, Shi W, Ding G, Li S, Liu M, Gao H. Ring-shaped corneoscleral lamellar keratoplasty guided by high-definition optical coherence tomography and Scheimpflug imaging for severe

Terrien's marginal corneal degeneration. Graefes Arch Clin Exp Ophthalmol. 2012;250:1795–801.
21. Accorinti M, Gilardi M, Giubilei M, De Geronimo D, Iannetti L. Corneal and scleral dellen after an uneventful pterygium surgery and a febrile episode. Case Rep Ophthalmol. 2014;28:111–5.
22. Lin SB, Li ZY, Zhan XK, Cai JH. A case of corneoscleral dellen after medial rectus recession combined with pterygium resection in right eye. Eye Sci. 2012;27(4):198–201.
23. Fenzl CR, Moshirfar M, Gess AJ, Muthappan V, Goldsmith J. Dellen-like keratopathy associated with glaucoma drainage devices. World J Clin Cases. 2014;2(1):1–4.
24. Ono T, Yuki K, Ozeki N, Shiba D, Tsubota K. Ocular surface complications after trabeculectomy: incidence, risk factors, time course, and prognosis. Ophthalmologica. 2013;230(2):93–9.
25. Kymionis GD, Plaka A, Kontadakis GA, Astyrakakis N. Treatment of corneal dellen with a large diameter soft contact lens. Cont Lens Anterior Eye. 2011;34(6):290–2.
26. Das S, Link B, Seitz B. Salzmann's nodular degeneration of the cornea: a review and case series. Cornea. 2005;24:772–7.
27. Farjo AA, Halperin GI, Syed N, Sutphin JE, Wagoner MD. Salzmann's nodular corneal degeneration clinical characteristics and surgical outcomes. Cornea. 2006;25:11–5.
28. Graue-Hernandez E, Mannis MJ, Eliasieh K, Greasby TA, Beckett LA, Bradley JC, Schwab IR. Salzmann nodular degeneration. Cornea. 2010;29:283–9.
29. Lim MC, Chan WK. Salzmann nodular degeneration after laser in situ keratomileusis. Cornea. 2009;28:577.
30. Lange AP, Bahar I, Sansanayudh W, Kaisermann I, Slomovic AR. Salzmann nodules: a possible new ocular manifestation of Crohn's disease: a case report. Cornea. 2009;28:85–6.
31. Roszkowska AM, Spinella R, Aragona P. Recurrence of Salzmann nodular degeneration of the cornea in a Crohn's disease patient recurrence. Int Ophthalmol. 2013;33(2):185–7.
32. Humeric V, Yoo SH, Karp CL, Galor A, Vajzovic L, Wang J, Dubovy SR, Forster RK. In vivo morphologic characteristics of Salzmann nodular degeneration with ultra-high-resolution optical coherence tomography. Am J Ophthalmol. 2011;151:248–56.
33. Meltendorf C, Bühren J, Bug R, Ohrloff C, Kohnen T. Correlation between clinical in vivo confocal microscopic and ex vivo histopathologic findings of Salzmann nodular degeneration. Cornea. 2006;25:734–8.
34. Ku JY, Grupcheva CN, McGhee CN. Microstructural analysis of Salzmann's nodular degeneration by in vivo confocal microscopy. Clin Experiment Ophthalmol. 2002;30:367–8.
35. Roszkowska AM, Aragona P, Spinella S, Pisani A, Puzzolo D, Micali A. Morphological and confocal investigation on Salzmann nodular degeneration of the cornea. Invest Ophthalmol Vis Sci. 2011;52:5910–9.
36. Stone DU, Astley RA, Shaker RP, Chodosh J. Histopathology of Salzmann nodular corneal degeneration. Cornea. 2008;27:148–51.
37. Vannas A, Hogan MJ, Wood I. Salzmann's nodular degeneration of the cornea. Am J Ophthalmol. 1975;79:211–9.
38. Roszkowska AM, Colosi P, De Grazia L, Mirabelli E, Romeo G. One year outcome of manual alcohol assisted removal of Salzmann's nodular degeneration. Graefes Arch Clin Exp Ophthalmol. 2009;247:1431–4.
39. Germundsson J, Fagerholm P. Phototherapeutic keratectomy in Salzmann's nodular degeneration. Acta Ophthalmol Scand. 2004;82(2):148–53.
40. Sinha R, Chhabra MS, Vajpayee RB, Kashyap S, Tandon R. Recurrent Salzmann's nodular degeneration: Report of two cases and review of literature. Indian J Ophthalmol. 2006;54:201–2.
41. Severin M, Kirchhof B. Recurrent Salzmann's corneal degeneration. Graefes Arch Clin Exp Ophthalmol. 1990;228:101–4.
42. Yoon KC, Park YG. Recurrent Salzmann's nodular degeneration. Jpn J Ophthalmol. 2003;47:401–4.
43. Jhanji V, Rapuano CJ, Vajpayee RB. Corneal calcific band keratopathy. Curr Opin Ophthalmol. 2011;22(4):283–9.

44. Najjar DM, Cohen EJ, Rapuano CJ, Laibson PR. EDTA chelation for calcific band keratopathy: Results and long-term follow-up. Am J Ophthalmol. 2004;137:1056–64.
45. Anderson DF, Prabhasawat P, Alfonso E, Tseng SC. Amniotic membrane transplantation after the primary surgical management of band keratopathy. Cornea. 2001;20(4):354–61.
46. Croxatto JO, Dodds CM, Dodds R. Bilateral and massive lipoidal infiltration of the cornea (secondary lipoidal degeneration). Ophthalmology. 1985;92(12):1686–90.
47. Levy J, Benharroch D, Lifshitz T. Bilateral severe progressive idiopathic lipid keratopathy. Int Ophthalmol. 2005;26:181–4.
48. Johnson GJ. Aetiology of spheroidal degeneration of the cornea in Labrador. Br J Ophthalmol. 1981;65(4):270–83.
49. Gray RH, Johnson GJ, Freedman A. Climatic droplet keratopathy. Surv Ophthalmol. 1992;36(4):241–53.
50. Johnson GJ, Overall M. Histology of spheroidal degeneration of the cornea in Labrador. Br J Ophthalmol. 1978;62(1):53–61.
51. D'Amico DJ, Kenyon KR, Ruskin JN. Amiodarone keratopathy: drug-induced lipid storage disease. Arch Ophthalmol. 1981;99(2):257–61.
52. Hollander DA, Aldave AJ. Drug-induced corneal complications. Curr Opin Ophthalmol. 2004;15(6):541–8.
53. Falke K, Büttner A, Schittkowski M, Stachs O, Kraak R, Zhivov A, Rolfs A, Guthoff R. The microstructure of cornea verticillata in Fabry disease and amiodarone-induced keratopathy: A confocal laser-scanning microscopy study. Graefes Arch Clin Exp Ophthalmol. 2009;247(4):523–34.
54. Ciancaglini M, Carpineto P, Zuppardi E, Nubile M, Doronzo E, Mastropasqua L. In vivo confocal microscopy of patients with amiodarone-induced keratopathy. Cornea. 2001;20(4):368–73.
55. Rose GE, Lavin MJ. The Hudson-Stahli line. I: An epidemiological study. Eye (Lond). 1987;1:466–70.
56. Rose GE, Lavin MJ. The Hudson-Stahli line. III: Observations on morphology, a critical review of aetiology and a unified theory for the formation of iron lines of the corneal epithelium. Eye (Lond). 1987;1:475–9.
57. Hansen A, Norn M. Astigmatism and surface phenomena in pterygium. Acta Ophthalmol (Copenh). 1980;58(2):174–81.
58. Iwamoto T, DeVoe AG. Electron microscopical study of the Fleisher ring. Arch Ophthalmol. 1976;94(9):1579–84.
59. Romero-Jiménez M, Santodomingo-Rubido J, Wolffsohn JS. Keratoconus: a review. Cont Lens Anterior Eye. 2010;33(4):157–66.
60. Bricola G, Scotto R, Mete M, Cerruti S, Traverso CE. A 14-year follow-up of photorefractive keratectomy. J Refract Surg. 2009;25(6):545–52.
61. Probst LE, Almasswary MA, Bell J. Pseudo-Fleischer ring after hyperopic laser in situ keratomileusis. J Cataract Refract Surg. 1999;25(6):868–70.
62. Yeung L, Chen YF, Lin KK, Huang SC, Hsiao CH. Central corneal iron deposition after myopic laser-assisted in situ keratomileusis. Cornea. 2006;25(3):291–5.

Chapter 3
Corneal Epithelial Nuclear Ferritin and Its Transporter Ferritoid Afford Unique Protection to DNA from UV Light and Reactive Oxygen Species

Thomas F. Linsenmayer, Kelly E. Beazley, Cindy X. Cai, James P. Canner, John M. Fitch, James K. Kubilus, John M. Millholland, Maria Nurminskaya, Christopher Talbot, and Naomi B. Zak

3.1 Introduction

Ultraviolet (UV) light constitutes a major environmental hazard to all exposed tissues of the body, including the cornea. UV light can damage a wide variety of macromolecules ranging from DNA to proteins and lipids. This damage can result directly from the radiation itself, or it can be indirect through the radiation-mediated generation of reactive oxygen species (ROS). The resulting cellular changes can be severe. In skin, for example, UV-induced damage to DNA is thought to be a major factor in the increasing incidence of epidermal cancers—if not the major one [23]. However, cells of the corneal epithelium (CE) seem to be refractory to such damage, as primary cancers of these cells are extraordinarily rare—even though this tissue is transparent and is constantly exposed to ROS-generating UV light and O_2 [44]. This suggests that CE cells have evolved defense mechanisms that prevent damage to their DNA. Studies in our laboratory suggest that one such mechanism involves having the iron-sequestering molecule ferritin in a nuclear localization—rather than the cytoplasmic location it typically has in other cell types [11]. As will be discussed, this seems to greatly diminish the damage of UV-generated ROS to DNA—and possibly other nuclear components—most likely by sequestering free iron and thus inhibiting the Fenton reaction (Fig. 3.1). So, in this way it serves a unique protective function for CE cells.

Numerous studies have shown that DNA is a major target of UV-induced cellular damage. Damage to DNA can be direct, involving for example, DNA strand

T.F. Linsenmayer, Ph.D. (✉) • J.K. Kubilus • C. Talbot
Department of Integrated Physiology and Pathobiology, Tufts University Medical School, 136 Harrison Avenue, Boston, MA 02111, USA
e-mail: Thomas.Linsenmayer@tufts.edu

© Springer Science+Business Media New York 2015   39
M.A. Babizhayev et al. (eds.), *Studies on the Cornea and Lens*,
Oxidative Stress in Applied Basic Research and Clinical Practice,
DOI 10.1007/978-1-4939-1935-2_3

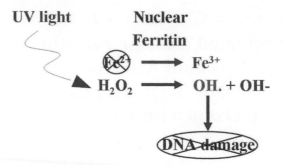

Fig. 3.1 In our model for the protection of the DNA of CE cells from UV-induced damage the ROS, here H_2O_2, is produced by UV light. If free iron is present, this catalyzes the formation of hydroxyl radicals (OH·) by the Fenton reaction. And these hydroxyl radicals, being the most energetic ROS can damage DNA. If, however, nuclear ferritin is present to block the Fenton reaction, such damage will be prevented

breakage and DNA–protein cross-linking. Or, it can be indirect, resulting from UV-induced formation of ROS, which, in turn, cause damage to DNA through alterations in the bases and deoxyribose moieties, strand breaks, and DNA cross-linking [21]. These changes can be mutagenic and carcinogenic [13]. ROS can also damage lipids and proteins. Thus, cells have evolved protective mechanisms against such oxidative damage [2] A growing body of evidence suggests that the ferritin-mediated sequestration of iron is one of these [3]—and that other iron-binding proteins, such as lactoferrin, can function similarly [42].

3.2   Free Iron and UV-Induced Damage to DNA

The deleterious effects of UV-induced ROS are greatly exacerbated by free iron, which catalyzes the Fenton reaction [46]. In the Fenton reaction (Fig. 3.1) Fe^{2+} catalyzes the conversion of hydrogen peroxide (H_2O_2) and superoxide (O_2^-) to the hydroxyl radical (OH·)—which, although acting over a short distance, is the most active ROS [28]. The Fenton reaction has been shown to increase H_2O_2-produced DNA breaks in preparations of nuclei [35,36] and rat hepatocytes [43]. At least part of the Fenton-reaction-mediated damage to nuclear DNA results from the presence of free iron in the nucleus [37]. The function of iron in the nucleus is unknown; however, its deleterious effects on DNA have been demonstrated [37]. It has also been determined that free iron can bind to specific sites on DNA and that this can generate Fenton-reaction-derived OH-radicals which specifically cleave the DNA at these sites [25,27,32].

Thus, we hypothesize that it is important to keep free iron in the nucleus at a low level—especially true for CE cells which are constantly exposed to environmental

Ferritin Subunit   Ferritin Molecule

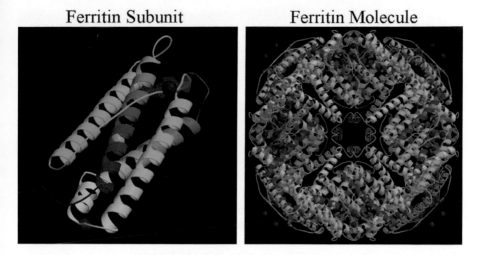

Fig. 3.2 Ferritin is a ubiquitous iron-storage molecule that is capable of sequestering up to 4,500 atoms of iron in its core. Ferritin subunits are comprised of five helical domains, and the "typical" ferritin molecule is a supramolecular complex comprised of 24 of these subunits

ROS-generating UV light and O_2. Our studies suggest that one mechanism through which this is achieved is by having ferritin in the nucleus where it would be most effective in sequestering free iron.

3.3   Cytoplasmic Ferritin

Cytoplasmic ferritin has been extensively characterized. It is a highly conserved protein composed of 24 subunits (Fig. 3.2) with an aggregate molecular weight of >450 kD. It has a central core that can sequester as many as 4,500 molecules of iron in the ferric state. Each ferritin subunit (Fig. 3.2) has a molecular weight of ~19 kD and is comprised of five α-helices connected by loops of varying lengths [22]. In most mammalian species ferritin has heavy (H) and light (L) chain subunits, however, in the chicken, the species we have utilized for our work, only the H-chain has been detected—suggesting that the molecule is homopolymeric.

A number of the characteristics of the cytoplasmic ferritin molecule have been elucidated. This includes its supramolecular assembly, which involves interactions of the ferritin subunits, through specific molecular regions and sites in their primary sequence. To generate this complex, individual ferritin subunits associate with one another through interactions of their helical domains—which is a concept that is part of our proposed mechanism for the nuclear transport of ferritin (described later).

Assembly starts with the formation of dimers and ends as the 24-mer complex [29]. Functionally, molecular assembly is required for the importation of free iron and its sequestration within the core of the molecule—which removes free iron from participation in the generation of Fenton reaction-derived OH· radicals (Fig. 3.1) [17].

3.4   The Nuclear Ferritin of Corneal Epithelial Cells

In most cells ferritin is cytoplasmic—presumably due to the large size of the assembled molecule preventing its passive diffusion into the nucleus. However, our studies show that in avian CE cells, the ferritin is largely nuclear.[1]

The initial observation suggesting the possibility of ferritin being a nuclear component of CE cells arose serendipitously while screening—by IHC—a "whole-corneal" monoclonal antibody library for antibodies of potential interest for studying corneal development [49]. One of these antibodies (6D11), when tested on sections of anterior eyes, showed staining only of the CE. And subsequent, more extensive and detailed analyses using this antibody [10] showed its IHC staining to be localized exclusively to the nuclei of the CE cells (Figs. 3.3 and 3.12). Also, when this antibody was used to immunoscreen a cDNA expression library—constructed with cDNAs from chicken CE tissue—all of the positive clones had inserts whose sequence was "typical" cytoplasmic ferritin.

Thus, avian CE cells have nuclear ferritin, and this is a "typical ferritin", so it is not a different isoform that is modified to undergo nuclear transport. Instead, there is considerable evidence that CE cells have a tissue-specific nuclear transporter that associates with ferritin and carriers it into the nucleus. This transporter has been termed ferritoid—for its similarities to a ferritin subunit [38] and described later. In addition, evidence will be presented that following their nuclear transport, ferritin and ferritoid remain associated with one another as a unique, heteropolymeric complex(es) that has characteristics that differentiate it from all other known vertebrate ferritins [39].

3.5   The Role of Nuclear Ferritin in Protecting CE Cell DNA from Damage by UV Radiation

Based on the model presented in Fig. 3.1, studies were initiated to determine whether nuclear ferritin might be involved in preventing the DNA of CE cells from UV damage.

---

[1] However, these in vitro studies also showed that ferritoid undergoes upregulation at the translational level—as the stimulation of mRNA synthesis, in itself, does not account quantitatively for the magnitude of the increase in ferritoid protein. Thus, for ferritoid it seems that following synthesis of its RNA, translational upregulation regulation also becomes important. (For a more detailed discussion of this see [6]).

6D11   Hoechst

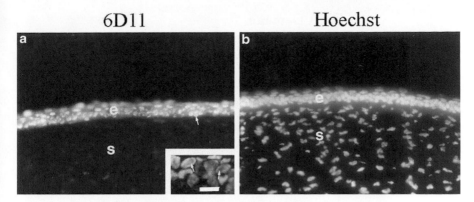

Fig. 3.3 (a) Fluorescence IHC for ferritin on a section of mature, embryonic cornea. The CE is labeled "e", the stroma is labeled "s". The inset shows a higher magnification of the CE. The only detectible antibody staining is within the CE cell nuclei. There it is distributed throughout—except for the nucleoli (see inset). (b) Hoechst dye that identifies nuclei. This research was originally published in the Journal of Biological Chemistry; Cai CX, Birk DE, Linsenmayer TF. Ferritin Is a Developmentally Regulated Nuclear Protein of Avian Corneal Epithelial Cells. 1997;272(19):12831–12839, © The American Society for Biochemistry and Molecular Biology

One approach was to UV-irradiate primary cultures of CE cells, and then evaluating the extent of strand breakage of their nuclear DNA—using fluorescence 3'-in situ-*end labeling* (ISEL) (as can be seen in Fig. 3.4).[2]

This assay was used in a number of studies (e.g., [11], and reviewed in [30]). One study compared the relative effects of UV-induced damage to CE cells (with their nuclear ferritin)—versus cell types that do not have nuclear ferritin (e.g., corneal fibroblasts, and epidermal and dermal cells). Of these cell types, CE cells were by far the most resistant to DNA damage. Then, it was determined whether prevention of iron-mediated damage was involved—implicating the Fenton reaction. If so, the UV-induced damage should be exacerbated by elevated iron in the cell types without nuclear ferritin, but CE, with their normal elevated iron CE cells should be refractory to elevated iron. To test this, the effect of elevated iron on UV-induced damage was examined for CE cells versus corneal fibroblasts. For this, both types of cells were cultured in medium with the normal concentration of iron (normal-iron medium) versus medium in which the concentration of iron was elevated (high-iron medium). Then, the cells were irradiated and the damage was accessed by ISEL. In high-iron medium, the irradiated fibroblasts showed a large increase in

---

[2] As primary cultures of pre-ferritin stage CE cells sometimes do not provide sufficient material for certain types of analyses, whole corneal organ cultures were employed (i.e., corneas floating in medium). The behavior of the CE of such corneas was identical to that of CE cells in primary culture, and the quantities of CE harvested from such cultured corneas [using treatment with the enzyme Dispase [45] easily provides sufficient CE tissue for assays most assays (e.g., qRT-PCR, protein determinations, and microarrays.

DNA Damage   Hoechst

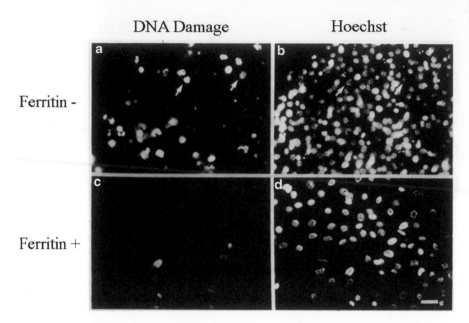

Ferritin -

Ferritin +

Fig. 3.4 All of our results are consistent with the nuclear ferritin of CE cells being protective against damage from UV irradiation and H_2O_2. In the experiment shown here, when CE cells are cultured in the presence of the iron chelator DFX (**a**, **b**), which prevents their synthesis of ferritin (ferritin−), and then are treated with UV light, extensive damage occurs (as determined by ISEL) when compared to cells with normal ferritin synthesis (**c**, **d**; ferritin+). Modified from [11]

damage to their DNA. However, for the CE cells increasing the iron produced little increase, if any, over their normal low level of damage.

These results are consistent with the CE cells having a mechanism(s) that prevents Fenton-reaction-mediated damage. They are also consistent with this mechanism involving ferritin in a nuclear location, as the nuclear ferritin of the CE cells was able to prevent UV-induced DNA damage—even in elevated iron—but the cytoplasmic ferritin of the fibroblasts did not confer protection.

Then to test directly whether the nuclear ferritin is responsible for preventing the UV-induced damage, experiments were performed in which the nuclear ferritin itself was manipulated. These experiments were based on the observations that pre-ferritin stage embryonic CE cells, when placed in culture, will initiate (precociously) the production of nuclear ferritin. However, if the iron chelator deferoxamine (DFX) is included in the medium, the initiation of ferritin synthesis is blocked (shown in detail in Fig. 3.15). Consistent with this nuclear ferritin being responsible for conferring protection, when the cultures with nuclear ferritin were irradiated (Fig. 3.4, ferritin+) the only cells showing an ISEL signal were occasional small rounded cells that were most likely naturally occurring, apoptotic cells—as these were also seen in un-irradiated cultures. However, in the cultures in which the nuclear synthesis was blocked (Fig. 3.4, ferritin−) most cells showed strong ISEL signals.

A modification of this experimental approach is based on the observation that the DFX blockage of nuclear ferritin synthesis can be reversed by replacing the DFX-containing medium with iron-containing medium (without DFX) (shown in Fig. 3.16). In addition, during the recovery from DFX there is a "window in time" in which some cells have initiated the production of nuclear ferritin, whereas others have not. Thus, irradiating cultures during this period makes it possible, in the same culture, to compare UV-induced damage to cells that have nuclear ferritin with ones that do not (data not presented, but see [11] and the review by Linsenmayer et al. [30]. Again, the cells that had nuclear ferritin (identified by IHC) showed little damage (by ISEL), and those that lacked nuclear ferritin showed damage. Also in the small population of cells that exhibited signals for both nuclear ferritin and damage to DNA, the ones with the least ferritin showed the most damage to DNA, and vice versa.

3.5.1   Protection from Damage by Reactive Oxygen Species (ROS)

If the protection afforded by CE nuclear ferritin involves the prevention of Fenton reaction-derived OH· radicals, protection should also be conferred against damage by ROS (e.g., H_2O_2) (as diagrammed in Fig. 3.1). So this was examined [9].

Similar to the UV-induced damage to DNA, the damage induced by H_2O_2 (determined by ISEL) was concentration and time-dependent. Of the four cell types tested (CE cells, corneal stromal cells, epidermal cells, and dermal cells), at all concentrations of H_2O_2, CE cells by far had the least damage. Also, CE cells showed a delayed temporal response to the appearance of damage, as compared to the other cell types.

And again, a protective role for nuclear ferritin was observed when DFX was used to inhibit the appearance of nuclear ferritin—followed by exposure to H_2O_2. Also consistent with the protection being effected through iron sequestration, for CE cells elevating iron in the medium did not increase damage by H_2O_2, whereas for epidermal cells it did.

It is also known that CE cells have other mechanisms that potentially may afford protection from oxidative damage—such as enzymatic scavenge pathways that exist for degrading H_2O_2 (e.g., catalase and glutathione peroxidase). So, in these studies it was also determined whether the protection afforded by nuclear ferritin is independent of these scavenge pathways. For these studies, the uptake and clearance of H_2O_2 was determined in CE cells versus several other cell types. This involved Fluorescence Activated Cell Sorting of cells that had been pre-labeled with an H_2O_2-sensitive dye and then were exposed to H_2O_2. Of the cell types examined, CE cells not only showed the highest uptake of the H_2O_2, but also the most rapid clearance. This clearance was most likely due to catalase and/or glutathione peroxidase—which is consistent with CE cells having scavenger enzymes that may have a role in protection against oxidative damage [12,19,34].

However, when H_2O_2-mediated damage was evaluated in the presence and absence of inhibitors of catalase and glutathione peroxidase there was no detectible difference in the damage to DNA. Thus, the protection afforded by nuclear ferritin is independent of cytoplasmic detoxifying enzymes.

3.6   Ferritoid: The Nuclear Transporter for CE Cell Ferritin

3.6.1   Isolation of a cDNA for Ferritoid and its Tissue Specificity

Ferritoid was initially detected as a "ferritin-like" component in a cDNA library of restriction enzyme-fragments of "up-regulated" corneal genes—using the subtractive hybridization procedure of [48].

As it was known from studies by others [18,41] that ferritin subunits associate with one another during the self-assembly of a supramolecular ferritin complex, we pursued the possibility that this clone might encode part of a ferritin-binding domain of the hypothesized nuclear transporter. If so, the encoded molecule should be specific for the CE—which was the next property that was examined and confirmed—both by in situ hybridization (Fig. 3.5a, b) and by IHC (Fig. 3.5c) using a ferritoid antibody.

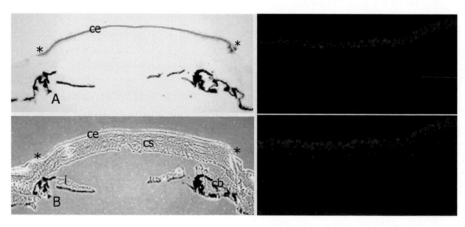

Fig. 3.5 If the tissue specificity of the nuclear ferritin transport is due to ferritoid, then ferritoid itself should be tissue-specific. As can be see in this figure, at both the mRNA and protein levels, ferritoid is specific for the CE. In (**a**) the in situ signal for ferritoid (*purple*) is found only in the CE—ending at the corneal-scleral junction (*asterisks*). **b** is a light micrograph of **a** and shows that the black regions in **a** and **b** are pigmented epithelia. The top right panel is IHC for ferritoid protein (*red*), and the bottom right panel is stained with Hoescht (*blue*). This research was originally published in the Journal of Biological Chemistry. John M. Millholland, John M. Fitch, Cindy X. Cai, Eileen P. Gibney, Kelly E. Beazley and Thomas F. Linsenmayer. Ferritoid, a Tissue-specific Nuclear Transport Protein for Ferritin in Corneal Epithelial Cells. 1997;278:23963–23970. © The American Society for Biochemistry and Molecular Biology

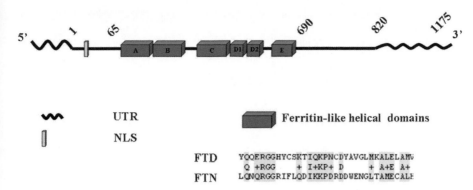

Fig. 3.6 The predicted structure of ferritoid includes: (1) a "ferritin-like" region that consists of all five helical domains of a ferritin, (2) a nuclear localization signal (NLS), and (3), 5' and 3' untranslated regions (UTR). Also shown in this figure is a conceptual amino acid sequence of a portion of one of the helical domains of ferritoid (FTD), and its comparison to the corresponding sequence in ferritin (FTN). The amino acids that are conserved between FTD and FTN are highlighted in *yellow*. From [38]

To obtain a full-length ferritoid cDNA, ferritin-like restriction-fragment clones from the subtracted library were used to screen another cDNA library, constructed using much longer cDNAs from chicken CE tissue. This screening identified other clones. The largest—diagrammed in Fig. 3.6—is likely to be full-length as it contains an uninterrupted reading frame (ORF) flanked by a 5' consensus translation start site and a 3' poly-A tail. Also, its conceptual translation product (shown in Fig. 3.6) contains two regions that fit with its proposed role as the nuclear transporter for ferritin.

The first of these regions—which we propose is involved in the interaction/binding between ferritoid–ferritin—is "ferritin-like". Consistent with this possibility, its conceptual amino acid sequence has 55 % similarity to ferritin, and it has most of the amino acids that studies by others [24] have shown in ferritin to be involved in the assembly of the supramolecular ferritin complex. Also, computer modeling [1,40] confirmed this region to have all five helical domains that are characteristic of a ferritin subunit (Fig. 3.6).

However, in ferritoid these helical domains are unique, as they are comprised of alternating short amino acid sequences that are highly conserved between ferritoid and ferritin (in Fig. 3.6 highlighted in yellow), and intervening among these sequences are equally short ones that have no similarity. The unique nature of this arrangement was extended by 3-D structural modeling (Fig. 3.7), in which the sequence of the ORF of ferritoid was inserted into the sequence of the known crystal structure of the H-chain subunit of human ferritin. This showed in ferritoid that the short "ferritin-like" amino acid sequences (shown in red in Fig. 3.7) are not distributed randomly along the lengths of the helices. Instead they are located preferentially along the outer surface of the helical domains—where we propose they interact with/bind to a ferritin subunit.

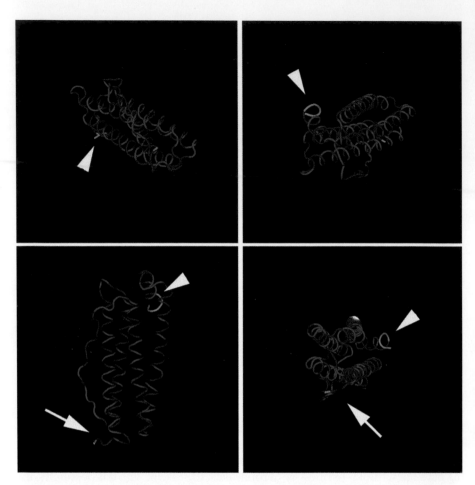

Fig. 3.7 In this figure, the crystal structure is of human ferritin-H is shown in four different planes. Into this structure, the deduced amino acid sequence of ferritoid has been superimposed. Also the amino acids of ferritoid that are conserved with those of ferritin are colored red, and the dissimilar amino acids are colored blue. As can be seen, in ferritoid, the conserved amino acids are preferentially located along the outer faces of the helices—where they would be available for interactions with ferritin. This research was originally published in the Journal of Biological Chemistry. John M. Millholland, John M. Fitch, Cindy X. Cai, Eileen P. Gibney, Kelly E. Beazley and Thomas F. Linsenmayer. Ferritoid, a Tissue-specific Nuclear Transport Protein for Ferritin in Corneal Epithelial Cells. 1997;278:23963–23970. © The American Society for Biochemistry and Molecular Biology

The second functional region—which is located just upstream of the ferritin-like region—contains a consensus SV40-like nuclear localization signal (Fig. 3.6, NLS) that experimental evidence shows is necessary for the nuclear transport of both ferritoid and ferritin (described next).

3.6.2 Functional Analysis of the Nuclear Transport of Ferritin by Ferritoid

To examine functionally whether ferritoid can transport ferritin into the nucleus, co-transfections were performed with epitope-tagged constructs for ferritoid (V5-tagged) and ferritin (myc-tagged) and the results were analyzed by double-labeled fluorescence IHC for the epitope tags.[3] For these experiments, Cos-1 cells were used, as this is a cell type that had been used previously for transfection studies of ferritin [15]. In co-cultures transfected with full-length constructs for ferritin and ferritoid, in the cells that have both molecules, they are both exclusively nuclear [Fig. 3.8, NLS(+)]. Consistent with the NLS of ferritoid being functional and required, when the co-transfections were performed with the ferritoid deletion construct from which the NLS had been removed [Fig. 3.8, NLS(−)], the ferritin and ferritoid are distributed uniformly throughout the cell.[4]

Then, to examine directly the association between ferritoid and ferritin, two approaches were employed. One was co-immunoprecipitation, and the other was a yeast two-hybrid binding assay.

The co-immunoprecipitations were performed on lysates of CE cells that had been co-transfected with epitope-tagged, full-length constructs for ferritoid (V5) and ferritin (myc). Then, the immunoprecipitates were analyzed by immunoprecipitation with antibodies for the epitope tags. The results showed that immunoprecipitates with either antibody contained both ferritoid and ferritin. Thus, ferritin and ferritoid associate with one another with an affinity strong enough to undergo co-immunoprecipitation.

For the yeast two-hybrid assays, the system employed not only shows qualitatively whether components interact with one another—as determined by the activation of a gene that is required for growth of a reporter yeast strain—it also provides quantitative information on relative strengths of these interactions. This latter parameter is based on the studies of [7] which showed that the stronger the interaction between two components, the greater the "relative plating efficiency" (RPE)—with RPE being defined as the number of colonies formed on medium that requires the interaction for growth, versus the number of colonies formed by a strain that grows constitutively.

---

[3] The CAM is comprised of all the extraembryonic membranes of the developing embryo. It supports the full-range development of an explanted cornea—even if the explants are from early stage embryos [50].

[4] In these studies, another potentially interesting observation concerns the cellular relationship between ferritoid and cytokeratin K3 (a marker for CE cell differentiation). In the central cornea—where all the CE cells are mature—ferritoid and K3 are both found in all cells. However, at the periphery, where the CE cells are undergoing their initial differentiation, ferritoid is present in basal CE cells that are not yet synthesizing detectible K3. Thus, the synthesis of ferritoid may represent an early event in the differentiation of CE cells. In addition, this observation reinforces the potential importance of ferritoid in the protection of the DNA of cells from damage—especially if the cells that have ferritoid, but no K3, are CE stem cells.

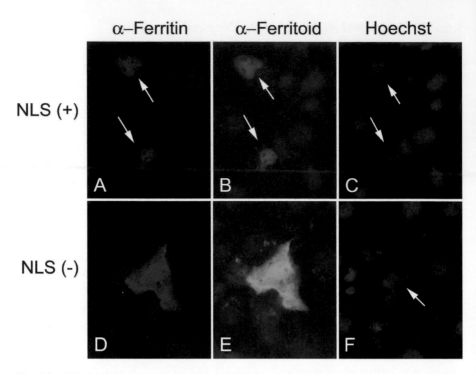

Fig. 3.8 COS-1 cells were co-transfected with a construct for ferritin, and constructs for ferritoid that are either full-length [with an intact NLS(+)] or ones from which the NLS has been deleted [NLS(−)]. The cells were analyzed by double-label IHC for ferritin (*red*) and ferritoid (*green*). Hoechst dye is also shown (*blue*). When the ferritin is co-transfected with the full-length ferritoid construct (NLS+), both ferritin and ferritoid are co-localized within the nucleus (arrows). When the co-transfections are done with the ferritoid construct that lacks the NLS(−) the ferritoid and ferritin are distributed uniformly throughout the cell. Modified from research originally published in the Journal of Biological Chemistry. John M. Millholland, John M. Fitch, Cindy X. Cai, Eileen P. Gibney, Kelly E. Beazley and Thomas F. Linsenmayer. Ferritoid, a Tissue-specific Nuclear Transport Protein for Ferritin in Corneal Epithelial Cells. 1997;278:23963–23970. © The American Society for Biochemistry and Molecular Biology

For the ferritoid–ferritin analyses, the RPE used for normalization was the homopolymeric ferritin–ferritin interaction that occurs during assembly of a ferritin supramolecular complex. This was assigned a value of 100 % (Fig. 3.9). Then, the heteropolymeric ferritoid–ferritin interactions were evaluated, which gave an RPE of ~70 %. This result confirmed that ferritoid and ferritin interact with one another, and it also suggested that this heteropolymeric interaction is somewhat weaker than that which occurs between ferritin monomers. However, this heteropolymeric interaction is strong enough to produce complexes that within the nuclei of CE cells in situ are stable (as described next).

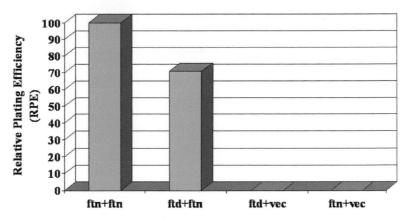

Fig. 3.9 The relative plating efficiencies (RPEs) of a yeast two-hybrid assay in which of ferritoid–ferritin interactions (ftd–ftn) are compared to ferritin–ferritin interactions (ftn–ftn; which are assigned a value of 100 %) shows that the ferritoid–ferritin interaction is slightly weaker—giving an RPE of ~70 %. In this assay, neither the ferritoid nor the ferritin, when inserted singly into the activation domain-containing vector, showed detectible growth

3.6.3   CE Nuclear Ferritin: A Stable Heteropolymeric Ferritin–Ferritoid Complex with Unique Properties

The studies already presented show that (1) ferritin and ferritoid interact with one another, and (2) this interaction is required for the nuclear transport of ferritin. However, these observations raised a number of additional questions concerning the ferritin and ferritoid that are present within CE cells in situ: including (1) the type(s) of complexes that are formed between these two components, (2) the subcellular localization(s) of these complexes, and (3) whether the complexes are transient—being present only during the transport process—or whether, once formed they remain as stable complexes. Also, if the complex(es) are stable, do they have unique characteristics/properties that distinguish them from other ferritin complexes? Certain of these questions have now been addressed.

3.6.4   Size(s) of the Ferritin–Ferritoid Complex(es)

For an initial characterization of the endogenous ferritin–ferritoid complexes of CE cells, extracts of CE tissue (from mature embryos) were fractionated by FPLC size-exclusion chromatography, followed by identification of the ferritoid and ferritin by Western blot.[5] Using this approach, the fractions with ferritoid (FTD) eluted as a

---

[5] By the time these analyses were undertaken, we had an anti-ferritoid antibody [6].

single peak (Fig. 3.10b), which by comparison with molecular weight markers (Fig. 3.10a), had a median molecular weight ~260 kDa (fraction #8). This is approximately half the size of a "typical" ferritin which has a molecular weight of ~470 kDa and is a 24-mer. Considerably lesser amounts of ferritoid were present in the highest molecular weight fractions (indicative of a "typical" 24-mer ferritin) and in lower molecular weight fractions that most likely are precursors of the higher molecular weight complexes (~50–100 kDa).

Western blot with the ferritin antibody (Fig. 3.10b, FTN) showed that the ~260 kDa peak also contained most of the ferritin, and immunoprecipitation of the fractions with this antibody (Fig. 3.10c) brought down both ferritin and ferritoid, showing that both components are co-assembled with one another.

To analyze further the subcellular distributions of the ferritin–ferritoid complexes, nuclear-cytoplasmic fractionation of CE was performed, and the extracts were analyzed as just described. The results (not presented) showed that both the nuclear and cytoplasmic compartments contained ferritin–ferritoid, complexes.

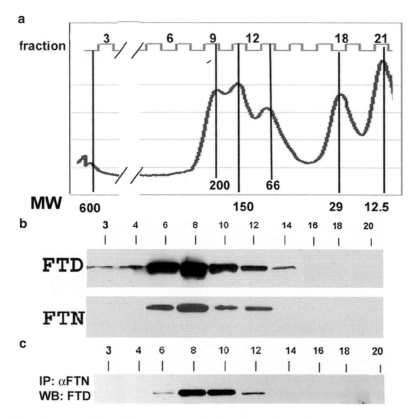

Fig. 3.10 Analyses of CE cell lysates enriched for ferritin and ferritoid—by a heat treatment—and then fractionated by gel filtration chromatography. (**a**) A profile of the molecular weight markers. Fractions were collected (**b**) and analyzed by Western blot for ferritoid (FTD) and ferritin (FTN), and by co-immunoprecipitation followed by immunoblot (**c**). Modified from [39]

However, the nuclear complex(es) were the 260 kDa form, and the cytoplasmic complexes were predominantly the ~470 kDa form.

Taken together, these results suggest that the CE cell "nuclear ferritin" is a stable, heteropolymeric ferritin–ferritoid complex that is half the size of a "typical" cytoplasmic ferritin (most likely a dodecamer). Thus, ferritoid serves not only as the nuclear transporter for ferritin, but once in the nucleus, it remains a component of a unique supramolecular complex.

3.6.5   Similarities of the CE Nuclear Ferritin–Ferritoid Complexes to the Bacterial DNA-Binding Proteins of Starvation (Dps)

To our knowledge, it is unique for a eukaryotic ferritin to be the size of the CE nuclear ferritin–ferritoid. However, certain bacteria have "ferritin-like molecules" that, by a number of criteria, are half the size of a "typical" ferritin (i.e., they are dodecamers). These bacterial proteins—termed the DNA-binding proteins of starvation (Dps)—bind to the DNA of environmentally stressed bacteria, which is thought to be one way they prevent damage to their DNA [33]. In addition, the Dps are low in iron—containing only 5–10 atoms per supramolecular complex, but they have the capability of incorporating up to 500 atoms of iron per complex [47]—which may make them especially efficient in sequestering iron and contributing to the prevention of oxidative damage.

In addition, studies by others [26,27] have shown that iron can bind to DNA with high affinity, and that this iron can catalyze the oxidative cleavage of DNA (through the Fenton reaction).

Therefore, we examined whether the CE nuclear ferritin–ferritoid complexes might also bind to nuclear DNA, where they could directly protect it from damage. And, we also examined whether these complexes were inherently of low iron—which might facilitate their ability to sequester free iron.

To determine whether CE ferritin–ferritoid complexes bind DNA, an electrophoretic mobility shift assay (EMSA) was employed—using as a source of DNA a mixture of double-stranded, random 23-mers. The results showed a shifted band only with the 260 kDa ferritin–ferritoid complexes, and none with either the larger ~470 kDa cytoplasmic complexes, or the lower molecular weight ones. Therefore, DNA binding seems to be another property of the nuclear ferritin–ferritoid complexes.

For iron, we have not yet tested directly the sequestering capacity of the CE nuclear ferritin–ferritoid complexes. However, two observations suggest that compared to the cytoplasmic ferritin of other tissues, the nuclear ferritin complexes are low in iron. First, when tissue sections were examined histochemically for iron (using Prussian Blue staining), a time when nuclear ferritin was present at high levels (as determined by IHC), none was detectable in the CE. However, in liver and heart, iron was clearly detectable (by Prussian Blue), even though the IHC signal for ferritin was weak. Second, when the FPLC column fractions of the CE

Fig. 3.11 The relative concentrations of ferritoid and ferritin mRNA in the CE of developing embryos—determined by qRT-PCR at 2-day intervals—are shown between E6 and E14. From [6]

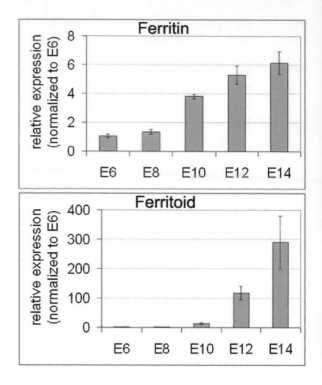

ferritin–ferritoid complexes were assayed for iron (using the ferrozine method), none was detectible in the 260 kDa nuclear complexes, but it was present in the ~470 kDa, cytoplasmic ferritin.

3.7 Developmental Analyses of Ferritin and Ferritoid

3.7.1 Ferritin and Ferritoid mRNAs During Development

Previous studies [10] suggested that during development, the production of CE nuclear ferritin is effected chiefly at the translational level—as ferritin mRNA (detected by Northern blot) is present as early as E6. However, the protein is not detected until approximately 5 days later. These results for the mRNA have been confirmed by qRT-PCR. They have also been extended to include ferritoid mRNA—which is predominantly under transcriptional control (Fig. 3.11) as the mRNA is not detectible until the protein also becomes detectible (~E10.5). Then, over the next several days ferritin mRNA undergoes a modest, progressive increase (of ~fivefold), while ferritoid mRNA increases several hundred fold (Fig. 3.11). Thus, the production of these two closely related family members—whose functions depend on the presence of each other—are controlled largely by different mechanisms.

Fig. 3.12 Double-labeling of a mature (E15) chicken cornea with the ferritoid antibody (*red*) and the ferritin antibody (*green*) show their colocalization in merged images (*yellow*) within the nuclei (*stained blue* with Hoechst) of the CE cells. From [6]

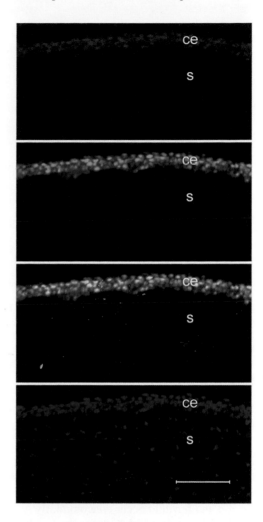

3.7.2   Ferritin and Ferritoid Proteins During Development

To elucidate the developmental relationships between ferritin and ferritoid proteins, single and double-label IHC was performed using the anti-ferritin monoclonal antibody and a newly produced anti-ferritoid antibody which was made against a peptide sequence in ferritoid that shares little homology with ferritin [6].

In initial analyses, these antibodies were used for doubling labeling analyses on sections of fully differentiated CE cells in the central cornea of late-stage embryos. An example of this staining can be seen in Fig. 3.12, in which both the ferritoid staining (red), and the ferritin staining (green) are found exclusively in the nuclei of CE cells—as verified by comparison with Hoechst staining. In addition, the merged images (yellow) showed that all cells with nuclear ferritin

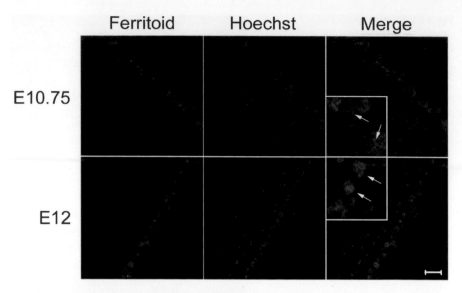

Fig. 3.13 Confocal IHC for ferritoid (*red*) in the CE at E10.75 before the initiation of ferritin synthesis shows ferritoid in a cytoplasmic, perinuclear rather than nuclear localization. IHC for ferritoid at E12, when both ferritoid and ferritin are being synthesized shows ferritoid in the nucleus. Comparison of the insets in the merged images at E10.75 and E12 suggests that ferritin must be present for the nuclear transport function of ferritoid to be active. Scale bar in bottom right panel = 10 μm. From [6]

also had nuclear ferritoid—which again is consistent with ferritoid being the nuclear transporter in vivo.

In previous developmental studies on CE cell ferritin—which employed single-label IHC (as no ferritoid antibody was yet available)—when ferritin became detectable it was always nuclear. This suggested that either: (1) ferritoid was already present at the time the synthesis of ferritin was initiated, or (2) that the syntheses of ferritin and ferritoid occur concomitantly.

By the time these studies were undertaken a ferritoid antibody was available. So by IHC, we were able to distinguish between these possibilities. We were also able to uncover other relationships between ferritin and ferritoid that became apparent during these analyses.

First, the appearance of these two proteins was examined—by IHC performed at closely spaced, temporal intervals [6]. The results showed that ferritoid becomes detectable at E10.75—approximately six hours earlier than ferritin, which is not detected until E11.

In these analyses, another observation that is relevant in elucidating the mechanisms involved in nuclear transport is that when ferritoid is first detected (i.e., in the absence of ferritin), it is cytoplasmic rather than nuclear. This can be seen in Fig. 3.13 (E10.75)—especially in the merged image of ferritoid with Hoechst dye, and in the associated insert. However, at E12 (not included here), when both ferritin and ferritoid are present, the staining for ferritoid is now nuclear—co-localizing with the Hoechst dye (as can be seen in the merged image and insert). These observations suggest that the nuclear transport activity of ferritoid, to be functional, requires an interaction with ferritin.

In the cornea of mature embryos, a similar sequence of events for the appearance of ferritoid and ferritin occurs in the peripheral CE (Fig. 3.14). This a region where CE cells continuously undergo progressive differentiation—starting as stem cells (in the limbus, and then migrating centripetally into the CE, where they continuously undergo differentiation [16]. In this region, for ferritoid and ferritin, their is a progression that consists of: (1) the cells at the extreme periphery having ferritoid that is cytoplasmic, and little if any ferritin (Fig. 3.14d); (2) a centripetal progression in which the cells acquire the synthesis of ferritin, but both the ferritoid nor ferritin are exclusively cytoplasmic (Fig. 3.14e), and eventually (3), the cells have ferritoid and ferritin—both of which are nuclear (Fig. 3.14f). The one notable difference in this progression—versus that which occurs during early development—is the stage in which cells have both ferritoid and ferritin, but neither has undergone nuclear transport.

Thus, while nuclear ferritin requires that a CE cell has both ferritoid and ferritin, this is not sufficient per se to initiate nuclear transport. Instead, at least one additional regulatory step is required; as described below this is likely to involve phosphorylation.[6]

3.7.3 Developmental Regulation of Ferritin and Ferritoid

Developmentally there are two general mechanisms by which the temporal initiation of ferritin and ferritoid synthesis could be regulated. One is that the regulation resides within the developing cornea itself (i.e., it is autonomous); the other is that it involves factor(s) produced elsewhere in the embryo (e.g., by systemic factors).

3.7.3.1 Systemic Factors

To distinguish between autonomous and systemic regulation, corneas from different stages of embryos—all pre-ferritin/ferritoid—were explanted (grafted) to the chorioallantoic membrane (CAM) of host embryos. The host embryos were also pre-ferritin–ferritoid but their ages differed from the donor cornea [4].[7]

---

[6] As this uniform distribution is also found in single transfections for ferritin, we think that this reflects the over-expression of ferritin by the transfected construct is more rapid than the assembly of the supramolecular ferritin complexes. Therefore some of the newly-synthesized monomers, and low molecular weight complexes are able to diffuse into the nucleus.

[7] At the time these studies were done, there was no antibody against ferritoid—thus the need to use epitope tags.

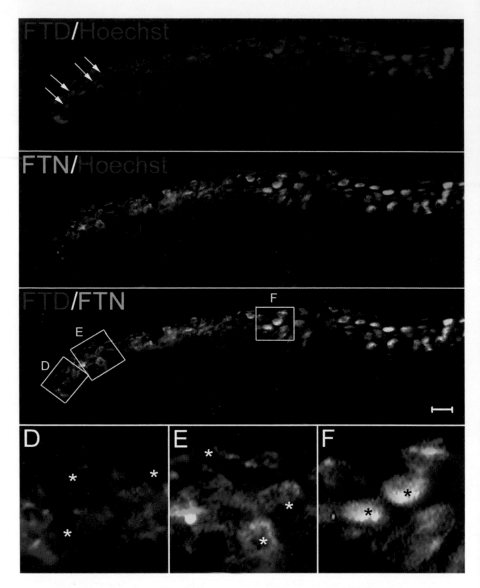

Fig. 3.14 IHC of sections of E15 cornea at the corneal–scleral junction—where cells from the limbus migrate centripetally into the cornea and differentiate into mature CE cells)—immunostained for ferritoid (*red*) and ferritin (*green*), and counterstained with Hoechst (*blue*) for nuclei show the nuclear localization of ferritin and ferritoid occurs with maturation. Enlarged areas of the merged images for ferritoid and ferritin (*D–F*) are confocal micrographs of the regions demarcated by the corresponding boxes in which nuclei are designated by asterisks show that closer to the limbus, panel *D*, ferritin and ferritoid are cytoplasmic, then become perinuclear as seen in "*E*" and then nuclear as seen in "*F*" in more mature CE cells closer to the central cornea. Scale bar = 10 μm. From [6]

As tissues explanted to the CAM are nourished by vessels from host embryo, this should provide the explant with any systemic factor(s) produced by the host. Thus, if systemic factors are responsible for regulating the temporal synthesis of ferritin and ferritoid, these should appear in the explant concomitant with their appearance in the host. The results (evaluated by IHC for ferritin and ferritoid) showed this to occur– thus providing evidence for the involvement of systemic factors from the host [4].

3.7.3.2   Thyroxine as a Regulatory Factor

Then, to begin to determine what systemic factor(s) might be involved in this regulation, thyroxine was chosen for examination as: (1) studies on hepatocytes had shown that thyroxine amplifies the synthesis of cytoplasmic ferritin, (2) thyroxine has been shown to influence the developmental expression of other corneal genes, and (3) formation of the thyroid gland temporally occurs shortly before the appearance of the CE nuclear ferritin and ferritoid (see description in [4]). For these studies, organ cultures of pre-ferritin stage embryonic corneas were employed.[8]

In such cultures, when the medium is supplemented with fetal bovine serum (FBS—which could contain a myriad of factors), the CE of the cultured corneas precociously initiates the synthesis of ferritoid and ferritin (as determined by IHC). For example, a cultured E8 cornea, which is chronologically three days pre-ferritin/ferritoid, will initiate synthesis of ferritin and ferritoid after one day in culture. However, when such corneas are cultured in serum-free medium, only an occasional CE cell will initiate synthesis of ferritoid and ferritin—even after a prolonged period of time in culture. Thus, whole corneal organ cultures were deemed suitable for examining the developmental regulation of ferritin and ferritoid synthesis. These results also suggested that the factor(s) involved in this regulation are present in FBS.

To determine whether thyroxine is one such factor, serum-free cultures were supplemented with T3, the active form of thyroxine. In these cultures, IHC showed that the addition of T3 induced the synthesis of both ferritoid and ferritin in the CE. Also, immunoblot showed the increase for ferritin to be ~70 % of that affected by FBS; for ferritoid the increase was ~50 %. Thus, a large part of the stimulatory activity in serum can be attributed to thyroxine. So, it was determined that the in vitro stimulation by both FCS and T3 involve the same mechanisms that operate in vivo—with ferritin being largely translational (as there was at most a slight increase in its mRNA) and ferritoid being largely transcriptional (as there is a large increase in its mRNA).[9]

---

[8] The ISEL method [8, 14] is based on UV-induced ROS damage to DNA activating an excision-repair system [20, 28]. This temporarily produces DNA breaks that can be detected by 3′-end labeling. Also, when necessary the fluorescent signals can be quantified—as the percentage of cells showing a positive signal, and the strength of the signal produced [11].

[9] Several other studies have reported ferritin in a nuclear location in other cell types, including nucleated red blood cells and cells in developing rat brain, as well as astrocytoma and glial cell lines, and cells subjected to iron overloading and other pathological conditions.

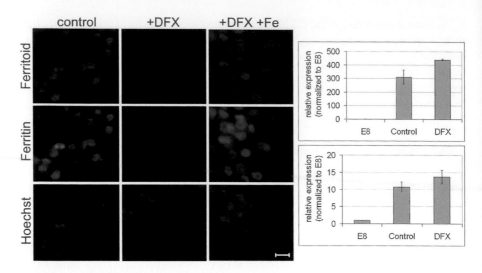

Fig. 3.15 Fluorescence IHC of E8 CE cell cultures in which CE cells were cultured for 48 h in normal medium (control), medium supplemented with 100 µM deferoxamine (+DFX) or equimolar concentrations of deferoxamine and ferrous sulfate (+DFX +FeSO$_4$) and then labeled for ferritoid and ferritin (nuclei counterstained with Hoechst) show that both ferritin and ferritoid synthesis are regulated by iron. Scale bar = 10 µm. In addition, by qRT-PCR the mRNAs for both ferritoid and ferritin are not decreased DFX consistent with the block being translational. From [6]

3.7.4 Iron-Mediated Translational Regulation of Ferritin and Ferritoid

As described previously for ferritin, when pre-ferritin stage CE cells are put in vitro (either as primary cells or whole corneal organ cultures) ferritin protein synthesis is initiated precociously [4,10]. Also, regulation of this appearance involves iron—as the addition of the iron chelator DFX to the medium, inhibits this synthesis [Fig. 3.15, ferritin (+DFX)]. Likewise, the appearance of ferritoid is also blocked [Fig. 3.15, ferritoid (+DFX)]. That the action of DFX on both ferritin and ferritoid involves iron sequestration is demonstrated by including, as a control equimolar iron—which negates the action of DFX [Fig. 3.15 (+DFX +FeSO$_4$)].

As iron is a known translational regulator of cytoplasmic ferritin, it seems likely that the DFX-mediated inhibition of CE nuclear ferritin and ferritoid, also involves translational regulation. However, this conclusion is predicated on DFX not inhibiting the synthesis of mRNAs for ferritoid and ferritin. This assumption was verified by qRT-PCR analyses of corneal cultures [Fig. 3.15, right panels) which showed that DFX treatment did not decrease the mRNAs for ferritin and ferritoid—if anything, it slightly elevated them.

An additional observation from these studies was that intracellularly regulation occurs between ferritin and ferritoid. Following release from DFX inhibition (Fig. 3.16)—which can be achieved by replacing the DFX-containing medium with DFX-free medium that also contains iron—ferritoid and ferritin reappear. And, this is

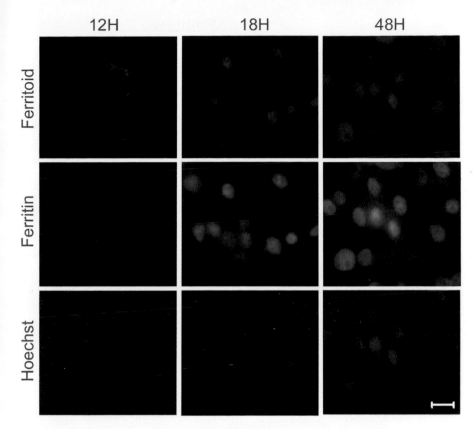

Fig. 3.16 When DFX-treated CE cultures are allowed to recover, most cells synthesize both ferritin and ferritoid by 18 h after removal of DFX. And by 48 h the cultures are identical with the controls. However, after as early as at 12 h of recovery, a small population of cells synthesize ferritoid (but not yet ferritin), and in these the ferritoid is cytoplasmic/perinuclear, consistent with what is observed in vivo. From [6]

in the same temporal sequence as occurs during development in vivo (i.e., ferritoid is synthesized before ferritin). Also as observed in vivo, during the period when the CE cells are synthesizing ferritoid—but not yet ferritin—the ferritoid is cytoplasmic (Fig. 3.16, ferritoid, 12 h) However, once ferritin synthesis is also initiated, both ferritoid and ferritin undergo nuclear transport (Fig. 3.16, ferritoid, 18H and ferritin, 18H).

3.7.5 The Ferritin–Ferritoid Interactions Required for Nuclear Transport, Involve the Phosphorylation of Ferritoid

At least some of this regulation may involve phosphorylation—as the conceptual ferritoid sequence has four consensus phosphorylation sites in its COOH-terminal region (depicted in Fig. 3.17). And, phosphorylation is a regulatory process that has

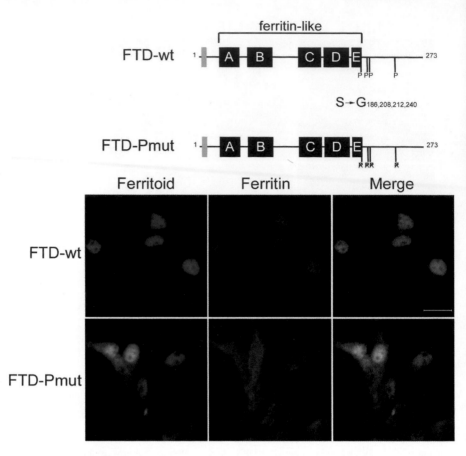

Fig. 3.17 To examine the role for C-terminal phosphorylation of ferritoid in ferritin-binding/ transport, a ferritoid construct was generated in which all of the C-terminal phosphorylation sites are mutated to glycines that cannot be phosphorylated. Then COS-1 cells were co-transfected with full-length constructs for ferritin—and for ferritoid that is either full-length wild type (FTD-wt), or in which the four consensus phosphorylation sites were mutated. (FTD-Pmut). Only with the FTD-wt were the ferritoid and ferritin nuclear. From [5]

been shown to affect a wide variety of cellular events—including protein–protein interactions and subcellular localizations. Therefore, we examined the possibility that phosphorylation may be involved in regulating the association of ferritoid with ferritin—which is the initial step required for nuclear transport.

We first determined whether the ferritin–ferritoid complex(es) are phosphorylated—and if so, which ones. For this, CE cell lysates were fractionated by gel filtration chromatography (as described earlier and shown in Fig. 3.10). Then, the resulting fractions were analyzed for phosphorylation, by immunoblot with an anti-phosphoserine antibody). The results showed that the ferritoid in the 260 kDa (nuclear) complex(es) is phosphorylated (confirmed by tandem mass spectrometry), but no phosphorylation was detected in the ~470 kDa (cytoplasmic) complexes, or in the lower molecular weight (putative precursor) fractions.

As described earlier, when COS-1 cells are co-transfected with full-length constructs for ferritoid and ferritin, nuclear transport of both occurs [Fig. 3.17 (FTD-wt)]. However, if the co-transfections were performed with a ferritoid construct in which the four consensus phosphorylation sites were mutated to non-phosphorylatable glysines (FTD-Pmut), only the ferritoid undergoes preferential, nuclear localization—while the ferritin is distributed uniformly throughout the cell (i.e., in both the cytoplasm and nucleus as can also be seen in [38]. Also co-immunoprecipitation showed that the mutated ferritoid no longer associates with ferritin. Taken together these results show that the mutated ferritoid is still capable of undergoing nuclear transport; however, it is no longer capable of transporting ferritin—most likely due to a lack of their association with one another.

To examine whether these analyses employing COS-1 cells reflect what actually occurs in CE cells, cultures of CE cells were treated with a variety of kinase inhibitors. The results are consistent with the results obtained with the COS-1 cells. In addition they suggest that the naturally occurring effect of the phosphorylation of ferritoid in CE cells is phosphokinase C (PKC)—as the only inhibitors that interfered with nuclear transport were ones that inhibit this enzyme. Consistent with this conclusion, the cultures treated with the PKC inhibitors showed a reduction in the amount of ferritoid immunoprecipitated with the phosphoserine antibody, and they also showed decreased binding of ferritoid to ferritin (as determined by co-immunoprecipitation).

Therefore, these studies, when taken together, suggest that phosphorylation of ferritoid is involved in regulating formation of the ferritin–ferritoid complex(es), and that these complexes, once formed, undergo subsequent nuclear translocation.

3.7.6 Does the CE of the Adult Cornea Have Ferritin–Ferritoid Complexes, and Why Are They in the Embryo

Our working hypothesis is that a major function of nuclear ferritin is to prevent UV-mediated damage to DNA. However, all of the in vivo work thus far presented here has been from embryonic corneas, and this a period when it is unlikely that UV radiation would pose much of a problem. However, we have also examined CE tissue from adult corneas [39]—which is when such protection would be needed. These analyses showed that in adult CE both ferritin and ferritoid mRNAs are present at levels similar to those in the mature embryo. And protein analyses showed that the molecular complexes in the adult CE are indistinguishable from those in the embryo—being present mostly in the 260 kDa (nuclear) form.

Conversely then, why do ferritin–ferritoid complexes even exist in the CE of the developing embryo? For this, at least two possibilities exist. One is that their embryonic appearance prepares the CE for future oxidative insults that will be encountered following hatching. However, another possibility is that the embryonic nuclear ferritin may have a role in protecting the CE from ROS produced by the embryo itself. It has been reported that amniotic fluid has ROS that are high enough they can be detected by the reaction products they produce [31]. And, our studies have shown that the nuclear ferritin can directly protect embryonic CE cells from ROS (e.g., H_2O_2). If this latter explanation is correct, the nuclear ferritin–ferritoid complexes are likely to have different protective roles at different times in life (i.e., in ovo from endogenous ROS produced by the embryo itself, and in the adult from ROS environmentally produced by UV irradiation and molecular O_2.)

Acknowledgement This work was supported by NIH Grant R01EY013127 to TFL.

References

1. Alexandrov NN, Nussinov R, Zimmer RM. Fast protein fold recognition via sequence to structure alignment and contact capacity potentials. In: Hunter L, Klein TE, editors. Pacific symposium on biocomputing. Singapore: World Scientific Publishing Co.; 1995. p. 53–72.
2. Audic A, Giacomoni PU. DNA nicking by ultraviolet radiation is enhanced in the presence of iron and of oxygen. Photochem Photobiol. 1993;57:508–12.
3. Balla J, Jacob HS, Balla G, Nath K, Eaton JW, Vercellotti GM. Endothelial-cell heme uptake from heme proteins: induction of sensitization and desensitization to oxidant damage. Proc Natl Acad Sci U S A. 1993;90:9285–9.
4. Beazley KE, Canner JP, Linsenmayer TF. Developmental regulation of the nuclear ferritoid–ferritin complex of avian corneal epithelial cells: roles of systemic factors and thyroxine. Exp Eye Res. 2009;89:854–62.
5. Beazley KE, Nurminskaya M, Linsenmayer TF. Phosphorylation regulates the ferritoid–ferritin interaction and nuclear transport. J Cell Biochem. 2009;107:528–36.
6. Beazley KE, Nurminskaya M, Talbot CJ, Linsenmayer TF. Corneal epithelial nuclear ferritin: developmental regulation of ferritin and its nuclear transporter ferritoid. Dev Dyn. 2008;237:2529–41.
7. Brent R, Finley Jr RL. Understanding gene and allele function with two-hybrid methods. Annu Rev Genet. 1997;31:663–704.
8. Bromidge TJ, Howe DJ, Johnson SA, Phillips MJ. Adaptation of the TdT assay for semi-quantitative flow cytometric detection of DNA strand breaks. Cytometry. 1995;20:257–60.
9. Cai C, Ching A, Lagace C, Linsenmayer T. Nuclear ferritin-mediated protection of corneal epithelial cells from oxidative damage to DNA. Dev Dyn. 2008;237:2676–83.
10. Cai CX, Birk DE, Linsenmayer TF. Ferritin is a developmentally regulated nuclear protein of avian corneal epithelial cells. J Biol Chem. 1997;272:12831–9.
11. Cai CX, Birk DE, Linsenmayer TF. Nuclear ferritin protects DNA from UV damage in corneal epithelial cells. Mol Biol Cell. 1998;9:1037–51.
12. Cejkova J, Stipek S, Crkovska J, Ardan T. Changes of superoxide dismutase, catalase and glutathione peroxidase in the corneal epithelium after UVB rays. Histochemical and biochemical study. Histol Histopathol. 2000;15:1043–50.
13. Cerutti PA. Prooxidant states and tumor promotion. Science. 1985;227:375–81.
14. Coates PJ, Save V, Ansari B, Hall PA. Demonstration of DNA damage/repair in individual cells using in situ end labelling: association of p53 with sites of DNA damage. J Pathol. 1995;176:19–26.
15. Corsi B, Perrone F, Bourgeois M, Beaumont C, Panzeri MC, Cozzi A, Sangregorio R, Santambrogio P, Albertini A, Arosio P, Levi S. Transient overexpression of human H- and L-ferritin chains in COS cells. Biochem J. 1998;330:315–20.

16. Cotsarelis G, Cheng SZ, Dong G, Sun TT, Lavker RM. Existence of slow-cycling limbal epithelial basal cells that can be preferentially stimulated to proliferate: implications on epithelial stem cells. Cell. 1989;57:201–9.

17. Cozzi A, Santambrogio P, Levi S, Arosio P. Iron detoxifying activity of ferritin. Effects of H and L human apoferritins on lipid peroxidation in vitro. FEBS Lett. 1990;277:119–22.

18. Gerl M, Jaenicke R, Smith JM, Harrison PM. Self-assembly of apoferritin from horse spleen after reversible chemical modification with 2,3-dimethylmaleic anhydride. Biochemistry. 1988;27:4089–96.

19. Green K. Free radicals and aging of anterior segment tissues of the eye: a hypothesis. Nucleic Acids Res. 1995;27:143–9.

20. Grossman L, Caron PR, Mazur SJ, Oh EY. Repair of DNA-containing pyrimidine dimers. FASEB J. 1988;2:2696–701.

21. Halliwell B, Aruoma OI. DNA damage by oxygen-derived species. Its mechanism and measurement in mammalian systems. FEBS Lett. 1991;281:9–19.

22. Harrison PM, Ford GC, Rice DW, Smith JM, Treffry A, White JL. Structural and functional studies on ferritins. Biochem Soc Trans. 1987;15:744–8.

23. Hart RW, Setlow RB, Woodhead AD. Evidence that pyrimidine dimers in DNA can give rise to tumors. Proc Natl Acad Sci U S A. 1977;74:5574–8.

24. Hempstead PD, Yewdall SJ, Fernie AR, Lawson DM, Artymiuk PJ, Rice DW, Ford GC, Harrison PM. Comparison of the three-dimensional structures of recombinant human H and horse L ferritins at high resolution. J Mol Biol. 1997;268:424–48.

25. Henle ES, Han Z, Tang N, Rai P, Luo Y, Linn S. Sequence-specific DNA cleavage by Fe^{2+}-mediated fenton reactions has possible biological implications. J Biol Chem. 1999;274:962–71.

26. Henle ES, Linn S. Formation, prevention, and repair of DNA damage by iron/hydrogen peroxide. J Biol Chem. 1997;272:19095–8.

27. Henle ES, Luo Y, Linn S. Fe^{2+}, Fe^{3+}, and oxygen react with DNA-derived radicals formed during iron-mediated Fenton reactions. Biochemistry. 1996;35:12212–9.

28. Janssen YM, Van HB, Borm PJ, Mossman BT. Cell and tissue responses to oxidative damage. Lab Invest. 1993;69:261–74.

29. Lavoie DJ, Ishikawa K, Listowsky I. Correlations between subunit distribution, microheterogeneity, and iron content of human liver ferritin. Biochemistry. 1978;17:5448–54.

30. Linsenmayer TF, Cai CX, Millholland JM, Beazley KE, Fitch JM. Nuclear ferritin in corneal epithelial cells: tissue-specific nuclear transport and protection from UV-damage. Prog Retin Eye Res. 2005;24:139–59.

31. Longini M, Perrone S, Vezzosi P, Marzocchi B, Kenanidis A, Centini G, Rosignoli L, Buonocore G. Association between oxidative stress in pregnancy and preterm premature rupture of membranes. Clin Biochem. 2007;40:793–7.

32. Luo Y, Henle ES, Linn S. Oxidative damage to DNA constituents by iron-mediated fenton reactions. The deoxycytidine family. J Biol Chem. 1996;271:21167–76.

33. Martinez A, Kolter R. Protection of DNA during oxidative stress by the nonspecific DNA-binding protein Dps. J Bacter. 1997;179:5188–94.

34. Martins EA, Chubatsu LS, Meneghini R. Role of antioxidants in protecting cellular DNA from damage by oxidative stress. Mutat Res. 1991;250:95–101.

35. Mello-Filho AC, Hoffmann ME, Meneghini R. Cell killing and DNA damage by hydrogen peroxide are mediated by intracellular iron. Biochem J. 1984;218:273–5.

36. Mello-Filho AC, Meneghini R. Iron is the intracellular metal involved in the production of DNA damage by oxygen radicals. Mutat Res. 1991;251:109–13.

37. Meneghini R. Iron homeostasis, oxidative stress, and DNA damage. Free Radic Biol Med. 1997;23:783–92 [Review] [74 refs].

38. Millholland JM, Fitch JM, Cai CX, Gibney EP, Beazley KE, Linsenmayer TF. Ferritoid, a tissue-specific nuclear transport protein for ferritin in corneal epithelial cells. J Biol Chem. 2003;278:23963–70.

39. Nurminskaya MV, Talbot CJ, Nurminsky DI, Beazley KE, Linsenmayer TF. Nuclear ferritin: a ferritoid–ferritin complex in corneal epithelial cells. Invest Ophthalmol Vis Sci. 2009;50:3655–61.

40. Rost B. PHD: predicting one-dimensional protein structure by profile-based neural networks. Methods Enzymol. 1996;266:525–39.
41. Santambrogio P, Pinto P, Levi S, Cozzi A, Rovida E, Albertini A, Artymiuk P, Harrison PM, Arosio P. Effects of modifications near the 2-, 3- and 4-fold symmetry axes an human ferritin renaturation. Biochem J. 1997;322:461–8.
42. Shimmura S, Suematsu M, Shimoyama M, Tsubota K, Oguchi Y, Ishimura Y. Subthreshold UV radiation-induced peroxide formation in cultured corneal epithelial cells: the protective effects of lactoferrin. Exp Eye Res. 1996;63:519–26.
43. Shires TK. Iron-induced DNA damage and synthesis in isolated rat liver nuclei. Biochem J. 1982;205:321–9.
44. Smolinand G, Thoft RA. The cornea: scientific foundations and clinical practice. Boston: Little, Brown and Company; 1987.
45. Spurr SJ, Gipson IK. Isolation of corneal epithelium with dispase II or EDTA: effects on the basement membrane zone. Invest Ophthalmol Vis Sci. 1985;26:818–27.
46. Stohs SJ, Bagchi D. Oxidative mechanisms in the toxicity of metal ions. Free Radic Biol Med. 1995;18:321–36.
47. Su MH, Cavallo S, Stefanini S, Chiancone E, Chasteen ND. The so-called *Listeria innocua* ferritin is a Dps protein. Iron incorporation, detoxification, and DNA protection properties. Biochemistry. 2005;44:5572–8.
48. Wang Z, Brown DD. A gene expression screen. Proc Natl Acad Sci U S A. 1991;88:11505–9.
49. Zak NB, Linsenmayer TF. Monoclonal antibodies against developmentally regulated corneal antigens. Dev Biol. 1983;99:373–81.
50. Zak NB, Linsenmayer TF. Analysis of corneal development with monoclonal antibodies. II. Tissue autonomy in cornea-skin recombinants. Dev Biol. 1985;108:455–64.

Chapter 4
Excitatory Amino Acid Transporters, Xc⁻ Antiporter, γ-Glutamyl Transpeptidase, Glutamine Synthetase, and Glutathione in Human Corneal Epithelial Cells

Marlyn P. Langford, Thomas B. Redens, and Donald E. Texada

Abbreviations

Cys	Cystine
EAAT	Excitatory amino acid transporter
GGT	Gamma–glutamyl transpeptidase
Gln	Glutamine
Glu	Glutamate
GS	Glutamine synthetase
HCEC	Human conjunctival epithelial cells
NAC	N-acetylcysteine
ROS	Reactive oxygen species
xCT	Xc-exchanger C-terminal light chain

4.1 Introduction

The 0.5–0.8 mm thick, transparent avascular cornea domes the anterior external surface of the eye and focuses light toward the lens and retina. Three phenotypically and phylogenically differentiated cell layers are found in the normal cornea; a single long-lived endothelial cell layer lines the posterior corneal surface bathed by aqueous humor, multiple layers of interdigitating stromal keratocytes (fibroblasts)

M.P. Langford, Ph.D. (✉) • T.B. Redens • D.E. Texada
Department of Ophthalmology, Louisiana State University Health Sciences Center, Shreveport, LA 71130-3932, USA
e-mail: mlangf@lsuhsc.edu; treden@lsuhsc.edu; dtexada@lsuhsc.edu

© Springer Science+Business Media New York 2015
M.A. Babizhayev et al. (eds.), *Studies on the Cornea and Lens*,
Oxidative Stress in Applied Basic Research and Clinical Practice,
DOI 10.1007/978-1-4939-1935-2_4

of neural crest origin provide strength and structure, and 4–6 layers of epithelial cells cover the anterior surface bathed by tear fluid. The epithelium is composed of a basal layer of metabolically active columnar epithelial cells, 2–3 layers of tightly adherent outward-migrating wing cells, and 1–2 layers of flat superficial epithelial cells at the cornea–tear interface that are progressively displaced by the underlying wing cells.

The cornea and intraocular tissues are exposed constantly to reactive oxygen species (ROS) generated by environmental factors, metabolism, and disease [1–4]. Glutathione (GSH; a tripeptide of glutamate, cysteine, and glycine) is a potent intracellular antioxidant found throughout the cornea, but predominantly in the endothelium and epithelium [5, 6]. The importance of GSH to corneal endothelium function, viability, and protection from ROS is supported by the observations that oxidative stress depletes corneal GSH leading to decreased corneal barrier function, hydration, cell membrane integrity, and xenobiotic detoxification [7–12]. Loss of GSH and related enzymes due to irradiation, drying, diabetes, inflammation, aging, and administration of some drugs predisposes cornea, conjunctiva, and intraocular tissues to oxidative stress [1–4, 13, 14].

The immunohistochemical mapping of GSH, glutamate (Glu), cystine (Cys) glycine, and glutamine (Gln), their respective transporter systems, γ-glutamyltranspeptidase (GGT; catalyzes the cleavage of extracellular glutathione into its constituent amino acids which can then be transported into the cell), and glutamine synthetase (GS catalyzes the condensation of Glu and ammonia to form Gln) in different lens regions [15, 16], ciliary body epithelial [17–19] and cornea endothelium and epithelium [5, 20] provide anatomical support for contributory roles in regulating intracellular Glu and Cys, maintaining GSH, and protecting the cell against oxidative stress. The purpose of the current study was (i) to determine the cellular distribution of Glu, EAATs, Xc$^-$ antiporter, GGT, and GS in human corneal epithelial cells (HCEC) and (ii) to determine their importance to GSH homeostasis utilizing specific transporter and metabolic inhibitors. The differential expression of EAAT, Xc$^-$ antiporter, GGT, and GS by HCEC in culture was similar to that previously reported in human cornea [20]. Notably, inhibition of EAAT, Xc$^-$ antiporter, GGT and GS depleted glutathione (GSH) levels in HCEC suggesting their activities are important to GSH homeostasis and protection from oxidative stress.

4.2   Materials and Methods

Transformed human corneal epithelial cells (HCEC). Transformed HCEC were provided by Dr. Zan Pan (Department of Ophthalmology, Dyson Vision Research Institute, Weill Cornell Medical College, New York, NY) and maintained at low passage using standard tissue culture techniques. The cells were grown in Dulbecco's minimum essential media (DMEM; Sigma, St Louis, MO) supplemented with 5 % fetal bovine serum (Atlanta Biologicals, Lawrenceville, GA) and insulin (1 unit/mL, Novolin®, Novo Nordisk Pharmaceutical Industries, Inc. Clayton, NC) as previously described [21]. For experiments, HCEC sub-stocks were propagated in flask cultures and trypsinized cells were suspended (5×10^6 cells/mL) in a 1:1 mixture of DMEM and F12 medium (Sigma-Aldrich, St. Louis, MO) containing low levels of glucose (21 mg/dL), 2 % bovine calf serum (HyClone Laboratories; Logan, UT), insulin and antibiotics (100 U penicillin and 100 µg streptomycin/mL; GIBCO-BRL; Grand Island, NY). Two milliliters of cell suspension were pipetted into each 6-well dish culture (Sarstedt, Inc., Newton, NC). The cultures were incubated for 48 h at 37 °C in a 5 % CO_2 humidified atmosphere within a water-jacketed incubator (Forma-Scientific, Fredrick, MD).

Reagents. Inhibitors of excitatory amino acid transporters (EAAT) L- and D-glutamic acid (L- and D-Glu), non-metabolizable D-aspartic acid (D-Asp), D/L-THREO-β-hydroxyaspartate (THA; high-affinity inhibitor of EAAT1 and EAAT3), and kainic acid (KA; inhibitor of EAAT2) [22], and inhibitors of Cys/Glu exchanger (Xc⁻ antiporter) α-aminoadipic acid (AAA) [23], and N-acetylcysteine (NAC) were purchased from Sigma Chemical Company (St. Louis, MO). The non-transportable Glu/Asp transport inhibitor DL-threo-β-benzyloxyaspartic acid (TBOA) [24] was purchased from Tocris Bioscience (Tocris Cookson Ltd., Ellisville, Missouri). Acivicin (AT¹²⁵; 25 µg/mL) [l-((S,5S)-(-amino-3-chloro-4,5-dihydro-5-isoxazoleacetic acid; inhibitor of GGT)] [25], 5 mM L-methionine-D/L-sulfoximine (MSO; inhibitor of GS) [26], and 5 mM buthionine sulfoximine (BSO; potent inhibitor of GSH synthetase) [27] were purchased from Sigma Chemical Company (St. Louis, MO). Stock solutions were prepared in DMEM, aliquoted, and stored at −20 °C.

Radiolabeled amino acids. [¹⁴C]-L-Glu (specific activity; 254 mCi/mM), [³H]-D-Asp (specific activity; 14 Ci/mM), and [¹⁴C]-L-Cys (specific activity; 250 mCi/mM) were purchased from New England Nuclear (Boston, MA).

Na-dependent and Na-independent uptake of radiolabeled l-Glu, d-Asp, and l-Cys. The kinetic uptake of trace amounts (10–100 nCi/mL) of radiolabeled L-Glu, D-Asp and L-Cys in the presence and absence of Na^+ was performed in side-by-side parallel experiments using triplicate 48 h-old confluent cell monolayers in 6-well culture dishes (Sarstedt, Newton, NC). The medium was removed from triplicate cell monolayers and rinsed three times with 1 mL volumes of pre-warmed (37 °C) Na^+-free choline-buffered medium (CBM; 16.8 g $C_5H_{14}NO \cdot Cl$, 3.3 g $C_5H_{14}NO \cdot HCO_3$, 140 mg $CaCl_2$, 100 mg $MgSO_4$, and 350 mg KCl/L; pH 7.4) or phosphate-buffered saline (PBS). One mL of pre-warmed CBM or PBS containing trace levels of [¹⁴C]-L-Glu (~1 µCi/5nM/mL), [³H]-D-Asp (~1 µCi/700pM/mL), or [¹⁴C]-L-Cys (~1 µCi/5nM/mL) was added to replicate cultures and incubated at 37 ° C for 5, 10, and 20 min. The CBM or PBS containing the free radiolabeled amino acid was removed quickly and the culture plate placed on wet ice. The amount of radiolabeled amino acid (CPM) taken up by the cultured cells was determined as previously described [28]. Briefly, the cultures were rinsed with three 1.0 mL volumes of chilled CBM or PBS and 0.25 mL 5 % trichloroacetic acid (TCA; Fisher Scientific, Fair Lawn, NJ) was added to each culture, the cell monolayer scraped off with a plastic spatula and the TCA and cell precipitate pipetted into a vial and centrifuged at 5,000g for 5 min. The TCA fraction was harvested and pipetted into a scintillation vial containing 4 mL of ScintiVerse, II cocktail (Fisher Scientific, Fair Lawn, NJ), counted in a Beckman LS6500 scintillation counter (Beckman Instruments, Palo Alto, CA), and radioactivity expressed as mean CPM/culture.

Immunoreagents: Rabbit polyclonal anti-glutamate antibody (Cat# AB5018; Lot# 24070069) was purchased from Chemicon International (Temecula, CA, USA; www.chemicon.com/; see http://www.millipore.com). Rabbit anti-rat antisera to the C-terminal 14–21 amino acid oligopeptides of EAAT-1 (Cat# GLAST11-S), EAAT-2 (Cat# GLT-11-S), EAAT-3 (Cat# EAAC-11-S), EAAT-4 (Cat# EAAT-41-S), and rabbit antiserum to the 21 amino acid C-terminal oligopeptide of huamn EAAT-5 (Cat# EAAT51-S) proteins were purchased from Alpha Diagnostics International (San Antonio, TX; www.4adi.com). Polyclonal rabbit antibody to a synthetic N-terminal peptide (1-50 aa) of murine xCT (NB300-318) was purchased from Novus Biologicals (Littleton, CO). Rabbit anti-GGT was generously provided by Dr. J.D. (David) Castle (Department of Cell Biology, University of Virginia Health System School of Medicine, Charlottesville, VA) [29]. Goat polyclonal antibody to GS was purchased from Abcam, Inc. (ab6585; Cambridge, MA, USA). Fluorescein-isothiocyanate (FITC)-tagged anti-rabbit and anti-goat IgG antisera were purchased from Jackson ImmunoResearch Laboratories, Inc., West Grove, PA).

Immunofluorescent analyses for proteins and mitochondria detection. The immunofluorescent antibody (IFA) analysis of HCEC was performed using polyclonal antibodies to Glu, EAAT subtype-specific peptides and xCT-specific peptide, rabbit anti-GS, and rabbit anti-GGT as previously described [20]. Briefly, EAATs and xCT immunoreactive proteins were detected by IFA in confluent HCEC monolayers grown on Lab-Tek® 8 chamber glass slide systems (Nunc Nalgene Corporation) for 48 h at 37 °C. The media was removed and the slide cell cultures rinsed with PBS and fixed for 15 min with cold 1 % paraformaldehyde. The slides were rinsed six times with cold PBS and reacted with 1:100 dilutions of primary antibodies. After 16 h incubation at 4 °C, the primary antibodies were removed. The slides were washed three times with PBS. FITC-tagged secondary antibody in PBS (1:1,000) was added and the slide cultures incubated for 6 h at 37 °C before rinsing six times with PBS to remove unbound secondary antibody. Excess moisture was blotted from the slide, one drop of IFA mounting media (VWR International, Inc., West Chester, PA) applied/section, and a glass cover slip mounted on each slide. The processed slides were stored at 4 °C.

Mitochondria in unfixed HCEC were labeled using Mitotracker Green (M-7514; Invitrogen, Eugene, OR) as per the manufacturer's instructions. Briefly, HCEC grown in Lab-Tek® 2 chamber cover glass systems (Nunc Nalgene Corp.) were incubated in 1 mL of a 1 µM solution of Mitotracker Green in DMEM (serum-free, 15 mM HEPES) for 30 min. The medium containing the Mitotracker Green was removed and the cells rinsed three times with 1 mL volumes of 37 °C PBS. Fresh pre-warmed PBS was pipetted on to the cells before viewing by fluorescent microscopy.

Fluorescing cells and mitochondria were observed under equal exposure conditions using a ×40 .60 N.A. plan Fluor objective on a Nikon Eclipse TE300 inverted microscope with an epi-fluorescence attachment with single band exciters for FITC

(492 nm, ×18) and TRITC (572 nm, ×23). Digital images were captured using a Photometrics Cool SNAPfx monochrome CCD camera controlled with Scanalytics IPLab software.

Substrate-specific inhibition of Glu, Asp, and Cys transporter activity in HCEC. Na⁺-dependent EAAT and Na⁺-independent Xc-antiporter activities were evaluated by comparing inhibition of radiolabeled Glu, D-Asp, and Cys by L-Glu, D-Glu, L/D-Asp, threo-β-hydroxyaspartate (THA; high-affinity inhibitors of Glu transport by cloned EAAT1 and EAAT3), kainic acid (KA; potent inhibitor of cloned EAAT2) [22], L-α-aminoadipic acid (AAA) (inhibitor of EAAT3 uptake of Cys), and NAC [23, 30]. Briefly, the culture media was removed and the HCEC cultures rinsed twice with pre-warmed PBS. For concentration-dependent experiment, 1.0 mL of pre-warmed PBS or CBM containing 1 μCi/mL [¹⁴C]-L-Glu, 0.1 μCi/mL [³H]-D-Asp, or 1 μCi/mL [¹⁴C]-L-Cys with and without different concentrations (1–100 μM) of each inhibitor was added to triplicate cultures. After incubation at 37 °C for 20 min, the PBS was removed from the cultures and rinsed three times with 1 mL volumes of cold PBS or CBM to remove free radiolabeled amino acid. Cold 5 % TCA was added (0.3 mL/culture), the cell precipitate separated from the TCA soluble fraction and the amount of free intracellular [¹⁴C]-L-Glu, [³H]-D-Asp or [¹⁴C]-L-Cys determined for each inhibitor concentration by scintillation counting as above. The mean 50 % inhibitory concentration (IC₅₀) was calculated per inhibitor from the results of three independent experiments ran in triplicate. For substrate inhibition studies, the mean percent inhibition of radiolabeled amino acid uptake by a 100 μM concentration of each inhibitor was compared. The mean percent inhibition/experiment was calculated by dividing the mean counts per minute (CPM)/mg protein (Bio-Rad protein assay) in the three inhibitor-treated cultures by the mean CPM/mg protein in the three placebo-untreated control cultures.

Effect of transport and metabolic inhibitors on GSH in HCEC: To determine if acute inhibition of EAAT would reduce GSH levels, media containing L-Glu, D-Asp, THA, and TBOA (10 mM) [22, 24] was pipetted onto replicate 6-well plate cultures of HLEC and incubated for 6 h. The effect of AT¹²⁵ (25 μg/mL; GGT inhibitor) [25], 5 mM MSO (GS inhibitor) [26], and 5 mM BSO (potent inhibitor of GSH synthesis) [27] on intracellular GSH levels was determined for comparison. The concentration of reduced GSH was determined using a colorimetric/microplate reader assay [31]. Briefly, the culture media was removed and the cells washed three times with PBS. Phosphoric acid precipitating solution (PAPS; 1.67 g metaphosphoric acid, 0.2 g EDTA, 30 g NaCl/100 mL double-distilled H₂O) was added in (100 μL)/6-well culture. The cells were detached with a plastic spatula (Disposable Cell Lifter; Fisher Scientific), the precipitate solution clarified by centrifugation at 10,000*g* for 5 min and the PAPS harvested. Twenty microliters of each PAPS sample and serial dilutions of a 20 mM GSH standard solution were added to triplicate 96-well microtiter culture plate wells containing 180 μL of 0.3 M Na₂HPO₄ and 20 μL of a DTNB solution (20 mg 5,5'- dithiobis-2-nitrobenzoic acid/100 mL 1 % sodium citrate in double-distilled H₂O). The concentration of the yellow derivative (5'-thio-2-nitrobenzoic acid) was measured after incubation at room temperature for 10 min using a Cayman spectrophotometric plate reader (OD₄₀₅ nm; Cayman Chemical Autoreader;

Ann Arbor, MI). The mM GSH concentrations was calculated/culture and the mean of two experiments ran in triplicate was determined.

Statistical analysis. Means with standard deviations (SD) or standard errors (SEM) were calculated using Microsoft Excel software (Windows XP-2000, Professional). Statistical differences between control and treatment groups were considered statistically significant at p-values <0.05 as determined by unpaired Student t test analysis. Graphs were generated using Sigma Plot software.

4.3   Results

Na⁺-dependent uptake of L-Glu, D-Asp, and L-Cys in HCEC. The time-dependent uptake of [¹⁴C]-L-Glu, [³H]-D-Asp, and [¹⁴C]-L-Cys by HCEC in the presence and absence of Na⁺ is presented in Fig. 4.1. The amount of [¹⁴C]-L-Glu taken up by HCEC after 20 min was 5.6 fold higher in the presence of Na ($2,256 \pm 232$ CPM/20 min) than the absence of Na⁺ (405 ± 37 CPM/20 min; $p < 0.0001$) (Fig. 4.1a). Similarly, 3.1-fold more [³H]-D-Asp was taken up by HCEC in the presence of Na⁺ ($1,862 \pm 268$ CPM/20 min) than under Na⁺-free conditions (600 ± 73 CPM/20 min; $p < 0.0001$) (Fig. 4.1b). In comparison, [¹⁴C]-L-Cys uptake by HCEC was similar in the presence of Na⁺ (937 ± 88 CPM/20 min) and the absence of Na⁺ (843 ± 98 CPM/20 min; $p = 0.72$) (Fig. 4.1c). The uptake of [¹⁴C]-L-Glu and [³H]-D-Asp was Na⁺-dependent. In contrast, the uptake of [¹⁴C]-L-Cys in HCEC was similar in the presence and absence of Na⁺. Taken together, the results are consistent with Glu and Asp transport by high-affinity Na⁺-dependent EAAT and support L-Cys transport by Na⁺-independent Xc⁻ antiporter.

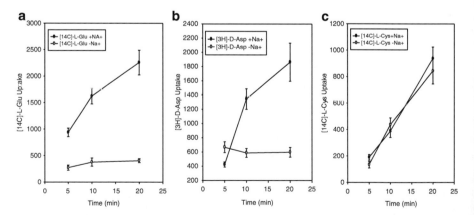

Fig. 4.1 Uptake of radiolabeled (**a**) L-Glu, (**b**) D-Asp, and (**c**) L-Cys by Na⁺-dependent and Na⁺-independent transport systems in HCEC. Each point represents the mean CPM/culture and SEM of three experiments

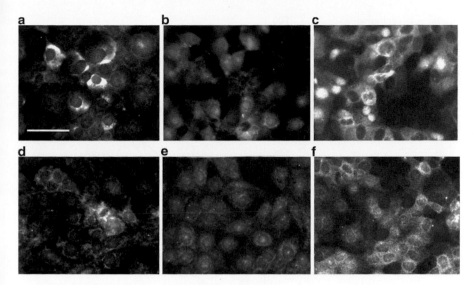

Fig. 4.2 EAATs and Xc⁻ antiporter (xCT) in HCEC cultures. Cellular localization of immunoreactive (**a**) EAAT1, (**b**) EAAT2, (**c**) EAAT3, (**d**) EAAT4, (**e**) EAAT5, and (**f**) xCT peptides in HCEC. Note the predominance of cytoplasmic EAAT1 and EAAT3 labeling, the weaker EAAT2, EAAT4, and EAAT5 labeling, and the surface labeling of xCT in HCEC. Bar = 50 µm

Cellular distribution of immunoreactive EAATs and xCT (light chain of Glu/Cys exchanger) [32] *in HCEC.* All Na⁺-dependent EAATs and xCT were detected in HCEC cultures (Fig. 4.2). The relative amount and location of the EAAT1-5 varied from undetectable to highly concentrated in HCEC within each monolayer. Notably, EAAT1 was expressed weakly in some cells, but was highly concentrated in the cytoplasm on one side of the nucleus in many positive HCEC (Fig. 4.2a). EAAT2 was absent in some cells, but was detected in the cytoplasm and nuclei of other HCEC (Fig. 4.2b). EAAT3 signal was detected in the plasma membrane, was highly concentrated in the cytoplasm of many cells, and was absent or very weakly expressed in some cells (Fig. 4.2c). EAAT4 was detected weakly in all HCEC, but was concentrated in the cytoplasm of a few cells (Fig. 4.2d). A weak EAAT5 signal was detected in the perinuclear cytoplasm of most cells (Fig. 4.2e). The immunoreactive signal for the light chain xCT of the Xc⁻ antiporter was detected on the membranes and cytoplasm of most cells (Fig. 4.2f). Thus, these results support the results of the Na-dependent uptake studies and suggest the relative levels of expression in HCEC cultures was EAAT3 > EAAT2 = xCT > EAAT1 > EAAT4 > EAAT5.

Concentration-dependent substrate inhibition of EAAT. To characterize the Na⁺-dependent EAAT activity in HCEC, [³H]-D-Asp uptake in the presence of different EAAT competitive inhibitor and blocker concentrations was determined. L-Glu, D-Asp, THA, and TBOA inhibited the uptake of [³H]-D-Asp in a concentration-dependent manner (Fig. 4.3). The IC₅₀ of the EAAT blocker TBOA (3.5 µM) was 2 to 8-fold lower than the IC₅₀s of the competitive EAAT inhibitors L-Glu, D-Asp, and

Fig. 4.3 Concentration-
dependent inhibition of
[³H]-ᴅ-Asp uptake in
HCEC. Each point represents
mean and SEM of three
experiments ran in triplicate

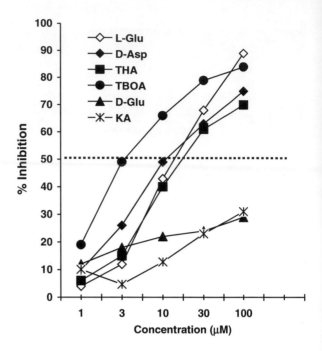

THA (~16 µM, 10 µM, and 20 µM, respectively). In contrast, the IC_{50} of KA was
>100 µM. The weak inhibition of KA and ᴅ-Glu suggests that the uptake of [³H]-ᴅ-
Asp in HCEC was not due to EAAT2. The strong inhibition of ᴅ-Asp uptake by ᴅ-
Asp, THA, and TBOA suggests EAAT1 and/or EAAT3 activity [22]. It should be
noted that the substrate specificity in a HCEC culture likely represents the
transporter(s) expressed by the majority of metabolically active cells. Further, the
EAAT1-5 and xCT (Xc⁻ antiporter light chain) expression levels and distribution
within cells of a culture suggests that more than one EAAT may be expressed by a
HCEC and that different EAATs may be expressed by Xc⁻ antiporter positive and
negative cells.

Substrate specificity of EAAT and Xc⁻ antiporter activities in HCEC.
Na-dependent EAAT and Na-independent Xc⁻ exchanger activities in HCEC were
determined by the uptake of radiolabeled ʟ-Glu, ᴅ-Asp, and ʟ-Cys in the presence
of 100 µM substrate-specific amino acid transporter inhibitors (Table 4.1).
The Na-dependent uptake of [¹⁴C]-ʟ-Glu by HCEC was inhibited strongly by Xᴀɢ
system substrate inhibitors of ʟ-Glu (76.7±2 %), ᴅ-Asp (78.3±5.3 %), and THA
(55.5±2 %), but was inhibited weakly by ᴅ-Glu (20.6±10.1 %) and KA (33.5±8 %).
Inhibition of [¹⁴C]-ʟ-Glu uptake by AAA or NAC was not detected (<10 %).
In comparison, Na⁺-dependent [³H]-ᴅ-Asp uptake was inhibited strongly by ʟ-Glu
(79±0.4 %), ᴅ-Asp (79.4±3.4 %), THA (74.2±5.9 %), and TBOA (91.6±4.1 %),
but was weakly inhibited by ᴅ-Glu (47±10.2 %), and KA (46±8.5 %) consistent
with the results of the concentration-dependent inhibition of ᴅ-Asp uptake (Fig. 4.3).

Table 4.1 Substrate-specific inhibition of radiolabeled Glu, D-Asp and Cys uptake in HCEC

Inhibitors	Percent uptake inhibition by 100 µM of:		
	$[^{14}C]$-L-Glu	$[^3H]$-D-Asp	$[^{14}C]$-L-Cys
L-Glu	76.7±2.0	79.0±0.4	64.6±0.7
D-Glu	20.6±10.1	47.0±10.2	78.6±2.9
D-Asp	78.3±5.3	79.4±3.4	
THA	55.4±2.0	74.2±5.9	
TBOA		91.6±4.1	
KA	33.5±8.0	46.0±8.5	
AAA	<10.0		54.3±5.8
NAC	<10.0		86.3±2.6
Cys			90.7±3.7

The weak inhibition Glu or Asp by D-Glu and KA suggests low or not EAAT2 uptake activity in HCEC. Taken together, the results of the Glu/Asp inhibition studies support EAAT1/3 uptake of Glu/Asp in HCEC, but cannot rule out Glu uptake by EAAT2, EAAT4, and EAAT5.

To define the substrate specificity of Cys transport by HCEC, the uptake of radiolabeled L-Glu and L-Cys was determined in the presence of inhibitors. Cys (90.7±3.7 %), NAC (86.3±2.6 %), and D-Glu (78.6±2.9 %) inhibited Na^+-dependent $[^{14}C]$-L-Cys uptake more than L-Glu (64.6±0.7 %) and AAA (54.3±5.8 %). These results support Na^+-independent Cys uptake and suggest that $[^{14}C]$-L-Cys uptake via Na^+-independent Xc⁻ exchanger, but do not rule out Na^+-dependent EAAT transport of Cys [33, 34].

Glu and Cys inhibition reduced GSH. Based upon the results of the substrate inhibition studies above, we next determined the effect of Glu and Cys transport inhibitors on GSH in HCEC. GSH levels were reduced most significantly by 6 h incubation in media containing L-Glu (24.3±21.6 %), D-Asp (21±11.7 %), and AAA (24±5.3 %) (Fig. 4.4). GSH levels were reduced less by THA (12±5.7 %) and KA (3±8 %). These results are consistent with the idea that EAAT and Xc⁻ antiporter uptake of Glu and Cys plays a significant role in maintaining GSH in HCEC.

Cellular distribution of mitochondria, Glu, GGT, and GS in HCEC. Mitochondria (MitoTracker Green dye positive) were concentrated in the cytoplasm around the nucleus of almost all HCEC in a monolayer (Fig. 4.5a). Glu was detected in all HCEC, but most cells contained highly concentrated Glu in the mitochondria-rich perinuclear cytoplasm (Fig. 4.5b). Some HCEC had concentrated Glu over the nucleus suggestive of apical localization. Punctate packets of highly concentrated GGT were detected between most HCEC and GGT was detected weakly on the nuclear membrane of some cells (Fig. 4.5c). Immunoreactive GS was detected in the perinuclear cytoplasm of many cells in the monolayer, but GS signal was absent or uniquely expressed in the cytoplasm of some HCEC (Fig. 4.5d). The results are consistent with high levels of Glu in mitochondria-rich perinuclear cytoplasm.

Fig. 4.4 Effect of Glu and
Cys transport inhibitors on
GSH in HCEC. Each point
represents mean and SEM of
three experiments ran in
triplicate

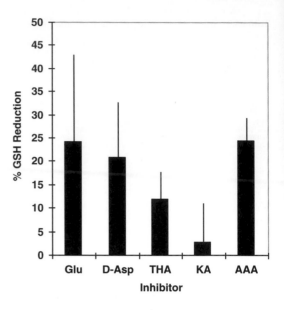

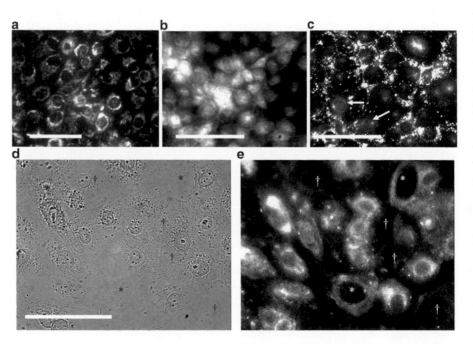

Fig. 4.5 Distribution of (**a**) mitochondria, (**b**) Glu, (**c**) GGT, and (**d, e**) GS in HCEC. Note the
punctate signals of highly concentrated GGT between HCEC and weak signal on nuclear
membranes (*arrows*) of a few HCEC. Also, note GS signals were strong, absent (†), or weak (*) in
some putative nucleophagic HCECs. Bar = 50 μm

Fig. 4.6 Percent GSH reduction in HCEC treated with AT^{125}, MSO, and BSO for 6 h. Mean and SEM of three experiments ran in duplicate

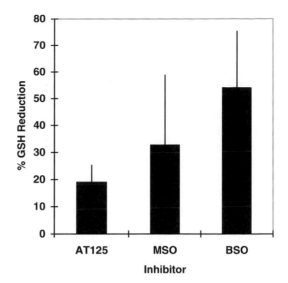

The variable GGT and GS expression by HCEC may suggest metabolic differences consistent with corneal cell differentiation from columnar to the wing epithelia cell phenotype.

Effect of GGT and GS inhibition on GSH levels in HCEC. To determine the relative importance of GGT and GS to GSH levels, HCEC cultures were treated with AT^{125} (5 μg/mL; inhibits >95 % GGT activity; data not shown) [25], 5 mM MSO (inhibitor of GS conversion of Glu to Gln) [26], or BSO (5 mM; inhibitor of GSH synthesis) [27]. The mean percent GSH reduction (±SEM) was determined from three experiments (Fig. 4.6). The GSH concentration in HCEC cultures was reduced by AT^{125} (19±6.4 %) and MSO (32.7±26.2 %), but was reduced about two-fold more by BSO (54±21.4 %). These results suggest the high GSH concentration in HCEC (39.7±2.4 mM) is dependent predominantly on GSH synthesis. Moreover, the results suggest GGT and GS metabolic activities play significant roles in maintaining the GSH levels in HCEC.

4.4 Discussion

All EAATs were detected in HCEC as in human cornea [20], but HCEC transport of Glu and D-Asp was via EAAT1/3 activity based upon the Na$^+$-dependent uptake, substrate specificity of D-Asp and Glu uptake, and predominant expression. However, low levels of Glu/Asp uptake, export or intracellular transport by EAAT2, EAAT4, and EAAT5 cannot be ruled out. GSH depletion following EAAT inhibition is consistent with the idea that GSH levels are dependent on Glu transporter activity in HCEC as in retinal cells [33, 34]. Recently, EAAT3 was shown to transport Cys [35, 36] suggesting an alternate pathway for Cys uptake. Moreover, EAAT3 knockout mice are GSH-deficient and susceptible to oxidative stress that is sensitive to antioxidant treatment [37–39]. Thus, the detection of EAAT3 activity in HCEC, as well as in human corneal endothelial and epithelial cells [20], suggests EAAT3 may contribute to GSH synthesis by transporting Cys and Glu in columnar epithelium and endothelium, but additional experiments are needed to confirm the role of EAAT3 Cys transport and protection from oxidative stress in corneal epithelium. Concomitantly, Cys uptake in HCEC was consistent with Xc^- antiporter activity based upon Na^+-independent uptake activity, sensitivity to Xc^- antiporter inhibitors, and xCT (Xc^- antiporter light chain) expression on most HCEC in the monolayer. The ≈25 % reduction in GSH following incubation of HCEC with inhibitors of Cys uptake (Glu and AAA) supports the importance of Xc^- antiporter in supplying Cys for GSH synthesis [40, 41]. That is, extracellular Glu taken up by EAAT can be utilized by the cell for metabolism and GSH synthesis or exchanged for an extracellular Cys by the Xc^- antiporter; providing the essential Cys for GSH synthesis. The coupling of EAAT and Xc^- antiporter activities has been shown to protect against GSH loss and oxidative stress [41–43]. The results also suggest that the coupling of EAAT and Xc^- antiporter activities in HCEC may protect against oxytosis; cell death due to Glu inhibition of Xc^- antiporter Cys uptake leading to a depletion of GSH and susceptibility to oxidative stress [42, 43]. In this regard, high levels of Glu and EAAT dysfunction have been noted in hypertensive and diabetic eyes [44–47]. Our results are consistent with the idea that Glu uptake by EAAT and exchange for Cys by the Xc^- antiporter are critical to HCEC Glu, Cys, and GSH homeostasis and protection against oxidative stress.

The expression of GGT on cell and nuclear membranes of HCEC was similar to that in human corneal tissue [13, 20]. The detection of GGT supports GSH recapture via the γ-glutamyl cycle [48]. That is, the catalytic cleavage of extracellular GSH by the ectoenzyme GGT generates free precursor amino acids that can then be taken up by HCEC. The function of intracellular and nuclear membrane-bound GGT in mitochondria-rich HCEC and corneal columnar epithelial cells is unknown, but suggests the presence of the glutaminase-GGT pathway [49]. The ≈20 % reduction in GSH following incubation of HCEC with acivicin (AT^{125}; GGT inhibitor) [24] was similar to the reduction in GSH by Glu and Cys uptake inhibition and was consistent with the idea that GGT activity contributes to the GSH levels. The >2-fold greater reduction in GSH levels by BSO than AT^{125} suggests that cytosolic/mitochondrial GSH synthesis [50] in HCEC cultures plays a greater role in maintaining the cellular GSH levels than GSH transport. The importance of GGT to GSH and protecting against cell death due to oxidative stress has been shown in vitro and in vivo [51–55]. Thus, loss of corneal GGT due to aging [13], diabetes [14] and other factors would reduce the capacity of the cornea epithelium to maintain GSH via the γ-glutamyl cycle and lessen the antioxidant potential of the cornea.

GS was detected in many of the HCECs in a 48 h culture, but was absent or weakly expressed in ≈25 % of the cells. GS converts intracellular Glu in the presence of ammonia to Gln [26]. The ≈30 % reduction in GSH caused by the GS inhibitor MSO was greater than the reduction caused by EAAT, Xc^- antiporter, and GGT inhibitors, but was less than by BSO (inhibitor of GSH synthesis). The reduction in

GSH by MSO suggests inhibition of GS activity leads to intracellular Glu accumulation which inhibits Cys uptake by Xc⁻ antiporter (and/or EAAT3) and predisposes the cell to Glu-mediated oxidative stress. This scenario is supported by previous studies in neuronal cells showing GS and Xc⁻ antiporter activities protect neuronal tissue from the toxic effects of Glu and ammonia by generating Gln and exporting Glu [41, 42]. It is not clear why GS is elevated in some HCECs in culture or predominantly expressed in Xc⁻ antiporter positive wing cells in human cornea [20]. However, GS has been detected in ciliary body epithelium [17–19] and retinal pigmented cells [26] where it likely functions to efflux Gln. In addition, GS activity has been shown to regulate Glu-mediated cell swelling [56], enhance clearance of extracellular Glu [57] and protect against neuronal cell death [58]. If GS exerts similar modulatory activities in the corneal epithelium then GS activity could help maintain GSH, protect against endogenous and exogenous oxidative stress, clear the high Glu concentration in columnar cells [20], provide Gln for other corneal cells, reduce Glu produced by dysfunctional mitochondria, and play a role in wing cell morphological changes. Other possibilities cannot be ruled out. Parallel studies using inhibitors of GSH, Glu, Cys, and Gln biogenesis in the corneal epithelium in vivo and in the HCEC model are needed to provide insights into the role of GS in corneal protection against oxidative stress due to endogenous and exogenous factors.

The transporter activity, immunocytological and pharmacological studies presented demonstrate the presence and importance of Glu and Cys transporter systems, GSH recapture, and Gln–Glu metabolic pathways to GSH homeostasis in human corneal epithelial cells. Further, our results and the expression of these systems in the cornea [6, 20] suggest each plays an important role in corneal health by protecting against ROS produced by ultraviolet light [59], Glu metabolism in corneal columnar epithelial cells [60, 61], and mitochondrial dysfunction as wing cells undergo mitophagy [62, 63]. Moreover, the results suggest that GSH and other antioxidant systems protect the corneal endothelium and epithelium and guard the stromal keratocytes [64] against oxidative toxicity due to environmental factors, aging, normal and abnormal metabolism and inflammation.

Acknowledgements The authors thank Christopher Duggan for excellent technical support. The authors have no conflicts of interest. The authors acknowledge the support of the LSUHSC, Department of Ophthalmology Faculty Improvement Fund.

References

1. Green K. Free radicals and aging of anterior segment tissues of the eye: a hypothesis. Ophthalmic Res. 1995;27 Suppl 1:143–9.
2. Buddi R, Lin B, Atilano SR, Zorapapel NC, Kenney MC, Brown DJ. Evidence of oxidative stress in human corneal diseases. J Histochem Cytochem. 2002;50:341–51.
3. Ganea E, Harding JJ. Glutathione-related enzymes and the eye. Curr Eye Res. 2006;31:1–11.
4. Shoham A, Hadziahmetovic M, Dunaief JL, Mydlarski MB, Schipper HM. Oxidative stress in diseases of the human cornea. Free Radic Biol Med. 2008;45:1047–55.
5. Whikehart DR. Total oxidized glutathione in bovine corneal epithelium and endothelium. Exp Eye Res. 1975;25:89–92.

6. Li B, Lee MS, Lee RS, Donaldson PJ, Lim JC. Characterization of glutathione uptake, synthesis, and efflux pathways in the epithelium and endothelium of the rat cornea. Cornea. 2012;31:1304–12.

7. Whikehart DR, Edelhauser HF. Glutathione in rabbit corneal endothelia: the effects of selected perfusion fluids. Invest Ophthalmol Vis Sci. 1978;17:455–64.

8. Anderson EL, Wright DD. The roles of glutathione reductase and γ-glutamyl transpeptidase in corneal transendothelial fluid transport mediated by oxidized glutathione and glucose. Exp Eye Res. 1982;35:11–9.

9. Riley MV. A role for glutathione and glutathione reductase in control of corneal hydration. Exp Eye Res. 1984;39:751–8.

10. Araie M, Shirasawa E, Hikita M. Effect of oxidized glutathione on the barrier function of the corneal endothelium. Invest Ophthalmol Vis Sci. 1988;29:1884–7.

11. Spector A. Oxidation and aspects of ocular pathology. CLAO J. 1990;16(1 Suppl):S8–10.

12. Nakamura M, Nakano T, Hikida M. Effects of oxidized glutathione and reduced glutathione on the barrier function of the corneal endothelium. Cornea. 1994;13:493–5.

13. Redmond P, Burnham JM, Langford MP, Misra RP, Redens TB, Texada DE. Age-related decrease in human corneal γ-glutamyltranspeptidase activity. Cornea. 2013;32:e121–6.

14. Burnham JM, Sakhalkar M, Langford MP, Liang C, Redens TB, Jain SK. Diabetic and non-diabetic human cornea and tear γ-glutamyl transpeptidase activity. Clin Ophthalmol. 2013;7:99–107.

15. Li B, Li L, Donaldson PJ, Lim JC. Dynamic regulation of GSH synthesis and uptake pathways in the rat lens epithelium. Exp Eye Res. 2010;90:300–7.

16. Lim JC, Lam L, Li B, Donaldson PJ. Molecular identification and cellular localization of a potential transport system involved in cystine/cysteine uptake in human lenses. Exp Eye Res. 2013;116:219–26.

17. Langford MP, Gosslee JM, Misra RP, Liang C, Redens TB, Welbourne TC. Apical accumulation of glutamate in GLAST-1, glutamine synthetase positive ciliary body non-pigmented epithelial cells. Clin Ophthalmol. 2007;1:43–53.

18. Hu RG, Lim JC, Kalloniatis M, Donaldson PJ. Cellular localization of glutamate and glutamine metabolism and transport pathways in the rat ciliary epithelium. Invest Ophthalmol Vis Sci. 2011;52:3345–53.

19. Li B, Umapathy A, Tran LU, Donaldson PJ, Lim JC. Molecular identification and cellular localisation of GSH synthesis, uptake, efflux and degradation pathways in the rat ciliary body. Histochem Cell Biol. 2013;139:559–71.

20. Langford MP, Redmond P, Chanis R, Misra RP, Redens TB. Glutamate, excitatory amino acid transporters, Xc⁻ antiporter, glutamine synthetase and γ-glutamyl transpeptidase in human corneal epithelium. Curr Eye Res. 2010;53:221–30.

21. Pan Z, Wang Z, Yang H, Zhang F, Reinach PS. TRPV1 activation is required for hypertonicity-stimulated inflammatory cytokine release in human corneal epithelial cells. Invest Ophthalmol Vis Sci. 2011;52:485–93.

22. Arriza JL, Fairman WA, Wadiche JI, Murdock GH, Kavanaugh MP, Amara S. Functional comparisons of three GLU transporter subtypes clone from human motor cortex. J Neurosci. 1994;14:5559–69.

23. Tsai MJ, Chang YF, Schwarcz R, Brookes N. Characterization of L-alpha-aminoadipic acid transport in cultured rat astrocytes. Brain Res. 1996;741:166–73.

24. Waagepetersen HS, Shimamoto K, Schousbe A. Comparison of effects of DL-threo-beta-benzyloxyaspartate (DL-TBOA) and 1-trans-pyrrolidine-2,4-dicarboxylate (t-2,4-PDC) on uptake and release of [³H]D-aspartate in astrocytes and glutamatergic neurons. Neurochem Res. 2001;26:661–6.

25. Stole E, Smith TK, Manning JM, Meister A. Interaction of gamma-glutamyl transpeptidase with acivicin. J Biol Chem. 1994;269:21435–9.

26. Reif-Lehrer L, Coghlin J. Conversion of glutamic acid to glutamine by retinal glutamine synthetase. Exp Eye Res. 1973;17:321–8.

27. Griffith OW. Mechanism of action, metabolism, and toxicity of buthionine sulfoximine and its higher homologs, potent inhibitors of glutathione synthesis. J Biol Chem. 1982;257:13704–12.
28. Langford MP, Stanton GJ, Johnson HM. Biological effect of staphylococcal enterotoxin A on human peripheral lymphocytes. Infect Immun. 1978;22:68–78.
29. Castle JD, Cameron RS, Patterson PL, Ma AK. Identification of high molecular weight antigens structurally related to gamma-glutamyl transferase in epithelial tissues. J Membr Biol. 1985;87:13–26.
30. Kato S, Ishita S, Sugawara K, Mawatari K. Cystine/glutamate antiporter expression in retinal Müller glial cells: implications for DL-alpha-aminoadipate toxicity. Neuroscience. 1993;57:473–82.
31. Rahman I, Kode A, Biswas SK. Assay for quantitative determination of glutathione and glutathione disulfide levels using enzymatic recycling method. Nat Protoc. 2006;1:3159–65.
32. Sato H, Tamba M, Ishii T, Bannai S. Cloning and expression of a plasma membrane cystine/glutamate exchange transporter composed of two distinct proteins. J Biol Chem. 1999;274:11455–8.
33. Reichelt W, Stabel-Brown J, Pannicke T, Weichert H, Heinemann U. The glutathione level of retinal Müller glial cells is dependent on high-affinity sodium-dependent uptake of glutamate. Neuroscience. 1997;77:1213–24.
34. Agostinho P, Duarte CB, Oliveira CR. Impairment of excitatory amino acid transporter activity by oxidative stress conditions in retinal cells: effect of antioxidants. FASEB J. 1997;11:154–63.
35. Zerangue N, Kavanaugh MP. Interaction of L-cysteine with a human excitatory amino acid transporter. J Physiol. 1996;493:419–23.
36. Hayes D, Wiessner M, Rauen T, McBean GJ. Transport of L-[¹⁴C]cystine and L-[¹⁴C]cysteine by subtypes of high-affinity glutamate transporters over-expressed in HEK cells. Neurochem Int. 2005;46:585–94.
37. Aoyama K, Suh SW, Hamby AM, Liu J, Chan WY, Chen Y, Swanson RA. Neuronal glutathione deficiency and age-dependent neurodegeneration in the EAAC1 deficient mouse. Nat Neurosci. 2006;9:119–26.
38. Cao L, Li L, Zuo Z. N-acetylcysteine reverses existing cognitive impairment and increased oxidative stress in glutamate transporter type 3 deficient mice. Neuroscience. 2012;220:85–9.
39. Berman AE, Chan YY, Brennan AM, Reyes RC, Adler BL, Suh SW, Kauppinen TM, Edling Y, Swanson RA. A-acetylcysteine prevents loss of dopaminergic neurons in the EAAC1⁻/⁻ mouse. Ann Neurol. 2011;69:509–20.
40. Bridges RJ, Natale NR, Patel SA. System xc⁻ cystine/glutamate antiporter: an update on molecular pharmacology and roles within the CNS. Br J Pharmacol. 2012;165:20–34.
41. Mysona B, Dun Y, Duplantier J, Ganapathy V, Smith SB. Effects of hyperglycemia and oxidative stress on the glutamate transporters GLAST and system xc⁻ in mouse retinal Müller glial cells. Cell Tissue Res. 2009;335:477–88.
42. Lewerenz J, Klein M, Methner A. Cooperative action of glutamate transporters and cystine/glutamate antiporter system Xc⁻ protects from oxidative glutamate toxicity. J Neurochem. 2006;98:916–25.
43. Tan S, Schubert D, Maher P. Oxytosis: a novel form of programmed cell death. Curr Top Med Chem. 2001;1:497–506.
44. Dreyer EB, Zurakowski D, Schumer RA, Podos SM, Lipton SA. Elevated glutamate levels in the vitreous body of humans and monkeys with glaucoma. Arch Ophthalmol. 1996;114:299–305.
45. Li Q, Puro DG. Diabetes-induced dysfunction of the glutamate transporter in retinal Müller cells. Invest Ophthalmol Vis Sci. 2002;43:3109–16.
46. Pulido JE, Pulido JS, Erie JC, Arroyo J, Bertram K, Lu MJ, Shippy SA. A role for excitatory amino acids in diabetic eye disease. Exp Diabetes Res. 2007;2007:36150.
47. Alves Mde C, Carvalheira JB, Módulo CM, Rocha EM. Tear film and ocular surface changes in diabetes mellitus. Arq Bras Oftalmol. 2008;71(6 Suppl):96–103.

48. Meister A, Tate SS. Glutathione and related gamma-glutamyl compounds: biosynthesis and utilization. Annu Rev Biochem. 1976;45:559–604.
49. Welbourne TC. Glutaminase-gamma-glutamyltransferase: subcellular localization and ammonia production in acidosis. Proc Soc Exp Biol Med. 1978;159:294–7.
50. Söderdahl T, Enoksson M, Lundberg M, Holmgren A, Ottersen OP, Orrenius S, Bolcsfoldi G, Cotgreave IA. Visualization of the compartmentalization of glutathione and protein-glutathione mixed disulfides in cultured cells. FASEB J. 2003;17:124–6.
51. Hill KE, Von Hoff DD, Burk RF. Effect of inhibition of gamma-glutamyltranspeptidase by AT-125 (acivicin) on glutathione and cysteine levels in rat brain and plasma. Invest New Drugs. 1985;3:31–4.
52. Cotgreave IA, Schuppe-Koistinen I. A role for gamma-glutamyl transpeptidase in the transport of cystine into human endothelial cells: relationship to intracellular glutathione. Biochim Biophys Acta. 1994;1222:375–82.
53. Karp DR, Shimooku K, Lipsky PE. Expression of gamma-glutamyl transpeptidase protects ramos B cells from oxidation-induced cell death. J Biol Chem. 2001;276:3798–804.
54. Carlisle ML, King MR, Karp DR. Gamma-glutamyl transpeptidase activity alters the T cell response to oxidative stress and Fas-induced apoptosis. Int Immunol. 2003;15:17–27.
55. Chevez-Barrios P, Wiseman AL, Rajas E, Ching-nan O, Lieberman MW. Cataract development in γ-glutamyl transpeptidase-deficient mice. Exp Eye Res. 2000;71:575–82.
56. Izumi Y, Matsukawa M, Benz AM, Izumi M, Ishikawa M, Olney JW, Zorumski CF. Role of ammonia in reversal of glutamate-mediated Müller cell swelling in the rat retina. Glia. 2004;48:44–50.
57. Shaked I, Ben-Dror I, Vardimon L. Glutamine synthetase enhances the clearance of extracellular glutamate by the neural retina. J Neurochem. 2002;83:574–80.
58. Gorovits R, Avidan N, Avisar N, Shaked I, Vardimon L. Glutamine synthetase protects against neuronal degeneration in injured retinal tissue. Proc Natl Acad Sci U S A. 1997;94:7024–9.
59. Shimmura S, Tadano K, Tsubota K. UV dose-dependent caspase activation in a corneal epithelial cell line. Curr Eye Res. 2004;28:85–92.
60. Lewerenz J, Dargusch R, Maher P. Lactacidosis modulates glutathione metabolism and oxidative glutamate toxicity. J Neurochem. 2010;113:502–14.
61. Welbourne T, Nissim I. Regulation of mitochondrial glutamine/glutamate metabolism by glutamate transport: studies with ^{15}N. Am J Physiol Cell Physiol. 2001;280:C1151–9.
62. Gottlieb RA, Carreira RS. Autophagy in health and disease. 5. Mitophagy as a way of life. Am J Physiol Cell Physiol. 2010;299:C203–10.
63. Wang CH, Wu SB, Wu YT, Wei YH. Oxidative stress response elicited by mitochondrial dysfunction: implication in the pathophysiology of aging. Exp Biol Med (Maywood). 2013;238:450–60.
64. Izzotti A, Saccà SC, Longobardi M, Cartiglia C. Sensitivity of ocular anterior chamber tissues to oxidative damage and its relevance to the pathogenesis of glaucoma. Invest Ophthalmol Vis Sci. 2009;50:5251–8.

Chapter 5
Transforming Growth Factor: β3 Regulates Cell Metabolism in Corneal Keratocytes and Fibroblasts

D. Karamichos, J.M. Asara, and J.D. Zieske

Abbreviations

ECM	Extracellular matrix
GNG	Glucogenesis
HCF	Human corneal fibroblasts
HCK	Human corneal keratocytes
TCA	Tricarboxylic acid cycle
TGF-β	Transforming growth factor-beta

5.1 Introduction

The cornea is a transparent, avascular, immunologically privileged tissue. In humans, it has a diameter of approximately 11.5 mm and a thickness of 0.5–0.6 mm in the center [1, 2]. About 90 % of the cornea's thickness is occupied by the stroma; a thick, transparent highly organized layer consisting of regularly arranged collagen

D. Karamichos (✉)
Department of Ophthalmology, University of Oklahoma Health Sciences Center,
608 Stanton L. Young Boulevard, DMEI PA-409, Oklahoma City, OK 73104, USA
e-mail: Dimitrios-karamichos@ouhsc.edu

J.M. Asara
Division of Signal Transduction and Mass Spectrometry Core,
Beth Israel Deaconess Medical Center and the Department of Medicine,
Harvard Medical School, Boston, MA, USA

J.D. Zieske
Schepens Eye Research Institute/Massachusetts Eye and Ear and the Department
of Ophthalmology, Harvard Medical School, Boston, MA, USA

© Springer Science+Business Media New York 2015
M.A. Babizhayev et al. (eds.), *Studies on the Cornea and Lens*,
Oxidative Stress in Applied Basic Research and Clinical Practice,
DOI 10.1007/978-1-4939-1935-2_5

fibers along with sparsely distributed keratocytes [1, 2]. Human corneal keratocytes (HCK) are a population of quiescent, mesenchymal derived cells and are found between the collagen lamellae of the mature corneal stroma. Despite the fact that HCKs only occupy 10 % of the total corneal stromal area, they are key players in long-term corneal transparency and integrity. Corneal stroma architecture is highly organized, consisting of aligned collagenous lamellae and extracellular matrix (ECM) components that are secreted and deposited by HCKs [3, 4].

In the event of an injury to the cornea, HCKs that survive apoptosis become activated and begin the synthesis and deposition of new ECM components, widely known as tissue remodeling. These activated cells are known to assume a "repair phenotype" and are called fibroblasts. Following more severe corneal injuries or at the advanced stages of healing process, fibroblasts transform into myofibroblasts; which is the stage where ECM components are actively secreted. This transformation is thought to be caused by transforming growth factor-β (TGF-β). There are three isoforms of TGF-β, identified in humans (TGF-β1, -β2, and -β3). TGF-β3 is an isoform that we and others have identified as a potential factor for resolving corneal scarring [5–7]. In an attempt to explore TGF-β3's effect on corneal cells we present here a novel study about the metabolism of corneal cells in vitro and how it can be altered using just a single factor; TGF-β3.

Despite the huge number of studies on corneal stroma and corneal cells, very little is known about metabolic activity of human corneal cells. Even less is known about the differences in metabolism between corneal keratocytes and fibroblasts. In this study, mass spectrometry-based metabolomics [8, 9], a technique developed to identify and quantify endogenous metabolites, was used to examine changes in metabolism. Metabolomics has been used successfully in ocular diseases [10–14] and we therefore wanted to investigate the metabolic differences and similarities of the HCKs and HCFs derived from normal individuals, using two established in vitro models: (a) conventional 2D cultures, and (b) self-assembled ECM-3D constructs.

This current chapter presents evidence for a metabolic role of TGF-β3 in HCK and corneal fibroblasts. Investigating the metabolic changes in two different systems (2D and 3D) allows us to compare cells in a monolayer culture and cells in a self-assembled ECM, which more closely resembles an in vivo-like condition.

5.2   Methods

5.2.1   Primary Cultures Established

HCKs and HCFs were isolated from human corneas ($n=3$) from healthy patients without ocular disease. All samples were obtained from NDRI (National Disease Research Interchange; Philadelphia, PA). All research adhered to the tenets of the Declaration of Helsinki. Tissue was processed, as previously described [15]. Briefly, corneal epithelium and endothelium were removed from the stroma by scraping with a razor blade. The stromal tissue was then cut into ~2×2 mm pieces and placed into T25 culture flasks. Explants were cultured in Eagle's Minimum Essential

Medium (EMEM: ATCC; Manassas, VA) with either 1 % (HCKs) or 10 % (HCFs) fetal bovine serum (FBS: Atlantic Biologicals, Miami, FL). Following 1–2 weeks of cultivation, the cells were passaged into 100 mm cell culture plates and allowed to grow to 100 % confluence before being used in our 2 systems (2D and 3D).

5.2.2 Conventional 2D Cultures

All cell types (HCKs and HCFs) were seeded in 6-well plates at 10^6 cells/well and cultured, for 4 weeks, in EMEM with either 1 % (HCKs) or 10 % FBS (HCFs), with VitC (Vit C: 0.5 mM 2-O-α-D-glucopyranosyl-L-ascorbic acid: Wako Chemicals USA, Inc.; Richmond, VA), and with or without T3 (0.1 ng/mL TGF-β3: R&D systems, Minneapolis, MN). Cultures without T3 served as Controls (C).

5.2.3 3D Constructs

For the development of the self-assembled 3D ECM constructs, as previously described [6, 16], cells were plated on transwell 6-well plates containing polycarbonate membrane inserts with 0.4 μm pores (Costar; Charlotte, NC) at a density of 10^6 cells/well. Cells were cultured in identical conditions as the 2D cultures; EMEM with either 1 % (HCKs) or 10 % FBS (HCFs)+VitC±T3 (0.1 ng/mL). The cultures were allowed to grow for 4 weeks. The optimal concentration of T3 was previously determined by comparing a concentration series ranging from 0.1 to 10 ng/mL [6, 16]. Cultures without T3 served as Controls (C).

5.2.4 Metabolite Extraction

All cultured cells and constructs were collected and processed as previously reported [17]. Briefly, a homogenizer was used to ground samples in ice-cold 80 % MeOH. Samples were centrifuged (14,000 g, 10 min, 4 °C) and supernatants incubated on dry ice. Pellets were further disrupted in ice-cold 80 % MeOH and combined with previous supernatants. Plasma metabolites were extracted twice in 80 % ice-cold MeOH. Metabolite extracts were vortexed and centrifuged (14,000 g, 10 min, 4 °C). Supernatants were dried and stored at −80 °C until further analysis.

5.2.5 Targeted Mass Spectrometry

Liquid chromatography-tandem mass spectrometry (LC-MS/MS) was used to analyze the samples as described previously [18]. Samples were re-suspended using 20 μL LC/MS grade water, and 5–7 μL was injected into a hybrid 5500 QTRAP triple quadrupole mass spectrometer (AB/SCIEX) coupled to a Prominence UFLC HPLC system (Shimadzu, Columbia, MD). Samples were analyzed via selected reaction monitoring (SRM) of a total of 256 unique endogenous water-soluble metabolites for steady-state analyses. Some metabolites were targeted in both positive and negative ion mode, for a total of 289 SRM transitions, using positive/ negative ion polarity switching. Approximately 10–14 data points were acquired per detected metabolite. Samples were delivered to the mass spectrometer via hydrophilic interaction chromatography (HILIC) using a 4.6 mm i.d × 10 cm Amide XBridge column (Waters) at 375 μL/min. Peak areas from the total ion current for each metabolite SRM transition were integrated using MultiQuant v2.0 software (AB/SCIEX).

5.2.6 Statistical and Pathway Analysis

We performed a two-step analysis of our findings. Firstly, we analyzed and compared HCFs and HKCs for their metabolic differences and secondly, the effect of TGF-β3 on their metabolism. The focus of this study was the regulation of metabolism by a single growth factor; TGF-β3. Therefore Table 5.1 shows our overall findings, where Figs. 5.1, 5.2, 5.3, and 5.4 concentrate on metabolic differences on HCFs and HKCs following TGF-β3 stimulation. All experiments were repeated at least three times and data were analyzed for significant variations ($p < 0.05$) using one way ANOVA. Metabolic pathways were investigated based on regulated metabolites using Metaboanalyst 2.0 free online software (www.MetaboAnalyst.ca) [19].

5.3 Results

5.3.1 Expression of Metabolites

We have identified 256 endogenous water-soluble metabolites of which more than 60 were significantly regulated between groups. We performed the analysis using only the metabolites that were up or downregulated by at least twofold. This cutoff

Table 5.1 Summary of the number of metabolites that were up or down-regulated following TGF-β3 stimulation on both 2D and 3D systems, for both cell types (HCKs and HCFs)

		# of up-regulated metabolites	# of down-regulated metabolites
2D	HCK + T3	3	20
	HCF + T3	15	1
3D	HCK + T3	56	3
	HCF + T3	4	3

Only the metabolites passing filtering criteria were included (at least 2:1 up or down-regulation)

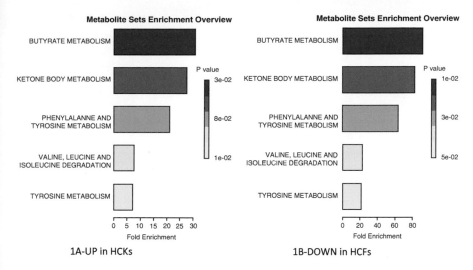

Fig. 5.1 Summary of pathway enrichment analysis in 2D system. Above is a display of the diversity of signaling pathways that are enriched on the basis of all the metabolites passing filtering criteria. The most significant *p*-values are in *red* while the least significant are in *yellow* and *white*. (**a**) Pathways affected based on metabolites that were at least twofold up-regulated in HCKs compared to HCFs, following TGF-β3 stimulation. (**b**) Pathways affected based on metabolites that were at least twofold down-regulated in HCFs compared to HCKs, following TGF-β3 stimulation

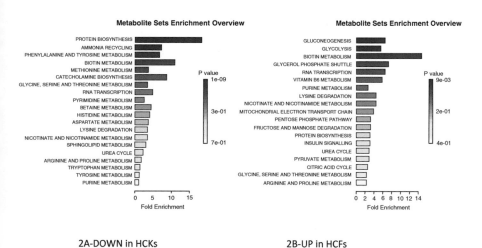

Fig. 5.2 Summary of pathway enrichment analysis in 2D system. Above is a display of the diversity of signaling pathways that are enriched on the basis of all the metabolites passing filtering criteria. The most significant *p*-values are in *red* while the least significant are in *yellow* and *white*. (**a**) Pathways affected based on metabolites that were at least twofold down-regulated in HCKs compared to HCFs, following TGF-β3 stimulation. (**b**) Pathways affected based on metabolites that were at least twofold up-regulated in HCFs compared to HCKs, following TGF-β3 stimulation

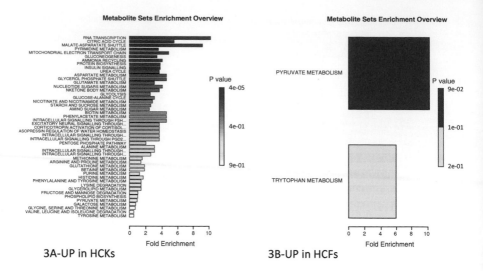

3A-UP in HCKs   3B-UP in HCFs

Fig. 5.3 Summary of pathway enrichment analysis in 3D system. Above is a display of the diversity of signaling pathways that are enriched on the basis of all the metabolites passing filtering criteria. The most significant *p*-values are in *red* while the least significant are in *yellow* and *white*. (**a**) Pathways affected based on metabolites that were at least twofold up-regulated in HCKs compared to HCFs, following TGF-β3 stimulation. (**b**) Pathways affected based on metabolites that were at least twofold up-regulated in HCFs compared to HCKs, following TGF-β3 stimulation

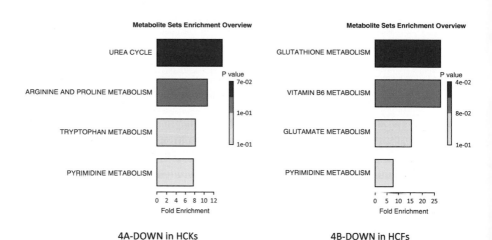

4A-DOWN in HCKs   4B-DOWN in HCFs

Fig. 5.4 Summary of pathway enrichment analysis in 3D system. Above is a display of the diversity of signaling pathways that are enriched on the basis of all the metabolites passing filtering criteria. The most significant *p*-values are in *red* while the least significant are in *yellow* and *white*. (**a**) Pathways affected based on metabolites that were at least twofold down-regulated in HCKs compared to HCFs, following TGF-β3 stimulation. (**b**) Pathways affected based on metabolites that were at least twofold down-regulated in HCFs compared to HCKs, following TGF-β3 stimulation

was chosen in order to ensure that we included only the vastly abundant metabolites. We investigated both a conventional 2D system and a 3D self-assembled ECM. More specifically, in 2D cultures, we found three metabolites significantly upregulated in HCKs when stimulated with TGF-β3 and 20 downregulated. In contrast, we found 15 upregulated and 1 downregulated in HCFs (Table 5.1). When the cells were stimulated to secrete their own ECM, using our 3D system, metabolic regulation was altered. In HCKs, 56 metabolites were upregulated and 3 were downregulated, while 5 were upregulated and 3 downregulated in HCFs. Surprisingly, in our 3D model there were seven metabolites that were not found at all, in HCKs, with or without TGF-β3 and four in HCFs. Furthermore, we found eight metabolites that were completely missing in HCK Controls and one in HCF Controls. Those were identified only upon TGF-β3 stimulation. While the missing metabolites are interesting, further investigation is needed before we can draw any conclusions. This chapter will discuss specific changes in cellular metabolism based on the above findings.

5.3.2   Conventional 2D Cultures: Metabolic Activity

Using the Metaboanalyst software, we were able to dissect which metabolic pathways are affected based on our raw data. In cells that are conventionally cultured (2D) we found significant differences between HCKs and HCFs. As shown in Fig. 5.1a the predicted metabolic pathways that are affected in 2D—HCKs based on the three metabolites that are upregulated following stimulation with T3 are: Butyrate, Ketone body, and Phenylalanine/Tyrosine metabolism. All of these are associated with energy production and suggest that T3 stimulates activation of HCKs. Butyrates are important for cells since they provide energy. Without them cells undergo apoptosis [20]. In cornea, butyrate was found to stimulate fibronectin synthesis in cultured rabbit corneal blocks in situ [21]. Ketone bodies are three different water-soluble biochemicals that are produced by the liver from fatty acids and are another source of energy for the cells. In cornea, not much is known, however levobunolol, a ketone-containing drug, has been used on rabbit eyes [22]. That study concluded that levobunolol had significant effects on energy metabolism of corneal epithelium, iris-ciliary body, and lens. Phenylalanine/Tyrosine metabolism is considered to be one of the key steps in signal transduction and regulation of enzymatic activity. Any defects in the Phenylalanine/Tyrosine metabolism may lead to a variety of clinical disorders, including Tyrosinemia type II which results in painful corneal lesions and scars [23].

Surprisingly, the same pathways were predicted to be affected with HCFs following stimulation with TGF-b3, only this time the metabolite was downregulated suggesting that HCFs and HCKs metabolism is differentially affected by T3 (Fig. 5.1b). This would suggest that T3 is acting on HCFs positively by reducing the stress levels and moving the cells back towards HCKs.

On the other hand, based on the 20 metabolites downregulated upon T3 stimulation of HCKs, the pathways affected were: protein biosynthesis, ammonia recycling, phenylalanine/tyrosine, and biotin metabolism. Ammonia recycling is important to be kept under control. Organisms that cannot easily and quickly remove ammonia usually have to convert it to some other substance, like urea which is much less toxic. Insufficient removal of ammonia is toxic. This might not be important on an in vitro model, such as this one, where toxins are released into the media but it definitely provides an initial indicator. In cornea, not much is known about ammonia metabolism; however exposure to gaseous ammonia can cause blindness [24]. Phenylalanine/tyrosine seems to be affected both from metabolites that are up and downregulated with T3 (Fig. 5.2a) suggesting a more complex mechanism.

Based on the 15 metabolites that were upregulated with HCFs that were stimulated with T3, the pathways affected included: glucogenesis (GNG), glycolysis, RNA transcription, and biotin metabolism (Fig. 5.2b). GNG and glycolysis are part of the same process and are vital in humans for maintaining blood glucose levels to normal. In cornea GNG is important as glucose is used as fuel source and has been linked to wound healing. The other two pathways RNA transcription is clearly important for gene expression [25, 26] and biotin metabolism is vital for cell growth and production of fatty acids [27]. Clearly T3 presence in HCFs has a predicted positive effect in terms of the cellular metabolism.

5.3.3 Self-Assembled 3D Cultures: Metabolic Activity

Not surprisingly, the metabolic pathways affected on cells that were cultured in our 3D self-assembled model were different between HCKs and HCFs. Figure 5.3a is representation of the affected metabolic pathways in 3D—HCKs based on the 56 metabolites that are upregulated following stimulation with T3. The most significantly affected pathway is the RNA transcription which is the first step of gene expression. In other words HCKs in 3D constructs can be stimulated by T3 to initiate gene expression which can be vital if we are to target specific genes. Furthermore, we found that citric acid cycle, also known as the tricarboxylic acid cycle (TCA cycle), or the Krebs cycle [28–30], is affected. TCA cycle is a key component of the metabolic pathway by which all aerobic organisms generate energy. TCA also provides precursors of certain amino acids as well as the reducing agent NADH (Nicotinamide adenine dinucleotide) that is key in numerous biochemical reactions. In fact enzymes that are known to be involved or affect NADH metabolism are normally targets for drug discovery. Later in this chapter, the TCA and specific metabolites altered in both HCKs and HCFs are shown. Another metabolic pathway that is affected and is linked to energy is the malate-aspartate shuttle. The malate-aspartate shuttle is a biochemical system that generates ATP via translocation of electrons across cellular membranes [31]. Even though in cornea there are currently no studies specifically reporting the activity of this pathway, it is easily understood that there is a key role of the malate-aspartate shuttle, at least in terms of cellular energy. Other notable pathways, as with the 2D system, include GNG and protein biosynthesis; both vital in human's blood glucose levels and in cornea's fuel supply.

Surprisingly though, the metabolic pathways upregulated by T3 stimulation in HCF—3D constructs were minimum. Only, four metabolites were significantly upregulated: Acetylphosphate, 2,3-dihydroxybenzoic acid, Xanthurenic acid, and Cholesteryl sulfate. Based on these, Metaboanalyst predicted that two pathways are affected: Pyruvate and Tryptophan metabolism. The significance, as indicated by the p value in Fig. 5.3b suggests that only pyruvate metabolic pathway was altered. Pyruvate metabolism however, is very important. In fact, it is a key intersection in the network of metabolic pathways. Pyruvate can be converted into carbohydrates via gluconeogenesis, to fatty acids or energy through acetyl-CoA, and to the amino acid alanine. It is also the output of the anaerobic metabolism of glucose [32] which we saw being affected by T3 stimulation on most of our systems. Pyruvate alone and/or in combination with Lactate is known to be indicators for oxidative stress in cornea as well as being regulated in cyclodiathermy [33].

T3 stimulation caused some metabolites to be downregulated on 3D constructs and Fig. 5.4a, b shows the predicted pathways affected. For HCKs the Urea cycle and the arginine/proline metabolism (Fig. 5.4a) was affected and for the HCFs the Glutathione and Vitamin B6 metabolism (Fig. 5.4b). Urea was also affected, indirectly, via the ammonia recycling on our 2D system indicating similar effect in 2D and 3D systems. The effect on arginine/proline pathway is expected since a key component of the arginine/proline metabolic pathway is ornithine; and ornithine is primarily an intermediate of the urea cycle. In cornea, arginine downregulation is important and has been reported as a key mechanism for corneal allografts immune privilege [12]. Finally, HCFs showed regulation of Vitamin B6 and Glutathione, with the two of them actually related. Glutathione is vital for various protective roles in the human body and the Vitamin B6 as pyridoxal-5′-phosphate (PLP) required in the formation of glutathione precursors [34]. Glutathione has been reported multiple times as an oxidative stress indicator while Vitamin B6 depletion have been shown to increase glutathione concentrations in rat liver [35, 36] and human plasma [37].

5.3.4 TCA Regulation

In this paragraph we present a schematic color-coded representation of the specific metabolites that were up- or downregulated within the TCA cycle on our 2D and 3D systems with the two different cell types. The schematic was generated based on the raw data obtained from the samples.

TCA cycle is a series of enzyme-catalyzed chemical reactions that form a key part of aerobic respiration in cells. This cycle is also known as the Krebs cycle. The greatly simplified schematic shown here starts with glucose (without actually showing the glycolysis cycle) and continues with pyruvate which is the end product of glycolysis and the first step of all types of cell respiration in the mitochondria. TCA cycle is not only part of the glucose pathway for the purpose of breaking down of glucose but is also part of all metabolites, including sugars, amino acids, and fatty acids. It is primarily fed from glutamine which is then converted to glutamate via glutaminase and then catalyzed to α-ketogluterate via glutaminate dehydrogenase. Within each of these groups of molecules there is always a pathway that leads into the TCA cycle. It is easily understood that in order for us to determine and characterize the metabolic differences in HCKs and HCFs we need to look at the TCA cycle regulation first. Figure 5.5 shows the metabolites within the TCA cycle that were regulated in our 2D system. Oxaloacetate products such as Asparagine, Lysine, Methionine, and Isoleucine were downregulated in HCKs, along with Arginine which is downstream from α-ketoglutarate, and Alanine. No metabolites were upregulated in HCKs. On the other hand, HCFs showed five upregulated metabolites; Alanine, Lactate, Citrate, Isocitrate, and Lysine. Note that Alanine and Lysine are the common and differentially regulated ones between HCKs and HCFs.

Figure 5.6 shows the metabolites within the TCA cycle that were regulated in our 3D system. Unexpectedly, only HCK-upregulated metabolites were within the TCA cycle. In other words there were no metabolites up- or downregulated in HCFs and

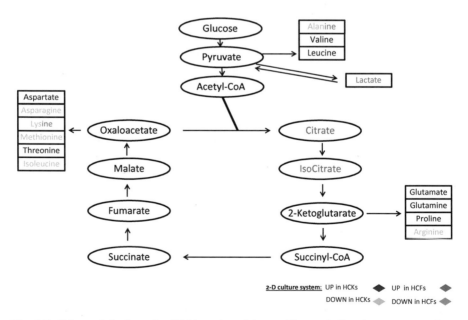

Fig. 5.5 Color-coded schematic of TCA cycle and the specific metabolites that were up or down regulated in the 2D system using HCKs and HCFs. The schematic was generated based on the metabolomics raw data. No metabolites were up-regulated in HCKs (◆) and none were down-regulated in HCFs (◆). Lysine and Alanine were up-regulated in HCFs and down-regulated in HCKs (◆). Asparagine, Methionine, Isoleucine, and Arginine were also down-regulated in HCKs (◆) while Lactate, Citrate, and Isocitrate were up-regulated in HCFs (◆)

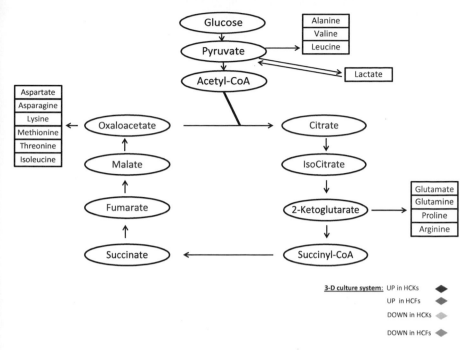

Fig. 5.6 Color-coded schematic of TCA cycle and the specific metabolites that were up or down regulated on the 3D system using HCKs and HCFs. The schematic was generated based on the metabolomics raw data. No metabolites were up-regulated in HCFs (◆), and none were down-regulated in HCFs (◆) or HCKs (◆). Asparagine, Lysine, Methionine, Oxaloacetate, Malate, Fumarate, Citrate, Isocitrate, Glutamate, and Arginine all were up-regulated in HCKs (◆)

no downregulated in HCKs. This would suggest that HCFs are probably as activated as they are going to be in terms of their metabolism. This may prove important in future studies on how to regulate HCFs responses. The metabolites that were upregulated were: Asparagine, Lysine, Methionine, Oxaloacetate, Malate, Fumarate, Citrate, Isocitrate, Glutamate, and Arginine. This is in complete contrast with the 2D system (Fig. 5.5) where no HCK upregulated metabolites were found. It is clear that T3 has an important impact on the TCA cycle of the HCKs in our 3D model that more closely represents what we see in vivo. This is true for both primary TCA metabolites (citrate, isocitrate, α-ketogluterate, succinate, fumerate, malate, and oxaloacetate) and secondary such as Lysine, Methionine, and Asparagine.

5.4 Discussion and Conclusions

Although generally in the literature metabolomics have not been utilized to its full potential, they hold great power in identifying and perhaps provide solutions to a variety of diseases [38]. Our data certainly supports this since we were able to investigate, identify, and analyze a large number of metabolites that were expressed in two different corneal cell types HCKs and HCFs. Both these cell types have been studied by us [6, 7, 16], and others [39–41], in an attempt to understand their characteristics and use their potential in vivo for treatment of corneal trauma.

Corneal trauma and injury can be caused by a number of ways including: chemical irritation or burn, scratches or scrapes, and foreign body in the eye [42]. Upon injury, by any of the above means, corneal wound healing and remodeling begins. Corneal wound healing is a process and a sequence of events which are dependent and controlled by many factors. Ultimately, most of these injuries will result in corneal scarring and some kind of vision impairment. What is important from clinical perspective is how the healing process can be manipulated or altered so the end result (scarring) is minimized. Unfortunately, much of our knowledge concerning corneal wound healing comes from experimental work with animals (mice, rats, and rabbits) [43]. There have been very limited amount of studies that involve human subjects. This is a problem because there are certain vital anatomical differences between these species, in particular the absence of Bowman's layer in rodents which can cause issues when animal model results are extrapolated to human clinical findings.

Understanding the mechanisms behind human corneal wound healing is important if we are going to treat and minimize corneal scarring. Our data here, using cells isolated from human subjects, indicates that there are multiple metabolic differences between HCKs and HCFs, two cell types that are widely known for different types of activities in the human corneal stroma. This is a novel approach for corneal scarring and it is certainly promising since the metabolic activities vary so much between them that we could potentially suggest a metabolic treatment for corneal wounds and scars. In fact, we have shown here that with the addition of a single growth factor, TGF-β3, we can modify important metabolic pathways such as the TCA cycle.

TCA cycle is a key component of a lot of metabolic pathways. It is involved in all three major groups: lipids, proteins, and carbohydrates. The end product of this cycle is cellular energy. TCA is a major indicator of oxidative stress and any inefficient cycling of TCA will lead to altered redox state. Our data clearly indicates that TGF-β3 can modulate the TCA cycle on both 2D and 3D systems. However, not all cell types are affected the same on both environments. TGF-β3 affected the HCF's TCA cycle in 2D and HCKs in 3D. More experimental work is required in order to be able to safely conclude on the TGF-β3's metabolic activity, but it is possible that one growth factor can alter cellular metabolic balance. More specifically, in our previous studies, we have reported that HCKs are not able to secrete and assemble an ECM without T3 stimulation [7]. This would correlate with our TCA cycle findings where HCKs metabolism is activated upon T3 stimulation and it might be part of the reason these cells are secreting ECM upon T3 stimulation. On the other hand HCFs seem to be at their peak of activation, in terms of their metabolism, and therefore are able to secrete ECM even without T3 stimuli [6].

Growth factors are rapidly emerging as a new generation of ophthalmic pharmaceuticals. More and more studies are suggesting that they will play a major role in wound healing management in the future. Of course, wound healing in cornea is affected by many factors including the size of the wound, its depth, and tear quality and therefore further investigations are required. Understanding the role of TGF-β3 in corneal cellular metabolism may prove to be key for the regulation of some or all corneal wounds.

Acknowledgments This work was supported by National Institutes of Health Grant EY023568 (D.K) and EY020886 (DK and JDZ)

References

1. Patel SV, McLaren JW, Hodge DO, Bourne WM. Normal human keratocyte density and corneal thickness measurement by using confocal microscopy in vivo. Invest Ophthalmol Vis Sci. 2001;42(2):333–9.
2. Reinstein DZ, Archer TJ, Gobbe M, Silverman RH, Coleman DJ. Stromal thickness in the normal cornea: three-dimensional display with artemis very high-frequency digital ultrasound. J Refract Surg. 2009;25(9):776–86.
3. Maurice DM. The structure and transparency of the cornea. J Physiol. 1957;136(2):263–86.
4. Muller LJ, Pels E, Schurmans LR, Vrensen GF. A new three-dimensional model of the organization of proteoglycans and collagen fibrils in the human corneal stroma. Exp Eye Res. 2004;78(3):493–501.
5. Shah M, Foreman DM, Ferguson MW. Neutralisation of TGF-beta 1 and TGF-beta 2 or exogenous addition of TGF-beta 3 to cutaneous rat wounds reduces scarring. J Cell Sci. 1995;108(Pt 3):985–1002.
6. Karamichos D, Hutcheon AE, Zieske JD. Transforming growth factor-beta3 regulates assembly of a non-fibrotic matrix in a 3D corneal model. J Tissue Eng Regen Med. 2011;5(8):e228–38.
7. Karamichos D, Rich CB, Zareian R, Hutcheon AE, Ruberti JW, Trinkaus-Randall V, et al. TGF-beta3 stimulates stromal matrix assembly by human corneal keratocyte-like cells. Invest Ophthalmol Vis Sci. 2013;54(10):6612–9.
8. Dettmer K, Aronov PA, Hammock BD. Mass spectrometry-based metabolomics. Mass Spectrom Rev. 2007;26(1):51–78.
9. Rochfort S. Metabolomics reviewed: a new "omics" platform technology for systems biology and implications for natural products research. J Nat Prod. 2005;68(12):1813–20.
10. Greiner JV, Kopp SJ, Glonek T. Phosphorus nuclear magnetic resonance and ocular metabolism. Surv Ophthalmol. 1985;30(3):189–202.
11. Risa O, Saether O, Lofgren S, Soderberg PG, Krane J, Midelfart A. Metabolic changes in rat lens after in vivo exposure to ultraviolet irradiation: measurements by high resolution MAS 1H NMR spectroscopy. Invest Ophthalmol Vis Sci. 2004;45(6):1916–21.
12. Fu H, Khan A, Coe D, Zaher S, Chai JG, Kropf P, et al. Arginine depletion as a mechanism for the immune privilege of corneal allografts. Eur J Immunol. 2011;41(10):2997–3005.
13. Klyce SD. Stromal lactate accumulation can account for corneal oedema osmotically following epithelial hypoxia in the rabbit. J Physiol. 1981;321:49–64.
14. Nguyen TT, Bonanno JA. Lactate-H$^+$ transport is a significant component of the in vivo corneal endothelial pump. Invest Ophthalmol Vis Sci. 2012;53(4):2020–9.
15. Guo X, Hutcheon AE, Melotti SA, Zieske JD, Trinkaus-Randall V, Ruberti JW. Morphologic characterization of organized extracellular matrix deposition by ascorbic acid-stimulated human corneal fibroblasts. Invest Ophthalmol Vis Sci. 2007;48(9):4050–60.
16. Karamichos D, Guo XQ, Hutcheon AE, Zieske JD. Human corneal fibrosis: an *in vitro* model. Invest Ophthalmol Vis Sci. 2010;51(3):1382–8.

17. Webhofer C, Gormanns P, Reckow S, Lebar M, Maccarrone G, Ludwig T, et al. Proteomic and metabolomic profiling reveals time-dependent changes in hippocampal metabolism upon paroxetine treatment and biomarker candidates. J Psychiatr Res. 2013;47(3):289–98.

18. Yuan M, Breitkopf SB, Yang X, Asara JM. A positive/negative ion-switching, targeted mass spectrometry-based metabolomics platform for bodily fluids, cells, and fresh and fixed tissue. Nat Protoc. 2012;7(5):872–81.

19. Xia J, Mandal R, Sinelnikov IV, Broadhurst D, Wishart DS. MetaboAnalyst 2.0—a comprehensive server for metabolomic data analysis. Nucleic Acids Res. 2012;40(Web Server issue):W127–33.

20. Donohoe DR, Garge N, Zhang X, Sun W, O'Connell TM, Bunger MK, et al. The microbiome and butyrate regulate energy metabolism and autophagy in the mammalian colon. Cell Metab. 2011;13(5):517–26.

21. Tanaka H, Nishida T. Butyrate stimulates fibronectin synthesis in cultured rabbit cornea. J Cell Physiol. 1985;123(2):191–6.

22. Lee VH, Chien DS, Sasaki H. Ocular ketone reductase distribution and its role in the metabolism of ocularly applied levobunolol in the pigmented rabbit. J Pharmacol Exp Ther. 1988;246(3):871–8.

23. Tsai CP, Lin PY, Lee NC, Niu DM, Lee SM, Hsu WM. Corneal lesion as the initial manifestation of tyrosinemia type II. J Chin Med Assoc. 2006;69(6):286–8.

24. Mahmoud SS. Corneal toxicity associated with ammonia exposure investigated by Fourier transform infrared spectroscopy. Biophys Rev Lett. 2009;04(4):331.

25. Nelson DL, Cox MM. Lehninger principles of biochemistry. New York: Worth Publishers; 2000. p. 724.

26. Young JW. Gluconeogenesis in cattle: significance and methodology. J Dairy Sci. 1977;60:1–15.

27. George SJ, Bernard AW, Albers RW, Stephen FK, Michael UD, editors. Basic neurochemistry. 6th ed. Philadelphia: Lippincott-Raven; 1999.

28. Lowenstein JM. Methods in enzymology, volume 13: citric acid cycle. Boston: Academic; 1969.

29. Krebs HA, Weitzman PDJ. Krebs' citric acid cycle: half a century and still turning. London: Biochemical Society; 1987.

30. Lane N. Life ascending: the ten great inventions of evolution. New York: W. W. Norton & Co; 2009.

31. Monty K, Matthew PS, Matsudaira PT, Lodish HF, Darnell JE, Lawrence Z, et al. Molecular cell biology. 5th ed. San Francisco: W. H. Freeman; 2003. p. 973.

32. Lehninger AL, Nelson DL, Cox MM. Principles of biochemistry. 5th ed. New York, NY: W.H. Freeman and Company; 2008. p. 528.

33. Schutte E, Schulz I, Reim M. Lactate and pyruvate levels in the aqueous humor and in the cornea after cyclodiathermy. Albrecht Von Graefes Arch Klin Exp Ophthalmol. 1972;185(4):325–30.

34. Lamers Y, O'Rourke B, Gilbert LR, Keeling C, Matthews DE, Stacpoole PW, et al. Vitamin B-6 restriction tends to reduce the red blood cell glutathione synthesis rate without affecting red blood cell or plasma glutathione concentrations in healthy men and women. Am J Clin Nutr. 2009;90(2):336–43.

35. Lima CP, Davis SR, Mackey AD, Scheer JB, Williamson J, Gregory 3rd JF. Vitamin B-6 deficiency suppresses the hepatic transsulfuration pathway but increases glutathione concentration in rats fed AIN-76A or AIN-93G diets. J Nutr. 2006;136(8):2141–7.

36. Hsu JM, Buddemeyer E, Chow BF. Role of pyridoxine in glutathione metabolism. Biochem J. 1964;90(1):60–4.

37. Davis SR, Quinlivan EP, Stacpoole PW, Gregory 3rd JF. Plasma glutathione and cystathionine concentrations are elevated but cysteine flux is unchanged by dietary vitamin B-6 restriction in young men and women. J Nutr. 2006;136(2):373–8.

38. Robertson DG, Watkins PB, Reily MD. Metabolomics in toxicology: preclinical and clinical applications. Toxicol Sci. 2011;120 Suppl 1:S146–70.

39. Chan AA, Hertsenberg AJ, Funderburgh ML, Mann MM, Du Y, Davoli KA, et al. Differentiation of human embryonic stem cells into cells with corneal keratocyte phenotype. PLoS One. 2013;8(2):e56831.
40. Petroll WM, Lakshman N, Ma L. Experimental models for investigating intra-stromal migration of corneal keratocytes, fibroblasts and myofibroblasts. J Funct Biomater. 2012;3(1):183–98.
41. Jester JV, Budge A, Fisher S, Huang J. Corneal keratocytes: phenotypic and species differences in abundant protein expression and in vitro light-scattering. Invest Ophthalmol Vis Sci. 2005;46(7):2369–78.
42. Khaw PT, Shah P, Elkington AR. Injury to the eye. BMJ. 2004;328(7430):36–8.
43. Netto MV, Mohan RR, Ambrosio Jr R, Hutcheon AE, Zieske JD, Wilson SE. Wound healing in the cornea: a review of refractive surgery complications and new prospects for therapy. Cornea. 2005;24(5):509–22.

Chapter 6
Corneal Stem Cells: A Source of Cell Renewal with Therapeutic Potential

Ana J. Chucair-Elliott, Michael H. Elliott, Alex Cohen, and Daniel J.J. Carr

Abbreviations

CSSCs	Corneal stromal stem cells
LSCD	Limbal stem cell deficiency
LSCs	Limbal stem cells
MIP-1α	Macrophage inflammatory protein-1
MMP-2	Metalloproteinase-2
MSCs	Mesenchymal stem cells
SJS	Steven-Johnson syndrome
TNFα	Tumor necrosis factor α
TSG-6	Factor (TNF)-α-stimulated gene/protein 6
TSP-1	Thrombospondin-1
UMSCs	Umbilical MSCs
VEGF	Vascular endothelial growth factor

A.J. Chucair-Elliott (✉) • M.H. Elliott • A. Cohen
Department of Ophthalmology, University of Oklahoma Health Sciences Center,
608 Stanton L. Young Boulevard, Oklahoma City, OK 73104, USA
e-mail: ana-chucair@ouhsc.edu; michael-elliott@ouhsc.edu; alex-cohen@ouhsc.edu

D.J.J. Carr
Departments of Ophthalmology and Microbiology and Immunology,
University of Oklahoma Health Sciences Center, Oklahoma City, OK, USA
e-mail: dan-carr@ouhsc.edu

© Springer Science+Business Media New York 2015   99
M.A. Babizhayev et al. (eds.), *Studies on the Cornea and Lens*,
Oxidative Stress in Applied Basic Research and Clinical Practice,
DOI 10.1007/978-1-4939-1935-2_6

6.1   Introduction: the Cornea and Its Niche of Stem Cells

The cornea is a tough, avascular tissue that forms a barrier between the contents of the eye and the outside world. Its transparency provides two-thirds of the refractive power required to focus light on the retina and sustain visual function. Three cellular components separated by Bowman's and Descemet's membranes work together in order to maintain clarity. The outermost anterior layer consists of squamous epithelium up to seven cell layers deep followed by a thick corneal stroma populated by keratocytes and composed mainly of water and collagen type I and type IV, and a single-layer corneal endothelium which functions to pump water from the cornea. Surrounding the corneal tissue is the vascular conjunctiva, which provides moisture and lubrication to the ocular surface. At the junction between the cornea and conjunctiva resides the limbus, a transitional region that possesses 10–12 layers of epithelium without the presence of Bowman's or Descemet's membranes. Here, the endothelial cells display a larger, flatter morphology compared to the endothelium residing in the central cornea. The limbus acts as a "reservoir" for limbal stem cells (LSCs) which maintain and regenerate the transparent state of the corneal surface. The deficiency of LSCs results in the loss of corneal clarity compromising the visual axis that can lead to unilateral or bilateral corneal blindness [24]. Understanding the function of corneal stem cells is critical for the development of cell-based therapies to treat diseased cornea.

6.2   Stem Cells in the Cornea

The corneal stem cell niche maintained at the limbus is composed of epithelial [24, 40] and stromal [4, 25, 67] stem cells. Although the term LSCs exclusively refers to the corneal epithelial progenitors, the limbal location has been described as common to epithelial and stromal stem cells [31]. The proximity between LSCs and corneal stromal stem cells (CSSCs) is thought to provide symbiotic support for their maintenance [65].

6.2.1   Limbal Epithelial Stem Cells

In 1971, Davanger and Evensen described the palisades of Vogt [17], a microenvironment of cellular and extracellular components where the LSC niche was generally accepted to reside (Fig. 6.1). Later work supported the hypothesis that this microenvironment regulates the fate of LSCs [73, 85]. The close proximity of these undulating structures to the limbal blood vessels provides oxygen and nutrient supply to LSCs [83] and protects them from shearing forces [28]. Underneath the basement membrane, the limbal stroma appears heavily innervated, vascularized,

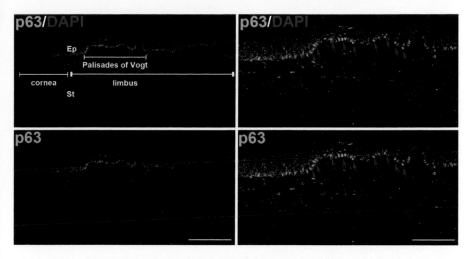

Fig. 6.1 Expression of p63 in the basal epithelium of the human limbus. Representative confocal microscope images of human corneas immunolabeled with antibody specific to the putative stem cell marker p63 (*green*). Nuclei were counterstained with DAPI (*blue*). Note the detection of p63-positive signal is stronger at the basal limbal epithelium located within the palisades of Vogt and becomes weaker toward the central cornea (*left panels*), where this signal is no longer present. *Red arrows* depict P63-positive cells at the limbus (*right panels*). *Ep* corneal epithelium, *St* cornea stroma. Scale bar left panel: 200 μm and right panel: 100 μm

and populated by heterogeneous and poorly defined cells [29, 44, 47]. Although the anatomic features in the limbus suggest that LSCs intimately interact with cells in the underlying limbal stroma, the dependence of LSCs on such close interactions has not been demonstrated [47].

The discovery of robust markers of LSCs would greatly help in the isolation and expansion techniques for more efficient cell-based therapies. Despite a growing list of phenotypic markers [18, 72, 86], the search for a unique marker to identify LSCs is an ongoing challenge [6, 18]. In 2001, Pellegrini et al. highlighted a nuclear protein called p63 [63], originally described with epidermal development [88], and depicted in Fig. 6.1 in a representative image from a human cornea. Specifically, positive staining for an isoform of p63, ΔNp63, was found in human basal epithelium and not in the central epithelium of human corneas [63]. Later, evidence further supported ΔNp63α as the main p63 isoform located within the basal to intermediate layers of the human limbal and conjunctival epithelia [37]. The presence of several isoforms of ΔNp63 was noted during corneal wounding, in correlation with limbal cell migration and corneal regeneration and differentiation [21]. Another group reported the expression of an ATP-binding cassette transporter group 2 protein, ABCG2, expressed by stem cells from a wide variety of sources [96]. This protein was associated with a group of cells called "side population" cells that facilitated the transport of the DNA-binding dye Hoechst. Additional work by other research groups located the expression of ABCG2 to a side population of corneal limbal epithelial cells; therefore, identifying ABCG2 as a new putative marker for LSCs [5, 84]. Beside p63 and ABCG2, a potential candidate for LSC marker was reported by Di Girolamo et al., who showed selective expression of the low-affinity nerve growth factor receptor p75 to the limbal basal epithelium [20]. In addition to p75, previous groups have noted that limbal basal epithelial cells are positive for cytokeratin 19 [36], and integrin α9 [77]. Negative indirect markers associated with terminal differentiation, such as cytokeratins 3 [71] and 12 [8, 41], involucrin [9], and connexin 43 [53] have been described that help distinguish the LSC phenotype.

The described functional characteristics of LSCs include high proliferative capacity and slow-cycling [12], aspects shared with epidermal stem cells [43] that might signify a strategy for DNA protection [61] and to maintain stemness and quiescence until signaling toward their activation and proliferation is triggered [18].

Although LSCs are widely believed to actively support the renewal of continuously shedding epithelial cells at the surface of the cornea, questions remain as to whether this area is the exclusive source of stem cells for the cornea [25, 82]. Work by Majo et al., introduced the controversial concept that in addition to the limbus, the cornea itself can harbor "stemness" [51]. Specifically, this study showed that serially transplanted mouse corneal epithelium could be self-maintained, and contained oligopotent stem cells. It was noted that these cells were also present throughout the whole cornea surface of the pig [51]. In addition to their findings, the authors proposed that the limbus is a zone of equilibrium where the expanding conjunctiva and corneal epithelium meet. A disequilibrium that can result from, for example, extensive corneal wound would elicit the migration of LSCs into the cornea.

6.2.2 Stromal Stem Cells

Funderburgh et al. [25] first isolated stromal progenitor cells from bovine corneas under the rationale that stem and progenitor cells exhibit clonal growth. Although such clones did not show multipotency, they expressed markers common to mesenchymal stem cells (MSCs), non-hematopoietic cells with the ability for self-renewal and the potential for differentiation into a variety of cell lineages of mesenchymal origin [66]. Markers expressed by stromal progenitor cells included Bmi1, CD90 (Thy1), CD73, CD166, ABCG2, Fhl1, stem cell factor (kit ligand), and Notch1 [25]. The presence of multipotent cells in the stroma was evidenced by several cornea stem cell researchers. A side population of cells from mouse and rabbit stroma were expanded clonally in attachment-free cultures as floating "neurospheres" [2, 91, 92] resembling characteristics of neural stem cells (NSCs) in vitro. Consistent with the rodent findings, work by Du et al. identified stem cells in the stroma of the human cornea [23]. The stromal side population cells termed as corneal stromal stem cells (CSSCs) could be clonally expanded and expressed astrocyte/neuronal-specific markers.

Polisetty et al. [67] used a model of stromal cell culture from human limbal explants to propose that the limbal stroma supporting the limbal epithelium contains a unique population of spindle-shaped cells similar to bone-marrow-derived MSC relative to colony forming efficiency, population doubling capacity, and low immunogenicity. Recent work consolidated the MSC-nature of stem cells in the human stromal limbus [4, 26] and showed that these cells were able to form colonies [26]. Moreover, these cells exhibited immunosuppressant features including inhibition of T cell proliferation in a TGFβ-dependent manner [26].

Similar to what has been found for LSCs, no consensus has been reached regarding a *bonafide* marker to identify stromal stem cells. As an example, the cell surface glycoprotein CD34 is a controversial marker for stem cells in the stroma [65]. In human cornea, CD34 was reported to be expressed by most keratocytes [35, 76, 80], and also by MSCs as they differentiate into keratocytes after transplantation into the corneas of mice [49]. Although keratocytes no longer express the eye-specific protein PAX6, the corneal stromal stem cells conserve this protein. As a consequence, the expression was reported to be unique for stem cells in the stroma [23, 25].

6.2.3   Maintenance of the Limbal Stem Cell Niche

As introduced earlier, stem cells in the cornea reside within a LSC niche. In the last few years, there has been a concerted effort to understand how the LSC niche is maintained in order to facilitate corneal integrity [7, 56, 60, 87]. As an example, in an effort to elucidate how adult lineage-committed epithelial cells are regulated, Xie et al. [87] isolated both LSCs/limbal progenitor cells and their native niche cells and allowed both cell types to reunite in culture to generate sphere growth. In their model, blocking CXCR4 receptor either with an antagonist or neutralizing antibody, the investigators found a lower capacity of LSCs to form growth spheres as a result of a commitment toward a more differentiated state. These findings suggested the limbal location of corneal LSCs within the LSC niche is critical for the maintenance of their stemness and function, which is mediated by the stromal cell-derived factor-1 (SDF-1) and its receptor CXCR4.

6.3   Stem Cell Deficiencies in the Cornea

Most of the current knowledge about stem cell deficiencies involves the LSC pool which can be partially or totally depleted due to genetic defects, injury or infectious disease [16, 64, 74]. By definition, limbal stem cell deficiency (LSCD) refers to a heterogenous group of pathologies with deleterious effects on corneal integrity and wound healing [70]. As a consequence of stem cell depletion, an invasion of peripheral and central cornea by conjunctival epithelium may take place [61]. Such abnormal tissue localization induces neovascularization of the normally avascular ocular area and corneal opacification leading to impaired visual acuity.

The main conditions that determine LSCD can be grouped according to their hereditary or acquired origins. One of the most common forms of hereditary LSCD, called aniridia is characterized by mutations in the PAX6 gene, a highly evolutionary conserved transcription factor that controls the morphogenesis of all tissues in the eye [34]. In aniridia, there is a "miscommunication" between the developing corneal epithelium and the microenvironment of the anterior compartment of the eye that results in the development of a significant keratopathy not apparent at birth, but progressive from late childhood [11]. The role of PAX6 as a regulator for cell proliferation at the limbus was highlighted in recent work [32]. In this study, PAX 6, along with other putative transcription factors, was able to bind the promoter site of ΔNp63 leading to increased cell proliferation of limbal epithelial cells.

Among the common causes of acquired LSCD are chemical burns (i.e. alkali and acid), thermal burns, contact lens misuse, and direct instilled topical drugs [32, 48, 64]. In addition, acquired forms of LSCD include immune diseases, such as Steven-Johnson syndrome (SJS), characterized by cell apoptosis and necrosis resulting in epithelial detachment, greatly due to adverse drug reactions [64].

6.4 Management of Limbal Stem Cell Deficiency

Eye diseases that result from LSCD translate into a difficult management problem. From the patient's perspective, the vision loss is partial or complete in either one or both eyes, and often cosmetically unsatisfactory, uncomfortable, or painful [75]. Clinical consequences of LSCD are recurrent or persistent ulceration of the corneal surface with the risk of thinning, vascularization, or infection. Under such circumstances, an effective treatment would offer a significant improvement in quality of life [19, 39, 52]. Conservative treatment options for patients with LSCD include the use of anti-inflammatory drugs, intensive lubrication with non-preserved artificial drops, the use of contact lenses to protect the corneal surface and provide pain relief, and the use of 10–20 % autologous serum eye drops to reduce inflammation and promote epithelial healing [27, 50, 68, 75, 93].

Regarding surgical options, earlier work by Kenyon and Tseng recommended limbal autograft transplantation for treatment of widespread ocular surface damage with loss of limbal epithelial stem cells and, specifically, for ocular surface defects such as those derived from chemical or thermal burns, and contact lens-induced keratopathy [38]. This method is currently in use and offers advantages such as the maintenance of the transplanted LSC in their natural niche, alternative sources of the limbal tissue (deceased donor, autologous tissue from the fellow healthy eye or limbal tissue donated by a living relative), and low cost [75]. An alternative technique, the transplantation of ex vivo expanded limbal epithelial sheets, is a widely used therapy for LSC deficiency [62]. Briefly, this therapeutic approach consists of placing a small limbal biopsy removed from either the patient or a donor onto transplantable carriers such as denuded human amniotic membrane (AM) to support the migration of limbal cells to form a limbal-like epithelial sheet [62, 81]. The success of this type of LSC transplantation is obscured by the depletion of LSC in culture, and requires the optimization of surgical procedures, control of the microenvironment onto which cells are supposed to engraft, and proper pharmacological support [62, 64].

Whether using allotransplantation or autotransplantation, the availability of LSCs seems to always be the limiting factor. For this reason, emphasis has been applied on the identification of potential sources of stem cells from other organs that might serve as autologous sources of replacement in ocular surface diseases, as has been shown for example, in the case of MSCs [50], adult epidermal [10, 54, 55], or oral mucosal [3, 13, 20] stem cells. In the next section of this chapter, we will discuss the use of MSCs for the development of cell-based therapies in the eye.

6.5 Therapeutic Potential of Mesenchymal Stem Cells in Treating LSCD

MSCs, originally isolated from bone marrow, are a type of multipotent cell population that now are shown to reside in almost every type of connective tissue [15] including adipose tissue [94], cord blood [59], and oral tissue [78, 95]. MSCs are characterized as a heterogenous population of cells that proliferate in vitro as plastic adherent cells with spindle-shaped morphology [66]. Based on surface markers, these cells are distinguished from hematopoietic cells by being negative for CD11b, CD14, CD34, CD45, and human leukocyte antigen (HLA)-DR, but expressing CD73, CD90, and CD105. Along with phenotypic markers, the capacity to differentiate into multiple mesenchymal lineages including osteocytes, chondrocytes, and adipocytes is used as criteria to define MSCs [22].

Due to their immunosuppressive properties, MSCs are extensively studied as new therapeutic tools for a variety of clinical applications, such as cases of graft-versus-host disease [46, 45] and autoimmune disorders [79]. In the last few years, encouraging results have highlighted the potential of MSCs in the treatment of ocular surface defects in animal models [33, 50, 57, 90]. However, the potential of these cells to differentiate into epithelial cells as needed to treat LSCD in human patients requires to be further examined.

Since chemical burns are a common cause of LSCD, animal models using this type of injury have been developed in order to test the beneficial effects of bone-marrow-derived adult MSCs on corneal surface repair. For instance, Ma et al. presented a study comparing the therapeutic efficacy of transplantation of human

MSCs and LSCs in a rat model [50]. In their work, transplantation of human MSCs on amniotic membrane, like LSCs in amniotic membrane, could reconstruct severely damaged rat corneal surface suggesting a new method to treat corneal disorders. Specifically, it is tempting to speculate based on these results that one could use the patient's own MSCs to treat their unique corneal disorder without the complications due to allotransplantation-induced immunorejection. Experimentally, the engraftment of systemically transplanted MSCs into locally injured rabbit cornea was shown by Ye et al. [90]. According to this study, MSCs could be home to the alkali-burned cornea, differentiate in response to the local microenvironment, and stimulate local repair response, promoting wound healing upon injury. Along with their repair effect, the presence of MSCs stimulated the proliferation of LSCs which we interpret to suggest that the MSCs that migrated into the wounded limbal and corneal areas exerted a positive influence on the native cornea cell proliferation.

Regarding the mechanism/s of action of MSCs, one group reported the topical application of either MSCs or MSC-derived conditioned media restored the stem cell niche in the chemically burned rat cornea through anti-angiogenic and anti-inflammatory activities of MSCs [57]. When analyzing the levels of soluble factors related to angiogenesis and inflammation, the authors found thrombospondin-1 (TSP-1), a powerful anti-angiogenic factor, was elevated while metalloproteinase-2 (MMP-2), a pro-angiogenic factor, was decreased upon MSC treatment. In addition, MSCs increased the levels of IL-10 and TGFβ resulting in reduced infiltration of inflammatory cells in the chemically burned tissue. Consistent with the restoration effect mediated by soluble factors, further studies revealed other potential therapeutic factors as part of the mechanism(s) of action of MSCs in the damaged cornea, such as the multipotent anti-inflammatory protein tumor necrosis factor (TNF)-α-stimulated gene/protein 6 (TSG-6) [58, 69]. Of interest, and controversial with the idea that MSCs require engraftment into the injured cornea to promote wound repair, another group reported that intraperitoneal and intravenous administration of human MSCs suppressed the development of corneal inflammation and opacity after chemical injury without engraftment of the cells in the cornea [69]. In support of the anti-inflammatory and angiostatic effects of MSCs, downregulation of the macrophage inflammatory protein-1 (MIP-1α), tumor necrosis factor α (TNFα), and vascular endothelial growth factor (VEGF) mRNA expression was found upon subconjunctival injection of MSCs into wounded corneas [89].

The impact of corneal injury on the homeostasis of endogenous MSCs is poorly understood. A recent report presented evidence that as a result of corneal injury, an increase in the frequency of circulating endogenous MSCs and elevated levels of the stem cell-specific chemoattractants SDF-1 and substance P was found in the peripheral blood [42]. These results implied that one or more of the soluble factors such as SDF-1 or substance P may be key to mobilize MSCs to home specifically to the injured cornea. An important question to address when considering the therapeutic application of MSCs in treating LSCD is whether the transplanted cells are able to differentiate into corneal epithelial cells. In this regard, a study found MSCs could be induced to transdifferentiate into corneal epithelial cells by corneal stromal cells in a transwell system and furthermore, such transdifferentiated MSC could then be seeded on a human amniotic membrane in an LSCD rat model [33]. The outcome of this approach was the reconstruction of the corneal epithelium with improved corneal opacity and neovascularization [33].

Beside the use of MSCs originated from the bone marrow to treat damaged corneas, recent work demonstrated intrastromal transplantation of human umbilical MSCs (UMSCs) was successful to treat ocular defects of mucopolysaccharidosis VII mice, a model of human corneal congenital metabolic disease [14]. The findings included human UMSCs survived rejection and differentiated into resident stromal cells, ameliorating the corneal clouding due to abnormal accumulation of glycosaminoglycans in this disease model.

Unlike animal models, evidence based on the efficacy of MSCs to treat human corneal disease is limited. A recent case report described the topical application of autologous adipose-derived MSCs to promote corneal epithelial healing in a patient with post-traumatic persistent corneal epithelial defect refractory to other treatment options [1]. Although these results suggest MSC transfer might be a feasible approach for corneal regeneration, further studies are required to establish the therapeutic efficacy and safety of MSC application in the treatment of corneal disorders.

6.6 Concluding Remarks

Despite some controversial findings it is widely accepted the cornea does not contain stem cells and its epithelial cells are renewed by transient amplifying cells generated by asymmetric divisions of LSCs, residing at the limbus of the eye [30]. Failure to maintain a functional pool of LSCs due to injury, genetic, or infectious disease has detrimental effects on vision by causing LSCD, characterized by invasion of conjunctival cells onto the cornea, and resulting neovascularization and inflammation of the normally avascular ocular surface. Surgical interventions to treat such disease include the successful use of limbal grafts and ex vivo expanded epithelial sheets for transplantation. However, whether using allotransplantation or autotransplantation, the caveat these procedures face is the limited availability of LSCs. The use of stem cells from different sources other than the limbus, such as MSCs, has emerged as a promising therapeutic alternative that in the last decade has been tested beneficial in a variety of animal models of LSCD. Further studies are needed to expand our knowledge on the mechanisms of action of MSCs, and their potential and efficacy as stem cell sources of replacement in ocular surface diseases (see Fig. 6.2 for summary).

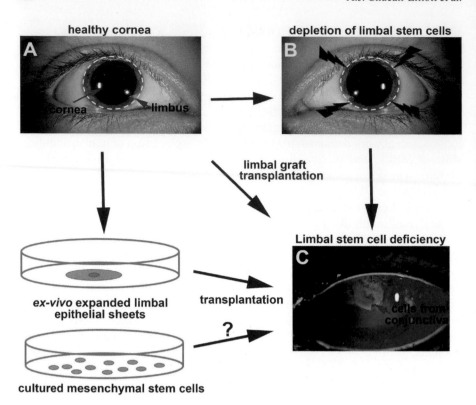

Fig. 6.2 Summary of surgical treatments for LSCD. (**a**) In a healthy cornea the renewal of central epithelial cells is sustained by LSCs at the limbus (depicted with *yellow discontinued lines*). Upon depletion of the LSC pool at the limbus (**b**) LSCD can develop, with clinical features that develop following the migration of cells of conjunctival origin onto the cornea. Such invading conjunctival cells onto the cornea are permeable to the dye fluorescein and can be observed under cobalt blue light by slit lamp examination (*green staining*) (**c**). Widely used surgical procedures to treat LSCD are limbal graft transplantations (from healthy fellow eye from the patient or donor eye) and the transplantation of ex vivo expanded limbal epithelial sheets. Ongoing studies are exploring the potential of MSCs as sources of stem cells to treat this disease

Acknowledgment This work was supported by a grant from the Oklahoma Center for Adult Stem Cell Research through the Oklahoma Tobacco Settlement Endowment Trust.

References

1. Agorogiannis GI, Alexaki VI, Castana O, Kymionis GD. Topical application of autologous adipose-derived mesenchymal stem cells (MSCs) for persistent sterile corneal epithelial defect. Graefes Arch Clin Exp Ophthalmol. 2012;250:455–7.
2. Amano S, Yamagami S, Mimura T, Uchida S, Yokoo S. Corneal stromal and endothelial cell precursors. Cornea. 2006;25:S73–7.

3. Blazejewska EA, Schlotzer-Schrehardt U, Zenkel M, Bachmann B, Chankiewitz E, Jacobi C, Kruse FE. Corneal limbal microenvironment can induce transdifferentiation of hair follicle stem cells into corneal epithelial-like cells. Stem Cells. 2009;27:642–52.
4. Branch MJ, Hashmani K, Dhillon P, Jones DR, Dua HS, Hopkinson A. Mesenchymal stem cells in the human corneal limbal stroma. Invest Ophthalmol Vis Sci. 2012;53:5109–16.
5. Budak MT, Alpdogan OS, Zhou M, Lavker RM, Akinci MA, Wolosin JM. Ocular surface epithelia contain ABCG2-dependent side population cells exhibiting features associated with stem cells. J Cell Sci. 2005;118:1715–24.
6. Chee KY, Kicic A, Wiffen SJ. Limbal stem cells: the search for a marker. Clin Experiment Ophthalmol. 2006;34:64–73.
7. Chen SY, Hayashida Y, Chen MY, Xie HT, Tseng SC. A new isolation method of human limbal progenitor cells by maintaining close association with their niche cells. Tissue Eng Part C Methods. 2011;17:537–48.
8. Chen WY, Mui MM, Kao WW, Liu CY, Tseng SC. Conjunctival epithelial cells do not transdifferentiate in organotypic cultures: expression of K12 keratin is restricted to corneal epithelium. Curr Eye Res. 1994;13:765–78.
9. Chen Z, de Paiva CS, Luo L, Kretzer FL, Pflugfelder SC, Li DQ. Characterization of putative stem cell phenotype in human limbal epithelia. Stem Cells. 2004;22:355–66.
10. Clausen H, Vedtofte P, Moe D, Dabelsteen E, Sun TT, Dale B. Differentiation-dependent expression of keratins in human oral epithelia. J Invest Dermatol. 1986;86:249–54.
11. Collinson JM, Chanas SA, Hill RE, West JD. Corneal development, limbal stem cell function, and corneal epithelial cell migration in the Pax6(+/-) mouse. Invest Ophthalmol Vis Sci. 2004;45:1101–8.
12. Cotsarelis G, Cheng SZ, Dong G, Sun TT, Lavker RM. Existence of slow-cycling limbal epithelial basal cells that can be preferentially stimulated to proliferate: implications on epithelial stem cells. Cell. 1989;57:201–9.
13. Cotsarelis G, Sun TT, Lavker RM. Label-retaining cells reside in the bulge area of pilosebaceous unit: implications for follicular stem cells, hair cycle, and skin carcinogenesis. Cell. 1990;61:1329–37.
14. Coulson-Thomas VJ, Caterson B, Kao WW. Transplantation of human umbilical mesenchymal stem cells cures the corneal defects of mucopolysaccharidosis VII mice. Stem Cells. 2013;31:2116–26.
15. da Silva ML, Chagastelles PC, Nardi NB. Mesenchymal stem cells reside in virtually all post-natal organs and tissues. J Cell Sci. 2006;119:2204–13.
16. Daniels JT, Dart JK, Tuft SJ, Khaw PT. Corneal stem cells in review. Wound Repair Regen. 2001;9:483–94.
17. Davanger M, Evensen A. Role of the pericorneal papillary structure in renewal of corneal epithelium. Nature. 1971;229:560–1.
18. Davies SB, Di GN. Corneal stem cells and their origins: significance in developmental biology. Stem Cells Dev. 2010;19:1651–62.
19. Di GN, Bosch M, Zamora K, Coroneo MT, Wakefield D, Watson SL. A contact lens-based technique for expansion and transplantation of autologous epithelial progenitors for ocular surface reconstruction. Transplantation. 2009;87:1571–8.
20. Di GN, Sarris M, Chui J, Cheema H, Coroneo MT, Wakefield D. Localization of the low-affinity nerve growth factor receptor p75 in human limbal epithelial cells. J Cell Mol Med. 2008;12:2799–811.
21. Di IE, Barbaro V, Ruzza A, Ponzin D, Pellegrini G, De LM. Isoforms of DeltaNp63 and the migration of ocular limbal cells in human corneal regeneration. Proc Natl Acad Sci U S A. 2005;102:9523–8.
22. Dominici M, Le BK, Mueller I, Slaper-Cortenbach I, Marini F, Krause D, Deans R, Keating A, Prockop D, Horwitz E. Minimal criteria for defining multipotent mesenchymal stromal cells. The International Society for Cellular Therapy position statement. Cytotherapy. 2006;8:315–7.
23. Du Y, Funderburgh ML, Mann MM, SundarRaj N, Funderburgh JL. Multipotent stem cells in human corneal stroma. Stem Cells. 2005;23:1266–75.

24. Dua HS, Azuara-Blanco A. Limbal stem cells of the corneal epithelium. Surv Ophthalmol. 2000;44:415–25.
25. Funderburgh ML, Du Y, Mann MM, SundarRaj N, Funderburgh JL. PAX6 expression identifies progenitor cells for corneal keratocytes. FASEB J. 2005;19:1371–3.
26. Garfias Y, Nieves-Hernandez J, Garcia-Mejia M, Estrada-Reyes C, Jimenez-Martinez MC. Stem cells isolated from the human stromal limbus possess immunosuppressant properties. Mol Vis. 2012;18:2087–95.
27. Geerling G, Maclennan S, Hartwig D. Autologous serum eye drops for ocular surface disorders. Br J Ophthalmol. 2004;88:1467–74.
28. Gipson IK. The epithelial basement membrane zone of the limbus. Eye (Lond). 1989;3(Pt 2):132–40.
29. Goldberg MF, Bron AJ. Limbal palisades of Vogt. Trans Am Ophthalmol Soc. 1982;80:155–71.
30. Haddad A. Renewal of the rabbit corneal epithelium as investigated by autoradiography after intravitreal injection of 3H-thymidine. Cornea. 2000;19:378–83.
31. Hashmani K, Branch MJ, Sidney LE, Dhillon PS, Verma M, McIntosh OD, Hopkinson A, Dua HS. Characterization of corneal stromal stem cells with the potential for epithelial transdifferentiation. Stem Cell Res Ther. 2013;4:75.
32. Hsueh YJ, Kuo PC, Chen JK. Transcriptional regulators of the DeltaNp63: their role in limbal epithelial cell proliferation. J Cell Physiol. 2013;228:536–46.
33. Jiang TS, Cai L, Ji WY, Hui YN, Wang YS, Hu D, Zhu J. Reconstruction of the corneal epithelium with induced marrow mesenchymal stem cells in rats. Mol Vis. 2010;16:1304–16.
34. Jordan T, Hanson I, Zaletayev D, Hodgson S, Prosser J, Seawright A, Hastie N, van Heyningen V. The human PAX6 gene is mutated in two patients with aniridia. Nat Genet. 1992;1:328–32.
35. Joseph A, Hossain P, Jham S, Jones RE, Tighe P, McIntosh RS, Dua HS. Expression of CD34 and L-selectin on human corneal keratocytes. Invest Ophthalmol Vis Sci. 2003;44:4689–92.
36. Kasper M. Patterns of cytokeratins and vimentin in guinea pig and mouse eye tissue: evidence for regional variations in intermediate filament expression in limbal epithelium. Acta Histochem. 1992;93:319–32.
37. Kawasaki S, Tanioka H, Yamasaki K, Connon CJ, Kinoshita S. Expression and tissue distribution of p63 isoforms in human ocular surface epithelia. Exp Eye Res. 2006;82:293–9.
38. Kenyon KR, Tseng SC. Limbal autograft transplantation for ocular surface disorders. Ophthalmology. 1989;96:709–22.
39. Kolli S, Ahmad S, Lako M, Figueiredo F. Successful clinical implementation of corneal epithelial stem cell therapy for treatment of unilateral limbal stem cell deficiency. Stem Cells. 2010;28:597–610.
40. Kulkarni BB, Tighe PJ, Mohammed I, Yeung AM, Powe DG, Hopkinson A, Shanmuganathan VA, Dua HS. Comparative transcriptional profiling of the limbal epithelial crypt demonstrates its putative stem cell niche characteristics. BMC Genomics. 2010;11:526.
41. Kurpakus MA, Maniaci MT, Esco M. Expression of keratins K12, K4 and K14 during development of ocular surface epithelium. Curr Eye Res. 1994;13:805–14.
42. Lan Y, Kodati S, Lee HS, Omoto M, Jin Y, Chauhan SK. Kinetics and function of mesenchymal stem cells in corneal injury. Invest Ophthalmol Vis Sci. 2012;53:3638–44.
43. Lavker RM, Sun TT. Epidermal stem cells: properties, markers, and location. Proc Natl Acad Sci U S A. 2000;97:13473–5.
44. Lawrenson JG, Ruskell GL. The structure of corpuscular nerve endings in the limbal conjunctiva of the human eye. J Anat. 1991;177:75–84.
45. Le BK, Frassoni F, Ball L, Locatelli F, Roelofs H, Lewis I, Lanino E, Sundberg B, Bernardo ME, Remberger M, Dini G, Egeler RM, Bacigalupo A, Fibbe W, Ringden O. Mesenchymal stem cells for treatment of steroid-resistant, severe, acute graft-versus-host disease: a phase II study. Lancet. 2008;371:1579–86.

46. Le BK, Rasmusson I, Sundberg B, Gotherstrom C, Hassan M, Uzunel M, Ringden O. Treatment of severe acute graft-versus-host disease with third party haploidentical mesenchymal stem cells. Lancet. 2004;363:1439–41.
47. Li W, Hayashida Y, Chen YT, Tseng SC. Niche regulation of corneal epithelial stem cells at the limbus. Cell Res. 2007;17:26–36.
48. Lichtinger A, Pe'er J, Frucht-Pery J, Solomon A. Limbal stem cell deficiency after topical mitomycin C therapy for primary acquired melanosis with atypia. Ophthalmology. 2010;117:431–7.
49. Liu H, Zhang J, Liu CY, Wang IJ, Sieber M, Chang J, Jester JV, Kao WW. Cell therapy of congenital corneal diseases with umbilical mesenchymal stem cells: lumican null mice. PLoS One. 2010;5:e10707.
50. Ma Y, Xu Y, Xiao Z, Yang W, Zhang C, Song E, Du Y, Li L. Reconstruction of chemically burned rat corneal surface by bone marrow-derived human mesenchymal stem cells. Stem Cells. 2006;24:315–21.
51. Majo F, Rochat A, Nicolas M, Jaoude GA, Barrandon Y. Oligopotent stem cells are distributed throughout the mammalian ocular surface. Nature. 2008;456:250–4.
52. Mangione CM, Lee PP, Gutierrez PR, Spritzer K, Berry S, Hays RD. Development of the 25-item National Eye Institute Visual Function Questionnaire. Arch Ophthalmol. 2001;119:1050–8.
53. Matic M, Petrov IN, Chen S, Wang C, Dimitrijevich SD, Wolosin JM. Stem cells of the corneal epithelium lack connexins and metabolite transfer capacity. Differentiation. 1997;61:251–60.
54. Nakamura T, Inatomi T, Sotozono C, Amemiya T, Kanamura N, Kinoshita S. Transplantation of cultivated autologous oral mucosal epithelial cells in patients with severe ocular surface disorders. Br J Ophthalmol. 2004;88:1280–4.
55. Nishida K, Yamato M, Hayashida Y, Watanabe K, Yamamoto K, Adachi E, Nagai S, Kikuchi A, Maeda N, Watanabe H, Okano T, Tano Y. Corneal reconstruction with tissue-engineered cell sheets composed of autologous oral mucosal epithelium. N Engl J Med. 2004;351:1187–96.
56. Notara M, Shortt AJ, Galatowicz G, Calder V, Daniels JT. IL6 and the human limbal stem cell niche: a mediator of epithelial-stromal interaction. Stem Cell Res. 2010;5:188–200.
57. Oh JY, Kim MK, Shin MS, Lee HJ, Ko JH, Wee WR, Lee JH. The anti-inflammatory and anti-angiogenic role of mesenchymal stem cells in corneal wound healing following chemical injury. Stem Cells. 2008;26:1047–55.
58. Oh JY, Roddy GW, Choi H, Lee RH, Ylostalo JH, Rosa Jr RH, Prockop DJ. Anti-inflammatory protein TSG-6 reduces inflammatory damage to the cornea following chemical and mechanical injury. Proc Natl Acad Sci U S A. 2010;107:16875–80.
59. Oh W, Kim DS, Yang YS, Lee JK. Immunological properties of umbilical cord blood-derived mesenchymal stromal cells. Cell Immunol. 2008;251:116–23.
60. Ordonez P, Chow S, Wakefield D, Di GN. Human limbal epithelial progenitor cells express alphavbeta5-integrin and the interferon-inducible chemokine CXCL10/IP-10. Stem Cell Res. 2013;11:888–901.
61. Osei-Bempong C, Figueiredo FC, Lako M. The limbal epithelium of the eye—a review of limbal stem cell biology, disease and treatment. Bioessays. 2013;35:211–9.
62. Pellegrini G, De LM, Arsenijevic Y. Towards therapeutic application of ocular stem cells. Semin Cell Dev Biol. 2007;18:805–18.
63. Pellegrini G, Dellambra E, Golisano O, Martinelli E, Fantozzi I, Bondanza S, Ponzin D, McKeon F, De LM. p63 identifies keratinocyte stem cells. Proc Natl Acad Sci U S A. 2001;98:3156–61.
64. Pellegrini G, Rama P, Di RA, Panaras A, De LM. Hurdles in a successful example of limbal stem cell-based regenerative medicine. Stem Cells. 2013;32:26–34.
65. Pinnamaneni N, Funderburgh JL. Concise review: stem cells in the corneal stroma. Stem Cells. 2012;30:1059–63.

66. Pittenger MF, Mackay AM, Beck SC, Jaiswal RK, Douglas R, Mosca JD, Moorman MA, Simonetti DW, Craig S, Marshak DR. Multilineage potential of adult human mesenchymal stem cells. Science. 1999;284:143–7.
67. Polisetty N, Fatima A, Madhira SL, Sangwan VS, Vemuganti GK. Mesenchymal cells from limbal stroma of human eye. Mol Vis. 2008;14:431–42.
68. Poon AC, Geerling G, Dart JK, Fraenkel GE, Daniels JT. Autologous serum eyedrops for dry eyes and epithelial defects: clinical and in vitro toxicity studies. Br J Ophthalmol. 2001;85:1188–97.
69. Roddy GW, Oh JY, Lee RH, Bartosh TJ, Ylostalo J, Coble K, Rosa Jr RH, Prockop DJ. Action at a distance: systemically administered adult stem/progenitor cells (MSCs) reduce inflammatory damage to the cornea without engraftment and primarily by secretion of TNF-alpha stimulated gene/protein 6. Stem Cells. 2011;29:1572–9.
70. Santos MS, Gomes JA, Hofling-Lima AL, Rizzo LV, Romano AC, Belfort Jr R. Survival analysis of conjunctival limbal grafts and amniotic membrane transplantation in eyes with total limbal stem cell deficiency. Am J Ophthalmol. 2005;140:223–30.
71. Schermer A, Galvin S, Sun TT. Differentiation-related expression of a major 64K corneal keratin in vivo and in culture suggests limbal location of corneal epithelial stem cells. J Cell Biol. 1986;103:49–62.
72. Schlotzer-Schrehardt U, Kruse FE. Identification and characterization of limbal stem cells. Exp Eye Res. 2005;81:247–64.
73. Schofield R. The stem cell system. Biomed Pharmacother. 1983;37:375–80.
74. Shortt AJ, Secker GA, Notara MD, Limb GA, Khaw PT, Tuft SJ, Daniels JT. Transplantation of ex vivo cultured limbal epithelial stem cells: a review of techniques and clinical results. Surv Ophthalmol. 2007;52:483–502.
75. Shortt AJ, Tuft SJ, Daniels JT. Corneal stem cells in the eye clinic. Br Med Bull. 2011;100:209–25.
76. Sosnova M, Bradl M, Forrester JV. CD34+ corneal stromal cells are bone marrow-derived and express hemopoietic stem cell markers. Stem Cells. 2005;23:507–15.
77. Stepp MA, Zhu L, Sheppard D, Cranfill RL. Localized distribution of alpha 9 integrin in the cornea and changes in expression during corneal epithelial cell differentiation. J Histochem Cytochem. 1995;43:353–62.
78. Su WR, Zhang QZ, Shi SH, Nguyen AL, Le AD. Human gingiva-derived mesenchymal stromal cells attenuate contact hypersensitivity via prostaglandin E2-dependent mechanisms. Stem Cells. 2011;29:1849–60.
79. Sun L, Akiyama K, Zhang H, Yamaza T, Hou Y, Zhao S, Xu T, Le A, Shi S. Mesenchymal stem cell transplantation reverses multiorgan dysfunction in systemic lupus erythematosus mice and humans. Stem Cells. 2009;27:1421–32.
80. Toti P, Tosi GM, Traversi C, Schurfeld K, Cardone C, Caporossi A. CD-34 stromal expression pattern in normal and altered human corneas. Ophthalmology. 2002;109:1167–71.
81. Tsai RJ, Li LM, Chen JK. Reconstruction of damaged corneas by transplantation of autologous limbal epithelial cells. N Engl J Med. 2000;343:86–93.
82. Uchida S, Yokoo S, Yanagi Y, Usui T, Yokota C, Mimura T, Araie M, Yamagami S, Amano S. Sphere formation and expression of neural proteins by human corneal stromal cells in vitro. Invest Ophthalmol Vis Sci. 2005;46:1620–5.
83. Van Buskirk EM. The anatomy of the limbus. Eye (Lond). 1989;3(Pt 2):101–8.
84. Watanabe K, Nishida K, Yamato M, Umemoto T, Sumide T, Yamamoto K, Maeda N, Watanabe H, Okano T, Tano Y. Human limbal epithelium contains side population cells expressing the ATP-binding cassette transporter ABCG2. FEBS Lett. 2004;565:6–10.
85. Watt FM, Hogan BL. Out of Eden: stem cells and their niches. Science. 2000;287:1427–30.
86. Wolosin JM. Cell markers and the side population phenotype in ocular surface epithelial stem cell characterization and isolation. Ocul Surf. 2006;4:10–23.
87. Xie HT, Chen SY, Li GG, Tseng SC. Limbal epithelial stem/progenitor cells attract stromal niche cells by SDF-1/CXCR4 signaling to prevent differentiation. Stem Cells. 2011;29:1874–85.

88. Yang A, Kaghad M, Wang Y, Gillett E, Fleming MD, Dotsch V, Andrews NC, Caput D, McKeon F. p63, a p53 homolog at 3q27-29, encodes multiple products with transactivating, death-inducing, and dominant-negative activities. Mol Cell. 1998;2:305–16.
89. Yao L, Li ZR, Su WR, Li YP, Lin ML, Zhang WX, Liu Y, Wan Q, Liang D. Role of mesenchymal stem cells on cornea wound healing induced by acute alkali burn. PLoS One. 2012;7:e30842.
90. Ye J, Yao K, Kim JC. Mesenchymal stem cell transplantation in a rabbit corneal alkali burn model: engraftment and involvement in wound healing. Eye (Lond). 2006;20:482–90.
91. Yoshida S, Shimmura S, Nagoshi N, Fukuda K, Matsuzaki Y, Okano H, Tsubota K. Isolation of multipotent neural crest-derived stem cells from the adult mouse cornea. Stem Cells. 2006;24:2714–22.
92. Yoshida S, Shimmura S, Shimazaki J, Shinozaki N, Tsubota K. Serum-free spheroid culture of mouse corneal keratocytes. Invest Ophthalmol Vis Sci. 2005;46:1653–8.
93. Young AL, Cheng AC, Ng HK, Cheng LL, Leung GY, Lam DS. The use of autologous serum tears in persistent corneal epithelial defects. Eye (Lond). 2004;18:609–14.
94. Zannettino AC, Paton S, Arthur A, Khor F, Itescu S, Gimble JM, Gronthos S. Multipotential human adipose-derived stromal stem cells exhibit a perivascular phenotype in vitro and in vivo. J Cell Physiol. 2008;214:413–21.
95. Zhang Q, Shi S, Liu Y, Uyanne J, Shi Y, Shi S, Le AD. Mesenchymal stem cells derived from human gingiva are capable of immunomodulatory functions and ameliorate inflammation-related tissue destruction in experimental colitis. J Immunol. 2009;183:7787–98.
96. Zhou S, Schuetz JD, Bunting KD, Colapietro AM, Sampath J, Morris JJ, Lagutina I, Grosveld GC, Osawa M, Nakauchi H, Sorrentino BP. The ABC transporter Bcrp1/ABCG2 is expressed in a wide variety of stem cells and is a molecular determinant of the side-population phenotype. Nat Med. 2001;7:1028–34.

Chapter 7
New Agents for Treating Dry Eye Syndrome

Simon E. Skalicky, Con Petsoglou, Avinash Gurbaxani, Clare L. Fraser, and Peter McCluskey

7.1 Introduction

The trilaminar tear film consists of a mucinous basal layer, an aqueous midphase secreted from the lacrimal gland and accessory lacrimal glands and a superficial Meibomian gland-derived lipid layer [1]. Dry eye syndrome (DES) describes a group of disorders characterized by an inadequate quality and/or volume of tears resulting in chronic corneal and conjunctival dryness together with deficiencies of all layers of the tear film [2, 3].

DES is common with 5–30 % of people aged 50 years and older suffering from DES symptoms; approximately two-thirds are women [4, 5]. The prevalence is increasing due to an aging population and increased exposure to exacerbating factors; these include contact lens wear, dry or humid indoor environments, prolonged computer use and prescribed medications such as antidepressants, antihistamines, and antihypertensives [6–11].

DES is characterized by symptoms of foreign body sensation, irritation, dryness, light sensitivity, and itching. Biomicroscopic examination, enhanced with fluorescein staining to detect ocular surface irregularities, may reveal deficient tear volume and/or distribution, superficial punctate erosions, epithelial defects, and corneal filaments. In severe cases this may lead to corneal ulcers, thinning, or perforation [12, 13]. DES significantly interferes with normal visual function, daily activities, social and physical functioning, workplace productivity and quality of life [14, 15].

DES may be caused by decreased aqueous tear production or excessive tear evaporation; the two frequently coexist [16]. Decreased tear production, known as

S.E. Skalicky (✉) • C. Petsoglou • A. Gurbaxani • C.L. Fraser • P. McCluskey
Save Sight Institute, University of Sydney, Sydney, NSW, Australia

Sydney Eye Hospital, Macquarie St, Sydney, NSW, Australia, 2000
e-mail: seskalicky@gmail.com

© Springer Science+Business Media New York 2015
M.A. Babizhayev et al. (eds.), *Studies on the Cornea and Lens*,
Oxidative Stress in Applied Basic Research and Clinical Practice,
DOI 10.1007/978-1-4939-1935-2_7

aqueous insufficiency or keratoconjunctivitis sicca, is characterised by lacrimal gland hyposecretion and may be due to inflammatory conditions, including Sjogren's syndrome. It is treated with surface lubrication, punctual occlusion, and anti-inflammatory agents. Excessive tear evaporation is due to dysfunction of the lipid layer. This results in a rapid tear film breakup despite normal aqueous production; treatment involves improving the flow and quality of Meibomian gland secretions.

The treatments outlined in this chapter predominantly focus on decreased aqueous tear production (aqueous insufficiency); however successful treatment invariably involves concordantly addressing disturbances in the lipid layer.

DES has a multifactorial pathophysiology involving osmotic, mechanical, and inflammatory insults to the tear film, epithelium, and subepithelial nerve plexus. Aqueous deficiency results in increased tear osmolarity. This leads to ocular surface epithelial cell desiccation, stimulating osmotic stress-induced pathways: mitogen-activated protein kinases c-Jun N-terminal kinase, extracellular signal-related kinase, and p38 [17–19]. Key transcription factors (nuclear factor κB and activator protein 1) are stimulated. This leads to increased production of TNF-α, IL-1 and IL-6 (pro-inflammatory cytokines), chemokines, and matrix metalloproteases. This pro-inflammatory signal activates antigen presenting cells [19, 20]. The subsequent immune response predominantly involves CD4+ helper T-cell (T_H) subsets 1 and 17 [21]. In response to the immunological drive there is epithelial dysfunction, squamous metaplasia, decreased epithelial barrier function and reduced goblet cell number and secretory function that increases surface instability [22–25]. Immune drive leads to vascular endothelial growth factor expression, with limbal lymphangiogenesis and further recruitment of inflammatory cells and mediators [26]. The inflammatory and mechanical insults cause exposure of subepithelial corneal nerve plexuses, resulting in altered structure and number of the sub-basal nerves [27, 28].

Aqueous deficiency is most commonly treated with lubricating eye drops, ointments, and punctual occlusion. These measures are used initially and provide symptomatic relief. There has been significant research in recent years into topical and systemic treatments that target ocular surface inflammation to promote a healthy ocular surface and tear complex [3]. Tear film and ocular surface epithelial cell cytokines, chemokines, major histocompatibility complex antigens, cell adhesion molecules, and matrix metalloproteases are novel potential molecular targets, as are pro-secretory adenosine receptor stimulators [29–35]. This chapter outlines some of the new and potential future treatments available for treating DES, including their advantages and limitations.

7.2   Agents used to treat Dry Eye Syndrome

7.2.1   Ocular Lubricants

Artificial tears are commonly used to lubricate the ocular surface; these control symptoms in most cases of DES. The active ingredient of these solutions is a hydrophilic molecule with polymerised polysaccharide additives that determine hydration

Table 7.1 Commonly used artificial tear lubricants and commonly used preservatives

Commonly used lubricating agents and preservatives	
Low Viscosity	Sodium hyaluronate
	Polyvinyl alcohol
	Hydroxypropyl methylcellulose
	Glycerin
	Polyethylene glycol
Medium Viscosity	Carbopol
	Carboxylmethylcellulose sodium 1 %
High Viscosity	Mineral oil
	Paraffin
	White petrolatum
Preservative	Benzalkonium chloride
	Sodium chlorate
	Sodium perborate
	Purite ® (Allergan, Irvine, CA, USA)
	Polyquad ® (Alcon, Fort Worth, TX, USA)

and viscosity (Table 7.1). In addition the solutions contain water, electrolytes, and buffers. The more viscous solutions have a longer duration of effect but can blur patients' vision [36, 37].

Most available solutions contain a preservative. The preservative prevents bacterial colonization and growth within the bottle; however, the emulsifying effect of the preservative can destabilize the ocular surface and cause additional discomfort. Formulations of non-preserved eye drops (dispensed in small plastic minims) are available for severe cases of DES, patients with preservative allergy, and for cases requiring frequent drop instillation (more than four drops daily) to avoid preservative toxicity. Various products have been specifically formulated to address specific deficiencies in the tear film. Some are designed to improve surface wettability as hypotonic solutions counteracting hypertonicity [38], or to dilute and physically remove noxious surface substances [39], prevent tear evaporation by supplementing the lipid layer [40], or prevent tear film breakup by supplementing the mucin layer [41].

7.2.2 Corticosteroids

Corticosteroids (Table 7.2) bind to and activate the cytoplasmic glucocorticoid receptor, inducing a wide range of effects including inhibition of pro-inflammatory and promotion of anti-inflammatory gene transcription in leukocytes as well as promoting lymphocytic apoptotic pathways [42]. Corticosteroids improve both the symptoms and signs of DES. Their beneficial effect has been documented in a variety of clinical and laboratory studies [43–46]. However their use is not routine owing to a number of ocular complications, most notably cataract, ocular

Table 7.2 Commonly used topical corticosteroid and nonsteroidal anti-inflammatory agents

Topical Corticosteroids	
Potency: mild	Hydrocortisone acetate 0.5 % or 1 %
	Fluoromethalone 0.1 %
Potency: moderate	Fluoromethalone acetate 0.1 %
	Prednisolone phosphate 1 %
	Rimexolone 1 %
	Loteprednol etabonate 0.2 % or 0.5 %
Potency: strong	Dexamethasone 0.1 %
	Prednisolone acetate 1%[a]
Topical nonsteroidal anti-inflammatory drugs	
Agent	Ketorolac tromethamine 0.5 %
	Diclofenac sodium 0.1 %
	Flurbiprofen sodium 0.03 %

[a]Occasionally formulations are combined with epinephrine; this combination is generally not recommended for DES

hypertension that can lead to glaucoma, and opportunistic infection; consequently they are unsuited to long-term management strategies. They are useful in brief courses to manage acute exacerbations of DES. Generally more potent topical corticosteroid agents are associated with a greater incidence and severity of ocular complications. Certain milder agents, or agents with poor intraocular penetration have a reduced effect on cataractogenesis and raised intraocular pressure but may still be useful in controlling ocular surface inflammation in DES [47].

7.2.3 Nonsteroidal Anti-Inflammatory Drugs

Topical nonsteroidal anti-inflammatory drugs (Table 7.2) can be effective in some ocular inflammatory diseases but have not been consistently shown to be effective in the management of DES [45, 48]. They can induce ocular complications such as reduced corneal sensitivity that can lead to persistent epithelial defects and corneal thinning; rarely they are associated with progressive corneal lysis. For these reasons they are not commonly used in the management of DES [49].

7.2.4 Tetracycline Derivatives

Tetracyclines have a significant effect on Meibomian gland dysfunction, improving flow and constitution of secretions primarily through inhibition of matrix metalloprotease activity [50]. In addition to stabilizing the lipid layer they have a modest anti-inflammatory effect on the ocular surface that may be beneficial in aqueous deficiency [20].

Pro-inflammatory epithelial cell surface antigens, including MHC Class II molecules, are downregulated by minocycline [51]. Osmotic stress-related cell-signaling pathways in epithelial cells, including c-Jun N-terminal kinase, extracellular signal–related kinase and mitogen-activated protein kinase in epithelial cells, are inhibited by doxycycline [52]. Consequently there is reduced production of pro-inflammatory cytokines by these cell-signaling pathways.

Tetracyclines are most commonly administered orally, however, topical formulations are available; experimental evidence suggests that liposomal-bound topical doxycycline may have increased bioavailability than non-liposomal topical preparations [53]. Dry eye related to Meibomian gland disease and contact lens wear can be successfully treated with topical azithromycin at 1 % concentration [54, 55].

7.2.5 Vitamin A

Vitamin A (retinol) is an important tear film component that maintains a moist mucosal ocular surface [56]. Stored and reflexively secreted from the lacrimal gland, retinol influences gene transcription controlling proliferation and differentiation of corneal and conjunctival epithelial cells, preventing keratinization and squamous metaplasia [57]. A rabbit model of dry eye suggested that topical retinoic acid may be protective for osmotically stressed keratoconjunctival epithelial cells [58].

Topical vitamin A (retinol palmitate) is readily available for the treatment of dry eye (® AFT Pharmaceuticals, Australia). As an emulsion it predominantly improves signs rather than symptoms of DES [59]. However, a recent clinical trial compared topical retinoic acid administered four times daily to cyclosporine administered twice daily; both had similar positive effects on the signs and symptoms of DES [60].

7.2.6 Secretagogs

Mucin is an essential component of the tear film basal layer critical for lubrication of the corneal and conjunctival epithelial surfaces. Secretagogs stimulate secretion of mucin; this reduces instability of the tear film commonly associated with DES.

7.2.6.1 Bromhexine

Bromhexine, an oral mucolytic agent, reduces the viscosity of mucous secretions of the respiratory tract. It has also been shown to improve symptoms and objective signs of ocular surface disease in Sjogren's Syndrome [61].

7.2.6.2 Diquafosol tetrasodium

Diquafosol tetrasodium is a mucin secretagog that has been shown to ameliorate symptoms and ocular surface features of DES [62]. It is a topical $P2Y_2$ receptor agonist; activation of the $P2Y_2$ receptor, a Gq protein-coupled adenosine triphosphate receptor, causes chloride channel opening resulting in nonglandular secretion of water and mucin [63]. Topical diquafosol at 3 % concentration (Diquas® (Inspire Pharmaceuticals, Raleigh, NC, USA) gained approval for the treatment of DES in Japan in 2010.

7.2.6.3 IB-MECA (CF101)

IB-MECA (CF101) is an oral purine receptor agonist that has been evaluated as a potential future treatment for DES. Adenosine receptor binding of this agent initiates signal transduction to downregulate protein kinases and TNF-α production downstream [64]. A phase II multicenter, double-masked, randomized placebo-controlled trial involving CF101 administered to 68 subjects with DES resulted in improved tear film breakup time, corneal epithelial staining and volume of the tear meniscus after 12 weeks [65].

7.2.7 Autologous Serum

Autologous serum (AS) contains crucial tear film components such as neurotrophic growth factor, epidermal growth factor, vitamin A, fibronectin, and lysozyme. These key molecular and trophic factors optimize the ocular surface; accordingly AS is used relatively commonly in ocular surface disease and DES unresponsive to topical immunosuppressive agents [66]. Compounded from patients' own blood, AS is an important therapeutic option for DES as well as other ocular surface conditions such as superior limbic keratoconjunctivitis, persistent epithelial defects and neurotrophic keratopathy [67]. AS can be prepared at a dedicated blood-bank service or with the use of an office-based centrifuge. 1–2 units of blood is donated by the patient each time AS is formulated; diluted or undiluted serum is extracted and stored in small plastic minims. These can be used immediately or frozen. The process is repeated as required, generally 6–12 monthly; newer preparation and storage techniques have been developed to optimize stability of growth factors and other ingredients [68]. Several randomized clinical studies have reported AS to have a superior effect over non-preserved artificial tears in the treatment of DES [69, 70].

7.2.8 Cyclosporine A

Topical cyclosporine A (CsA) has been commonly used in the treatment of DES since FDA approval was granted in 2003 for a 0.05 % ophthalmic emulsion (Restasis; Allergan, Inc., Irvine, CA) [71]. As a potent immunosuppressive CsA exerts its effect by binding to intracytoplasmic chaperone proteins cyclophilins A and D. The cyclosporine–cyclophilin A complex inhibits calcineurin phosphatase, a pro-inflammatory transcription factor that stimulates T lymphocyte cytokine production including IL-2 [72]. This is complementary to the influence of the cyclosporine–cyclophilin D complex which reduces the permeability of mitochondria, preventing pro-apoptotic enzyme release such as cytochrome c and other mediators of pro-apoptotic caspase pathways [73]. Topical CsA administered for 3–6 months reduces surface expression of apoptotic markers including CD40, CD40 ligand, and Fas in addition to immunologic markers such as HLA-DR [32, 74]. These are expressed at increased ocular surface levels in patients with DES. Other markers of CsA-induced anti-inflammatory activity include increased density of mucin-secreting goblet cells, reduced squamous cell metaplasia, and increased tear flow; the latter may be mediated by release of parasympathetic-associated neurotransmitters [75, 76]. In clinical evaluation a randomized, double-masked, placebo-controlled, dose-ranging clinical trial demonstrated CsA to provide significant improvement in symptoms and signs with the 0.05 % dosage most effective [77, 78].

CsA has very low solubility in aqueous solutions and the tear film, resulting in highly variable and incomplete absorption from conventional topical formulations [3, 79]. This is largely due to its structure consisting of a rigid cyclic peptide with four intramolecular hydrogen bonds that is highly hydrophobic [80]. To counteract its low bioavailability several methods of drug delivery have been developed. The preservative benzalkonium chloride enhances corneal penetration by disrupting epithelial cell barriers; however, it also increases surface irritation [81]. Microspheres, micelles, cyclodextrins, nanoparticles, and liposomes have been used—each system working by surrounding the hydrophobic CsA to enhance corneal penetration; however, these are typically associated with rapid clearing of CsA from the tear film [82–85].

7.2.9 Calcineurin Phosphate Inhibitors

Calcineurin phosphate inhibitors exert their anti-inflammatory effect by forming a complex with intracellular macrophilin-12, which inhibits calcineurin-related transcription of T lymphocyte pro-inflammatory cytokines [86]. Commercially available topical calcineurin phosphate inhibitors include tacrolimus (Protopic,

Astellas, Tokyo, Japan) and pimecrolimus (Elidel, Novartis, Basel, Switzerland). Although initially developed for the treatment of atopic dermatitis [87], topical tacrolimus has good efficacy in the treatment of DES [88]. A more recently developed topical calcineurin phosphate inhibitor, LX-214 (voclosporin), was shown in a phase I study to be well tolerated [89]. It has been suggested that in addition to their anti-inflammatory effect, topical skin preparations applied on the eyelids may improve meibomian gland function. No topical calcineurin phosphate inhibitors have yet received FDA approval for DES.

When systemically administered, calcineurin phosphate inhibitors may have a contributory role in oncogenesis; this trend has been mostly noticed when used post organ-transplantation associated with non-melanoma skin cancers and lymphoproliferative malignancy [90]. Whether this risk translates to topical administration is unclear; however, a certain amount of systemic absorption occurs with the use of all topical ophthalmic preparations and can be minimized by digital occlusion of the lacrimal sac and puncta or use of punctual plugs.

7.2.10   Essential Fatty Acids

Dietary supplements of oral essential polyunsaturated fatty acids (FAs) omega-6 (linolenic acid) and omega-3 (α-linolenic acid) have been shown to be beneficial in various inflammatory conditions including ulcerative colitis, Sjogren's syndrome, and rheumatoid arthritis [91]. Essential fatty acids are precursors essential for synthesizing eicosanoids. Eicosanoids include prostacyclins, prostaglandins, leukotrienes, and thromboxanes which are all involved in key modulatory functions of immune activation. In general, omega-3 is anti-inflammatory and omega-6 is pro-inflammatory [92]. Omega-3 FAs reduce levels of key pro-inflammatory eicosanoids including prostaglandin E2, leukotriene B4, and cytokines IL-1 and TNF [93, 94].

Topical application of omega-3 has a clinically and experimentally demonstrated anti-inflammatory effect [93, 95]. Resolvin E1, a topically administered omega-3 FA derivative, was evaluated in a murine model of DES, and promoted corneal epithelial integrity, tear volume and decreased COX-2 expression [96]. It has recently been investigated clinically and has completed Phase II investigation [97].

7.2.11   Mucoprotective agents

Rebamipide is a mucoprotective agent that stimulates mucus secretion by inducing prostaglandin E2 production [98]. It is occasionally used to treat gastric ulcers. It has been evaluated as an ocular surface mucoprotective agent. Experimental work suggests that it improves ocular surface mucin secretion; current Phase III investigation is underway evaluating it as a treatment for DES [99]. Ecabet sodium, also used for ulcers of the gastrointestinal tract, is another medication that upregulates prostaglandin E2 pathway and mucous production. It is also currently being evaluated for the treatment of DES with phase III investigation [99].

7.3 Future Directions

With ongoing discovery of new molecular targets, increased understanding of the immunopathogenesis of DES and the success of new systemic anti-inflammatory agents including biological agents in treating inflammatory disease there is likely to be several new therapeutic agents to treat DES in the near future. A comprehensive review of therapeutic targets currently under investigation is beyond the scope of this chapter; however, a few will be discussed.

Tasocitinib (CP-690,550, Pfizer, New York, NY) is a Janus kinase 3 (JAK) inhibitor that has shown promise for treating rheumatoid arthritis and increasingly there is supportive evidence for its role in other inflammatory diseases [100]. JAKs are receptor-associated kinases involved in pro-inflammatory cytokines signal transduction that are specific to immune cells [101]. Aquaporins influence lacrimal gland fluid secretion, and their expression in Sjogren's disease is altered; [102] aquaporin may have a role as a potential therapeutic target in DES but this is not yet elucidated. Other molecular targets implicated in DES immunopathogenesis include integrin $\alpha 4\beta 1$ and the $T_H 17$ subset of lymphocytes; these may have a role in treating DES [103, 104].

7.4 Conclusion

DES is a common, chronic ocular surface disorder that causes reduced visual function, surface discomfort and can affect workplace productivity and quality of life. Treatment of DES predominantly involves a combination of ocular surface lubricants and topical anti-inflammatory agents. Some agents such as corticosteroids and cyclosporine A have an established role in the treatment of DES; more are currently undergoing evaluation in clinical trials. As knowledge of the immunopathogenesis of DES grows, there is the potential for new therapeutic targets to be identified and evaluated as therapeutic agents. The process of translation of experimental to clinical evaluation is key to this process; hopefully new immunomodulatory and tear-film enhancing agents will lead to improved management for patients with DES.

Conflicts of Interest None

Sources of Funding None

References

1. Reidy JJ. Basic and clinical science course. Section 8. External disease and cornea. San Francisco: American Academy of Ophthalmology; 2008.
2. The definition and classification of dry eye disease: report of the Definition and Classification Subcommittee of the International Dry Eye WorkShop. Ocul Surf. 2007; 5(2):75–92.
3. Yavuz B, Bozdağ Pehlivan S, Unlü N. An overview on dry eye treatment: approaches for cyclosporin a delivery. Scientific World J. 2012;2012:194848; * A concise overview of the pharmacokinetics of topical cyclosporine and new drug delivery systems.
4. The epidemiology of dry eye disease: report of the Epidemiology Subcommittee of the International Dry Eye WorkShop. Ocul Surf. 2007; 5(2): 93–107.
5. Schaumberg DA, Sullivan DA, Buring JE, Dana MR. Prevalence of dry eye syndrome among US women. Am J Ophthalmol. 2003;136(2):318–26.
6. Moss SE, Klein R, Klein BE. Prevalence of and risk factors for dry eye syndrome. Arch Ophthalmol. 2000;118(9):1264–8.
7. Brewitt H, Sistani F. Dry eye disease: the scale of the problem. Surv Ophthalmol. 2001;45 Suppl 2:S199–202.
8. Ding J, Sullivan DA. Aging and dry eye disease. Exp Gerontol. 2012;47(7):483–90.
9. Guillon M, Maissa C. Contact lens wear affects tear film evaporation. Eye Contact Lens. 2008;34(6):326–30.
10. Yee RW, Sperling HG, Kattek A, Paukert MT, Dawson K, Garcia M, Hilsenbeck S. Isolation of the ocular surface to treat dysfunctional tear syndrome associated with computer use. Ocul Surf. 2007;5(4):308–15.
11. Pflugfelder SC. Prevalence, burden, and pharmacoeconomics of dry eye disease. Am J Manag Care. 2008;14(3 Suppl):S102–6.
12. Lemp MA. Report of the National Eye Institute/Industry workshop on Clinical Trials in Dry Eyes. CLAO J. 1995;21(4):221–32.
13. Pult H, Purslow C, Murphy PJ. The relationship between clinical signs and dry eye symptoms. Eye (Lond). 2011;25(4):502–10; * A 47-patient cohort study comparing clinical symptoms of dry eye with new and traditional clinical tests. When used in combinations these tests correlate with symptomatic severity of dry eye syndrome.
14. Li M, Gong L, Sun X, Chapin WJ. Anxiety and depression in patients with dry eye syndrome. Curr Eye Res. 2011;36(1):1–7.
15. Miljanovic' B, Dana R, Sullivan DA, Schaumberg DA. Impact of dry eye syndrome on vision-related quality of life. Am J Ophthalmol. 2007;143(3):409–15.
16. McGinnigle S, Naroo SA, Eperjesi F. Evaluation of dry eye. Surv Ophthalmol. 2012; 57(4):293–316.
17. Luo L, Li DQ, Corrales RM, Pflugfelder SC. Hyperosmolar saline is a proinflammatory stress on the mouse ocular surface. Eye Contact Lens. 2005;31(5):186–93.
18. De Paiva CS, Pangelinan SB, Chang E, et al. Essential role for c-Jun N-terminal kinase 2 in corneal epithelial response to desiccating stress. Arch Ophthalmol. 2009;127(12):1625–31.
19. Li DQ, Luo L, Chen Z, Kim HS, Song XJ, Pflugfelder SC. JNK and ERK MAP kinases mediate induction of IL-1β, TNF-α and IL-8 following hyperosmolar stress in human limbal epithelial cells. Exp Eye Res. 2006;82(4):588–96.
20. Stevenson W, Chauhan SK, Dana R. Dry eye disease: an immune-mediated ocular surface disorder. Arch Ophthalmol. 2012;130(1):90–100; ** An excellent overview of the immunopathogenesis of dry eye and the associated molecular targets for anti-inflammatory treatment of dry eye syndrome.
21. El Annan J, Chauhan SK, Ecoiffier T, Zhang Q, Saban DR, Dana R. Characterization of effector T cells in dry eye disease. Invest Ophthalmol Vis Sci. 2009;50(8):3802–7.
22. De Paiva CS, Chotikavanich S, Pangelinan SB, et al. IL-17 disrupts corneal barrier following desiccating stress. Mucosal Immunol. 2009;2(3):243–53.

23. Chen YT, Nikulina K, Lazarev S, et al. Interleukin-1 as a phenotypic immunomodulator in keratinizing squamous metaplasia of the ocular surface in Sjogren's syndrome. Am J Pathol. 2010;177(3):1333–43.
24. De Paiva CS, Villarreal AL, Corrales RM, et al. Dry eye–induced conjunctival epithelial squamous metaplasia is modulated by interferon-γ. Invest Ophthalmol Vis Sci. 2007;48(6):2553–60.
25. Yeh S, Song XJ, Farley W, Li DQ, Stern ME, Pflugfelder SC. Apoptosis of ocular surface cells in experimentally induced dry eye. Invest Ophthalmol Vis Sci. 2003;44(1):124–9.
26. Goyal S, Chauhan SK, El Annan J, Nallasamy N, Zhang Q, Dana R. Evidence of corneal lymphangiogenesis in dry eye disease: a potential link to adaptive immunity? Arch Ophthalmol. 2010;128(7):819–24.
27. Benıtez-Del-Castillo JM, Acosta MC, Wassfi MA, et al. Relation between corneal innervation with confocal microscopy and corneal sensitivity with noncontact esthesiometry in patients with dry eye. Invest Ophthalmol Vis Sci. 2007;48(1):173–81.
28. Hosal BM, Ornek N, Zilelioglu G, Elhan AH. Morphology of corneal nerves and corneal sensation in dry eye: a preliminary study. Eye (Lond). 2005;19(12):1276–9.
29. Yoon KC, Jeong IY, Park YG, Yang SY. Interleukin-6 and tumor necrosis factor-α levels in tears of patients with dry eye syndrome. Cornea. 2007;26(4):431–7.
30. Enrıquez-de-Salamanca A, Castellanos E, Stern ME, et al. Tear cytokine and chemokine analysis and clinical correlations in evaporative-type dry eye disease. Mol Vis. 2010;16:862–73; * An evaluation of cytokines and other inflammatory molecules in the tear films of patients with mild-to-moderate dry eye syndrome compared to healthy subjects. Interleukin (IL) 1-receptor antagonist, IL-6, IL-8/CXCL-8 and epidermal growth factor levels correlated with symptoms and signs of dry eye syndrome.
31. Chotikavanich S, de Paiva CS, Li Q, et al. Production and activity of matrix metalloproteinase-9 on the ocular surface increase in dysfunctional tear syndrome. Invest Ophthalmol Vis Sci. 2009;50(7):3203–9.
32. Brignole F, Pisella PJ, Goldschild M, De Saint JM, Goguel A, Baudouin C. Flow cytometric analysis of inflammatory markers in conjunctival epithelial cells of patients with dry eyes. Invest Ophthalmol Vis Sci. 2000;41(6):1356–63.
33. Gao J, Morgan G, Tieu D, et al. ICAM-1 expression predisposes ocular tissues to immune-based inflammation in dry eye patients and Sjogrens syndrome-like MRL/lpr mice. Exp Eye Res. 2004;78(4):823–35.
34. Zheng X, de Paiva CS, Li DQ, Farley WJ, Pflugfelder SC. Desiccating stress promotion of Th17 differentiation by ocular surface tissues through a dendritic cell mediated pathway. Invest Ophthalmol Vis Sci. 2010;51(6):3083–91.
35. Chauhan SK, Dana R. Role of Th17 cells in the immunopathogenesis of dry eye disease. Mucosal Immunol. 2009;2(4):375–6.
36. Lee JH, Ahn HS, Kim EK, Kim TI. Efficacy of sodium hyaluronate and carboxymethylcellulose in treating mild to moderate dry eye disease. Cornea. 2011;30(2):175–9.
37. McDonald CC, Kaye SB, Figueiredo FC, Macintosh G, Lockett C. A randomised, crossover, multicentre study to compare the performance of 0.1% (w/v) sodium hyaluronate with 1.4% (w/v) polyvinyl alcohol in the alleviation of symptoms associated with dry eye syndrome. Eye (Lond). 2002;16(5):601–7.
38. Brignole F, Pisella PJ, Dupas B, et al. Efficacy and safety of 0.18% sodium hyaluronate in patients with moderate dry eye syndrome and superficial keratitis. Graefes Arch Clin Exp Ophthalmol. 2005;243(6):531–8.
39. Lemp MA. Management of dry eye. Am J Manag Care. 2008;14:S088–101.
40. Scaffidi RC, Korb DR. Comparison of the efficacy of two lipid emulsion eyedrops in increasing tear film lipid layer thickness. Eye Contact Lens. 2007;33(1):38–44.
41. Benelli U. Systane lubricant eye drops in the management of ocular dryness. Clin Ophthalmol. 2011;5:783–90.

42. Coutinho AE, Chapman KE. The anti-inflammatory and immunosuppressive effects of gluco-corticoids, recent developments and mechanistic insights. Mol Cell Endocrinol. 2011; 335(1):2–13.
43. De Paiva CS, Corrales RM, Villarreal AL, et al. Apical corneal barrier disruption in experimental murine dry eye is abrogated by methylprednisolone and doxycycline. Invest Ophthalmol Vis Sci. 2006;47(7):2847–56.
44. Lekhanont K, Leyngold IM, Suwan-Apichon O, Rangsin R, Chuck RS. Comparison of topical dry eye medications for the treatment of keratoconjunctivitis sicca in a botulinum toxin B–induced mouse model. Cornea. 2007;26(1):84–9.
45. Avunduk AM, Avunduk MC, Varnell ED, Kaufman HE. The comparison of efficacies of topical corticosteroids and nonsteroidal anti-inflammatory drops on dry eye patients: a clinical and immunocytochemical study. Am J Ophthalmol. 2003;136(4):593–602.
46. Pflugfelder SC, Maskin SL, Anderson B, et al. A randomized, double-masked, placebo controlled, multicenter comparison of loteprednol etabonate ophthalmic suspension, 0.5%, and placebo for treatment of keratoconjunctivitis sicca in patients with delayed tear clearance. Am J Ophthalmol. 2004;138(3):444–57.
47. Pavesio CE, Decory HH. Treatment of ocular inflammatory conditions with loteprednol etabonate. Br J Ophthalmol. 2008;92:455–9.
48. Zhu L, Zhang C, Chuck RS. Topical steroid and nonsteroidal anti-inflammatory drugs inhibit inflammatory cytokine expression on the ocular surface in the botulinum toxin B-induced murine dry eye model. Mol Vis. 2012;18:1803–12; * Botulin toxin B-induced mouse tear-deficiency dry eye model has been shown to mimic human non-Sjogren's disease. In this study topical fluoromethalone and not NSAIDs reduced staining of tumour necrosis factor α and IL-1β in corneal and conjunctival epithelia.
49. Congdon N, Schein O, Kulajta P, Lubomski L, Gilbert D, Katz J. Corneal complications associated with topical ophthalmic use of nonsteroidal anti-inflammatory drugs. J Cataract Refract Surg. 2001;27:622–31.
50. De Paiva CS, Corrales RM, Villarreal AL, Farley WJ, Li DQ, Stern ME, Pflugfelder SC. Corticosteroid and doxycycline suppress MMP-9 and inflammatory cytokine expression, MAPK activation in the corneal epithelium in experimental dry eye. Exp Eye Res. 2006;83:526–35.
51. Luo L, Li DQ, Doshi A, Farley W, Corrales RM, Pflugfelder SC. Experimental dry eye stimulates production of inflammatory cytokines and MMP-9 and activates MAPK signaling pathways on the ocular surface. Invest Ophthalmol Vis Sci. 2004;45:4293–301.
52. Nikodemova M, Watters JJ, Jackson SJ, Yang SK, Duncan ID. Minocycline down-regulates MHC II expression in microglia and macrophages through inhibition of IRF-1 and protein kinase C (PKC) alpha/beta II. J Biol Chem. 2007;282(20):15208–16.
53. Shafaa MW, El Shazly LH, El Shazly AH, El Gohary AA, El Hossary GG. Efficacy of topically applied liposome-bound tetracycline in the treatment of dry eye model. Vet Ophthalmol. 2011;14(1):18–25.
54. Veldman P, Colby K. Current evidence for topical azithromycin 1% ophthalmic solution in the treatment of blepharitis and blepharitis-associated ocular dryness. Int Ophthalmol Clin. 2011;51(4):43–52.
55. Nichols JJ, Bickle KM, Zink RC, Schiewe MD, Haque RM, Nichols KK. Safety and efficacy of topical azithromycin ophthalmic solution 1.0% in the treatment of contact lens-related dry eye. Eye Contact Lens. 2012;38(2):73–9.
56. Ubels JL, MacRae SM. Vitamin A is present as retinol in the tears of humans and rabbits. Curr Eye Res. 1984;3(6):815–22.
57. Ubels JL, Foley KM, Rismondo V. Retinol secretion by the lacrimal gland. Invest Ophthalmol Vis Sci. 1986;27(8):1261–8.
58. Odaka A, Toshida H, Ohta T, et al. Efficacy of retinol palmitate eye drops for dry eye in rabbits with lacrimal gland resection. Clin Ophthalmol. 2012;6:1585–93.

59. Kim EC, Choi JS, Joo CK. A comparison of vitamin a and cyclosporine a 0.05% eye drops for treatment of dry eye syndrome. Am J Ophthalmol. 2009;147(2):206–13.
60. Selek H, Unlu N, Orhan M, Irkec M. Evaluation of retinoic acid ophthalmic emulsion in dry eye. Eur J Ophthalmol. 2000;10(2):121–7.
61. Frist-Larsen K, Isager H, Manthorpe R. Sjogren's syndrome treated with bromhexine: a randomized clinical study. Br Med J. 1978;1:1579–81.
62. Tauber J, Davitt WF, Bokosky JE, et al. Double-masked, placebo controlled safety and efficacy trial of diquafosol tetrasodium (INS365) ophthalmic solution for the treatment of dry eye. Cornea. 2004;23:784–92.
63. Matsumoto Y, Ohashi Y, Watanabe H, Tsubota K. Diquafosol Ophthalmic Solution Phase 2 Study Group. Efficacy and safety of diquafosol ophthalmic solution in patients with dry eye syndrome: a Japanese Phase 2 clinical trial. Ophthalmology. 2012;119(10):1954–60; *A randomized, double-masked, multicenter clinical trial comparing topical diquafosol (1%, n=96, 3%, n=96) to placebo (n=94). There was a dose-dependent improvement in fluorescein staining and an improvement in symptom score in both treatment groups.
64. Peral A, Domínguez-Godínez CO, Carracedo G, Pintor J. Therapeutic targets in dry eye syndrome. Drug News Perspect. 2008;21(3):166–76.
65. Avni I, Garzozi HJ, Barequet IS, et al. Treatment of dry eye syndrome with orally administered CF101: data from a phase 2 clinical trial. Ophthalmology. 2010;117(7):1287–93.
66. Kojima T, Higuchi A, Goto E, Matsumoto Y, Dogru M, Tsubota K. Autologous serum eye drops for the treatment of dry eye diseases. Cornea. 2008;27 Suppl 1:S25–30.
67. Geerling G, Maclennan S, Hartwig D. Autologous serum eye drops for ocular surface disorders. Br J Ophthalmol. 2004;88:1467–74.
68. Bradley JC, Simoni J, Bradley RH, McCartney DL, Brown SM. Time- and temperature-dependent stability of growth factor peptides in human autologous serum eye drops. Cornea. 2009;28(2):200–5.
69. Kojima T, Ishida R, Dogru M, et al. The effect of autologous serum eye drops in the treatment of severe dry eye disease: a prospective randomized case-control study. Am J Ophthalmol. 2005;139:242–6.
70. Urzua CA, Vasquez DH, Huidobro A, Hernandez H, Alfaro J. Randomized double-blind clinical trial of autologous serum versus artificial tears in dry eye syndrome. Curr Eye Res. 2012;37(8):684–8.
71. Donnenfeld E, Pflugfelder SC. Topical ophthalmic cyclosporine: pharmacology and clinical uses. Surv Ophthalmol. 2009;54(3):321–38.
72. Eckstein LA, Van Quill KR, Bui SK, et al. Cyclosporin a inhibits calcineurin/nuclear factor of activated T-cells signaling and induces apoptosis in retinoblastoma cells. Invest Ophthalmol Vis Sci. 2005;46:782–90.
73. Waldmeier PC, Zimmermann K, Qian T, et al. Cyclophilin D as a drug target. Curr Med Chem. 2003;10:1485–506.
74. Brignole F, Pisella PJ, De Saint JM, et al. Flow cytometric analysis of inflammatory markers in KCS: 6- month treatment with topical cyclosporin A. Invest Ophthalmol Vis Sci. 2001;42:90–5.
75. Kunert KS, Tisdale AS, Gipson IK. Goblet cell numbers and epithelial proliferation in the conjunctiva of patients with dry eye syndrome treated with cyclosporine. Arch Ophthalmol. 2002;120:330–7.
76. Yoshida A, Fujihara T, Nakata K. Cyclosporin A increases tear fluid secretion via release of sensory neurotransmitters and muscarinic pathway in mice. Exp Eye Res. 1999; 68(5):541–6.
77. Sall K, Stevenson OD, Mundorf TK, et al. CsA Phase 3 Study Group. Two multicenter, randomized studies of the efficacy and safety of cyclosporine ophthalmic emulsion in moderate to severe dry eye disease. Ophthalmology. 2000;107:631–9.
78. Stevenson D, Tauber J, The RBL, Cyclosporin A. Phase 2 Study Group. Efficacy and safety of cyclosporin A ophthalmic emulsion in the treatment of moderate-to-severe dry eye disease: a dose-ranging, randomized trial. Ophthalmology. 2000;107:967–74.

79. Liu H, Wang Y, Li S. Advanced delivery of ciclosporin A: present state and perspective. Expert Opin Drug Deliv. 2007;4(4):349–58.
80. El Tayar N, Mark AE, Vallat P, et al. Solvent-dependent conformation and hydrogen-bonding capacity of cyclosporin A: evidence from partition coefficients and molecular dynamics simulations. J Med Chem. 1993;36(24):3757–64.
81. Skalicky SE, Goldberg I, McCluskey P. Ocular surface disease and quality of life in glaucoma patients. Am J Ophthalmol. 2012;153(1):1–9.
82. Kanai A, Alba RM, Takano T, et al. The effect on the cornea of alpha cyclodextrin vehicle for cyclosporin eye drops. Transplant Proc. 1989;21(1 pt 3):3150–2.
83. Schechter BA. Ketorolac during the induction phase of cyclosporin-A therapy. J Ocul Pharmacol Ther. 2006;22(2):150–4.
84. Milani JK, Pleyer U, Dukes A, et al. Prolongation of corneal allograft survival with liposome-encapsulated cyclosporine in the rat eye. Ophthalmology. 1993;100(6):890–6.
85. Dillen K, Bozdag S, Vandervoort J, Ludwig A. Evaluation of the physicochemical characteristics and activity of various kinds of ciprofloxacin HCl-loaded cationic nanoparticles. J Drug Del Sci Technol. 2007;17(1):49–56.
86. Sárdy M, Ruzicka T, Kuhn A. Topical calcineurin inhibitors in cutaneous lupus erythematosus. Arch Dermatol Res. 2009;301(1):93–8.
87. Russell JJ. Topical tacrolimus: a new therapy for atopic dermatitis. Am Fam Physician. 2002;66(10):1899–903.
88. Moscovici BK, Holzchuh R, Chiacchio BB, Santo RM, Shimazaki J, Hida RY. Clinical treatment of dry eye using 0.03% tacrolimus eye drops. Cornea. 2012;31(8):945–9.
89. Products and technologies. Lux BioSciences. 2011. Available at: http://www.luxbio.com/LX214.htm. Accessed Jul 2012.
90. Ormerod AD. Topical tacrolimus and pimecrolimus and the risk of cancer: How much cause for concern? Br J Dermatol. 2005;153:701–5.
91. Aragona P, Bucolo C, Spinella R, Giuffrida S, Ferreri G. Systemic omega-6 essential fatty acid treatment and pge1 tear content in Sjogren's syndrome patients. Invest Ophthalmol Vis Sci. 2005;46(12):4474–9.
92. Rosenberg ES, Asbell PA. Essential fatty acids in the treatment of dry eye. Ocul Surf. 2010;8(1):18–28; *A well-written review of the literature regarding essential fatty acids for the prevention or treatment of dry eye syndrome. All the studies demonstrated that essential fatty acid supplementation resulted in some improvement in dry eye syndrome yet the evidence is not strong.
93. Rashid S, Jin Y, Ecoiffier T, Barabino S, Schaumberg D, Dana R. Topical omega-3 and omega-6 fatty acids for treatment of dry eye. Arch Ophthalmol. 2008;126:219–25.
94. James MJ, Gibson RA, Cleland LG. Dietary polyunsaturated fatty acids and inflammatory mediator production. Am J Clin Nutr. 2000;71(1 suppl):343S–8.
95. Barabino S, Rolando M, Camicione P, et al. Systemic linoleic and gamma-linolenic acid therapy in dry eye syndrome with an inflammatory component. Cornea. 2003;22:97–101.
96. Li N, He J, Schwartz CE, Gjorstrup P, Bazan HE. Resolvin E1 improves tear production and decreases inflammation in a dry eye mouse model. J Ocul Pharmacol Ther. 2010; 26(5):431–9.
97. US National Institutes of Health Clinical Trials. 2011. Available at: http://clinicaltrials.gov/ct2/show/NCT00799552. Accessed Jul 2012.
98. Takeji Y, Urashima H, Aoki A, Shinohara H. Rebamipide increases the mucin-like glycoprotein production in corneal epithelial cells. J Ocul Pharmacol Ther. 2012;28(3):259–63.
99. Fahmy AM, Hardten DR. Treating ocular surface disease: new agents in development. Clin Ophthalmol. 2011;5:465–72; *A review of new agents in the treatment of ocular surface disease.
100. Coombs JH, Bloom BJ, Breedveld FC, et al. Improved pain, physical functioning and health status in patients with rheumatoid arthritis treated with CP-690,550, an orally active Janus

kinase (JAK) inhibitor: results from a randomised, double-blind, placebo-controlled trial. Ann Rheum Dis. 2010;69:413–6.

101. Schindler C, Levy DE, Decker T. JAK-STAT signaling: from interferons to cytokines. J Biol Chem. 2007;282(28):20059–63.

102. Ding C, Nandoskar P, Lu M, Thomas P, Trousdale MD, Wang Y. Changes of aquaporins in the lacrimal glands of a rabbit model of Sjögren's syndrome. Curr Eye Res. 2011;36(6):571–8; * Protein and mRNA levels of aquaporin-4 and -5 were assessed in the lacrimal glands of rabbits with induced autoimmune dacryoadenitis compared to age-matched controls. Levels were altered in the diseased lacrimal glands, being greater or less than normal depending on the site (ductal or acinar cells) and aquaporin type (-4 or -5).

103. Ecoiffier T, El Annan J, Rashid S, Schaumberg D, Dana R. Modulation of integrin $\alpha4\beta1$ (VLA-4) in dry eye disease. Arch Ophthalmol. 2008;126(12):1695–9.

104. Chauhan SK, El Annan J, Ecoiffier T, et al. Autoimmunity in dry eye is due to resistance of Th17 to Treg suppression. J Immunol. 2009;182(3):1247–52.

Chapter 8
Investigating Carcinine Transport and the Expression Profile of Transporter Genes in Human Corneal Epithelial Cells

Anne Kasus-Jacobi, Okhil K. Nag, Vibudhuta Awasthi, Mark A. Babizhayev, and H. Anne Pereira

Abbreviations

4-HNE	4-Hydroxynonenal
ABC	ATP-binding cassette
ATA2	Amino acid transporter A2
ATP	Adenosine triphosphase
COPT1	Copper transporter 1
Ct	Cycle threshold
HCEC	Human corneal epithelial cell
HPLC	High-performance liquid chromatography
HPRT1	Hypoxanthine phosphoribosyltransferase
LAT1	Large neutral amino acid transporter 1
MCT1	Monocarboxylate transporter 1

A. Kasus-Jacobi (✉)
Department of Pharmaceutical Sciences and Oklahoma Center for Neuroscience,
University of Oklahoma Health Sciences Center, 1110 N. Stonewall Avenue,
Oklahoma City, OK 73117-1223, USA
e-mail: anne-kasus-jacobi@ouhsc.edu

O.K. Nag • V. Awasthi
Department of Pharmaceutical Sciences, University of Oklahoma
Health Sciences Center, Oklahoma City, OK USA

M.A. Babizhayev
Innovative Vision Products Inc., New Castle, DE, USA

Innovative Vision Products Inc., Moscow, Russia

H.A. Pereira
Departments of Pharmaceutical Sciences, Pathology, and Cell Biology
and Oklahoma Center for Neuroscience, University of Oklahoma
Health Sciences Center, Oklahoma City, OK, USA

© Springer Science+Business Media New York 2015
M.A. Babizhayev et al. (eds.), *Studies on the Cornea and Lens*,
Oxidative Stress in Applied Basic Research and Clinical Practice,
DOI 10.1007/978-1-4939-1935-2_8

MRP	Multidrug resistance-associated protein
MS	Mass spectrometry
PEPT	Peptide transporter
PMP70	Peroxisomal membrane protein of 70 kDA
RT-PCR	Reverse transcription polymerase chain reaction
SLC	Solute carrier
VDAC	Voltage-dependent anion channel

8.1   Introduction

The major barrier to developing safe and effective new treatments for ocular diseases is the identification and implementation of effective methods of drug delivery [1]. Anatomic barriers and physiologic clearance mechanisms restrict the penetration of drugs into the eye globe [2]. Current delivery options are far from ideal because they involve repeated injections to deliver drugs directly inside the eye globe [1]. Intravitreal injection is the delivery method leading to the best intraocular bioavailability [2]. Therapeutic concentrations are easily reached with this method but intraocular clearance mechanisms cause the drug concentration to decrease over time, requiring repeated injections [2]. It is the most invasive method and is associated with the highest complication risks such as cataract, infection, or retinal detachment [2]. Thus, without an effective and safe drug delivery system, many compounds that show promise at the bench or in preclinical testing are failing clinically.

Topical delivery would have clear advantages over intraocular injection because it is noninvasive and it is compatible with a long-term treatment plan that could be conveniently auto-administered by patients. However, by contrast to intraocular injection, this method leads to the lowest intraocular bioavailability for most molecules [2]. To enter the eye globe after topical administration, drugs have to cross the blood-aqueous barrier formed by the tight junctions of the corneal epithelial cells. Alternatively, drugs can penetrate through the conjunctival/scleral route. The barrier for penetration through this route is the conjunctiva, which is lined with a stratified columnar epithelium with tight junctions, resting on a continuous basal lamina, forming a tight epithelial barrier. These corneal and conjunctival barriers, combined with the precorneal tear turnover and drainage mechanisms seem to pose insurmountable problems for efficient topical drug delivery to the inside of the eye. Indeed, the intraocular bioavailability of most compounds following topical delivery ranges from 0 to 0.0004 % [2].

Interestingly, a small number of topically administered drugs have been shown to efficiently reach the inside of the eye [3–5]. These drugs might be transported by endogenous transport mechanisms, which physiological role is the transport of nutrients through biological barriers. The identification of these transport mechanisms would be an important step to rationalize any efforts to improve ocular penetration and bioavailability of topical ophthalmic drugs. They could be specifically designed to take advantage of existing transport mechanisms, allowing effective penetration [6].

The dipeptide derivative carcinine (β-alanyl-histamine) is a multifunctional antioxidant combining interesting biological properties such as the scavenging of reactive oxygen species, the reduction of lipid hydroperoxides, and the quenching of toxic end-products of lipid peroxidation [7]. We have previously demonstrated in vitro that carcinine can quench 4-hydroxynonenal (4-HNE), one of the most abundant and most toxic product of lipid peroxidation [8]. The formation of an adduct between carcinine and 4-HNE prevents and even reverses the adduction of proteins by 4-HNE [8]. In cells, modification of proteins by 4-HNE typically inactivates their function and is followed by their rapid degradation. Since oxidative damage is associated with multiple ocular diseases, the possibility to use carcinine as a therapeutic to prevent and reverse oxidative damage on proteins in ocular tissues is promising. In the current study, we show that carcinine is one of the few compounds that can penetrate the eye globe and efficiently reach posterior segment tissues upon topical administration in the form of eye drops. We then investigated the transport route allowing carcinine to efficiently penetrate the eye. We used the immortalized human corneal epithelial cells (HCECs) to investigate carcinine uptake in these cells with the rational that if a carcinine transporter was identified, it could be exploited to increase the penetration of other ophthalmic drugs. We further reasoned that any other transporter identified on HCECs could potentially be exploited, so we investigated the expression profile of 84 transporter genes in these cells. We found that several genes encoding membrane transporters mediating the influx of amino acids (*SLC38A2*, *SLC3A2*, and *SLC7A5*), monocarboxylate (*SLC16A1*), and copper (*SLC31A1*) were expressed at high levels in HCECs. These corneal transporters are promising targets for the design of modified ophthalmic drugs with increased ocular penetration and bioavailability.

8.2 Carcinine Efficiently Penetrates the Eye Globe

8.2.1 Retinal Levels of Carcinine Following Topical Administration

We used our previously described HPLC/MS method to quantify the amount of carcinine in the mouse retina, following topical administration of 0.2 M carcinine eye drops [8, 9]. All animals were maintained and handled according to institutional guidelines and the ARVO Statement for the Use of Animals in Ophthalmic and Vision Research. A time-dependent accumulation of carcinine was found in the retina after administration of 1 drop (Fig. 8.1a). A peak of ~100 ng of carcinine/retina was found 30 min after administration, and this level rapidly decreased to less than ~25 ng/retina. When we administered a second eye drop of 0.2 M carcinine, 1 h after the first, the peak of retinal carcinine reached ~100 ng/retina in 30 min again

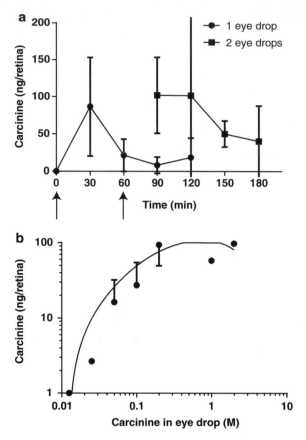

Fig. 8.1 Carnicine is efficiently transported to the retina following topical administration. Mice received carnicine dissolved in eye drop formulation (glycerol 1 %, carboxymethylcellulose 0.3 %, benzyl alcohol 0.3 %, in water, pH adjusted to 6.8 with potassium borate and potassium bicarbonate) [25]. (**a**) 0.2 M carnicine was administered at 0 min (*arrow*) and mice were euthanized 30, 60, 90, and 120 min later to collect retinas. Another set of mice received 0.2 M carnicine twice, at 0 and 60 min (indicated by *arrows*) and were euthanized every 30 min following second administration to collect retinas. (**b**) Mice received indicated concentrations of carnicine in eye drop formulation and were euthanized 30 min later to collect retinas. HPLC/MS was used to quantify the amount of carnicine per retina in each retinal sample [8]. The means of three retinal samples and error bars representing the standard error from the mean (SEM) were plotted

but this time the carnicine level remained constant for the following 30 min and stayed higher (~50 ng/retina) during at least 1 h. After only 1 application, a dose-dependent accumulation of carnicine was found in the retina when using between 0.0125 and 0.2 M carnicine in the eye drop formulation (Fig. 8.1b). There was no further increase in retinal carnicine when the concentration in eye drop increased from 0.2 to 2 M. This result suggests that a saturable transport of carnicine is taking place, rather than a passive diffusion.

8.2.2 Carcinine Lowers 4-HNE-Protein Adduct in Retinal Cells

We previously demonstrated the scavenging effect of carcinine on 4-HNE in vitro [8]. To determine if carcinine had this effect in vivo, we measured the amount of 4-HNE-protein adduct in mouse retinas, following administrations of 0.1 M carcinine eye drops, every hour for 6 h [9]. After treatment with carcinine eye drops, a significant ($P < 0.05$) 20 % decrease of 4-HNE-protein adduct was found in mouse retinas when compared with vehicle-treated eyes [9]. Low amount of 4-HNE is constantly generated in cells and reacts with surrounding proteins to form 4-HNE-protein adduct within the intracellular compartment. Thus, our result supports the idea that carcinine is still active after migrating through the eye globe to the retina, and suggests that carcinine is entering the retinal cells, perhaps through a transport mechanism, where it prevents and/or reverses adduct formation.

8.2.3 Retinal Bioavailability of Carcinine

As shown in Fig. 8.1b, after application of 1 eye drop (10 µL) of 0.2 M carcinine corresponding to 2,000 nmol of carcinine on the mouse cornea, 100 ng of carcinine is found in the retina corresponding to 0.5 nmol. Thus, 0.025 % of the topically applied carcinine reaches the retina, which is at least 60-fold higher than the 0–0.0004 % intraocular bioavailability of most compounds [2]. In addition, topical applications of carcinine lead to therapeutic concentrations of carcinine, as demonstrated by the significant decrease of 4-HNE-protein adduct in the retina [9]. This result is interesting because if identified, the transport route allowing carcinine to efficiently penetrate the eye and reach the retina could be exploited for the delivery of other drugs to the posterior segment of the eye. Drugs can be transported through the corneal epithelium or through the conjunctival epithelium [2]. We used the HCECs to investigate the uptake of carcinine with the goal of identifying the transporter(s) involved.

8.3 Carcinine in HCECs

8.3.1 Carcinine Uptake in HCECs

We first hypothesized that the peptide transporters PepT1 (encoded by the *SLC15A1* gene) and PepT2 (encoded by the *SLC15A2* gene) could be involved in the transport of carcinine through the corneal epithelium because expression of *SLC15A1* and *SLC15A2* genes were detected in human cornea and HCECs [10, 11]. In addition, a recent study demonstrated that carnosine and anserine, two dipeptide derivatives structurally similar to carcinine, are transported with high affinity by PEPT2 and with moderate affinity by PEPT1 in a human intestinal cell line and a rat renal cell line [12]. We used immortalized HCECs [13] to determine if carcinine could be transported inside the cells. We first incubated HCECs with 0.2 M carcinine for 30 min, and found carcinine in the cell extract (Fig. 8.2).

To determine if there was an influx of carcinine in these cells, we prepared a radiolabeled ^{14}C carcinine. ^{14}C carcinine was synthesized according to a published procedure [14]. The identity of intermediate ^{14}C-N-tert-butoxycarbonyl-b-alanylhistamine and final ^{14}C-β-alanylhistamine compounds were verified by their retention times in HPLC, using unlabeled standards. In the first experiment, a time-course of carcinine accumulation was performed using a concentration of 1 mM carcinine. A fast accumulation of carcinine was observed, peaking at 5 min followed by a very slow decrease over the following 15 min (Fig. 8.3a).

In the second experiment, we quantified the uptake of carcinine after incubation of HCECs for 20 min with concentrations of carcinine ranging from 0.01 to 100 mM. A linear dose-dependent accumulation of carcinine was found in the cells at the concentrations tested (Fig. 8.3b). This result suggests that there could be a passive, non saturable, influx of carcinine in these cells, or could indicate that carcinine simply binds the plasma membrane of the cells but does not penetrate the intracellular compartment. To determine whether or not carcinine penetrates the cells, we next investigated the biological effect of carcinine in the intracellular compartment by quantifying intracellular 4-HNE-protein adduct in HCECs.

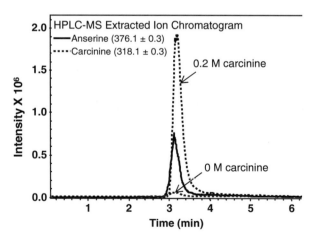

Fig. 8.2 Carcinine is found in HCECs after incubation with 0.2 M exogenous carcinine. SV-40 adenovirus immortalized HCECs were grown to 90 % confluence in 10 cm plates as previously described [13, 17, 18]. Cells were incubated with fresh basal medium containing 0 or 0.2 M carcinine for 30 min. Cells were then washed three times in PBS, harvested, and centrifuged. Cell pellet was extracted with acetonitrile, anserine was used as the internal standard, and HPLC-MS analysis of cell extracts was performed as previously described [8]. Extracted Ion Chromatogram using m/z: 318.1±0.3 for carcinine and m/z: 376.1±0.3 for anserine is shown. Carcinine was found in the cell extract after treatment with 0.2 M carcinine

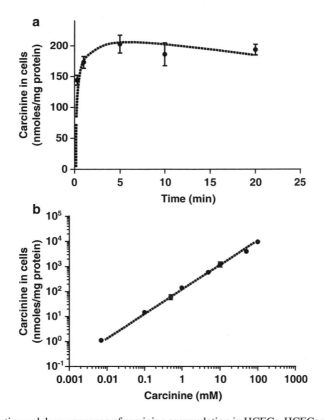

Fig. 8.3 Kinetics and dose–response of carcinine accumulation in HCECs. HCECs were grown to confluence in 24-well plates. (**a**) Cells were washed with basal medium once and incubated in basal medium containing 1 mM of unlabeled carcinine and a negligible concentration of ^{14}C carcinine (0.007 mM final, 0.16 µCi/µL final) for 0, 1, 5, 10 and 20 min. Cells were washed once in PBS and harvested in 0.3 M NaOH. ^{14}C associated with the cells was quantified in scintillation buffer. Protein concentrations were used to normalize the results. Each time point was performed in triplicate, mean and SEM were plotted. (**b**) Cells were washed with basal medium once and incubated in basal medium containing indicated concentrations of unlabeled carcinine and a negligible concentration of ^{14}C carcinine (0.007 mM final, 0.16 µCi/µL final) for 20 min. Cells were washed once in PBS and harvested in 0.3 M NaOH. ^{14}C associated with the cells was quantified in scintillation buffer. Protein concentrations were used to normalize the results. Each concentration of carcinine was performed in triplicate and error bars representing standard deviation (SD) were plotted. In most cases, they are too small to be visible on the graph

8.3.2 Carcinine Effect in HCECs

We first incubated HCECs with 75 µM of exogenous 4-HNE for 0, 1, and 3 h, washed the cells with PBS to eliminate extracellular 4-HNE, and quantified 4-HNE-protein adduct in cell homogenates as previously described [15]. A time-dependent formation of 4-HNE-protein adduct is shown in Fig. 8.4a. We then preincubated HCECs with 0, 1, or 10 mM carcinine for 1 and 3 h, removed the media containing

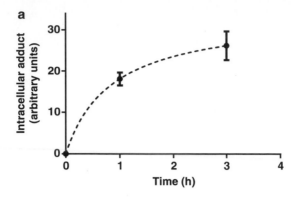

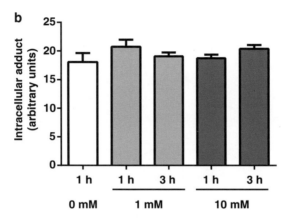

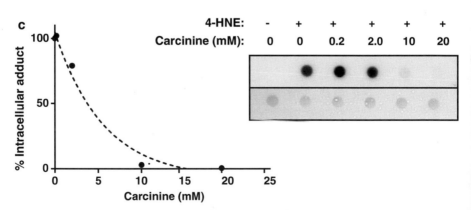

Fig. 8.4 Carcinine can quench 4-HNE in the cell medium but not in the HCEC intracellular compartment. HCECs were grown to confluence in 6-well plates. (**a**) Cells were washed with basal medium once and then incubated with basal medium containing 75 μM 4-HNE for 0, 1, and 3 h. (**b**) Cells were preincubated with basal medium containing 0, 1, or 10 mM carcinine for 1 or 3 h. At the end of the preincubation time, cells were washed three times with PBS and incubated with basal medium containing 75 μM 4-HNE for 1 h. (**c**) Cells were incubated with basal medium

carcinine, washed the cells with PBS to eliminate extracellular carcinine, and then treated the cells with 75 μM of 4-HNE for 1 h. If carcinine was entering the intracellular compartment, it would inhibit the formation of 4-HNE-protein adduct in a dose-dependent manner. As shown in Fig. 8.4b, there was no significant decrease of 4-HNE-protein adduct formed in any of the cells preincubated with carcinine, suggesting that carcinine was not efficiently transported inside the cells.

We then treated HCECs with carcinine and 4-HNE, added simultaneously to the cell culture medium. Increasing concentrations of carcinine were used in combination with 75 μM of 4-HNE. After 2 h incubation with medium, cells were harvested and intracellular 4-HNE-protein adducts were quantified. As shown in Fig. 8.4c, carcinine completely inhibited the formation of adducts with intracellular proteins when used at concentrations of 10 mM or higher. These results show the rapid quenching effect of 4-HNE by carcinine in the cell-culture medium, inhibiting the passive influx of 4-HNE and the formation of intracellular 4-HNE-protein adduct.

Taken together, these results show that carcinine is active outside but not inside the cells, demonstrating that it does not enter the HCECs. The linear dose-dependent accumulation of carcinine found in HCEC extracts in Fig. 8.3b was probably due to an interaction of carcinine with the plasma membranes. Carcinine is a dipeptide derivative (β-alanyl-histamine), formed by a hydrophobic alanyl and a positively charged histamine. Such compound is likely to bind and be inserted in the plasma membrane but might not be able to freely cross it because of the positive charge. Our results demonstrate that no passive diffusion is taking place across cell membranes and thus suggests that the intracellular quenching effect of carcinine previously found in retinal cells [9] was probably due to a transporter-mediated uptake. The *SLC15A2* mRNA has been found in the human retina [10] so it is possible that PepT2 was involved in the transport of carcinine in mouse retinal cells.

Our results suggest that the penetration route of carcinine into the eye globe is not through the corneal epithelium. Further experiments will investigate the alternative conjunctival/scleral route and the transport of carcinine through the conjunctival epithelium.

Fig. 8.4 (continued) containing 75 μM 4-HNE and 0, 0.2, 2, 10 or 20 mM carcinine for 2 h. At the end of incubation times, cells were washed three times with PBS and harvested in T-PER reagent containing protease inhibitors as previously described [15]. Whole cell homogenates are prepared and equal aliquots (5 μg) of protein are analyzed by dot blot as previously described [15]. Dot blot results are quantified using ImageJ and normalized to the amount of protein in each dot quantified with Ponceau red staining of the membranes as previously described [8]. Graph shown in (**a**) and (**b**) are representative results of triplicates for each experimental condition. Means and SD are plotted. Graph shown in (**c**) is representative of one sample per experimental condition. The percent of intracellular adduct is plotted using 0 mM carcinine as 100 % of intracellular adduct. The dot blot showing detection of 4-HNE-protein adduct (*top panel*) and Ponceau staining (*bottom panel*) is also shown

8.4   Expression Profile of Endogenous Transporters in HCECs

We reasoned that any transporter, expressed at sufficient level at the corneal blood-aqueous barrier, could be exploited to transport topical ophthalmic drugs. Two expression profile studies of transporter genes have been previously conducted, looking specifically at dissected human corneas [10, 16]. The cornea contains three cell types; the corneal epithelial cells forming the blood-aqueous barrier of the cornea, the keratinocytes, and the corneal endothelial cells. A recent study describes more specifically the expression profile of 12 transporter genes in the HCECs (cHCE) immortalized by infection with an amphotropic recombinant retrovirus containing HPV-16 ED/E7 genes [11]. In the current study, we investigated the expression profile of 84 transporter genes in the HCECs immortalized by infection with a recombinant simian virus 40-adenovirus vector [13]. Any transporter highly expressed in corneal epithelial cells is likely to form a gate through which modified drugs might be able to cross the corneal epithclium barrier.

HCECs immortalized with SV-40 adenovirus were maintained as previously described [13, 17, 18]. Total RNAs were prepared from HCECs at passage 10. Total RNA was used for reverse transcription cDNA synthesis. The Human Drug Transporters RT2 Profiler PCR Array (Qiagen) was used for quantification of gene expression. Two independent experiments were conducted using two different batches of HCECs.

Gene expression levels are shown in Fig. 8.5, relative to the *HPRT1* gene, a housekeeping gene with relatively low expression in these cells. The Ct number of *HPRT1* was 23 in both experiments and by definition its relative expression level is 1. The two major superfamilies of transporters identified in the human genome are represented on the PCR array; the ATP-binding cassette (ABC) superfamily (Fig. 8.5a) and the solute carrier (SLC) superfamily (Fig. 8.5b). The other types of transporters represented on the PCR array are shown in Fig. 8.5c.

The ABC superfamily includes 7 families and 48 members [19], of which 6 families and 27 members are quantified on the PCR array. Most of the ABC transporters are active efflux pumps driven by ATP hydrolysis [19]. As shown in Fig. 8.5a, most of them have negligible expression levels compared to *HPRT1* (set as 1). Only *ABCD3*, encoding a peroxisomal transporter of fatty acids (peroxisomal membrane protein of 70 kDa, PMP70) was expressed at a level close to *HPRT1*. *ABCC1* and *ABCC3*, encoding for the multidrug resistance-associated proteins MRP1 and MRP3, were previously detected in the human cornea and cHCE cells [11]. They were also found in HCECs but at negligible levels of 0.1 and 0.04 respectively, relative to *HPRT1*.

The SLC transporters are usually facilitated influx pumps [20]. This superfamily consists of 52 families and at least 298 members, of which 20 families and 45 members are quantified in this study. As shown in Fig. 8.5b, only five SLC transporters are expressed at higher levels than *HPRT1*. Fifteen members of the SLC superfamily are expressed at intermediate levels, below 1 but higher than 0.1,

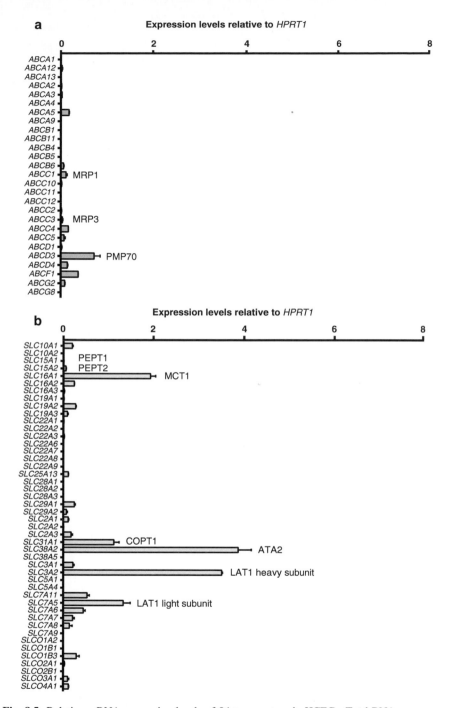

Fig. 8.5 Relative mRNA expression levels of 84 transporters in HCECs. Total RNAs were prepared from HCECs at passage 10 using the RNeasy Plus Universal Mini Kit from Qiagen. Total RNA (0.5 µg) was used for reverse transcription cDNA synthesis using the RT2 First Strand Kit (Qiagen). The Human Drug Transporters RT2 Profiler PCR Array (Qiagen) was used for quantification of gene expression. (**a**) Efflux transporters of the ABC superfamily, (**b**) influx transporters of the SLC superfamily, and (**c**) other transporters. Delta Ct values are calculated using the Ct value of *HPRT1* and converted to relative expression levels. Data are presented as the mean of two experiments with the SD

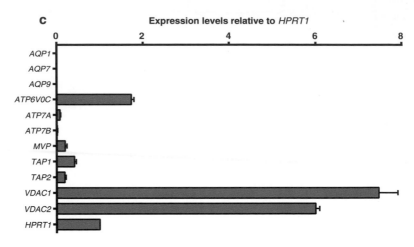

Fig. 8.5 (continued)

relative to *HPRT1*. The other 25 members, including *SLC15A1* (PEPT1) and *SLC15A2* (PEPT2) are expressed at negligible levels lower than 0.1. This is likely to explain the absence of carcinine uptake in HCECs.

The five highly expressed transporters are influx pumps, located on the plasma membrane. Three of them are amino acid transporters: the amino acid transporter A2 (ATA2), and the heavy and light subunits of the heterodimeric large neutral amino acid transporter (LAT1). One is a monocarboxylate transporter (MCT1) allowing influx of lactate, pyruvate, and other monocarboxylates to provide energy to the cells. The fifth one is a copper transporter (COPT1). To our knowledge, none of them has been previously described as transporters that are highly expressed in the cornea or in corneal epithelial cells.

Other transporters represented on the PCR array are shown in Fig. 8.5c. Three organelle transporters *ATP6V0C* (encoding a component of the vacuolar ATPase H⁺ transporter), and *VDAC1* and *VDAC2* (mitochondrial voltage-dependent anion channels) have relatively higher expression than *HPRT1*.

8.5 Taking Advantage of the Existing Transport Mechanisms

The high expression of the copper transporter gene is particularly interesting because a growing body of evidences shows that it facilitates the cellular intake of platinum-derived anticancer drugs such as cisplatin [21–23]. Recent advances in the understanding of how the copper transporter COPT1 works as a biochemical pump to specifically bind copper or platinum atoms outside the cell and facilitates their passage, through an otherwise impermeable cell membrane, into the cytoplasm can be used to develop new platinum-containing cancer drugs [21, 23, 24]. By extension, this knowledge could also be exploited for the development of topical ophthalmic drugs with increased intraocular bioavailability.

8.6 Conclusions

An intracellular excess of reactive oxygen species and lipid oxidation products induces oxidative damage. This toxic process is involved in diverse ophthalmic disorders affecting various regions of the eye, including the retina, the lens, and the cornea. Drugs such as carcinine that inhibit oxidative damage or induce repair mechanisms/detoxification and cannot freely cross cell membranes or epithelial barriers need transport mechanisms. Identifying such ocular transport mechanisms is important to design ophthalmic drugs with optimal bioavailability. Two approaches were used in this study to identify ocular transporters. The first one was to use the multifunctional antioxidant carcinine, which can efficiently penetrate the eye globe and inhibit oxidative damage within retinal cells, to identify its transporter(s). This approach led us to the conclusion that carcinine is not transported by the corneal epithelial cells, which suggests that its penetration takes place in the conjunctiva and not in the cornea. More studies will be necessary to identify the carcinine transporter(s) in ocular tissues. The second approach was to screen the expression profile of a large number of transporter genes in corneal epithelial cells to identify the most abundantly expressed ones. This approach led us to conclude that only a few influx transporters are highly expressed in corneal epithelial cells, including a copper transporter that could be hijacked for drug transport.

Acknowledgements This work was supported by grants HR10-152 from the Oklahoma Center for the Advancement of Science and Technology and R01EY015534 from the National Institute of Health. We would like to thank the Laboratory for Molecular Biology and Cytometry Research at the University of Oklahoma Health Sciences Center for the HPLC-MS analysis of unlabeled carcinine.

References

1. Behar-Cohen F. Drug delivery to the eye: current trends and future perspectives. Ther Deliv. 2012;3:1135–7.
2. Edelhauser HF, Rowe-Rendleman CL, Robinson MR, et al. Ophthalmic drug delivery systems for the treatment of retinal diseases: basic research to clinical applications. Invest Ophthalmol Vis Sci. 2010;51:5403–20.
3. Lambiase A, Tirassa P, Micera A, Aloe L, Bonini S. Pharmacokinetics of conjunctivally applied nerve growth factor in the retina and optic nerve of adult rats. Invest Ophthalmol Vis Sci. 2005;46:3800–6.
4. Williams KA, Brereton HM, Farrall A, et al. Topically applied antibody fragments penetrate into the back of the rabbit eye. Eye (Lond). 2005;19:910–3.
5. Furrer E, Berdugo M, Stella C, et al. Pharmacokinetics and posterior segment biodistribution of ESBA105, an anti-TNF-alpha single-chain antibody, upon topical administration to the rabbit eye. Invest Ophthalmol Vis Sci. 2009;50:771–8.
6. Mitra AK. Role of transporters in ocular drug delivery system. Pharm Res. 2009;26:1192–6.
7. Babizhayev MA, Guiotto A, Kasus-Jacobi A. N-Acetylcarnosine and histidyl-hydrazide are potent agents for multitargeted ophthalmic therapy of senile cataracts and diabetic ocular complications. J Drug Target. 2009;17:36–63.

8. Marchette LD, Wang H, Li F, Babizhayev MA, Kasus-Jacobi A. Carcinine has 4-hydroxynonenal scavenging property and neuroprotective effect in mouse retina. Invest Ophthalmol Vis Sci. 2012;53:3572–83.
9. Kasus-Jacobi A, Marchette LD, Xu C, Li F, Wang H, Babizhayev MA. Mechanisms of RDH12-induced Leber congenital amaurosis and therapeutic approaches. Adv Ophthalmol. 2012;1:473–96.
10. Zhang T, Xiang CD, Gale D, Carreiro S, Wu EY, Zhang EY. Drug transporter and cytochrome P450 mRNA expression in human ocular barriers: implications for ocular drug disposition. Drug Metab Dispos. 2008;36:1300–7.
11. Xiang CD, Batugo M, Gale DC, et al. Characterization of human corneal epithelial cell model as a surrogate for corneal permeability assessment: metabolism and transport. Drug Metab Dispos. 2009;37:992–8.
12. Geissler S, Zwarg M, Knutter I, Markwardt F, Brandsch M. The bioactive dipeptide anserine is transported by human proton-coupled peptide transporters. FEBS J. 2010;277:790–5.
13. Araki-Sasaki K, Ohashi Y, Sasabe T, et al. An SV40-immortalized human corneal epithelial cell line and its characterization. Invest Ophthalmol Vis Sci. 1995;36:614–21.
14. Bonomo RP, D'Alessandro F, Grasso G, et al. Carcinine-beta-cyclodextrin derivatives as scavenger entities of (OH)-O-center dot radicals and SOD-like properties of their copper(II) complexes. Inorg Chim Acta. 2008;361:1705–14.
15. Marchette LD, Thompson DA, Kravtsova M, Ngansop TN, Mandal MN, Kasus-Jacobi A. Retinol dehydrogenase 12 detoxifies 4-hydroxynonenal in photoreceptor cells. Free Radic Biol Med. 2010;48:16–25.
16. Dahlin A, Geier E, Stocker SL, et al. Gene expression profiling of transporters in the solute carrier and ATP-binding cassette superfamilies in human eye substructures. Mol Pharm. 2013;10:650–63.
17. Pereira HA, Ruan X, Gonzalez ML, Tsyshevskaya-Hoover I, Chodosh J. Modulation of corneal epithelial cell functions by the neutrophil-derived inflammatory mediator CAP37. Invest Ophthalmol Vis Sci. 2004;45:4284–92.
18. Griffith GL, Russell RA, Kasus-Jacobi A, et al. CAP37 activation of PKC promotes human corneal epithelial cell chemotaxis. Invest Ophthalmol Vis Sci. 2013;54:6712–23.
19. Holland IB. ABC transporters, mechanisms and biology: an overview. Essays Biochem. 2011;50:1–17.
20. Hediger MA, Clemencon B, Burrier RE, Bruford EA. The ABCs of membrane transporters in health and disease (SLC series): introduction. Mol Aspects Med. 2013;34:95–107.
21. Wang E, Xi Z, Li Y, et al. Interaction between platinum complexes and the C-terminal motif of human copper transporter 1. Inorg Chem. 2013;52:6153–9.
22. Chen HH, Kuo MT. Overcoming platinum drug resistance with copper-lowering agents. Anticancer Res. 2013;33:4157–61.
23. Cai H, Wu JS, Muzik O, Hsieh JT, Lee RJ, Peng F. Reduced 64Cu uptake and tumor growth inhibition by knockdown of human copper transporter 1 in xenograft mouse model of prostate cancer. J Nucl Med. 2014;55:622–8.
24. Tsigelny IF, Sharikov Y, Greenberg JP, et al. An all-atom model of the structure of human copper transporter 1. Cell Biochem Biophys. 2012;63:223–34.
25. Babizhayev MA, Micans P, Guiotto A, Kasus-Jacobi A. N-acetylcarnosine lubricant eyedrops possess all-in-one universal antioxidant protective effects of L-carnosine in aqueous and lipid membrane environments, aldehyde scavenging, and transglycation activities inherent to cataracts: a clinical study of the new vision-saving drug N-acetylcarnosine eyedrop therapy in a database population of over 50,500 patients. Am J Ther. 2009;16:517–33.

Part II
Basic Science in Lens

Chapter 9
Basic Review of the Oxidative Stress Role in Age-Related Cataractogenesis

Lepša Žorić, Dijana Miric, and Bojana Kisic

9.1 Introduction

Age-related cataract is the most common curable cause of blindness in the world. At the same time, it is the most common disease in the clinical ophthalmologic practice. Methods of treatment of cataracts through its extraction and installation of artificial lenses, although highly efficient, are a burden on the budget of each country [1, 2].

Like the aging process in the body, for decades, long accumulated oxidative damage to the lens is considered a key factor in senile cataract. In that process responsible is a balance of oxidative stress and antioxidant protection at the system level and at the local, in the eye. Mortality due to different causes is elevated in the population of elderly people with cataracts, compared to people of the same age group without cataracts, or with other eye diseases [3].

9.2 Lens of the Eye

Lens is a part of the dioptric apparatus of the eye. Thanks to the transparency and elasticity, it allows penetration of light rays to the retina and their focus in the macula. Transparency of the lens results from the optical homogeneity of its elements and depends on their structure, layout, composition, and biochemical processes in them.

L. Žorić (✉)
Ophthalmology Department, University of Pristina, Kosovska Mitrovica, Serbia
e-mail: zoriclepsa@gmail.com

D. Miric • B. Kisic
Institute of Biochemistry, University of Pristina, Kosovska Mitrovica, Serbia

© Springer Science+Business Media New York 2015
M.A. Babizhayev et al. (eds.), *Studies on the Cornea and Lens*,
Oxidative Stress in Applied Basic Research and Clinical Practice,
DOI 10.1007/978-1-4939-1935-2_9

In addition to refractive power, the lens has an important protective role. Through the absorption of ultraviolet part of the spectrum it protects the retina from the harmful effects of this part of the solar radiation. In this way, lens suffers photodynamic damage.

Since it is avascular, its metabolism is dependent on the composition of the aqueous humor, which surrounds it. From aqueous humor the lens receives a necessary amino acid for synthetic processes, antioxidants, and necessary hormones and glucose as an energy source. At the same time, the composition of the aqueous humor reflects the character and intensity of the biochemical processes in the lens. A high lactate level in aqueous humor was primarily a result of the anaerobic metabolism of glucose in the lens. Hydrogen peroxide in aqueous humor may be the cause, but also the consequence of oxidative processes in the lens [4, 5].

The lens is epithelial organ. However, during the life, all its cells, even those incurred during intrauterine development, remain within lens capsule, after transforming into lens fibers. Part of the energy produced in the lens is spent on their maintenance and a part on the slow mitotic cycle, which enables a very slow but steady growth of lens throughout life.

9.3 Oxidative and Photooxidative Stress

Oxygen is the condition of life of aerobic organisms, but the production of free oxygen radicals is its inevitable consequence. Hence, oxygen has a dual nature—it is necessary for life, but it is also toxic. Even normal atmospheric oxygen concentrations have slowly manifesting adverse effects.

Under normal conditions, the balance between the oxidative process and antioxidative capacity maintains stability of the composition and function of living cells. If this balance is disturbed due to increased production of radicals, reduced is the capacity of antioxidant defense system, or both processes at the same time, occurs the state of oxidative stress.

Oxidative stress has been performed physiologically in all of the breathing cells. Free radicals are created during normal cell homeostatic and defense mechanisms. Radicals which are formed during pathological processes are responsible for the damage at almost all biomolecules and manifestations are visible in the form of a number of diseases [6, 7].

The Sun, like oxygen, is a requirement for life on the Earth. Man is a "daily" organism and in his everyday activities follows a diurnal rhythm. Sunlight is necessary for a complete physical and mental development of human. Without sunlight, the development of visual function is not possible.

The human eye is exposed to ambient radiation for decades and a full spectrum of solar light has a wavelength with significant damaging potential. Photooxidative stress strats with the light absorbing organic molecules. Absorption of electromagnetic radiation of any tissue depends on its molecule excitability. Low-energy photons interact with electrons in the organic molecules and they produce a higher energy level in the outer orbitals. Thus the resulting excited state allows the formation of reactive oxygen radicals [5].

It is difficult to determine the ocular dose of ultraviolet light photodynamic damage to the human lens. In fact, it is the cumulative dose effect over a long period, which depends on solar radiation, its reflection from the stone, sand, snow, ozone depletion in the atmosphere, latitude and altitude, occupation [8], and the absorption spectrum of primary lens and its changes over a lifetime, antioxidant capacity of lenses and other endogenous factors in the eye, and the whole organism [9, 10]. It is known that artificial sources of ultraviolet rays also carry some risk of damaging the lens [8, 11].

Along with adaptation to aerobic conditions, during evolution have evolved protective, antioxidant mechanisms against the toxicity caused by free radicals. Antioxidant elements make a vast array of diverse biochemical molecules of endogenous or exogenous origin, which are activated depending on the actual mechanism of oxidative damage, their availability, and properties of the medium [5].

9.4   Cataractogenesis

Age-related cataract is a progressive opacification of the lens in people older than 45 years occurred without any known cause such as trauma, inflammation, hypocalcemia, medications or congenital factors. Cataract affects the quantity and quality of vision in various ways, depending on its type, maturity, or pigmentation.

Mitotic-capable epithelial cells are located on the anterior lens capsule and they, through the pupillary aperture, are the first lens cells facing to the electromagnetic radiation. Absorbing chromophores in the lens may initially be DNA bases and tryptophan. On this occasion arise photoproducts of DNA and subsequent changes in the type and amounts of mRNAs. Cells possess reparative enzymes of resulting damages. Incomplete repair, however, leads to permanent damage of DNA, synthesis of aberrant proteins, mutations and cell death [4, 12, 13]. Such changes are actually registered in human cataract and cataract in animal models caused by ultraviolet radiation [14, 15].

Tryptophan residues in the alpha and beta crystalline lens are easily oxidized in a hydrophilic environment. In a hydrophobic environment they are exposed to hydrogen peroxide and superoxide anion. In proteins and between them, disulfide bonds are formed, as well as with glutathione and cysteine. In this way, aggregations of proteins with altered tertiary structure are formed [14, 16]. In patients with diabetes, people with myopia and other eye diseases that process is more intense [14, 16, 17].

A moderate degree of oxidation makes proteins vulnerable to proteolysis, whereas more extensive and more intense oxidative stress consequences is a formation of aggregates of, which includes a reduction in protease activity also [18, 19]. Such a non-enzymatic oxidation is directly proportional to the concentration of hydrogen peroxide and the oxidized glutathione and reduced glutathione amount inversely.

Human lens and other lens of day organisms contain two groups of chromophores. The first is a product of tryptophan of low molecular weight having an absorption spectrum between 300 and 400 ηm and its presence is detected in the lens even before birth. Another group of pigments is related to lens proteins, appears in the second decade of life and its concentration increases with age, and the absorption spectrum extends to 500 ηm [20, 21]. Lens chromophores have a protective role basically, but those that appear in the cataractogenesis can further promote the oxidative process [22, 23].

Damage of the lipid cell membrane violates one of the conditions of cell integrity. Lipid peroxidation, according to the respective authors, is considered even as an essential causal factor of cataract [24, 25]. At the same time, the structure of membrane proteins gets disturbed [1, 25]. This leads to disruption of barrier function of the cell membrane, and membranes of cell organelles. Subjected to the oxidative modification are, as well, the calcium-adenosine triphosphatase (Ca^{2+}-ATPase), and sodium–potassium adenosine triphosphatase (Na^+-K^+-ATPase) which lead to a change in the level of intracellular calcium.

Formation and accumulation of the aforementioned colored and fluorescent oxidation products of protein intensify further photodynamic damage. Formation of protein aggregates of high molecular weight, which increases the scattering of light, is considered as the substance of the occurrence of nuclear cataracts [4, 5].

Changes in the tertiary structure of proteins, reducing the activity of antioxidant enzymes, loss of function of ion pumps and ion imbalance that, together, lead to changes in water content within the lens and the refractive index of the lens, happen in a formation of cortical cataracts [25].

9.5   The Lens Protection from Oxidation

Peptide glutathione is synthesized in the epithelium and in the cortex of the lenses in high concentrations. During the effect of light wavelength of 302 nm, it is a good cleaner (scavenger) of reactive radicals, while at higher wavelengths glutathione is a donor for glutathione peroxidase [19].

He maintains ascorbate and sulfhydryl groups of crystallines in a reduced state, and to some extent provides work of $Na+$-$K+$-ATP pump [26]. Adding glutathione to the medium in which the cells are cultured, and cataract showed a positive effect in maintaining their opacity [27, 28].

The lens crystallines, although the primary structural proteins of the lens, have a high antioxidant potential. It is primarily derived from its free sulfhydryl groups, but amino acids of a different composition can be targeted to oxidative stress [29].

In the lens there are other enzymatic and non-enzymatic antioxidants: superoxide dismutase, catalase, tioreduksin T, choline, taurine, metalloproteinases, ubiquinone, ascorbate, tocopherol, and carotenoids [26, 30–32]. Their antioxidant activity is dependent on the mechanism of oxidative damage, which is dependent on the polarity of the medium and the type of damage caused.

There are data that suggest that the decline in activity of certain enzymes in up to 95 % does not necessarily mean a significant disadvantage to their functions. It is, of course, only applies to the optimum conditions of functioning of the antioxidative protection [33].

Measuring the concentration of oxygen in the rabbit eye showed a dramatic decrease in its concentration of the cornea to the aqueous humor and, finally, the lens [34]. Pressure of oxygen in the lens is only about 20 % of that in the blood. Low concentration of oxygen in the lens is an important factor for preventing damage of oxidation, since it provides low level of the oxidation stress and eliminates it from the very beginning. Use of high concentration of oxygen for therapeutic purposes can have a negative impact on lens transparency [35]. Lipid peroxidation and changes in hydrocarbon chains are evident in guinea pigs chronically exposed to high concentrations of oxygen [36].

The intensity of the metabolism in the lens cells is slow. The energy production is dominated by anaerobic processes. Mitotic cycle in epithelial cells is rare, synthetic processes are slowed down.

A potent antioxidant, vitamin C, through active transport mechanisms against the concentration gradient, comes to the aqueous humor, and hence in the lens. There reaches a high concentration, up to several tens of times higher than in serum. By comparing the concentration of vitamin C in the humor of eyes of daily and nocturnal animals, it was observed that this feature belongs to daily organisms [37]. It is believed that this is due to adaptation mechanisms of lenses protection against photooxidative stress.

The integrity of cell membranes in the lens substantially preserves vitamin E [38]. However, the composition of membrane lipids of the lens cells reflects an evolutionary adaptive mechanism to reduce oxidative stress and longer life, in general. Camel, which even in old age rarely develops cataract, in their lens, cell membranes have a high content of sphingolipids in relation to the phosphatidylcholines. Similar system has been observed in man, who as compared to other mammals, relatively late in life develop cataract [39].

Hormone of the pineal gland, melatonin and structurally related tryptophan metabolites show antiperoxidation effect and acts as scavenger for highly toxic hydroxyl radicals. In experimental conditions, supplementation with melatonin significantly slows the development of selenium and X-radiation-induced cataracts [40, 41].

9.6   Recommendations for the Prevention of Cataracts

Long-term results of a prospective multicenter North American study for the study of eye diseases in selenium-Age-Related Eye Disease Study Group (AREDS) denied benefit from intake of megadoses of multivitamin preparations and trace elements in well-nourished elderly subjects, as well as any possible damage from them [42]. Authors of other similarly designed early studies-Roche European American

Cataract Trial (REACT) actually performed at approximately the same calendar period [43], however, found a small but statistically significant retardation of cataract in their patients with a long supplementation with vitamin C, tocopherol, and various carotenoids. Investigation of the influence of individual antioxidants does not always show the expected results [44].

Megadosing of liposoluble antioxidant vitamins can have adverse effects, and a caution is necessary in their application [4, 43]. Most moderate recommendations relate to the increased intake of vitamins and trace elements important in antioxidant processes in food [27], reduction of oxidative stress through changes in lifestyle protect your eyes from ultraviolet rays. Eye protective lenses are recommended whenever the sun comes out and they need to completely block UV rays and protect from blue light. Corrective lenses and Intraocular Implants (IOL) are produced with filters. Wide brim hats and caps are protecting to some extent. Protection at work and in sports is equally, if not more important.

9.7   Conclusion

Recent studies indicate that cataract is not only a local eye disease which operational methods successfully solve in a very high percentage. The occurrence of senile cataracts, especially some of its forms, can point to the existence of systemic pathological processes in the body.

Advanced age is the main risk factor identified with the appearance of age-related cataracts. Individual, family, gender, and racial diversity in its expression also suggest that there are several cataractogenesis causes. Identifying risk factors that may be moderated offers hope for the possibility of preventing cataract. In this sense, the most promising is better understanding of components of oxidative stress and antioxidant defense.

Today, the mainly known and still valid recommendations for the prevention of cataract are related mostly to general advice on lifestyle protection from ultraviolet rays and avoiding harmful habits.

References

1. Babizhayev MA, Deyev AI, Yermakova VN, Brikman IV, Bours J. Lipid peroxidation and cataracts: N-acetylcarnosine as a therapeutic toll to manage age-related cataracts in human and canine eyes. Drugs RD. 2004;5(3):125–39.
2. Latinović S, Čanadanović V, Babić N, Lješević LJ, Grković D, Žikić Z, Babović S, et al. Istraživanje kvaliteta života uslovljenog funkcijom vida: studija katarakte. Med Pregl. 2007;60(11–12):653–6.
3. Bernardino CR. AMD, cataracts and mortality. Arch Ophthalmol. 2007;125:917–24.
4. Spector A. Oxidative stress-induced cataract: mechanism of action. FASEB J. 1995; 9(12):1173–82.

5. Žorić L, Jovanović P. Oxidation stress in eye diseases pathogenesis (In Serbian). Niš: SVEN; 2006.
6. Čolak E. New markers of oxidative damage to macromolecules. J Med Biochem. 2008;27:1–16.
7. Mayne ST. Antioxidant nutrients and chronic disease: use of biomarkers of exposure and oxidative stress status in epidemiological research. J Nutr. 2003;133:933s–40.
8. Žorić L, Stojčić M. The influence of ultraviolet radiation on eye. Primary health care: open access.   http://www.omicsgroup.org/journals/the-influence-of-ultraviolet-radiation-on-eye-2167-1079.1000133.pdf
9. Delcourt C, Carriere I, Ponton-Sanchez A, Lacroux A, Covacho MJ, Papoz L. Light exposure and the risk of cortical, nuclear, and posterior subcapsular cataracts: the Pathologies Oculaires Liees a l'Age (POLA) study. Arch Ophthalmol. 2000;118(3):385–92.
10. Zoric L, Miric D, Novakovic T, Pavlovic A, Videnovic G, Trajkovic G. Age-related cataract and serum albumin concentration. Curr Eye Res. 2008;33(7):587–90.
11. Costagliola C, Balestrieri P, Fioretti F, Fumizio C, Rinaldi M. ArF193nm excimer laser corneal surgery as a possible risk factor in cataractogenesis. Exp Eye Res. 1994;58(4):453–73.
12. Spector A, Kliman N, Wang RR. Ultraviolet light induced DNA damage and repair in bovine lens epithelial cells. Invest Ophthalmol Vis Sci (Suppl). 1990;31:430–6.
13. Taylor HR. The biological effects of UV-B on the eye. Photochem Photobiol. 1989;50:489–95.
14. Boscia F, Grattagliano I, Vendemiale G, Micelli-Ferrari T, Altomare E. Protein oxidation and lens opacity in humans. Invest Ophthalmol Vis Sci. 2000;41(9):2461–5.
15. Zigler JS, Huang QL, Du XY. Oxidative modification of lens crystallyns by H_2O_2 and chelated iron. Free Rad Biol Med. 1989;7:499–505.
16. Žorić L. Some oxidation stress parameters in lens, aqueous humor and serum of patients with diabetes and age-related cataract. (In Serbian). Srp Arh Celok Lek. 2003;131(3–4):137–42.
17. Zoric L, Miric D, Milenkovic S, Jovanovic P, Trajkovic G. Pseudo exfoliation syndrome and its antioxidative protection deficiency as risk factors for age-related cataract. Eur J Ophthalmol. 2006;16(2):268–73.
18. Grosswiner LJ. Photochemistry of proteins: a review. Curr Eye Res. 1984;3:137–42.
19. Pau H, Graf P, Sies H. Glutathione levels in human lens: regional distribution in different forms of cataract. Exp Eye Res. 1990;50(1):17–20.
20. Roberts JE, Finley EL, Patat SA, Schey KL. Photooxidation of lens proteins with xanthurenic acid: a putative chromophore for cataractogenesis. Photochem Photobiol. 2001;74(5):740–4.
21. Cheng R, Lin B, Ortwerth BJ. Separation of the yellow chromophores in individual brunescent cataracts. Exp Eye Res. 2003;77(3):313–25.
22. Korlimbinis A, Hains PG, Truscott RJ, Aquilina JA. 3-Hydroxykynurenine oxidizes alpha-crystallyns: potential role in cataractogenesis. Biochemistry. 2006;45(6):1852–60.
23. Žorić L, Aleksić P, Koraćević D, Trajković G. Antioksidacioni kapacitet očne vodice kod različitih vrsta i pigmentacije senilne katarakte. Vojnosanit Pregl. 2005;62(12):909–15.
24. Mibu H, Nagata M, Hikida M. A study on lipid peroxide-induced lens damage in vitro. Exp Eye Res. 1994;58:85–90.
25. Transcott RJ. Age-related nuclear cataract-oxidation is the key. Exp Eye Res. 2005;80(5):709–25.
26. Lou MF. Thiol regulation in the lens. J Ocul Pharmacol Ther. 2000;16(2):137–48.
27. Bonnefoy M, Drai J, Kostka T. Antioxidants to slow aging, facts, and perspectives. Presse Med. 2002;32(25):1174–84.
28. Cronstein BN. Oxidative insults: sublethal injury to the endothelium by H_2O_2. J Lab Clin Med. 1991;117(1):6–7.
29. Chen SJ, Sun TX, Akhtar NJ, Liang JJ. Oxidation of human lens recombinant alpha A-crystallin and cysteine-deficient mutants. J Mol Biol. 2001;305(4):969–76.
30. John M, Jaworski C, Chen Z, Subramanian S, Ma W, Sun F, et al. Matrix metalloproteinases are down-regulated in rat lenses exposed to oxidative stress. Exp Eye Res. 2004; 79(6):839–46.

31. Fecondo JV, Augusteyn RC. Superoxide dismutase, catalase and glutathione peroxidase in the human cataractous lens. Exp Eye Res. 1983;36:15–8.
32. Cendella RJ, Neely AR, Sexton P. Concentration and distribution of ubiquinone (conezime Q), the endogenous lipid antioxidant, in the rat lens: effect of treatment with simvastatin. Mol Vis. 2005;11:594–602.
33. Goosey JD, Tuan WM, Garcia CA. A lipid peroxidative mechanism for posterior subcapsular cataract formation in rabbit: a possible model for cataract formation in tapetoretinal diseases. Invest Ophthalmol Vis Sci. 1984;25(2):608–12.
34. Shui YB, Fu JJ, Garcia C, Dattilo LK, Rajagopal R, McMillan S, et al. Oxygen distribution in the rabbit eye and oxygen consumption by the lens. Invest Ophthalmol Vis Sci. 2006;47(4):1571–80.
35. Palmquist D. Nuclear cataract and myopia during hyperbaric oxygen therapy. Br J Ophthalmol. 1984;68:113–6.
36. Borchman D, Giblin FJ, Leverenz VR, et al. Impact of aging and hyperbaric oxygen in vivo on guinea pig lens lipids and nuclear light scatter. Invest Ophthalmol Vis Sci. 2000; 41(10):3061–73.
37. Reiss GR, Werness PG, Zollman PE, Brubaker RF. Ascorbic acid levels in the aqueous humour of nocturnal and diurnal mammals. Arch Ophthalmol. 1986;104:753–5.
38. Ayala MN, Soderberg PG. Vitamin E can protect against ultraviolet radiation-induced cataract in albino rats. Ophthalmic Res. 2004;36(5):264–9.
39. Borchman D, Yappert MC, Afzal M. Lens lipids and maximum lifespan. Exp Eye Res. 2004; 79(6):761–8.
40. Yagci R, Aydin B, Erdurmus M, Karadag R, Gurel A, Durmus M, et al. Use of melatonin to prevent selenite-induced cataract formation in rat eyes. Curr Eye Res. 2006;31(10):845–50.
41. Karslioglu I, Ertekin MV, Taysi S, Kocer I, Sezen O, Gepdiremen A, et al. Radioprotective effect of melatonin on radiation-induced cataract. J Radiat Res. 2005;46(2):277–82.
42. Age-Related Eye Disease Study Group. A randomized, placebo-controlled clinical trial of high-dose supplementation with vitamins C, and E, and beta carotene for age-related cataract and vision loss: AREDS report no. 9. Arch Ophthalmol. 2001;119(10):1439–52.
43. Schalch W, Chylack LT. Antioxidant micronutrients and cataract: review and comparison of the AREDS and REACT cataract studies. Ophthalmologe. 2003;100(3):181–9.
44. Barcley L, Nghiem HT. Vitamin E may not reduce risk for age-related cataracts. http://www.medscape.com, May 2008.

Chapter 10
The Human Lens: A Living Biometric Indicator of Health Status and Successful Aging

Stuart Richer, William Stiles, Pejman Zargar, Mahsa Rezaei, Theresa Vo, Richard Bone, and Bill Sardi

10.1 Introduction

The human lens is the approximate size of a 325-g aspirin tablet. It is a slow metabolic tissue formed from the ectodermal embryonic layer, which develops into a biconvex, crystalline, and transparent structure [1]. With the help of the cornea, the lens refracts and focuses visible light upon the retina where it is converted to neuronal signals transmitted to the suprachiasmic nucleus and pineal gland (circadian rhythm), midbrain (pupillary function), the occipital cortex for vision, and the frontal cortex for eye movement coordination [2]. For image formation, the lens alters the focal distance of the eye (i.e., myopia, hyperopia, astigmatism), assisting in the formation of sharp images of perceived objects upon the retina. For near

S. Richer (✉)
Ocular Preventive Medicine, Captain James A. Lovell Federal Health Care Center, Chicago, IL, USA
e-mail: Stuart.Richer1@VA.Gov

W. Stiles
Ophthalmology, Captain James A. Lovell Federal Health Care Center, Chicago, IL, USA

P. Zargar • M. Rezaei
Chicago Medical School, Rosalind Franklin University of Medicine and Science, Chicago, IL, USA

T. Vo
Optometry, Captain James A. Lovell Federal Health Care Center, Chicago, IL, USA

R. Bone
Department of Physics, Florida International University, Miami, FL, USA

B. Sardi
Longevinex® Associates, Las Vegas, NV, USA

© Springer Science+Business Media New York 2015
M.A. Babizhayev et al. (eds.), *Studies on the Cornea and Lens*,
Oxidative Stress in Applied Basic Research and Clinical Practice,
DOI 10.1007/978-1-4939-1935-2_10

vision, an adjustment known as accommodation is modulated neuro-muscularly, via the ciliary body's release of tension upon lens fibril attachments. This process is degraded with age (presbyopia) and occurs earlier in smokers, diabetics, and those taking certain medications [1, 2]. Accommodation of the more rigid and oxidized, glycated, and cross-linked lens protein crystallins eventually results in aggregation, scattering centers, vision degradation, and ultimately cataract formation. Cataract extraction is the single largest component of the United States Medicare budget at 3.4 billion dollars and growing [3, 5]. Age-related cataracts (ARCs) must be differentially discriminated from congenital, ocular, systemic, and environmental assaults.

The human lens consists of three main compartments from external to internal: lens capsule, lens epithelium, and lens fibers (crystallins) [4]. The lens epithelium is composed of simple cuboidal epithelium cells and its core function is to maintain lens homeostasis [5]. The lens epithelium is also the source of precursors for both the lens capsule and lens fibers [4].

Cataract is classified as a protein aggregation disease in which the lens becomes cloudy and decreases visual acuity [6]. Cataract can occur in one or both eyes and it cannot spread from one eye to the other [7].

Life expectancy increased by the advancement of public hygiene and medical care has resulted in a higher prevalence of age-related disease, notably ARC, age-related macular degeneration (AMD), and glaucoma. Age-related cataract appears above 50 years of age. More than half of Americans by age 80 either suffer from ARC or have already had cataract surgery in one or both eyes. Cataracts are mostly seen in elderly people and aging cataract is most common. Yet, not all cataracts are age-related. Other subtypes include congenital, medical, traumatic, surgical (secondary cataracts), and those resulting from insults such as trauma, chemicals, or radiation [7] (see Table 10.1).

There are three types of aging cataracts: nuclear, cortical, and posterior subcapsular. Nuclear cataract, the most common type, is a central lens opacification where the nucleus loses its transparency, changing from clear to yellow and sometimes brown. This phenomenon occurs due to deamidation of lens proteins through oxidation, proteolysis, and glycation, which accumulates into high molecular-weight particles that scatter incoming light. Cortical cataract, the second most common type is associated with liquefaction of lens fibers due to electrolyte imbalance, overhydration, and excess carbohydrate intake. Finally, posterior subcapsular cataract, the third and the least common type, is typically formed by the death of lens fiber nuclei and epithelial cells clustering at the lens posterior pole. Cataract development has been attributed to metabolic dysglycemia, retinal diseases, alcoholism, drugs, malnutrition, and smoking [5]. Most common symptoms of cataract include: cloudy or blurry vision, poor night vision, dimmed color perception, perceived glare and halos from headlights, lamps, or sunlight [7] (see Table 10.2 and Fig. 10.1).

Table 10.1 Not all cataracts are related to age
Reprinted from [8] with permission from Dr. S. Richer

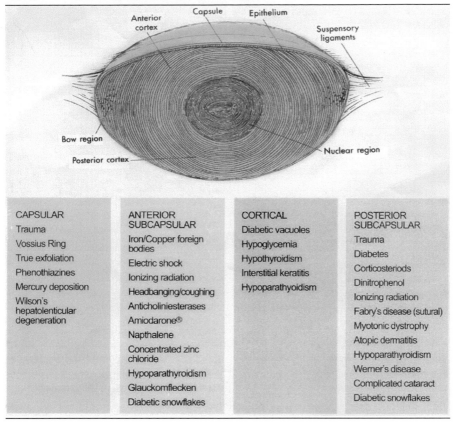

CAPSULAR	ANTERIOR SUBCAPSULAR	CORTICAL	POSTERIOR SUBCAPSULAR
Trauma	Iron/Copper foreign bodies	Diabetic vacuoles	Trauma
Vossius Ring	Electric shock	Hypoglycemia	Diabetes
True exfoliation	Ionizing radiation	Hypothyroidism	Corticosteriods
Phenothiazines	Headbanging/coughing	Interstitial keratitis	Dinitrophenol
Mercury deposition	Anticholiniesterases	Hypoparathyoidism	Ionizing radiation
Wilson's hepatolenticular degeneration	Amiodarone®		Fabry's disease (sutural)
	Napthalene		Myotonic dystrophy
	Concentrated zinc chloride		Atopic dermatitis
	Hypoparathyroidism		Hypoparathyroidism
	Glauckomflecken		Werner's disease
	Diabetic snowflakes		Complicated cataract
			Diabetic snowflakes

Table 10.2 Biochemistry of ARC (Age Related Cataract)

Damage to crystalline proteins/ proteinases with aging in older lenses
Oxidation
Glycation (nonenzymatic glycosylation)
Glyco-oxidation (oxidation of glycated proteins)
Addition of lipids
Cleavage/deamidation/racemization
Accumulation of ubiquitinated proteins
Inactivation of proteinases
Aggregation of lens crystallins (cytotoxicity)
Tissue dehydration/precipitation

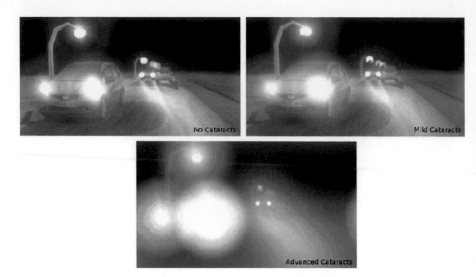

Fig. 10.1 Simulation of a cataract-affected view and its progression over time. Reprinted from [65] with permission from the Association of Computing Machinery, Inc.

10.2 Sunlight Protection

Sunlight consists of electromagnetic radiation, from infrared to visible light to ultraviolet. Eyes are protected by eyelid and facial structures such as eyebrows, cheekbones, and the nose. Yet eyes are still prone to high-energy ultraviolet (UV) radiation referred to as UVA, UVB, and UVC. UVB is blocked by the cornea and UVC, the most energetic with the shortest wavelength, is almost completely blocked by the ozone layer, UVA has the longest wavelength in the UV spectrum and penetrates deep within the skin and eye. Traversing the cornea, it can induce DNA damage of the anterior lens epithelial cells. Unprotected, exposure to such radiation damages the lens at all ages with infants and young children being particularly vulnerable. UVA damages human DNA indirectly by generating highly reactive chemical intermediates, such as hydroxyl and oxygen radicals [9].

Ecological studies suggest that cataract is more prevalent in countries with more sunlight exposure than in those with less sunlight exposure [10]. There are a higher number of cataract cases reported in countries like Israel and India compared with England [11]. An ecological study called *the Beaver Dam Eye Study* confirmed the association between UVB radiation exposure and cataracts in men only [12]. Another epidemiological study done by H. R. Taylor showed a significant association for cortical cataracts and UV radiation index. In this study, 838 watermen working on the Chesapeake Bay were surveyed. Taylor's UV radiation index was based upon number of outdoor hours worked, work location, and attenuation due to spectacle use and hat cover. The study found an association between cortical cataracts and higher UV exposure and suggested that children require more protection against sunlight, specifically UV radiation since they generally spend more time outdoors than adults [13].

10.3 Caloric Restriction

Lifestyle is an important factor in ARC. Many studies evaluated caloric restriction and its benefit(s) to human health. Caloric and dietary restriction is defined as a nutritious diet with 40 % fewer calories compared with an ad libitum diet [14]. Dietary caloric restriction has been shown to delay cataractogenesis. Wang et al. concluded that caloric restriction slowed protein insolubilization and blunted the decline of the total soluble thiols, protein thiols, reduced glutathione, and ascorbic acid levels in the lenses of old Brown Norway rats. In terms of lens protein aggregation, this study described cataract development in three stages: (1) the precipitation of gamma-crystallin, (2) the insolubilization of beta-crystallin, and (3) the final precipitation of alpha-crystallin which was saturated with other denatured lens proteins. A major risk factor in pathogenesis of human cataract is oxidative stress. Caloric restriction is ostensibly beneficial in retarding oxidative stress in the human lens via up-regulation of endogenous antioxidant pathways [15].

10.3.1 Diabetes Mellitus

Chiu et al. explored the association of carbohydrate intake and glycemic index in early cortical and nuclear lens opacity formation. Some 417 female subjects of 53–73 years of age from the Boston's Nurses' Health Study provided dietary information via 5 semi-quantitative food frequency questionnaires over a 14-year period. Lens opacities were evaluated using the Lens Opacity Classification System III (LOCS III). The authors concluded that there was a positive association between early cortical lens opacities and the quantity (but not the quality) of carbohydrate intake in middle-aged women. They did not find any significant association between quantity or quality of dietary carbohydrate intake and nuclear lens opacities in middle-aged women [16].

Significantly, cataract is up to five times more prevalent among patients suffering from diabetes mellitus (DM). Diabetics develop all three subtypes (cortical, posterior subcapsular, and nuclear). Risk factors are duration of diabetes and quality of glycemic control. Early insulin treatment has been shown to delay cataractogenesis in patients with both type 1 and type 2 diabetes.

DM is associated with several other diseases, such as heart and vascular complications, neuropathy, retinopathy and dyslipidemia. Hyperglycemia leads to a number of pathogenic mechanisms, including but not limited to increased aldose reductase (AR) activity, nonenzymatic glycation/glycooxidation, activation of protein kinase C, oxidative–nitrosative stress, and poly (ADP-ribose) polymerase (PARP) activation. The lens, in comparison to other tissues of the human body, is avascular and therefore dependent on anaerobic glucose metabolism pathways, such as glycolysis and the pentose phosphate pathway. The lens receives nutrients and oxygen through the blood-aqueous barrier as it takes up glucose in an insulin-independent pathway. In diabetic patients, excess blood glucose enters the sorbitol pathway, in which AR catalyzes NADPH-dependent reduction of glucose to sorbitol, and sorbitol dehydrogenase (SDH) catalyzes NAD-dependent oxidation of sorbitol to fructose. By increasing sorbitol pathway activity, intracellular sorbitol accumulates and results in osmotic stress. Increased AR activity contributes to oxidative–nitrosative stress, nonenzymatic glycation/glycoxidation, mitogen-activated protein kinase (MAPK) and cyclooxygenase-2 activation, accumulation of cytosolic Ca^{2+}, activation of NF-kappaB, and activator protein-1.

Although many mechanisms of diabetic cataractogenesis have been identified, there are no specific methods for prevention or management of diabetic cataracts apart from glucose control. Lifestyle changes, such as carbohydrate restriction, medication, smoking cessation, systemic and intraocular blood pressure management and nutrient supplementation, are all beneficial [17].

10.4   Plant Food: A Source of Nourishing Antioxidants and Cofactor Minerals/Vitamins

Diet-associated ARC is a crucial topic. Plant food containing antioxidants and cofactors fighting against free radicals that cause oxidative stress, are important to stay healthy and mitigate age-related processes. Exogenous and gene derived endogenous antioxidants protect the human body against aging and oxidative free radical assault. Free radicals, antioxidants and cofactors are the main focus of many human aging and supplementation studies. Due to new technologies and improvements in medicine, human life expectancy is increasing at a fast pace. This will not only raise the overall population, there will also be a larger older population in need of health and medical care. Therefore, having a balanced lifestyle in addition to a healthy diet with adequate nutrition is required for all of us.

Free radicals are paramagnetic molecules with unstable chemical configurations. Most free radicals are formed as a byproduct of carbohydrate, protein, and lipid metabolism. For example, Oxygen-free radicals (e.g., hydroxyl super-oxide) along with hydrogen peroxide are known as reactive oxygen species (ROS). Free radicals of oxygen and nitrogen - NOS are extremely and NOS reactive and react with macromolecules and nucleic acids. The amount of ROS being produced in the human body increases with age. This phenomenon in part explains age-associated increase in inflammation and myriad diseases such as neurodegenerative, cardiovascular, cataracts, DM, and cancer.

Antioxidants are any endogenous or exogenous molecules capable of stabilizing free radicals before they cause cellular damage. This protective system, particularly with endogenous antioxidants, decreases in power as humans age. Therefore, there is a greater need for the aged population to monitor their diet or take supplements of exogenous antioxidants. The most effective enzymatic antioxidants are glutathione peroxidase, catalase, and superoxide dismutase. Nonenzymatic antioxidants include vitamins E and C, thiol antioxidants (glutathione, thioredoxin, and lipoic acid), melatonin, carotenoids, and natural flavonoids.

Cofactors are molecules or ions that bind to the catalytic site of enzymes, accelerating biological reactions. Many antioxidant enzymes require metal and nonmetal cofactors such as copper (Cu), iron (Fe), selenium (Se), zinc (Zn), and vitamins B1 (thiamine), and B2 (riboflavin). Cofactors are essential for antioxidant functions, assisting in protection against oxidative damage by free radicals. Cofactor deficiency as well as unregulated and toxic levels of divalent metals can cause health complications, including cataract. Optimal cofactor intake is required for each individual to maintain optimal health.

Copper is a vital cofactor in many critical cellular reactions, including those involving cytochrome C oxidase and zinc-superoxide dismutase. Iron is an essential component of catalase enzymes for hemoglobin and myoglobin. As well as copper, excess iron and copper also has oxidative qualities, which, in contact with lipids, could cause oxidative stress. It has been reported that patients with high levels of iron and lipids are at higher risk of developing cancer. Selenium is another essential cofactor in multiform of glutathione peroxidases, which have been found to protect DNA and cellular oxidative damage. Selenium deficiency has been linked to DNA damage, neurodegenerative disease, cardiovascular diseases, cancer, and cataract. Zinc is another cofactor essential for antioxidant protection against vascular and immunological damage. Zinc has also been found to protect DNA and decrease the risk of cancer. It has also been shown to have a protective function for having a healthy skin, which is important in healthy aging. Copper and zinc are critical cofactors in bone matrix ingredients and synthesis, and the lack of these cofactors has been shown to be correlated with elderly osteoporosis and bone density loss. Vitamins such as thiamine (vitamin B1) and riboflavin (vitamin B2) are also considered as cofactors protecting against oxidative stress. The active form of vitamin B1, thiamine diphosphate, is used in important enzymes of oxidant stress defense. Vitamin B1 deficiency has been linked to neurodegenerative diseases and high levels of oxidative damage. The active form of vitamin B2, flavin dinucleotide, works as a co-enzyme for glutathione reductase and many other antioxidant enzymatic reactions. Evidence shows that vitamin B2 deficiency has been linked to cardiovascular complications and increased risk of developing cancer [18].

In terms of ocular tissue, Bruce H. Grahn et al. investigated the association between zinc and the eye. The authors found a significant concentration of zinc in the eye of most animal species, particularly the retina, retinal pigment epithelium, and choroid. Zinc has several known functions in the retina and retinal pigmented epithelium, such as altering synaptic transmission, regulating the rhodopsin reaction, modulating interactions between taurine and vitamin A, modifying photoreceptor plasma membranes, and acting as an antioxidant. Zinc deficiency can manifest in ocular complications such as altered vision, electroretinograms, oscillatory potentials, and, in severe cases, ultra-structural changes in the retina and retinal pigmented epithelium. The progression of degenerative retinal diseases are reduced by zinc supplements [19].

In another article, *Cataracts as an outcome of zinc deficiency in salmon*, the authors investigated zinc deficiency-related complications due to high intake of calcium, phosphorus, and phytate. This article concluded that a secondary deficiency of zinc has been associated with a high incidence of cataract in young salmon [20].

David L. Watts has explored the significance of deficiency and toxic levels of selenium in human and animal diets. The review paper, *the Nutritional Relationships of Selenium*, stated that selenium level constantly increased in the human lens from birth to old age. However, it also stated that lenses suffering from cataracts had sixfold less selenium content compared to that in healthy lenses in the same age group. It found that remarkably low selenium levels in the lens increased oxidative damage due to free radical species [21].

Endogenous antioxidants have recently been considered in cataract prevention, especially so for patients with frank and severe vitamin deficiency. Studies suggest that lens epithelial cell nuclear factor erythroid-2-related factor 2 (Nrf2) is an activator of endogenous antioxidants, preventing cataract. Notably, resveratrol (in red wine) and allicin (in garlic) are potent activators of Nrf2 [22].

From the oxidant stress side, homocysteine (related to hyperhomcystenuria, inborn errors of metabolism and dietary deficiency of abnormal methylation of B6, folic acid, and B12) within the endoplasmic reticulum (ER) was found to suppress NrF2, ostensibly resulting in cataract [23]. ER stress resulting from exposure of lens epithelial cell to homocysteine within fetal calf serum, resulted in significant degradation of Nrf2, within 24 h. The increased production of ROS decreased GSH reductase and catalase. Homocysteine exposure thus results in ER stress that suppresses Nrf2 dependent cellular antioxidant protection with subsequent death of lens epithelial cells. Nrf2 loss may also result from proteasomal degradation and increased cytoplasmic calcium. Thus high homocysteine concentrations induce ER stress, altering ROS generation, crystalline protein folding, and calcium balance. ROS regulation may help improve the lens epithelial cell Nrf2 pathway, reducing the rate of cataract formation [24].

10.4.1 Lutein and Zeaxanthin

There exist 600 known carotenoids in nature, with lutein and zeaxanthin as the two important dietary ocular xanthophylls found in both the lens and macula of the retina. Not surprisingly, the physiological concentration and function of these two nutrients are important in the prevention and management of both AMD and ARC. Lutein and zeaxanthin protect the eye from high energy blue light and possess powerful antioxidant activity [25].

Lutein and zeaxanthin are not endogenous and not synthesized by the human body. Therefore, there is great need for obtaining them through diet from dark leafy vegetables, such as spinach, kale, and broccoli, oranges, corn, and yellow and orange peppers [26].

The concentration of lutein and zeaxanthin in the macula of the retina can be quantified with instruments that measure macular pigment optical density (MPOD), which is a good indicator of visual function, eye health, and even overall systemic health [25]. The biochemistry of these two antioxidants is very similar. Zeaxanthin is a stereoisomer of lutein and it has a very similar mechanism of action to lutein. In the lens, they work by limiting lipid peroxidation, the primary cause for ocular and cardiovascular diseases.

Numerous studies illustrate the positive correlation between dietary supplements of lutein and zeaxanthin and reduced risk of AMD and ARC. A study by Seddon et al. with 356 subjects showed that the group who consumed 6 mg daily lutein and zeaxanthin had a 57 % decreased risk for developing AMD compared with those intaking 0.5 mg daily. Bernstein et al. showed that 4 mg daily supplementation of lutein increased macular pigment density, which is crucial for maintaining visual acuity and lowering risk of developing AMD [26].

The National Cancer Institute and U.S. Department of Agriculture recommend 5–6 mg daily lutein and zeaxanthin, which equals 5 servings of fruits and vegetables every day [25] (see Table 10.3).

10.4.2   Alcohol and Smoking

Smoking and alcoholism both have long-term destructive effects on human health, particularly in age-related diseases. The Blue Mountains Eye study investigated the effects of smoking and alcohol consumption on human lenses. The article, "Alcohol, Smoking, and Cataracts" by Robert G. Cumming et al. investigated the relationship between alcohol consumption, tobacco smoking, and cataractogenesis in a population-based cross-sectional study with 654 subjects. The authors concluded that smoking was associated with a higher rate of nuclear and posterior subcapsular cataract formation. Also, subjects abusing alcohol and smoking tobacco together showed a higher frequency of nuclear cataracts [27].

Another Blue Mountains study, "Alcohol Consumption and the Long-Term Incidence of Cataract and Cataract Surgery" by Gowri L. Kanthan et al. was a population-based prospective cohort study with 3,654 subjects aged 49 years and older. Interestingly, a U-shaped correlation between alcohol consumption and the long-term risk of cataract surgery was established. Subjects consuming over two standard drinks per day and subjects who abstained from alcohol consumption had a significantly higher prevalence of cataract surgery. In comparison, subjects consuming a moderate amount of alcohol (1 or less standard drink per day) had a 50 % lower incidence of cataract surgery [28].

Risk of smoking was investigated in the "Smoking and Risk of Age-Related Cataract: A Meta-Analysis." The study concluded an association between long-term smoking and developing ARC, particularly nuclear cataracts. However, the results and the underlying physiological mechanisms require more investigation, although it has been established that smoke contains toxins as well as depressing serum ascorbic acid, a water soluble antioxidant found in high concentrations in all ocular tissues [29].

Table 10.3 Nutrition, sources, roles and deficiencies in the United States

Nutrient	Main sources	Systemic role	Ocular role	RDA 19–50	>50	% deficient in RDA intake
B$_1$	Meat, fish, leafy vegetables, legumes, peas, whole grains	– Food and energy metabolism – Required for nervous system and muscle function	Cofactor for antioxidant enzymatic reactions	M: 1.2 mg/day F: 1.1 mg/day	M: 1.2 mg/day F: 1.1 mg/day	18.4 % (57 million)
B$_2$	Meat, milk, eggs, green leafy vegetables, legumes, nuts	Body growth, red blood cell formation, releasing energy from carbohydrates	Cofactor for antioxidant enzymatic reactions	M: 1.3 mg/day F: 1.1 mg/day	M: 1.3 mg/day F: 1.1 mg/day	10.9 % (34 million)
C	Cantaloupe, citrus fruits, kiwi fruit, mango, papaya, pineapple, broccoli, brussels sprouts, spinach	Growth and repair of tissues Repair and maintain cartilage, bones, and teeth immune system activity	Located in aqueous compartments of lens membranes as an antioxidant and protect lens proteases from photooxidative destruction	M: 90 mg/day F: 75 mg/day	M: 90 mg/day F: 75 mg/day	42.0 % (131 million)
E	Vegetable oils, nuts, seeds, green leafy vegetables	– Cellular health	Lipid soluble antioxidant concentrated in lens fibers, inhibit cataract formation by reducing photooperoxidation of lens lipids and stabilizing lens cell membranes	Adults: 15 mg/ day	Adults: 15 mg/ day	86.4 % (271 million)

		– Immune system – Assist fertility				
Cu	Oysters other shellfish, whole grains, beans, nuts, potatoes, dark leafy greens	Red blood cell formation, health of blood vessels, nerves system, immune system, and bones	Cofactor for antioxidant enzymatic reactions	Adults: 900 µg/day	Adults: 900 µg/day	15.8 % (50 million)
Fe	Meat, oyster, fish whole grains, beans, broccoli, spinach	Structure oxygen-carrying proteins such as hemoglobin and myoglobin	(1) Role in cornea: completion of the citric acid cycle, production of ATP as well as an essential component of the rate-limiting enzyme in DNA synthesis	M: 8 mg/day F: 18 mg/day	M: 8 mg/day F: 8 mg/day	10.5 % (33 million)
Se	Meat, poultry, fish, eggs, grains	DNA repair, apoptosis, and the endocrine and immune system activity	Cofactor of glutathione peroxidases, an antioxidant enzymatic reactions	Adults: 55 µg/day	Adults: 55 µg/day	8.5 % (26 million)
Zn	Beef, pork, lamb, fish, nuts, whole grains, legumes, and yeast	Cell division, cell growth, wound healing, breakdown of carbohydrates, and immune system activity	Altering synaptic transmission, regulation rhodopsin reaction, modifier of photoreceptor plasma membranes, and antioxidant	M: 11 mg/day	M: 11 mg/day	29.2 % (91 million)

(continued)

Table 10.3 (continued)

Nutrient	Main sources	Systemic role	Ocular role	RDA		% deficient in RDA intake
				19–50	>50	
				F: 8 mg/day	F: 8 mg/day	
Lutein zeaxanthin	Lutein: spinach, kale, broccoli; zeaxanthin: orange, corn, and yellow/orange peppers	Prevent/limit lipid peroxidation	Increase macular pigmentation density	Adults: 5.8 mg/day	Adults: 5.8 mg/day	Not Determined
			Biological antioxidant			

• (number of Americans based on population of 314 million)

Sources:

• DHHS. NIH. National Library of Medicine

• USDA NAL: United States Department of Agriculture, National Agriculture Library

• (1) Iron homeostasis and eye disease, Biochim Biophys Acta. 2009 July; 1790(7): 637–649. Published online 2008 November 14. doi: 10.1016/j. bbagen.2008.11.001

10.5   Systemic Diseases

10.5.1   Cardiovascular Complications

Several studies explored the association of aging cataractogenesis and systemic diseases such as cardiovascular complications. As mentioned, DM and high carbohydrate intake are recognized risk factors for ARC. However, hypertension and cardiovascular diseases are other systemic issues that have been investigated as well. Another Blue Mountains Study assessed the associations between DM, selected cardiovascular risk factors, and long-term incidence of cataract and cataract surgery. Three thousand six hundred and fifty-four Australians were initially selected for a 10-year longitudinal study. Age, gender, fasting glucose, blood pressure, and a few other factors were investigated separately. Results showed that baseline DM was associated with nuclear cataracts and impaired fasting glucose was associated with cortical cataracts. Subjects with higher body mass index (BMI) presented with posterior subcapsular cataract. Also, subjects taking anti-hypertensive medications had a higher incidence of cataract surgery with few associations also found between cardiovascular risk factors and long-term cataract development [30].

   Free radicals can cause oxidative damage to lipoproteins and increase the risk of atherosclerosis. In this regard, Frank B. Hu et al. hypothesized that cataracts from free radical oxidative stress could be a marker for atherosclerosis, particularly coronary heart diseases. During a 10-year study on women aged 45–63 years with no known coronary disease, stroke, or cancer, the authors confirmed that cataract extraction was significantly associated with higher risk of coronary heart diseases and overall mortality [31].

10.5.2   Mortality

ARC and visual impairment are associated with reduced life expectancy and quality of life. A survey of visual impairment and blindness between 1996 and 2000 recruited 10,293 individuals from urban and rural clusters in Andhra Pradesh, India. Participants were followed in an effort to determine ocular risks factors associated with visual disability, cataract, and mortality. From this cohort study, 799 of 4,188 participants with visual problems had died (19.1 %) and 308 participants had migrated. Mortality in males was greater than in females a decade later. The study indicated that any type of cataract or history of cataract surgery increased the mortality rate. It further suggests that cataract is a marker of aging [32].

   Another study comparing mortality risk in cataract patients, by degree of opacification, is instructive. One thousand eight hundred and sixty-four patients aged 64 years and older undergoing phacoemulsification surgery were followed for 5 years postoperatively. Of 901 patients with moderate and severe visual impairment before cataract surgery, 60.4 % showed no visual complaints, 15.5 % had mild problems, and

24.1 % of the patients remained with severe visual impairment 1 month postoperatively. While age-standardized mortality rates were not significantly lower in patients with mild or no visual impairment, correcting moderate and severe visual impairment in elderly patients with phacoemulsification surgery resulted in statistically significant lower mortality compared to that of un-operated patients [33].

The Blue Mountain Eye study cohort evaluated the relationship between cataract surgery and quality of life. In this study, 354 patients aged 49 years and above with visually significant cataract, underwent surgery and were examined 5 and 10 years later. Patients who underwent surgery had better vision and enjoyed significantly lower long-term mortality risk, than individuals with untreated cataract(s). Thus cataract surgery induced visual improvement associated with long-term survival. Other visual impairments were similarly associated with increased mortality [34].

However another study illustrates *increased mortality rate* among cataract surgery patients and cohorts following phacoemulsification. Data were collected for 933 consecutive patients in 2006 who underwent surgery between 2000 and 2001. The data were later negatively correlated with regional and standard mortality risk. This study is contrary to similar studies as cataract surgery is universally associated with improved survival and quality of life indices. Nonetheless, this illustrates the importance of continuous research and re-evaluation of medical knowledge in light of changes in population demographics and practice [35].

In summary, there is an increased mortality rate among patients with visual impairment in general and ARC specifically. Scientific evidence suggests that cataract is associated with deterioration of health status, as well as quality of life. Patients suffering from cataract and other visual impairments must seek medical care. Cataract surgery has the potential to permanently cure cataract related visual disability resulting in an improved quality of life.

10.5.3   Osteoporosis

Calcium is one of the essential minerals found abundantly in the human body. Calcium imbalance, particularly in the elderly, causes diverse physiological issues involving neurological, muscular, and skeletal tissues. Calcium assists electrolyte homeostasis in terms of cellular transport, renal function, enzymatic activation, and cellular division and communication. Hypercalcemia and hypocalcemia are related to symptoms such as nausea, vomiting, muscle weakness, sudden twitches, memory loss, depression as well as paresthesia in hands and feet, anxiety, seizure, and arrhythmias, respectively [36].

Osteoporosis, a complex multifactorial disorder, is a silent progressive bone disease more prevalent in females over 50 years of age. Calcium metabolism is an important factor in bone turnover, bone mass, and density. An observational case-control study from Israel found a significant association between cataract and osteoporosis among women of all age groups and men older than 75 years of age.

Serum calcium is maintained in a normal range via pituitary hormone secretion affecting bone resorption, gut calcium absorption, and renal calcium excretion. When serum calcium levels are low, bone resorption is the first line of compensation. When calcium resorption occurs at a faster rate than its deposition, bone mass and its density decrease. This is a simple explanation of osteoporosis explaining why serum calcium remains normal in most patients with osteoporosis [37].

Calcium homeostasis is an important factor in lens clarity. Increases in lens calcium have long been associated with cataract increasing on average 200–300 % compared to that of clear lenses. On the other hand, decreased serum calcium levels also result in lens swelling and opacification [38].

Calcium imbalance affects lens metabolism in terms of activation or inhibition of ion channels, electrolyte homeostasis, protein synthesis, protease activities, membrane permeability and finally aggregation of lipids and proteins along with actual calcium deposition in the lens which results in loss of lens clarity [37].

10.6   Postmenopausal Hormone Use: A Protective Effect Against Cataracts?

Research has broken new ground revealing the possibility of postmenopausal hormone use in having a protective role against cataracts. The meta-analysis research, though not definitive, led by Kairan Lai analyzed four cohorts and five case controls that are related to postmenopausal hormone therapy and cataract development. The Salisbury Eye Evaluation, the Framingham study, and Beaver Dam Eye studies were included in the analysis. The findings revealed a reduction in the risk of cataract associated with postmenopausal hormone use regardless of cataract type. In the findings, there was an association between hormone replacement therapy (HRT) use and nuclear cataract in case-control and cross-sectional studies. No association was found between cortical or posterior subcapsular cataracts in women who had use HRT at any point. Current users of HRT were shown to have reduced risk of any type of cataracts, but this cannot be proved in the past in the relation with HRT, the reason being that concentration and exposure time could be the cause of immunity in the current users of the estrogen. Another reason being posited is the antioxidant properties that preserve mitochondrial function and cell viability thus protecting the lens from being damaged by oxidative damage. Limitations of the study included the use of a self-reported questionnaire, the study design, and varying definitions, and for this reason a follow-up study was recommended. This gives reassurance to HRT users to continue therapy without fear of cataract development as suggested by prior research [39].

10.7 Age-Related Cataract Parallels Age-Related and Drug-Induced Nutrient Depletion

Normally, caloric intake decreases as one ages and therefore meticulous attention to nutrition and nutritional supplements become paramount. Micronutrients are essential traces of vitamins, antioxidants, and minerals, such as iron, calcium, vitamins B, C, and E, that are vital for growth and healthy aging. A large portion of the geriatric population in the United States and many other countries suffer from micronutrient deficiency. Therefore, intake of nutrient-rich food, in addition to taking daily multivitamin and mineral supplements, is required for vulnerable populations.

Different factors increasing the risk of malnutrition in the elderly include but are not limited to: decreased physical activity, reduced energy expenditure, poor appetite, and physical changes, such as menopause, various digestive system diseases, such as *H. pylori* infection and irritable bowel syndromes, and drug-induced malabsorption. Gastric pH is an important factor for absorbing nutrients in the intestine. More elderly people tend to suffer from hypochlorhydria and achlorhydria, which correlates with vitamin B_{12} and iron deficiency. Gastric hypochlorhydria or achlorhydria can occur as a result of certain clinical disorders or surgical procedures that cause reduced parietal cell mass or by unbalanced hormone and peptide action. They can also occur iatrogenically due to long-term administration of antacids and H_2 blockers or continuous administration of Na/K ATPase inhibiting drugs.

Gastric acid secretion decreases with aging [40]. Prescribed and over-the-counter medications amplify this phenomenon. Proton pump inhibitors (PPIs), such as Nexium® and H_2-receptor antagonists such as anti-histamines, are commonly prescribed medications used for dyspepsia, peptic ulcer disease, gastroesophageal reflux disease, and many other gastric acid complications. Side effects of long-term use of such medications include headache, nausea, diarrhea, abdominal pain, fatigue, and dizziness. Long-term adverse effects include hypomagnesemia, vitamin B_{12} deficiency, and iron, calcium, and zinc deficiency. Long-term use of PPIs and H_2-receptor antagonists could lead to chronic gastric complications, pneumonia, enteric infections, anemia, increased risk of bone fracture, and ARC [41].

10.8 Network Antioxidants, Not High Dose Antioxidants

Nutrition is an important factor in preventing a variety of pathogenic and age-related diseases. It has been suggested that proper nutrition could decelerate aging and aging-related diseases. Many investigations have studied the association between high dose vitamins C, E, and various physiological and age-related diseases, such as ARC and AMD. One study showed that taking high dose vitamins C and E had no appreciable effect on preventing ARC or AMD [42]. Another study stated that taking antioxidants as a network is more efficacious in preventing oxidative damage [43].

As ARC results from oxidative damage via free radical damage, scientists have focused more on dietary intake of antioxidants. Vitamin C (ascorbic acid) is located in the aqueous compartments of lens membranes, where it may function as an antioxidant and protect lens proteases from photooxidative destruction. Vitamin E is a lipid-soluble antioxidant concentrated in lens fibers and membranes and may inhibit cataract formation by reducing photoperoxidation of lens lipids and stabilizing lens cell membranes [42]. Recommended dietary allowance (RDA) of vitamin C for male and female adults are 90 and 75 mg, respectively. RDA for vitamin E for both male and female adults is 15 mg or 22.4 IU [44].

As a general thought, one would suppose that taking higher doses of vitamins and antioxidants would have a strong advantage on preventing disease and aging-related oxidative damage. However, many large scale long-term randomized controlled trials have disproven this hypothesis [45]. In fact, two population-based prospective cohort studies on men and women concluded that taking high doses of vitamin C (1,000 mg/day) and E (100 mg/day) supplements, but not low dose, was associated with higher risks of developing ARC [46]. A randomized, triple masked, placebo controlled, field based 5-year clinical trial from South India explored the effect of antioxidant supplements such as beta-carotene and vitamins C and E on slowing ARC progression. Results were negative, showing that there was no significant difference between groups taking the supplements versus placebo [47].

Dr. Lester Packer is one of the primary science researchers investigating the association of antioxidants and aging and aging-related diseases. His book, *The Antioxidant Miracle,* explores antioxidant activities, advantages, and side effects. He introduces the concept of an antioxidant network, in which antioxidants do not work separately or independently from each other, but rather interact with one another in complex ways. Among hundreds of antioxidants he refers to, only five antioxidants form this network: vitamins C and E, glutathione, lipoic acid, and coenzyme Q10 (CoQ10). The book's central thesis is that each antioxidant, through its normal course of activity, will become a pro-oxidant after neutralizing destructive free radical species. This being said, these new pro-oxidants are not as dangerous, but they still require elimination from the body [48].

10.9   Seminal Importance of the Glutathione and Ascorbate Redox State

Glutathione (GSH), the most predominant low molecular weight thiol in human cells, is a tripeptide containing the amino-acids cysteine, glycine, and glutamate. The sulfhydryl (R-SH, thiol) group of cysteine residues allows this molecule to be the most important endogenous intracellular antioxidant. Eighty-five to ninety percent of cellular GSH (0.5–10 mmol/L) exists in the cytosol and the remainder are

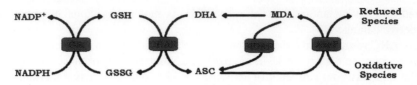

Fig. 10.2 Schematic of the glutathione–ascorbate cycle. *GR* glutathione reductase, *DHAR* dehydroascorbate reductase, *MDAR* mono-dehydroascorbate reductase, *AscP* ascorbate peroxidase

in organelles such as the mitochondria and nuclear matrix. Extra-cellular GSH can be found in concentrations as low as 2–20 μmol/L in plasma, with the exception of bile acid, which could have up to 10 mmol/L of GSH. Due to the thiol cysteine residue, GSH is capable of reducing free radicals and oxidative species to less dangerous compounds, becoming glutathione disulfide (GSSG) in the process. GSSG leaves the cell after being converted to its oxidized status and this contributes to intracellular loss of reducing power (see Fig. 10.2).

The GSH to GSSG concentration ratio is an important indicator of the cellular redox state and is associated with mortality risk. Under the healthy physiological state, GSH is about ten times more common than GSSG. This ratio reflects the oxidative and anti-oxidative status of cells. GSH production takes place in almost all cells of the human body through two pathways controlled by the cytosolic enzymes, gamma-glutamylcysteine synthetase (GCS), and GSH synthetase. The liver is the major producer and exporter for this natural antioxidant in the body [49]. However, GSH forms in the same manner within anterior lens epithelium and researchers have discovered that cataractous lenses have as much as 81 % less GSH concentration compared to that of clear lenses. There is no solid answer yet on whether this phenomenon is caused by a decrease in GSH production or defective GSSG recycling, or if there is an increase in GSH degradation in cataractogenesis. Researchers have also not ruled out the probability of direct loss of GSH from the lens by metabolic assault [50].

GSH can be found naturally in meat, eggs, milk, asparagus, watermelon, avocado, grapefruit, garlic, and onion. Precursor thiols can be taken in supplement form (i.e., *N*-acetyl cysteine).

10.9.1   The Glutathione-Ascorbate Cycle

GSH and vitamin C are both strong antioxidants. Ascorbate (ASC) reduces free radical species via ascorbate peroxidase (AscP) and converts to its oxidative state, mono-dehydroascorbate (MDA). MDA in turn is a radical itself and requires rapid reduction back to ascorbate or dehydroascorbate. MDA is reduced to ascorbic acid via mono-dehydroascorbate reductase (MDAR) by oxidizing NADH to NAD⁻. GSH is in charge of reducing dehydroascorbate to ascorbic acid via dehydroascorbate reductase and becomes GSSG in the process. GSSG is then recycled back to its reduced state via glutathione reductase using NADPH as an electron donor [51] (see Fig. 10.2).

10.10   Vision Degradation in Cataract: Clinicians Can Do Better

Given the lens's close association with healthy aging, as well as with the advent of pharmaceuticals and nutraceuticals to slow down and reverse ARC, new opportunities for vision assessment are now materializing. Metrics that accurately describe and record the structure and function of the lens are desperately needed. Reliance upon Snellen visual acuity in the twenty first century is questionable. In addition to patients' history, ocular examinations could include visual testing well beyond the simple high-contrast Snellen visual acuity and visual fields that are often not degraded, or provide a misleading representation of the extent of a patient's visual disability [52]. Such auxiliary tests can be divided into two categories: (1) tests that evaluate vision at different contrasts and (2) tests that evaluate the effect of light scatter, or disability glare. Disability glare is due to scattered intraocular strays of light that disturb normal vision and reduce contrast of the image formed on the retina [53]. Specifically, this would include a measure of the contrast sensitivity function (e.g., Stereo Optical Functional Vision Analyzer™) [54] or a measure of visual acuity at low contrast values of 10–25 % using inexpensive specially designed reading cards or modern optical electronic projection acuity devices [55]. The issue with measuring contrast degradation in ARC is that contrast sensitivity is a nonspecific metric that can also be degraded by AMD, glaucoma, and neurodegenerative disease such as Alzheimer's disease or Parkinson's.

There are various other office devices for evaluating lens transparency of the aging eye, such as GALILEI™, iTRACE™, and Pentacam™. GALILEI™ utilizes a dual revolving Scheimpflug camera that takes several images to display a three-dimensional analysis of the anterior chamber of the eye and topography of the cornea and the lens [56]. iTRACE™ takes advantage of ray tracing technology for analyzing corneal aberrations and visual function of the eye by employing binocular open field analysis of refraction that provide an accurate and complete wavefront map. This instrument also analyzes corneal diagnostic indices and calculates advanced intraocular lens parameters [57]. Pentacam® evaluates the anterior and posterior corneal surfaces, which provides true total refractive power of the cornea [58].

10.11   The Pentacam® PNS

The advantages of Pentacam® are the 5-function biometric capability of imaging the cornea, anterior chamber, and human lens in various planes and describing presurgical optical distortions resulting from keratoconus, prior refractive surgery, and ARC. The Pentacam® provides precise geometric information on structure, elevation, curvature, pachymetry (corneal thickness), depth of the anterior eye chamber, and derived topography of front and back corneal surface aberration.

The Pentacam® is a device consisting of a slit illumination system and a Scheimpflug camera capable of constructing three-dimensional images of the anterior segment of the eye by rotating and taking images in various planes. One important advantage of Pentacam® over other imaging devices is that its two special cameras compensate for eye movement. Also, its slit flash light is capable of transversing not fully transparent ocular tissue such as the lens. Illuminated biologic tissue scatters light, allowing for sectional pseudo-anatomic images to be captured [58]. The Pentacam® mathematically reconstructs the tomographic structure of the cornea and distance between anterior segment tissues and the lens. Recorded images allow for serial evaluation over time. With respect to the lens, the Scheimpflug camera takes several images by shining bright flashes through a dilated pupil, to measure density and depth. The Pentacam® is able to measure a cataract's location, density, thickness, and consistency. Another advantage of this device is its ability to evaluate anterior cortical and subcapsular cataract as well as posterior capsular opacification. It utilizes a precise densitometer that allows physicians to measure lens density variation [59]. Technicians in optometry or ophthalmology offices operate the Pentacam® and take only a few minutes to digitally image the human lens, with immediate densitometry data displayed.

Pentacam® densitometry is a novel clinical method for in office quantification of nuclear cataracts. Dr. Michael Belin, professor of ophthalmology at the University of Arizona demonstrated various clinical uses of the Pentacam® for corneal disease, refractive surgery, glaucoma, and cataract management. Regarding cataracts, it could be utilized for lens densitometry, post refractive intraocular lens calculation, and limbal relaxing incision planning. Pentacam® examination can be performed in normal and dilated pupils to evaluate nuclear density, or lens densitometry. This is a feature that scans various sections of the eye to provide a three-dimensional structure of the lens. This image can be used to quantify the density of cataracts and phacoemulsification settings in cataract surgery [60] (see Fig. 10.3).

10.12   The NASA Fiber Optic Imaging Device Using Dynamic Light Scattering (DLS)

There are many optical imaging techniques available currently for grading cataract, such as clinical Scheimpflug or slit-lamp photographic grading and various retro-illumination photographic devices. These methods could detect the cataracts graded equal or higher than 2 based on LOCS III [61]. Based on the American Optometric Association, a grade 2 cataract is already at a moderate stage [5]. Diagnosis of cataract in the early stages, before its progression to grade 2 or further, could be helpful for subclinical diagnosis and implementing early treatment [61].

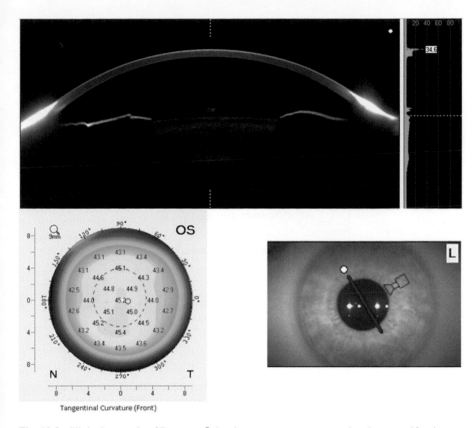

Fig. 10.3 Clinical example of Pentacam® densitrometry to measure nuclear lens opacification

Dr. Ansari et al. invented the fiber optic imaging probe using Dynamic Light Scattering (DLS) that can noninvasively detect and quantify the cataract in such early stages. This device can detect cataract two to three orders earlier in magnitude. The DLS system is also sensitive enough for early detection of vitreous modalities, corneal structures, and cataracts. In clinical settings, the volume of projected light on the lens is small and limited, and any minor deviation in opacification could alter the results significantly [62].

This device consists of a digital camera, automated fiber optic probe, a detector, and a digital correlator. By combining a corneal analyzer (Optikon 2000 Keratron, Italy) and DLS fiber optic imaging probe, a new clinical device for early detection of cataracts was created. This modified version of a corneal analyzer uses the power of DLS system to overcome issues related to projecting scattering light on the lens. An important feature of this instrument is the alignment program. It uses two fiber optic position sensors; one sending an infrared beam of light and the other collecting. This feature allows for infrared light instead of lasers for aligning the cameras. Once the infrared light passes through the cornea, the device is aligned with the corneal vertex and no more infrared light passes through the eye. Therefore, this feature

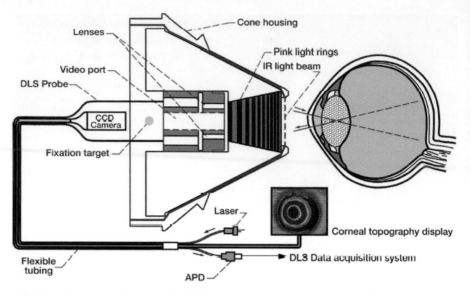

Fig. 10.4 Schematic diagram of the DLS optical system. Reprinted from [63] with permission from Dr. Manuel B. Datiles III

makes the system safe to evaluate human ocular tissue. As the instrument achieves its desired position, the DLS system begins projecting light for 2 or 5 s into the eye [63]. By the end of the procedure, the system saves and displays the DLS reflections and corneal map data (see Figs. 10.3 and 10.4).

Dr. Datiles et al. took advantage of the DLS method for clinical detection of pre-cataractous lens at the molecular level. They investigated changes of α-crystallins, which are molecular chaperones that prevent uncontrolled aggregation and unfolding of lens' proteins, keeping the lens clear from opacification. As age increases, there are less α-crystallin proteins available inside the lens, causing damaged proteins to aggregate and decrease transparency. Therefore, measuring the α-crystallin proteins in human lenses could be a significant reflection of protective reserve.

For this purpose, the DLS fiber optic imaging probe was placed inside a keratoscope (Keratron; Optikon 2000 SpA, Rome, Italy) with a three-dimensional aiming system. This instrument projects DLS light for 5 s into the eye and collects the scattered beams as random proteins inside the lens reflect light. Scattered light intensity is evaluated through mathematical measurements and time-autocorrelation show the amount of light scattered by α-crystallins and other aggregated proteins. Aggregated proteins have higher molecular weight than α-crystallin proteins and will scatter the light differently [61]. With DLS and previously mentioned devices, cataracts can be detected early and noninvasively (see Fig. 10.4).

10.13   CATRA: Cataract Maps from a Snap-on Eyepiece for Mobile Phones

Current diagnostic methods for cataract are based on back-scattering techniques, such as slit-lamp examination, requiring highly trained professionals or costly techniques like optical coherence tomography. Camera Culture Group at MIT developed CATRA, a compact self-evaluation device to detect and quantify human cataracts in third world countries using ordinary "smart phones." A modified parallax barrier is used to scan the crystalline lens and compute maps for opacity, attenuation, contrast, and point spread function (PSF). CATRA combines optics and mathematical knowledge with computer graphics to address an important vision problem through a user-based interactive technique.

CATRA has no moving parts and is built from low-cost components, which lets developing countries take advantage of this device. Cataracts scatter incoming light toward the retina and modify the eye's PSF. As the cataract progresses, the PSF loses its characteristic peak and the perceived image on the retina will be more blurry. CATRA generates collimated beams of light that go through the lens in order to project patterns on the fovea. The user sees these patterns and presses a few buttons to map the attenuation, opacity, contrast and other intrinsic parameters. This device is designed with a pair of stacked liquid crystal displays (LCDs), which constructs a programmable parallax barrier. Opacity and attenuation map are attained with a simpler design through a fixed pinhole on top an LCD. Existing cataract examination devices rely on back-scatter profiles but CATRA evaluates forward-scattering outlines to measure lens PSF with no coherent light sources or mechanical techniques. More importantly, CATRA solved the gazing issue by projecting a collimated beam of light directly onto the fovea.

In order to assess lens opacities, a fast scan is utilized as users perceive slight changes of brightness of a centered dot. Attenuation maps are attained by asking the user to match the brightness of beams that pass through various affected segments of lens against the reference point. CATRA by measuring the levels required to recognize the rotation of a pattern of each projected collimated path creates the contrast map. The PSF of each sub-aperture is attained by projecting a parametric pattern through a clear path that is matched with those naturally occluded paths [64] (see Table 10.4).

10.14   Computer-Aided Diagnosis System for Nuclear Cataract and Quantification via Slit-Lamp Images

Currently, ophthalmologists, optometrists, or trained technicians grade cataract via LOCS III or Oxford Clinical Cataracts Classification and Grading System (OCCCGS) [65]. In order to quantify the severity of cataracts, specialists compare slit-lamp examination findings with a set of standard photographs with increasing

Table 10.4 Comparison of CATRA technique against current available technologies and research tools

	Slit-lamp + visual acuity tests	Scheimpflug photography	Retro-illumination techniques	Shack–Hartmann wavefront aberrometer	Optical coherence tomography	CATRA
Scattering	Backward	Backward	Forward	Forward	0	Forward
Training	High	High	Medium	Low	High	Low
Data log	No	Computer	Computer	Computer	Computer	Cellphone
Mobility	>1 kg	>10 kg	>10 kg	>10 kg	>10 kg	<300 g
Speed	Slow	Slow	Fast	Fast	Medium	Fast
Scalability	Hard	Hard	Hard	Hard	Hard	Easy
AC need	Yes	Yes	Yes	Yes	Yes	No
Networked	No	No	No	No	No	Yes
Self-diagnose	No	No	No	No	No	Yes
Method	Subjective	Subjective	Objective	Objective	Objective	Subjective
Early cataracts	Hard to find	Hard to find	Hard to find	Yes	Yes	Yes
Accuracy	Personal skills	Personal skills	Personal skills	High	High	High
Cost	$5,000.00	$20,000.00	$5,000.00	$15,000.00	$10,000.00	$2.00

Reprinted from [65] with permission from the Association of Computing Machinery, Inc.

nuclear sclerosis, cortical opacification, and posterior subcapsular opacification. In this age, diagnosis of cataract and precise quantification is necessary to take place more efficiently. The current medical quantification is subjective, time consuming, and requires a large amount of training [66]. More importantly, each person grades cataract differently even with the standard grading system in place. For grading nuclear sclerosis, the interobserver agreement was found to be around 65 % and the intragrader agreement was between 70 and 80 % [66].

In order to have a consistent system that neither requires highly trained technicians nor being time consuming, a computer-aided diagnosis imaging program would be welcome. This novel system must be as precise and acceptable as educated practitioners' and trained technicians' quantifications. In this regard, the Computer-Aided Diagnosing System (CADS) for grading nuclear cataracts was introduced by Dr. Huiqi Li and his colleagues. Dr. Li is a Senior Research Fellow at the Institute for Infocomm Research, Agency for Science, Technology and Research in Singapore.

Via CADS, nuclear cataracts are automatically detected and graded through slit-lamp images. In the history of ophthalmology, there have been various attempts for automatic quantification of nuclear cataracts, such as depiction of anatomical structures on visual axes. However, due to low signal to noise ratio (SNR), these studies were unsuccessful. CADS for grading nuclear cataracts, the most common type, includes three fundamental components of structure detection, feature extraction, and grade prediction. The most important module is the anatomical structure detection factor, which must be highly accurate.

In this regard, bottom-up processing and top-down processing have been utilized for precise anatomical lens structure detection. Bottom-up strategy is used to locate lens position and top-down processing is used to hypothesize the results. Ultimately, lens structure detection takes place by a top-down strategy through a local searching method of Active Shape Model (ASM). ASM assists with corneal detection, horizontal and vertical profile clustering, and lens estimation through five steps of initialization: matching point detection, transformation between shape space and image space, model update, and convergence evaluation (see Fig. 10.5).

The next component is feature extraction, which is entirely based on detected anatomical lens structures by the prior level. Local features are extracted according to standard clinical protocols and identified landmarks. For the purpose of automatic grading, the Support Vector Machine (SVM) regression method is applied. Through various mathematical equations, the SVM regression method analyzes data provided by the last two components and grades the nuclear cataracts on the slit-lamp images. This system has been validated using over 5,000 clinical images. The success rate of structure detection has been shown to be 95 % and the average grading difference is 0.36 on a 5.0 scale.

The user interface of CADS is simple and user-friendly. The main menu consists of three features for loading images, feature extraction, and predicting cataract grades. In just a few seconds after uploading the slit-lamp image into the program, the nuclear cataract is quantified and its grade appears in a small box on the main page.

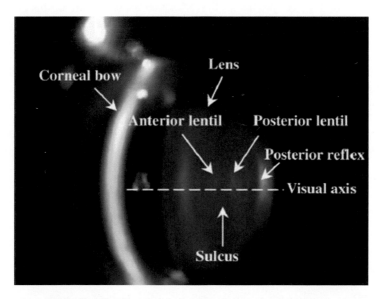

Fig. 10.5 Anatomical structure detection on a slit lamp image via Computer Aided Diagnosis System. Reprinted from [67] with permission from IEEE

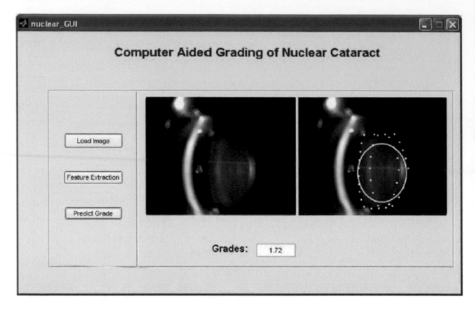

Fig. 10.6 User interface of the Computer Aided Diagnosis of Nuclear Cataract Software. Reprinted from [67] with permission from IEEE

Operation of this software does not require any specific educational level or past experience with similar equipment. This allows the system to be easily approachable in any ophthalmology or optometry clinic. Another useful feature of this program is the user intervention function. This feature allows users to fine-tune images that require correction before being graded. Special cases such as inaccurate focus, small pupil, or upper eyelid obscuration can be fixed prior to utilizing the grading program. This system is also able to process large numbers of images sequentially as it includes a batch processing mode [67]. Therefore, slit-lamp images could be uploaded and graded significantly faster than grading nuclear cataracts in an office setting.

The CADS for automatic grading of nuclear cataracts could improve the quality of health care by being more objective and consistent toward diagnosis. This system is based on ophthalmology standard protocols and saves time and workload for ophthalmologists, optometrists, and technicians. The user can utilize this system to grade a large number of slit lamp images in less time (see Figs. 10.5 and 10.6).

10.15 mapcat*SF*™: Dual Macular Pigment and Lens Opacification as a Broad Measure of Systemic Health

mapcat*SF*™ is a noninvasive, non-mydriatic optical instrument with a unique feature of simultaneously providing both lens and macular pigment optical densities. The instrument output includes the patient's "lens equivalent age" and the percentage

Fig. 10.7 mapcat*SF*™ measures macular pigment optical density and lens opacification. Reprinted from [69] with permission from Dr. Karen Thomas

of blue light blocked by the patient's macular pigment. The instrument operates on the principle of heterochromatic flicker photometry (HFP), generating visual stimuli that alternate rapidly between blue and green wavelengths. The patient adjusts the relative intensities of the two wavelengths to minimize perception of flicker. A large stimulus provides a measurement that is dominated by optical filtering by the lens, and a small stimulus provides a measurement that is, in addition, dependent on filtering by the macular pigment. The instrument is designed for ease of operation, particularly by elderly patients. The problem of Troxler fading of peripherally viewed stimuli experienced in traditional HFP is eliminated (see Fig. 10.7).

10.16   ClearPath DS-120: Measurement of Lens Autofluorescence Can Distinguish Subjects with DM from Those Without

Lens autofluorescence increases with age, which is consistent with the physiological mechanism of accumulation of advanced glycation end-products (AGEs). Lens autofluorescence has also been shown to increase with DM, duration of DM, and blood glucose levels, and has been shown to predict diagnosis of diabetes mellitus. This has been linked with pathology that accompanies DM, such as diabetic retinopathy. Interestingly, free reducing sugars initiate glycation and therefore diabetics usually show accelerated accumulation of AGEs. Thus, they are at higher risk for developing ocular, vascular, and other complications of DM compared to those without DM.

ClearPath DS-120 detects lens autofluorescence noninvasively in 6 s. With immediate results and no required blood draw, it is possible to diagnose prediabetic patients earlier than with traditional methods. ClearPath works by projecting a focused beam of blue light on the lens and measuring the autofluorescence of the

Fig. 10.8 ClearPath DS-120
measures lens
autofluorescence for early
detection of diabetes mellitus.
Reprinted from Freedom
Meditech website with
permission from Dr. Mark
Miehle (http://www.
freedom-meditech.com/)

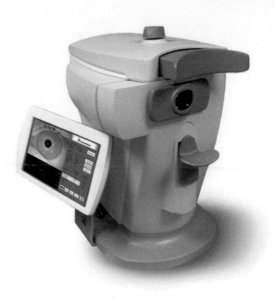

reflected light. The fluorescence ratio of the patient's lens is displayed on a touch
screen tablet and based on the ratio percentiles, further DM testing may be recommended. The results of the ClearPath evaluation can be recorded electronically and
printed for further analysis by referring physicians. A comparison of the lens autofluorescence with other screening methods in terms of specificity and sensitivity for
diagnosing DM has been promising. Lens autofluorescence has a specificity and
sensitivity of 94 and 67 %, while fasting blood glucose has 95 and 60 % and hemoglobin A1C has 79 and 49 %, respectively [68] (see Fig. 10.8).

10.17   Conclusion

The human lens is a biomarker of successful aging. Nuclear cataracts are a proxy
measure of the glutathione redox state as modulated by ascorbic acid and supporting micronutrient status. It is related to the general health of the individual. Cortical
cataracts are related to excessive sugar and carbohydrate intake, glycemic status,
and risk of diabetes. Posterior subcapsular cataract often reflects local retinal and
systemic metabolic pero-oxidative disturbances. Meaningful application of new
lens imaging technologies affords the opportunity for the clinician to quantify the
extent of lens opacification. Serial measurement of cataract with simple to highly
evolved computational digital optoelectronic instrumentation opens up the possibility for providing individualized therapeutic intervention. This is especially important for our most vulnerable citizens. These patients include smokers, alcoholics,
polypharmacy users, the elderly and malnourished populations who are not consuming or assimilating adequate plant food (vegetables and fruit). The Physicians

Health Study and AREDS II cataract sub-analysis provide evidence that these vulnerable populations, with low plant food intake, may benefit from dietary counseling and simple supplemental repletion of dietary antioxidants and supporting cofactor minerals and vitamins.

Acknowledgement   This chapter is based on original clinical work supported by the Optometry/ Ophthalmology sections of Captain James A Lovell Federal Health Care Facility, DVA-Naval Medical Center, North Chicago, IL, USA.

References

1. Duncan G. The aging human lens: structure, growth, and physiological behaviour. Br J Ophthalmol. 1997;81(10):818–23. http://bjo.bmj.com/content/81/10/818.full.
2. Brown N. The change in shape and internal form of the lens of the eye on accommodation. National Center for Biotechnology Information. U.S. National Library of Medicine. http://www.ncbi.nlm.nih.gov/pubmed/4702379
3. Optometric Clinical Practice Guideline. Care of the adult patient with presbyopia. American Optometric Association. http://www.aoa.org/documents/CPG-17.pdf
4. Stafford MJ. The histology and biology of the lens. Association of Optometrists—London. 2001.   http://www.optometry.co.uk/uploads/articles/0b3e55d71662f4e8381aea8637c48f4f_stafford20010112.pdf
5. Optometric Clinical Practice Guideline. Care of the adult patient with cataract. American Optometric Association. http://www.aoa.org/documents/CPG-8.pdf
6. Benedek GB. Cataract as a protein condensation disease. Association for Research in Vision and Ophthalmology. http://www.iovs.org/content/38/10/1911.full.pdf
7. Facts about cataract. National Eye Institute Health Information. http://www.nei.nih.gov/health/cataract/cataract_facts.asp#1a
8. Richer, et al. Etiology of non-congenital non-age related cataract. Optometry. 2001;72:767–78.
9. Svobodová AR, Galandáková A, Sianská J, Doležal D, Lichnovská R, Ulrichová J, Vostálová J. DNA damage after acute exposure of mice skin to physiological doses of UVB and UVA Light. National Center for Biotechnology Information. U.S. National Library of Medicine, July 2012. https://www.ncbi.nlm.nih.gov/pubmed/22271212
10. Mohan M, Sperduto RD, Angra SK, Milton RC, Mathur RL. India-US case-control study of age-related cataracts. India-US Case-Control Study Group. National Center for Biotechnology Information. U.S. National Library of Medicine, May 1989. Web. 2013. http://www.ncbi.nlm.nih.gov/pubmed/2818712
11. Hodge WG, Whitcher JP, Satariano W. Risk factors for age-related cataracts. National Center for Biotechnology Information. U.S. National Library of Medicine. Epidemiol Rev. 1995;17(2):336–46. http://www.ncbi.nlm.nih.gov/pubmed/8654515
12. Cruickshanks KJ, Klein BE, Klein R. Ultraviolet light exposure and lens opacities: The Beaver Dam Eye Study. National Center for Biotechnology Information. U.S. National Library of Medicine. http://www.ncbi.nlm.nih.gov/pmc/articles/PMC1694542
13. Taylor HR. Ultraviolet light and the eye: an epidemiologic study. National Center for Biotechnology Information. U.S. National Library of Medicine. http://www.ncbi.nlm.nih.gov/pmc/articles/PMC1298564/pdf/taos00012-0827.pdf
14. Anekonda TS. The benefits of calorie restriction and calorie restriction mimetics as related to the eye. National Center for Biotechnology Information. U.S. National Library of Medicine. http://www.ncbi.nlm.nih.gov/pmc/articles/PMC2939013
15. Wang K, Li D, Sun F. Dietary caloric restriction may delay the development of cataract by attenuating the oxidative stress in the lenses of brown Norway rats. National Center for

Biotechnology Information. U.S. National Library of Medicine. Jan–Feb 2004. http://www.ncbi.nlm.nih.gov/pubmed/14667836

16. Chiu C-J, Milton RC, Gensler G, Taylor A. Carbohydrate intake and glycemic index in relation to the odds of early cortical and nuclear lens opacities. The American Journal of Clinical Nutrition. http://ajcn.nutrition.org/content/81/6/1411.full

17. Obrosova IG, Chung SSM, Kador PF. Diabetic cataracts: mechanisms and management. National Center for Biotechnology Information. U.S. National Library of Medicine. Diabetes Metab Res Rev. 2010. http://www.ncbi.nlm.nih.gov/pubmed/20474067

18. Rahman K. Studies on free radicals, antioxidants, and co-factors. National Center for Biotechnology Information. U.S. National Library of Medicine. 2005. http://www.ncbi.nlm.nih.gov/pmc/articles/PMC2684512

19. Grahn BH, Paterson PG, Gottschall-Pass KT, Zhang Z. Zinc and the eye. National Center for Biotechnology Information. U.S. National Library of Medicine. 2001. http://www.ncbi.nlm.nih.gov/pubmed/11349933

20. Cataracts as an outcome of zinc deficiency in salmon. Nutr Rev. 1986;44(3):118–20.

21. Watts DL. The nutritional relationships of selenium. J Orthomol Med. 1994;9(2):111–7. http://orthomolecular.org/library/jom/1994/articles/1994-v09n02-p111.shtml.

22. Handa JT. How does the macula protect itself from oxidative stress? Mol Aspects Med. 2012;33(4):418–35.

23. Blundell M, Hunt L, Mayer E, Dick A, Sparrow J. Ophthalmol Scand. 1999;77(1):99–102.

24. Bek T. Ocular changes in heredo-oto-ophthalmo-encephalopathy. Br J Ophthalmol. 2000;84:1298–302.

25. Lutein & zeaxanthin. American Optometric Association. http://www.aoa.org/patients-and-public/caring-for-your-vision/diet-and-nutrition/lutein

26. Painter FM. Lutein and zeaxanthin. Alternat Med Rev. 2005;10(2):128–35. Thorne Research, Inc. http://www.thorne.com/altmedrev/.fulltext/10/2/128.pdf.

27. Cumming RG, Mitchell P. Alcohol, smoking, and cataracts: the Blue Mountains Eye Study. National Center for Biotechnology Information. U.S. National Library of Medicine. 1997. http://www.ncbi.nlm.nih.gov/pubmed/9338677

28. Kanthan GL, Mitchell P, Burlutsky G, Wang JJ. Alcohol consumption and the long-term incidence of cataract and cataract curgery: The Blue Mountains Eye Study. National Center for Biotechnology Information. U.S. National Library of Medicine. 2010. http://www.ncbi.nlm.nih.gov/pubmed/20619390

29. Ye J, He J, Wang C, Wu H. Smoking and risk of age-related cataract: a meta-analysis. National Center for Biotechnology Information. U.S. National Library of Medicine. 2012. http://www.ncbi.nlm.nih.gov/pubmed/22599585

30. Tan JS, Wang JJ, Mitchell P. Influence of diabetes and cardiovascular disease on the long-term incidence of cataract: The Blue Mountains Eye Study. National Center for Biotechnology Information. U.S. National Library of Medicine. 2008. http://www.ncbi.nlm.nih.gov/pubmed/18850468

31. Hu FB, Hankinson SE, Stampfer MJ, Manson JE, Colditz GA, Speizer FE, Hennekens CH. Prospective study of cataract extraction and risk of coronary heart disease in women. National Center for Biotechnology Information. U.S. National Library of Medicine. 2001. http://www.ncbi.nlm.nih.gov/pubmed/11323318

32. Fong CS, Mitchell P, Rochtchina E, Teber T, Hong T. Ophthalmology. 2009;93(3):290–5.

33. Sakata L, Sakata K, Malta RF. Cataract surgery and glaucoma. Ophthalmology. 2007;114(1):195–6.

34. Knudsen E, Baggesen K, Naeser K. Graefes Arch Clin Exp Ophthalmol. 2008;246(4):615–7.

35. Tabin G, Chen M, Espandar L. Cataract surgery for the developing world. Curr Opin Ophthalmol. 2008;19:55–9.

36. Catharine Ross A, Taylor CL, Yaktine AL, Del Valle HB, editors. Institute of Medicine (US) Committee to review dietary reference intakes for vitamin D and calcium. Washington, DC: National Academies Press; 2011.

37. Nemet AY, Hanhart J, Kaiserman I, Vinker S. Are cataracts associated with osteoporosis? Clin Ophthalmol. 2013;7:2079–84.
38. Tang D. Influence of age, diabetes, and cataract on calcium, lipid-calcium, and protein-calcium relationships in human lenses. Invest Ophthalmol Vis Sci. 2003;44(5):2059–66.
39. Vanessa Caceres: Eye world contributing author, April 2014. Post-menopausal hormone use: a protective effect against cataracts?
40. Holt PR, Russel RM. Chronic castritis and hypochlorhydria in the elderly. New York: CRC Press; 1993.
41. Sheen E, Triadafilopoulos G. Adverse effects of long-term proton pump inhibitor therapy. National Center for Biotechnology Information. U.S. National Library of Medicine. 2011. http://www.ncbi.nlm.nih.gov/pubmed/21365243
42. Christen WG, et al. Vitamins E and C and medical record-confirmed age-related macular degeneration in a randomized trial of male physicians. National Center for Biotechnology Information. U.S. National Library of Medicine. http://www.ncbi.nlm.nih.gov/pubmed/22503302
43. Packer L, Colman C. The antioxidant miracle: put lipoic acid, pycnogenol, and vitamins E and C to work for you. ISBN 13: 9780471353119.
44. Office of Dietary Supplements (ODS), National Institute of Health. http://ods.od.nih.gov/
45. Sesso HD, Buring JE, Christen WG. Vitamins E and C in the prevention of cardiovascular disease in men: the physicians' health study II randomized trial. http://www.ncbi.nlm.nih.gov/pmc/articles/PMC2586922/
46. Rautiainen S, Lindblad B, Morgenstern R. Vitamin C supplements and the risk of age-related cataract: a population-based prospective cohort study in women. National Center for Biotechnology Information. U.S. National Library of Medicine. http://www.ncbi.nlm.nih.gov/pubmed/23420353
47. Gritz D, Srinivasan M. Antioxidants for prevention of cataracts follow-up study. http://clinicaltrials.gov/show/NCT01664819
48. Packer L, Colman C. The antioxidant miracle. A brief review by Ivy Greenwell. Life Extension Magazine. 1999.
49. Wu G, Fang YZ. Glutathione metabolism and its implications for health. National Center for Biotechnology Information. U.S. National Library of Medicine. http://www.ncbi.nlm.nih.gov/pubmed/14988435
50. Head KA. Natural therapies for ocular disorders, part two: cataracts and glaucoma. National Center for Biotechnology Information. U.S. National Library of Medicine. 2006. http://www.ncbi.nlm.nih.gov/pubmed/11302779
51. Noctor G, Foyer CH. Ascorbate and glutathione: keeping active oxygen under control. National Center for Biotechnology Information. U.S. National Library of Medicine. http://www.ncbi.nlm.nih.gov/pubmed/15012235
52. Currie Z, Bhan A, Pepper I. Reliability of Snellen charts for testing visual acuity for driving: prospective study and postal questionnaire. Br Med J. 2000;321:990–2. http://www.bmj.com/content/321/7267/990.
53. Kaiser PK. Prospective evaluation of visual acuity assessment: a comparison of Snellen versus ETDRS charts in clinical practice. Am Ophthalmol Soc. 2009;107:311–24. US National Library of Medicine National Institutes of Health Search Term-Search Database. http://www.ncbi.nlm.nih.gov/pmc/articles/PMC2814576/?report=classic.
54. Mainster MA, Turner PL. Glare's causes, consequences, and clinical challenges after a century of ophthalmic study. Am J Ophthalmol. 2012;153(4):587–93. http://www.ajo.com/article/S0002-9394(12)00041-4/abstract.
55. The Optec® Functional Vision Analyzer™—contrast sensitivity tests with two glare levels under four testing conditions. http://www.stereooptical.com/products/vision-testers/functional-vision-analyzer
56. Optometric Clinical Practice Guideline. Care of the patient with visual impairment (low vision rehabilitation). American Optometric Association. http://www.aoa.org/documents/CPG-17.pdf

57. GALILEI™ Dual Scheimpflug Analyzer from Ziemer Group. http://www.ophthalmologyweb.com/5599-Computerized-Corneal-Analysis-Systems/53052-GALILEI-Dual-Scheimpflug-Analyzer/
58. iTRACE™ Combination Ray Tracing Aberrometer/Topographer. http://www.traceytechnologies.com/products_iTrace.htm
59. All Features: Pentacam™, the Gold Standard in Anterior Segment Tomography. http://www.pentacam.com/sites/hr_features.php
60. Belin MW. The Pentacam: precision, confidence, results, and accurate "Ks!". Cataract & Refractive Surgery Today. 2007. http://www.pentacam.com/downloads/studien/pentacam_aao_2006.pdf
61. Belin MW. Other clinical uses of the Pentacam: cataract, glaucoma, cornea. Applications of the Oculus Pentacam. http://oculus.cnpg.com/video/flatfiles/1342/
62. Datiles III MB, Ansari RR, Suh KI, et al. Clinical detection of precataractous lens protein changes using dynamic light scattering. Arch Ophthalmol. 2008;126(12):1687–93. archophthalmol.2008.507. http://archopht.jamanetwork.com/article.aspx?articleid=420932.
63. Ansari RR, Datiles MB III, King JF. A new clinical instrument for the early detection of cataract using dynamic light scattering and corneal topography. Optikon Men and Technology. NASA/CR—2000-209955. http://www.optikon.com/en/show-article.asp?section=archive&articleID=114
64. Ansari RR, King JF, Giblin FJ. Applications in bioastronautics and bioinformatics: early radiation cataracts detected by noninvasive, quantitative, and remote means. National Center for Microgravity Research. http://www.grc.nasa.gov/WWW/RT/RT2000/6000/6712ansari.html
65. Pamplona VF, Passos EB, Zizka J. CATRA: CATRA: interactive measuring and modeling of cataracts. Cataract maps with snap-on eyepiece for mobile phones. MIT Media Lab—CATRA. 2012. http://web.media.mit.edu/~pamplona/CATRA/
66. Cheung CY. Validity of a new computer-aided diagnosis imaging program to quantify nuclear cataract from slit-lamp photographs. National Center for Biotechnology Information. U.S. National Library of Medicine. 2011. http://www.ncbi.nlm.nih.gov/pubmed/21051727
67. Li H. A computer-aided diagnosis system of nuclear cataract. IEEE Xplore. 2010. http://ieeexplore.ieee.org/xpls/abs_all.jsp?arnumber=5415679
68. Cahn F. Measurement of lens autofluorescence can distinguish subjects with diabetes from those without. J Diabetes Sci Technol. 2014;8:43–9.
69. Bone RA, Mukherjee A. Innovative Troxler-free measurement of macular pigment and lens density with correction of the former for the aging lens. J Biomed Opt. 2013;18(10):107003. http://www.ncbi.nlm.nih.gov/pubmed/24114020.

Chapter 11
Oxidative Stress in Lens

Sam Young Yoon, Eunbi Kim, and Young Joo Shin

11.1 Oxidative Stress in Lens

11.1.1 Aging

Cataract is an opacity in the ocular crystalline lens, which results in visual disturbance if it is located on the visual axis. Cataract is a major cause of blindness globally. With aging, lens proteins undergo nonenzymatic, posttranslational modification and the accumulation of fluorescent chromophores, increasing susceptibility to oxidation and cross-linking and increased light scatter [1]. Because the human lens grows throughout life, the lens core is exposed for a longer period to such influences and the risk of oxidative damage increases in the fourth decade when a barrier to the transport of glutathione forms around the lens nucleus. Consequently, as the lens ages, its transparency falls and the nucleus becomes more rigid, resisting the change in shape necessary for accommodation. This is the basis of presbyopia. In some individuals, the steady accumulation of chromophores and complex, insoluble crystallin aggregates in the lens nucleus leads to the formation of a brown nuclear cataract. The process is homogeneous and the affected lens fibers retain their gross morphology. Cortical opacities are due to changes in membrane permeability and enzyme function and shear-stress damage to lens fibers with continued accommodative effort. Unlike nuclear cataract, progression is intermittent, stepwise, and nonuniform.

S.Y. Yoon • E. Kim • Y.J. Shin, M.D. (✉)
Department of Ophthalmology, Hallym University College of Medicine,
948-1 Daerim1-dong, Youngdeungpo-gu, Seoul 150-950, South Korea
e-mail: schinn@hanmail.net; schinn7@gmail.com

© Springer Science+Business Media New York 2015
M.A. Babizhayev et al. (eds.), *Studies on the Cornea and Lens*,
Oxidative Stress in Applied Basic Research and Clinical Practice,
DOI 10.1007/978-1-4939-1935-2_11

11.1.2   Cataract

11.1.2.1   Oxidative Stress of the Ocular Lens

Reactive oxygen species (ROS) play an important role in various biological processes. Oxidation of proteins has been associated with the aging process and many diseases. Free radicals can induce cumulative and irreversible damage to molecules, abnormality of cellular function and even cell death. Oxidative stress is the condition of the cellular damage from ROS exposure due to an imbalance between cellular ROS production and repairing damage caused by ROS [2]. We review the role of oxidative stress in the human ocular crystalline lens.

11.1.2.2   The Human Eye Lens and the Mechanism of Cataract

The lens is rare transparent structure which can be observed from the outside. It transmits light of various wavelengths, but filters almost ultra violet (UV) light which can damage the retina so that light hitting the retina is almost exclusively composed of visible wavelengths [3]. This UV filtering role of the lens is known to cause damage to lens proteins and may contribute to the development of cataract [4, 5]. To achieve the necessary refractive index for the lens to transmit light information, the protein concentration of the lens must be very high. The center of human lens contains a protein concentration as high as 450 mg/mL [6]. The lens epithelium, which covers the anterior surface of the organ, contains the majority of transporters and metabolic enzymes in the lens [7, 8] and is responsible for the growth and development of the entire lens [9]. Lens epithelial cells near the lens equator differentiate to form the fiber cells. This area of lens differentiation is called the superficial lens cortex. Mitochondrial respiration, carried out in the lens epithelium and superficial cortical fibers, accounts for 90 % of the oxygen consumed by the lens, although other oxygen consumers are found in the area beyond these mitochondria containing cells [10]. Oxygen enters the lens tissue via diffusion from the surrounding aqueous and vitreous humors but its consumption of oxygen keeps the partial pressure of oxygen in the lens fibers lower than at the surface of the lens [10]. In this way, potentially detrimental effects of oxygen on lens fiber cell proteins are minimized. This is important for the maintenance of lens homeostasis since the fiber cells are devoid of organelles and contain the majority of lens crystallin proteins required for light refraction [11]. Protein aggregation in the fibers as a result of oxidation, other protein modification and/or proteolysis results in loss of lens transparency and cataract formation [12–15]. In patients treated with hyperbaric oxygen (higher than normal oxygen tension), a high proportion go on to develop cataracts [16]. Treatment of guinea pigs with hyperbaric oxygen induced cross-linking of guinea pig lens nuclear crystallins into large disulfide-bonded aggregates capable of scattering light [17]. Collectively these properties of the lens make it uniquely susceptible to oxidative stress and consequently the lens has evolved a multitude of systems to combat ROS damage.

Sources of Lens ROS

Reactive oxygen species (ROS) are small readily diffusible molecules that contain unpaired electrons and include the hydroxyl radical ($\bullet$OH), the superoxide anion ($O_2 \cdot^-$), hydroperoxyl radicals (HO$\bullet$2), and peroxynitrite (OONO$^-$). Hydrogen peroxide (H_2O_2) and singlet oxygen ($^1O_2$) contain their full complement of electrons but exist in an unstable or reactive state. ROS form in cells as a consequence of both endogenous and exogenous oxidative stress initiators including altered cellular respiration, viral infection, immune activity, disease states, U.V.-light exposure, radiation exposure, smoke, metals, and drugs. Importantly, a fundamental source of endogenous ROS is produced in the mitochondria via the electron transport chain where inefficient electron coupling leads to the formation of $O_2 \cdot^-$ [18, 19]. It has been estimated that as much as 2 % of oxygen is converted to $O_2 \cdot^-$ under normal respiratory conditions [18, 20]. $O_2 \cdot^-$ is converted by the superoxide dismutases (detailed later) to form H_2O_2, H_2O_2 in the presence of metal ions in the Fenton reaction can form the highly reactive and damaging $\bullet$OH radical [21]. Interaction of $O_2 \cdot^-$ and nitric oxide (NO), a well characterized signaling molecule, gives rise to OONO$^-$, while $^1O_2$ may arise from absorption of photochemical energy by O_2 [22]. These many propagation reactions demonstrate that production of $O_2 \cdot^-$ can give rise to multiple damaging ROS species and thus the lens requires special antioxidant and enzyme systems to prevent damage to lens proteins exposed to ROS sources. Mitochondria produce less ATP and more ROS upon aging in mammalian tissues [22]. This leads to further mitochondrial damage and more ROS production. This gradual progression towards $O_2 \cdot^-$ production coupled with decreased antioxidant activity and levels in aging tissue is believed to contribute to a number of age-related degenerative disorders including age-related cataract. Evidence that mitochondrial ROS may play a role in lens protein oxidation and cataract formation comes from the fact that some inherited mitochondrial diseases are associated with an elevated risk of cataract [23], although the mechanisms underlying this phenomenon have not fully been elucidated. Several studies have provided evidence that loss of lens repair systems results in loss of mitochondrial combined with increased ROS levels in lens cells [24].

Defense Systems of Lens

ROS cause oxidative modifications to proteins resulting in loss or gain of function and protein aggregation which results in light scatter and cataract formation. Oxidative stress defense mechanisms operating to defend or protect the lens against ROS range from simple ROS scavenger systems to more advanced enzyme protective systems, free metal binding proteins, and actual repair proteins that can restore normal function to oxidatively damaged proteins. The lens is also home to an interesting family of chaperone proteins that act to prevent aggregation of oxidized proteins within the cell. Failures in antioxidant defense/detoxifying systems to protect proteins against oxidative stress-induced damage or subsequent failure of the protein repair systems to restore protein activity by reversing oxidative stress-induced damage requires the action of protein salvaging systems such as the proteasome [25] which degrade damaged proteins. Importantly, many of these antioxidant and repair systems also depend on the availability of specific reducing systems which may critically limit and regulate the function of these systems in the lens. The many layered systems of lens antioxidant defense and repair under normal circumstances function efficiently to prevent damage to lens proteins and prevent cataract formation. Oxidative stress has been implicated in a number of ocular diseases including age-related macular degeneration (AMD) [26], cataract [12–15, 27], Uveitis [28], corneal inflammation [29], glaucoma [30, 31], and Keratitis [32]. Understanding the roles and mechanisms that underlie the activities of these lens oxidative defense systems provides insight into the understanding of these and other oxidative stress-associated diseases. These systems and their functions are summarized in Fig. 11.1.

11.1.2.3   Oxidative Stress Defense and Repair Systems of the Ocular Lens

General ROS Scavengers of the Ocular Lens

General ROS scavengers can be roughly described as those molecules that nonenzymatically eliminate ROS by acting as chain breakers to directly prevent the propagation of ROS formation through direct binding and detoxification of ROS and/or ROS generators. In the lens, the primary reducing system or ROS scavenger is glutathione (GSH). Other important and well studied ROS scavengers in the lens are ascorbate (vitamin c) and vitamin E, while the cartotenoids, which are supplied solely in the diet, have been the subject of multiple supplementation studies to determine their ability to prevent or delay AMD and cataract.

Glutathione

GSH is the primary scavenger of ROS in the lens and the predominant reducing system in other ocular tissues, including the retina and cornea where it protects against chemical and oxidative stress-induced damage [33]. GSH is a dual function peptide that also participates in regulation of DNA and protein synthesis, cell-cycle control, signal transduction, and proteolysis. It is also involved in multiple metabolic pathways and the immune response [34]. GSH exists in unusually high levels in the lens. GSH functions to maintain protein thiol groups in their reduced form and therefore maintaining normal protein function [33]. GSH is maintained in its reduced form by the enzyme glutathione reductase. Under normal cellular conditions in the lens epithelium [35, 36] GSH is almost entirely found in its reduced state [37, 38] with barely detectable levels of the oxidized form of GSH (GSSG). GSH is also found at high levels in the lens fiber cells where it is likely transported from the epithelium to the fiber cells through lens connexins. GSH is completely broken down and resynthesized in the rabbit lens about every 48 h [39]. The epithelium of the lens also contains an active glutathione redox cycle. This pathway involves

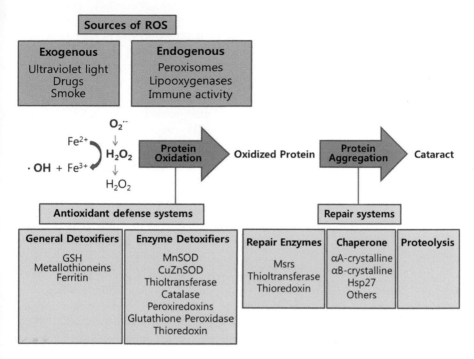

Fig. 11.1 Summary of oxidative stress defense system in eye lens. Multiple exogenous (ultraviolet light, drugs, metals, and smoke) and endogenous (peroxisomes, lipooxygenases, immune activity and altered mitochondrial respiration) sources of oxidative stress result in the formation of multiple forms of ROS such as $O_2^{.-}$, H_2O_2, and •OH. Two types of antioxidant systems (general and enzymatic detoxifiers) normally detoxify these reactive oxygen species to maintain lens homeostasis. These include GSH, MnSOD, CuZnSOD, catalase, the peroxiredoxins, and glutathioine peroxidase. Simultaneously, metallothioneins and ferritin function to limit access to free metals. Failure of these antioxidant systems leads to oxidation of proteins, loss of protein function, and protein aggregation. Specific repair systems including the Msrs and thioltransferase, thioredoxin can repair proteins restoring their normal function. Chaperone proteins such as αA-crystallin, αB-crystallin, and Hsp27 can remove toxic protein aggregates. Failure of these antioxidant defense and repair systems results in irreversible protein aggregation, loss of lens cell function, and ultimately cataract formation. *ROS* Reactive oxygen species, *GSH* glutathione, *MnSOD* Manganese superoxide dismutase, *CuZnSOD* copper zinc superoxide dismutase, *Msrs* methionine sulfoxide reductases, *Hsp27* Heat shock protein 27

glutathione reductase (GSH-Rx), NADPH, and the hexose monophosphate shunt (HMPS) which all function to reduce GSSG back to active GSH [40–42]. Cultured human lens cells challenged with H_2O_2 show a linear increase in HMPS activity while rabbit lenses subjected to t-butyl hydroperoxide (tBHP) stress show near complete oxidation of GSH to GSSG, but this ratio is reversed after recovery in complete medium [43] possibly as a consequence of increased shunt and glutathione redox cycle activity. Treatment of cultured lens epithelial cells with 0.1 mM H_2O_2 for up to 180 min showed depletion of the GSH pool over the first 60 min followed by full recovery [44], illustrating that lens cells react quickly to reduced GSH levels. Levels of reduced GSH have been shown to decrease in the human lens with increased aging [35] [45, 46] and upon cataract formation [36]. GSH was also observed to decrease in normal guinea pig lenses with aging, while slight decreases in guinea pig lens GSH were found following hyperbaric oxygen treatment [47]. Decreased concentrations of GSH or a shift from GSH to GSSG in the lens is believed to increase the rate of posttranslational modifications of crystallins [48], and to perpetuate damage to key proteins containing –SH groups including Na/K-ATPase, cytoskeletal proteins, and proteins associated with membrane permeability [43].

Ascorbate

Ascorbate (vitamin C) is found at high levels in the lens and is believed to be capable of consuming oxygen in the lens [49]. In the presence of redox available metal ions, ascorbate is oxidized to dehydroascorbate and H_2O_2 and in the process O_2 is consumed [10]. Dehydroascorbate is subsequently reduced back to ascorbate with reduced GSH and/or NADPH [50]. Ascorbate is transported from the plasma across the blood-aqueous barrier by the ciliary body into the aqueous humor where it is believed to supply the rest of the eye [51]. In addition to its oxidative role in H_2O_2 production, ascorbate can also act as a strong reductant and scavenger of ROS, particularly $O_2^{\cdot-}$. Ascorbate decreases membrane damage in diabetic rats [52], and photoperoxidation of lens membranes [53]. It also prevents riboflavin mediated light-induced damage to cation pumps in the lens [54, 55] and it plays a role in lens development and maintenance of transparency during development [56]. Ascorbate has been shown to protect the rat lens against selenite-induced cataract [57]. Incubation of mouse lenses with high levels of ascorbate also protected against ROS-mediated decreases in membrane transport, ATP production, and decreased GSH levels [58]. It is known that measured levels of ascorbate decline with age, possibly contributing to cataract formation. In fact, the ratio of ascorbate to dehydro-ascorbate decreased in cataractous lenses compared to clear lenses [59, 60], suggesting a requirement for ascorbate in lens homeostasis and prevention of cataract. These data provide evidence for a protective role for ascorbate in the lens, however, oxidized ascorbate–dehydroascorbate is likely damaging to the lens, suggesting that ascorbate can also contribute to lens oxidation and cataract formation, for instance, dehydroascorbate causes precipitation and cross-linking of bovine lens crystallins in vitro [61]. The interaction of ascorbate with light in the presence of metal ions, as described above, leads to the formation of H_2O_2 and subsequent interaction of H_2O_2 with metal ions, particularly Iron (Fe) that could produce the damaging species •OH. In this way ascorbate can actually become a pro-oxidant and lead to protein damage via both H_2O_2 and Fenton production of •OH. The addition of both Fe and ascorbate to cultured lens epithelial cells led to a significant increase in ferritin synthesis (the intracellular iron-storage protein) [62] and increased Fe loading into ferritin [63], therefore protecting against oxidative damage. A number of studies (reviewed by Chui & Taylor [64]) indicate that consuming elevated levels of antioxidants such as ascorbate, carotenoids, and tocopherol is associated with delayed development of various forms of cataract. Indeed the Blue Mountains Eye

Study found that participants with the highest quintile of total intake of vitamin C had a reduced risk of incident nuclear cataract [65]. Even more effective was an above-median intake of combined antioxidants (vitamins C and E, β-carotene, and zinc) which was also associated with a reduced risk of incident nuclear cataract [65].

Vitamin E

Vitamin E, the most important fat soluble antioxidant, is proposed to be an important chain breaking antioxidant for prevention of lipid peroxidation. Vitamin E is actually an umbrella term for a group of compounds called tocopherols and tocotrienols. α-tocopherol, the most active naturally occurring form, contains a chromane ring which reacts with organic peroxyl radicals and accounts for its antioxidant activity. It protects tissue lipids from free radical attack [66]. There is evidence to suggest that vitamin c acts to recycle the α-tocopheryl radical back to α-tocopherol, therefore implying a synergistic function between these two antioxidants [50]. Vitamin E supplementation has been shown to protect rat lenses against radiation-induced cataract [67]. The delaying or preventive effect of vitamin E on cataractogenesis has also been studied in in vivo sugar cataract models, for example galactose cataracts. Two studies showed that vitamin E supplementation showed no protective effect on galactosemic cataract in rat lenses [68, 69] but two other studies showed that direct topical instillation of vitamin E liposomes into eyes increased rat lens vitamin E and helped protect against galactosemic cataract [70, 71]. Just as the animal models for vitamin E supplementation and cataractogenesis produce mixed results so do the many human trials examining vitamin E supplementation and/or dietary/plasma levels and their association with cataract. Levels of lutein-zeaxanthin, retinoid, or α-tocopherol showed no differences between normal and cataractous human lenses [72] but in regular users of multivitamin supplements, the risk of nuclear opacification was reduced by one third. In regular users of vitamin E supplements and persons with higher plasma levels of vitamin E, the risk was reduced by approximately half [73]. A low plasma vitamin E level was associated with a 3.7-fold excess risk of the progression of early cortical lens opacities compared with the highest quartile in hypercholesterolemic eastern Finnish men [74]. In the Beaver Dam Eye Study, persons in with higher total serum tocopherol had a lower risk of cataract than those with a lower serum tocopherol [75] and lens opacities were associated with lower levels of riboflavin, vitamin E, Fe, and protein nutritional status [76]. By contrast, McNeil et al. [77] found that vitamin E supplemented for 4 years at a dose of 500 IU daily did not reduce the risk of nuclear, cortical, or posterior subcapsular.

Carotenoids

The best studied carotenoids in the eye are lutein and zeaxanthin. Carotenoids are made up of more than 600 structural variants that are all lipophilic pigments. They are synthesized in plants, fungi, bacteria, and algae and provided to mammalians only in their diet, particularly in egg yolk and dark leafy green vegetables such as spinach or kale. Some carotenoids, including β-carotene, can protect against photooxidative damage by scavenging $^1O_2$ and peroxyl radicals and can interact synergistically with other antioxidants [78]. They may lower the risk for several degenerative disorders, cardiovascular and ophthalmological diseases, and various types of cancer [79]. In the eye, lutein and zeaxanthin have been shown to filter high-energy wavelengths of blue visible light, primarily in the macula of the retina and to act as antioxidants that protect against the formation of ROS [3]. A study by Gale et al. [80] showed that the risk of nuclear cataract was lowest in people with the highest plasma concentrations of α- or β-carotene while the risk of cortical cataract was lowest in people with the highest plasma concentrations of lycopene. In addition the risk of posterior subcapsular cataract was lowest in those with higher concentrations of lutein but high plasma concentrations of vitamin C, vitamin E, or zeaxanthin and β-cryptoxanthin were not associated with decreased risk. In the Blue Mountain Eye Study, as mentioned above, an increased intake of vitamins C and E, β-carotene, and zinc was associated with a reduced risk of nuclear cataract but not with cortical or posterior subcapsular cataract [65]. The Beaver Dam Eye Study, also mentioned above, found that patients with the highest quintile of lutein intake were half as likely to develop nuclear cataract as those in the lowest quintile of intake [81]. In the Pathologies Oculaires Liées à l'Age (POLA) Study, the highest quintile of plasma zeaxanthin was significantly associated with reduced risk of nuclear or other cataract, among other carotenoids, only β-carotene showed a significant negative association with nuclear cataract [82]. In the Carotenoids in the Age Related Eye Disease Study (CAREDs), women in the group with high dietary levels of lutein and zeaxanthin had a 23 % lower prevalence of nuclear cataract compared with those with low levels [83]. Despite the conclusions of these studies, an analysis of a number of supplementation studies involving carotenoids using the FDA evidence review system, found no evidence to support the fact that lutein or zeaxanthin can protect against cataract formation [84].

Antioxidant Enzymes of the Eye Lens

The Superoxide Dismutases

The antioxidant enzymes of the eye catalytically remove ROS, in some cases generating other reactive or unstable species but generally resulting in a less toxic environment. Mitochondrial superoxide is dismutated to H_2O_2 and O_2 by the superoxide dismutases. Manganese superoxide dismutase (MnSOD, SOD 2) is present in the mitochondrial matrix while copper zinc superoxide dismutase (CuZnSOD, SOD 1) is localized to the intramembrane space of the mitochondria and in the cytosol [19]. In the absence of metal ions and thus the Fenton reaction, H_2O_2 produced in this reaction can then be detoxified by specialized enzymes including glutathione peroxidase (GSH-Px), the peroxiredoxins (Prxs), and catalase, described below. A recent study examined total SOD (TSOD), MnSOD, and CuZnSOD activities in lens epithelial cells derived from different types of cataract in patients having phacoemulsification cataract surgery. The highest level of all TSOD, MnSOD, and CuZnSOD activity was in patients 50 years or younger with the activity gradually declining with age. The level of TSOD activity increased in cortical cataract and the individual levels of CuZnSOD and MnSOD activities in lens epithelial cells were also higher in cortical cataracts [85]. MnSOD has been shown to protect human lens epithelial cells against oxidative stress using up- and down-regulation of the enzyme in lens epithelial cells [86]. The intrinsic apoptotic pathway was also significantly increased in MnSOD down-regulated lens epithelial cells compared to those with up-regulated MnSOD when exposed to superoxide [87]. CuZnSOD, which is not mitochondrial specific and makes up 90 % of total SOD [88], was shown to prevent H_2O_2-induced oxidative damage when overexpressed in whole rat lenses [89]. Treatment of human lens epithelial cells with a bolus of 17 β-estradiol, known to protect lens cells against oxidative stress by preserving the mitochondria, resulted in a significant rapid increase in the activity of MnSOD with no effect on levels of mRNA or protein [90].

Hydrogen Peroxide Detoxifiers

B.1 Catalase

Human catalase is normally found localized in the peroxisomes, where it functions to decompose H_2O_2 to H_2O. One study investigated the effect of over expressing catalase in both nucleus and mitochondria of mice, two areas normally devoid of catalase activity. They found that oxidative damage was reduced, H_2O_2 production and H_2O_2-induced aconitase (an enzyme involved in the TCA cycle) inactivation were attenuated, and importantly that cataract development, cardiac pathology, and the development of mitochondrial deletions were reduced [91]. Resveratrol (a polyphenol antioxidant found in plants) was found to protect human lens epithelial cells against H_2O_2-induced oxidative stress by increasing catalase, SOD-1, and heme oxygenase (HO-1) expression [92]. In rat lens epithelial explants, increased levels of GSH and catalase suppressed changes typically associated with TGFβ-induced transdifferentiation including wrinkling of the lens capsule, cell-surface blebbing, apoptotic cell loss, induction of αSMA, and loss of Pax6 expression [93]. However, lenses from mice deficient in catalase did not show any increased susceptibility to oxidative stress generated by photochemical reaction [94], suggesting that catalase is not the most important enzyme involved in decomposing H_2O_2.

B.2 The peroxiredoxins

Given its peroxisomal location and the evidence outlined above it is unlikely that catalase is the major H_2O_2 detoxification enzyme in the lens. One likely H_2O_2 detoxification system in the lens is the Prxs, which are peroxide scavengers that possess redox active cysteines and use the thioredoxin system as an electron donor to detoxify H_2O_2, $OONO^-$ and a wide range of hydroperoxides [95]. Six mammalians Prxs are known, with Prx 1, 2, and 6 found in the cytoplasm, Prx 3 in the mitochondria, Prx 4 in the endoplasmic reticulum, and Prx 5 found in various compartments in the cell including the peroxisomes and mitochondria [95–97]. Since the mitochondria is the major source of H_2O_2 formation, Prx3 is well studied. Significant levels of Prx 3 mRNA and protein have been detected in human lens epithelial and fiber cells and it has been shown that Prx 3 is inducible by H_2O_2 in human lens epithelial cells but not by tBHP or heat shock [97].

B.3. Glutathione peroxidase

Another potentially important H_2O_2 detoxification enzyme is GSH-Px which also reduces H_2O_2 to H_2O or alkyl peroxides to alcohols at the expense of reduced GSH [33]. In the lenses of H_2O_2-treated mice with deficient levels of GSH-Px, DNA strand breaks in the mice with elevated GSH-Px were 40 % of those with normal GSH-Px activity [98], while in the GSH-Px knockout mouse lens DNA damage was fivefold that of GSH-Px rich transgenic mice [98].

Free Metal Detoxifiers of the Eye Lens

Metallothioneins

Fenton-type reactions, where H_2O_2 reacts with free metal ions i.e., Iron (Fe^{2+}) or copper (Cu^{2+}) to produce the •OH radical, are a major source of oxidative stress initiated by transition metals [99] and are thought to be involved in the formation of cataract [100, 101]. Detoxification of these metals by metallothioneins in combination with careful control of free iron levels by ferritin is central to prevent ROS production. Human exposures to toxic metals such as Fe, Cu, cadmium (Cd), lead, aluminum, mercury and others may arise from fossil fuel emissions, industrial waste, cigarette smoke, and air pollution [102]. These metals have been associated with increased risk of cataract formation [103] and it has been proposed that detoxification of metals may play a major role in prevention of cataract. In support of this theory, increased Cd levels [104] and increased Cu levels [105] have been reported in cataract compared to clear lenses. Another study that compared corticonuclear and mature cataracts found higher Fe levels in the mature cataract [106] suggesting a role for these metals in cataract formation. Metallothioneins (MTs) serve to detoxify free metals; there are 16 known isoforms of MTs in humans, grouped into four classes: I, II, III, and IV. The human lens expresses MT classes I and II. Only one isoform, MTIIa, is specific for the lens epithelium, whereas the MTI isoforms are expressed at lower levels in both the lens epithelium and lens fibers [107]. In addition, MTIIa exhibits increased expression in age-related cataract compared with clear human lenses [108] suggesting a possible role for MTIIa in lens protection. Lens MTs (Ig, If, Ih, Ie, and IIa) were differentially induced by specific metals in human lens epithelial cells, specifically by Cd^{2+} and Zn^{2+}, but not Cu^{2+} [109]. Similar responses of the MTIIa gene were detected in identically treated primary human lens epithelial cells. Cd^{2+} and Zn^{2+} induced MTIIa to five times higher levels than MTIg [109]. Overexpression of MTIIa in lens epithelial cells has been shown to protect against Cd as well as TBHP-induced oxidative stress [110]. The same study also shows that MTIIa may play a role in regulating expression of other important antioxidant genes, HO-1, thioredoxin reductase, and MnSOD, antioxidant molecules that could further enhance protection against oxidative stress in lens cells, potentially delaying onset of cataract [110].

Ferritin

Fe is transported in the plasma using transferrin, the Fe transport protein, while ferritin is the intracellular protein responsible for iron binding and storage. Ferritin, which is found throughout the lens [111, 112], is a multimeric iron-storage protein consisting of 24 subunits of two types: heavy (H) and light (L). The ratio of these two chains is tissue specific and controls iron storage and availability [113]. Fe levels in the eye lens have been reported to be between 0.18 and 9.6 μg/g wet weight [114]. A number of studies have found increased Fe levels in cataractous lenses [106, 111] with redox active Fe (not bound to ferritin) also found at higher levels in cataractous compared to control lenses [100, 101]. Low levels of Fe have been found in both the aqueous and vitreous humors [115]. Levels are kept low by the blood ocular barrier preventing transferrin entrance into the eye [114]. Inflammation which causes a breakdown of this barrier results in large increases in iron concentration in both the aqueous and vitreous humors [116]. The lens tightly controls Fe levels and during inflammation the lens accumulates Fe by taking it up from increased levels in the aqueous and vitreous humors in both the transferrin and non-transferrin-bound forms. The lens Fe concentration returns to control levels following resolution of the inflammatory episode [117]. Why the lens would act as a sink for excess Fe is not known and is the subject of much research. In a recent study using knockdown of ferritin heavy chains in human lens epithelial cells, increased iron availability resulted in increased cystine uptake and GSH concentration and decreased nuclear translocation of hypoxia-inducible factor 1-α (a transcription factor that regulates vascular endothelial growth factor (VEGF) expression) and VEGF accumulation in the cell-conditioned medium [118].

Protein Repair Systems of the Eye Lens

Methionine Sulfoxide Reductases

While all amino acids are susceptible to oxidation, methionine and cysteine residues are among the most vulnerable due to the presence of a sulfur containing side chain that is sensitive to attack from ROS such as H_2O_2, •OH, hypochlorous acid, chloramines, and OONO$^-$ [119, 120]. In the eye lens protein methionine sulfoxide (PMSO), the oxidized form of methionine, levels increase upon aging [121] and in human cataractous lenses 60–70 % of total lens protein is found as PMSO [122]. ROS-mediated oxidation of methionine results in an asymmetric sulfur center and thus PMSO exists as two epimers, the S and R-epimers. Methionine oxidation in proteins can result in altered conformation, activity, sub-cellular localization patterns, and aggregation states which are associated with loss of cellular functions, apoptosis, and cell death [123]. Methionine oxidation is, however, reversible via a thioredoxin dependent reaction in which PMSO is converted to reduced methionine. This reaction is catalyzed by a family of enzymes called the methionine sulfoxide reductases (Msrs). The Msr family consists of MsrA (found in the cytosol and mitochondria) and three MsrBs; MsrB1 (localized in the cytosol and nucleus),

MsrB2 (localized in the mitochondria), and MsrB3 (localized in the endoplasmic reticulum and mitochondria) that act on the S- and R-epimers respectively [124]. The oxidized thioredoxin produced during the reduction of PMSO is subsequently reduced by thioredoxin reductase (TrxR) in an NADPH dependent reaction. It has been shown that in bovine lenses 40 % of Msr activity is due to MsrB while the remaining is MsrA [125]. MsrA has been shown to play an important role in protection of lens cells against oxidative damage and it has been shown to be required for the maintenance of lens transparency in vivo [126–128]. Gene silencing of MsrA decreases the resistance of lens epithelial cells to H_2O_2-induced oxidative stress resulting in increased mitochondrial ROS levels in human lens cells [126] and loss of lens cell mitochondrial function [24]. Similarly gene silencing of MsrB1, B2, and B3 results in decreased resistance to tBHP-induced oxidative stress and increased cell death in human lens epithelial cells [126]. Deletion of the MsrA gene in mice leads to oxidative stress-induced cataract [127]. By contrast, overexpression of MsrA in human lens cells protects against oxidative stress and preserves mitochondrial function [126]. Recently, both cytochrome c (cyt c) [127] and α-crystallin/sHSP [128] have been identified as key targets of MsrA function in the lens. Both proteins are critical for lens function. Cyt c is essential for mitochondrial electron transfer and is a key initiator of apoptosis in mammalian cells [129]. α-crystallin/sHSP is a molecular chaperone that is essential for the maintenance of lens transparency whose deletion has been shown to result in cataract formation [130–133] and is discussed in more detail below.

Thioltransferases

Thioltransferase, known interchangeably as glutaredoxin, is a GSH-dependent cytosolic protein and a member of the thiol-disulfide oxidoreductase enzyme family, containing a conserved CXXC active site [134]. Two isoenzymes of thioltransferase are known: cytosolic TTase-1 (Grx1) and mitochondrial TTase-2 (Grx2) [134]. The TTases use reduced GSH in their reaction to reduce protein thiols (s-thiolation) preventing disulfide bond formation and protein aggregation [134]. TTase activity was found to be higher in the lens epithelial layer than in the rest of the lens [135]. In cultured lens epithelial cells TTase 1 levels and activity were shown to be up-regulated by treatment with H_2O_2. Depletion of GSH and subsequent treatment with H_2O_2 also increased levels of TTase in human lens epithelial cells [136]. Porcine lenses cultured in H_2O_2 over a 24-h period showed similar trends for TTase, with slow transient increases in TTase activity, mRNA transcript, and protein levels in response to low level of H_2O_2 and a more rapid response to higher levels of H_2O_2 [137]. In a separate study, treatment of cultured lens epithelial cells with 0.1 mM H_2O_2 over 3 h inactivated the key glycolytic enzyme glyceraldehyde-3-phosphate dehydrogenase (G-3PD) by more than 50 % over the first 15 min, this returned to 80 % of normal activity by 180 min. Interestingly, the inactivated cellular G-3PD in the cell extract could be partially reactivated by human recombinant TTase but not GSH [138]. The absence of TTase 1 was also shown to increase lens cell susceptibility to UV-B radiation (UVRB), light scattering was increased in TTase−/− mouse lenses compared to wild type mouse lens after treatment with UVRB [139]. TTase-2 has been shown to protect against disruption of the mitochondrial transmembrane potential in lens epithelial cells during oxidative stress conditions [140] and more recently TTase-2 was shown to protect complex I of the electron transport chain and to prevent H_2O_2-induced apoptosis in lens epithelial cells [141].

Reducing Systems of the Eye Lens

Thioredoxin and Thioredoxin Reductase

This section could easily contain GSH which we have grouped under ROS scavengers. It could also contain a section on NADPH, which is required for Msr repair of PMSO, thioredoxin repair of disulfide bonds and the recycling of GSH but NADPH is mentioned specifically in many other sections in this review. Thioredoxin (Trx) is the major disulfide reductase responsible for maintaining proteins in a reduced state within the cell. It is a small, cysteine rich protein possessing a dithiol/disulfide active site (CGPC) that acts to reduce protein disulfides. Its action results in an internal disulfide on the thioredoxin protein itself and this must be reduced by TrxR in an NADPH dependent reaction in order to recycle Trx. Trx serves as an electron donor, via the thiol/disulfide exchange reaction, for a number of enzymes including Msrs, ribonucleotide reductases, as well the peroxiredoxins [142]. Reduction of disulfide bonds is thought to be an important regulatory step where Trx could control the redox state of critical SH groups involved in structure and catalytic function in proteins. There are two nuclear encoded isoforms of Trx, Trx 1 found in the cytosol, and Trx 2 which is localized solely to the mitochondria. Trx 1 not only is an important part of the antioxidant defensive against oxidative stress but it also plays a role in transcription, growth control, and immune function [142]. Early embryonic lethality results from knockout of either Trx 1 [143] or Trx 2 [144] and it has also been shown in mice that a haploinsufficiency (a single functional copy of a gene, where insufficient product is made) of Trx 2 results in the reduction of ATP production and increased ROS production [145]. Early work on mouse lenses revealed that in vivo photochemical oxidative stress to Emory mice resulted in a fivefold up-regulation of the Trx 1 gene in the lens at 3 weeks and a fourfold increase of the lens Trx 1 protein but that Trx 2 was unchanged [146]. In cultured lens epithelial cells, Trx 1 is up-regulated at both transcript and protein levels in response to H_2O_2 treatment [147]. Interestingly, human lens epithelial cells treated with exogenous human recombinant Trx 1 showed a simultaneous increase in mRNA expressions of mitochondrial MnSOD, TTase 1, TTase 2, and thioredoxin peroxidase IV (Prx3) [148]. Recently a study on 23 normal human lenses of 19–77 years which were grouped into 2nd, 3rd, 5th, 6th, and 7th decades, indicated that Trx 1 activity decreased with age. Activity of Trx 1 in the 7th decade appeared to be 30 % of the activity found in lenses from the 2nd decade, a significant drop in activity. Interestingly the corresponding protein levels of Trx 1 were unaffected and remained steady throughout the seven decades in the lenses studied [149].

Thioredoxin-Like Proteins

Recently a thioredoxin-like protein (TXNL) containing the conserved CXXC reducing motif was discovered in the retina of mice and termed rod derived cone viability factor (RdCVF) [150]. This protein known as TXNL6 in humans is a product of the NXNL1 gene. The discovery of a novel CXXC containing protein led to the hypothesis that other reducing systems could function in the eye in addition to or as an alternative to the better characterized reducing systems such as Trx. TXNL6 is part of a small family of thioredoxin-like proteins which consists of TXNL1, TXNL2, TXNL4, and TXNL5. TXNLs share the CXXC active site of Trx 1 and/or the thioredoxin fold which consists of a four stranded β sheets surrounding three α helices [151]. Little is known about the disulphide reducing action of many of the TXNLs but recent work has shown that TXNL6 is present in both the lens epithelium and lens fiber cells, that it is inducible by oxidative stress and that it can serve as a reducing agent for MsrA in the repair of essential lens proteins [152]. TXNL1 is known to be a redox sensor and part of the 26 s proteasome subunit [153, 154]. TXNL2 is also known as glutaredoxin 3 and may catalyze the reduction of glutathione mixed disulfides [155]. Little is known about the oxidoreductase role of TXNL4 if any but it is known to be essential for the G2/M transition of the cell cycle [156]. TXNL5 is known to have redox potential similar to that of Trx 1 [157] and it is known that it can itself be reduced by TxrR and that it inhibits TNF-α induced NFκB activation to a greater extent than Trx 1 [158].

Chaperone Proteins of the Eye Lens

α-Crystallin

Heat shock proteins (Hsps) are highly conserved proteins that are induced in response to various physiological and environmental stressors [159]; interestingly HSPs form essential partnerships with the proteasome and lysosomes in protein degradation processes [159]. The molecular chaperones expressed in lens epithelial cells include αA-crystallin, αB-crystallin, Hsp25/27, Hsp40, Hsc70, Hsp70, and TCP-1 [160–163]. In the lens the predominant Hsp is α-crystallin. α-crystallin is made up of two small HSPs (sHSPs); αA and αB-crystallin, α-crystallin makes up approximately 40–50 % of total lens protein and is crucial for maintaining lens transparency. In addition to its structural and refractive role in the lens, α-crystallin is a molecular chaperone [130, 132] that functions to prevent protein aggregation [131], and formation of high molecular weight aggregates in the lens. α-crystallin has also been implicated in apoptotic control and cell survival [9]. In its native state in the lens, α crystalline consists of two subunits called αA- and αB-crystallin that consist of 173 and 175 amino acids respectively. αA- and αB-crystallin share as much as 57 % homology and exist in the lens in a 3:1 ratio [164]. Deletion of αA-crystallin in mice results in smaller lenses compared to wild type, and the development of opacification that spreads with age [133]. Mutation of αA-crystallin (R116C), which causes loss of chaperone function, leads to cataract in humans [165].

Similarly a mutation of αB-crystallin (R120G) that causes cataract and desmin-related myopathy in humans, also leads to loss of chaperone function [166]. Importantly, the chaperone function of α-crystallin is also lost upon oxidative conditions [128, 167, 168]. Specific oxidation of bovine α-crystallin methionines to PMSO resulted in loss of chaperone function in vitro [128] and αA-crystallin methionines are found as PMSO in rat hereditary cataracts [169] and in MsrA knockout mice [128]. Oxidation by Fenton-type reaction of rat recombinant αA- and αB-crystallin was shown to result in higher molecular weight proteins that lacked chaperone function [170]. These results suggest that oxidation of α-crystallin results in loss of chaperone function and that loss of chaperone function plays a role in cataract formation. αA-crystallin and αB-crystallin as well as HSP27 can be induced by specific metals in SRA01/04 human lens epithelial cells. Cd^{2+} and Cu^{2+}, but not Zn^{2+}, induced αB-crystallin and HSP27 while αA-crystallin was induced by Cu^{2+} only [109].

References

1. Michael R, Bron AJ. The ageing lens and cataract: a model of normal and pathological ageing. Philos Trans R Soc Lond B Biol Sci. 2011;366(1568):1278–92.
2. Brennan LA, McGreal RS, Kantorow M. Oxidative stress defense and repair systems of the ocular lens. Front Biosci (Elite Ed). 2012;4:141–55.
3. Roberts RL, Green J, Lewis B. Lutein and zeaxanthin in eye and skin health. Clin Dermatol. 2009;27(2):195–201.
4. Abraham AG, Cox C, West S. The differential effect of ultraviolet light exposure on cataract rate across regions of the lens. Invest Ophthalmol Vis Sci. 2010;51(8):3919–23.
5. Hightower KR. The role of the lens epithelium in development of UV cataract. Curr Eye Res. 1995;14(1):71–8.
6. Fagerholm PP, Philipson BT, Lindstrom B. Normal human lens: the distribution of protein. Exp Eye Res. 1981;33(6):615–20.
7. Reddy VN. Metabolism of glutathione in the lens. Exp Eye Res. 1971;11(3):310–28.
8. Delamere NA, Tamiya S. Lens ion transport: from basic concepts to regulation of Na, K-ATPase activity. Exp Eye Res. 2009;88(2):140–3.
9. Andley UP. The lens epithelium: focus on the expression and function of the alpha-crystallin chaperones. Int J Biochem Cell Biol. 2008;40(3):317–23.
10. McNulty R, et al. Regulation of tissue oxygen levels in the mammalian lens. J Physiol. 2004;559(Pt 3):883–98.
11. Augusteyn RC. On the growth and internal structure of the human lens. Exp Eye Res. 2010;90(6):643–54.
12. Hejtmancik JF, Kantorow M. Molecular genetics of age-related cataract. Exp Eye Res. 2004;79(1):3–9.
13. Truscott RJ. Age-related nuclear cataract-oxidation is the key. Exp Eye Res. 2005;80(5):709–25.
14. Vinson JA. Oxidative stress in cataracts. Pathophysiology. 2006;13(3):151–62.
15. Lou MF. Redox regulation in the lens. Prog Retin Eye Res. 2003;22(5):657–82.
16. Palmquist BM, Philipson B, Barr PO. Nuclear cataract and myopia during hyperbaric oxygen therapy. Br J Ophthalmol. 1984;68(2):113–7.
17. Simpanya MF, et al. Aggregation of lens crystallins in an in vivo hyperbaric oxygen guinea pig model of nuclear cataract: dynamic light-scattering and HPLC analysis. Invest Ophthalmol Vis Sci. 2005;46(12):4641–51.

18. Cadenas E, Davies KJ. Mitochondrial free radical generation, oxidative stress, and aging. Free Radic Biol Med. 2000;29(3–4):222–30.
19. Murphy MP. How mitochondria produce reactive oxygen species. Biochem J. 2009;417(1):1–13.
20. Sas K, et al. Mitochondria, metabolic disturbances, oxidative stress and the kynurenine system, with focus on neurodegenerative disorders. J Neurol Sci. 2007;257(1–2):221–39.
21. Halliwell B, Gutteridge JM. Role of free radicals and catalytic metal ions in human disease: an overview. Methods Enzymol. 1990;186:1–85.
22. Wei YH, Lee HC. Oxidative stress, mitochondrial DNA mutation, and impairment of antioxidant enzymes in aging. Exp Biol Med (Maywood). 2002;227(9):671–82.
23. Finsterer J. Mitochondriopathies. Eur J Neurol. 2004;11(3):163–86.
24. Marchetti MA, et al. Silencing of the methionine sulfoxide reductase A gene results in loss of mitochondrial membrane potential and increased ROS production in human lens cells. Exp Eye Res. 2006;83(5):1281–6.
25. Shang F, Nowell Jr TR, Taylor A. Removal of oxidatively damaged proteins from lens cells by the ubiquitin-proteasome pathway. Exp Eye Res. 2001;73(2):229–38.
26. Hollyfield JG, et al. Oxidative damage-induced inflammation initiates age-related macular degeneration. Nat Med. 2008;14(2):194–8.
27. Spector A. Oxidative stress-induced cataract: mechanism of action. FASEB J. 1995;9(12):1173–82.
28. Gritz DC, et al. Histochemical localization of superoxide production in experimental autoimmune uveitis. Curr Eye Res. 1991;10(10):927–31.
29. Alio JL, et al. Antioxidant therapy in the treatment of experimental acute corneal inflammation. Ophthalmic Res. 1995;27(3):136–43.
30. Kumar DM, Agarwal N. Oxidative stress in glaucoma: a burden of evidence. J Glaucoma. 2007;16(3):334–43.
31. Tezel G. Oxidative stress in glaucomatous neurodegeneration: mechanisms and consequences. Prog Retin Eye Res. 2006;25(5):490–513.
32. Alio JL, et al. Effect of topical antioxidant therapy on experimental infectious keratitis. Cornea. 1995;14(2):175–9.
33. Ganea E, Harding JJ. Glutathione-related enzymes and the eye. Curr Eye Res. 2006;31(1):1–11.
34. Ballatori N, et al. Glutathione dysregulation and the etiology and progression of human diseases. Biol Chem. 2009;390(3):191–214.
35. Harding JJ. Free and protein-bound glutathione in normal and cataractous human lenses. Biochem J. 1970;117(5):957–60.
36. Bhat KS, et al. Effect of pigmentation on glutathione redox cycle antioxidant defense in whole as well as different regions of human cataractous lens. Exp Eye Res. 1991;52(6):715–21.
37. Giblin FJ, Chakrapani B, Reddy VN. Glutathione and lens epithelial function. Invest Ophthalmol. 1976;15(5):381–93.
38. Giblin FJ, et al. Exposure of rabbit lens to hyperbaric oxygen in vitro: regional effects on GSH level. Invest Ophthalmol Vis Sci. 1988;29(8):1312–9.
39. Reddy DV, Klethi J, Kinsey VE. Studies on the crystalline lens. XII. Turnover of glycine and glutamic acid in glutathione and ophthalmic acid in the rabbit. Invest Ophthalmol. 1966;5(6):594–600.
40. Giblin FJ, Nies DE, Reddy VN. Stimulation of the hexose monophosphate shunt in rabbit lens in response to the oxidation of glutathione. Exp Eye Res. 1981;33(3):289–98.
41. Giblin FJ, McCready JP, Reddy VN. The role of glutathione metabolism in the detoxification of H2O2 in rabbit lens. Invest Ophthalmol Vis Sci. 1982;22(3):330–5.
42. Giblin FJ, et al. Detoxification of H_2O_2 by cultured rabbit lens epithelial cells: participation of the glutathione redox cycle. Exp Eye Res. 1985;40(6):827–40.
43. Giblin FJ. Glutathione: a vital lens antioxidant. J Ocul Pharmacol Ther. 2000;16(2):121–35.
44. Xing KY, Lou MF. Effect of H_2O_2 on human lens epithelial cells and the possible mechanism for oxidative damage repair by thioltransferase. Exp Eye Res. 2002;74(1):113–22.

45. Lou MF, Dickerson Jr JE. Protein-thiol mixed disulfides in human lens. Exp Eye Res. 1992;55(6):889–96.
46. Bova LM, et al. Major changes in human ocular UV protection with age. Invest Ophthalmol Vis Sci. 2001;42(1):200–5.
47. Giblin FJ, et al. Nuclear light scattering, disulfide formation and membrane damage in lenses of older guinea pigs treated with hyperbaric oxygen. Exp Eye Res. 1995;60(3):219–35.
48. Reddy VN. Glutathione and its function in the lens: an overview. Exp Eye Res. 1990;50(6):771–8.
49. Eaton JW. Is the lens canned? Free Radic Biol Med. 1991;11(2):207–13.
50. Jacob RA, Burri BJ. Oxidative damage and defense. Am J Clin Nutr. 1996;63(6):985S–90.
51. Garland DL. Ascorbic acid and the eye. Am J Clin Nutr. 1991;54(6 Suppl):1198S–202.
52. Linklater HA, et al. Modelling cortical cataractogenesis. XI. Vitamin C reduces gamma-crystallin leakage from lenses in diabetic rats. Exp Eye Res. 1990;51(3):241–7.
53. Varma SD, Srivastava VK, Richards RD. Photoperoxidation in lens and cataract formation: preventive role of superoxide dismutase, catalase and vitamin C. Ophthalmic Res. 1982;14(3):167–75.
54. Varma SD, Kumar S, Richards RD. Light-induced damage to ocular lens cation pump: prevention by vitamin C. Proc Natl Acad Sci U S A. 1979;76(7):3504–6.
55. Varma SD, Richards RD. Ascorbic acid and the eye lens. Ophthalmic Res. 1988;20(3):164–73.
56. Brewitt B, Clark JI. A new method for study of normal lens development in vitro using pulsatile delivery of PDGF or EGF in HL-1 serum-free medium. In Vitro Cell Dev Biol. 1990;26(3 Pt 1):305–14.
57. Devamanoharan PS, et al. Prevention of selenite cataract by vitamin C. Exp Eye Res. 1991;52(5):563–8.
58. Hegde KR, Varma SD. Protective effect of ascorbate against oxidative stress in the mouse lens. Biochim Biophys Acta. 2004;1670(1):12–8.
59. Lohmann W, Schmehl W, Strobel J. Nuclear cataract: oxidative damage to the lens. Exp Eye Res. 1986;43(5):859–62.
60. Lohmann W, et al. Nuclear cataract and ascorbic acid. Naturwissenschaften. 1986;73(5):266–7.
61. Ortwerth BJ, Feather MS, Olesen PR. The precipitation and cross-linking of lens crystallins by ascorbic acid. Exp Eye Res. 1988;47(1):155–68.
62. McGahan MC, et al. Regulation of ferritin levels in cultured lens epithelial cells. Exp Eye Res. 1994;59(5):551–5.
63. Harned J, Grimes AM, McGahan MC. The effect of UVB irradiation on ferritin subunit synthesis, ferritin assembly and Fe metabolism in cultured canine lens epithelial cells. Photochem Photobiol. 2003;77(4):440–5.
64. Chiu CJ, Taylor A. Nutritional antioxidants and age-related cataract and maculopathy. Exp Eye Res. 2007;84(2):229–45.
65. Tan AG, et al. Antioxidant nutrient intake and the long-term incidence of age-related cataract: the Blue Mountains Eye Study. Am J Clin Nutr. 2008;87(6):1899–905.
66. Azzi A. Molecular mechanism of alpha-tocopherol action. Free Radic Biol Med. 2007;43(1):16–21.
67. Ross WM, et al. Radiation cataract formation diminished by vitamin E in rat lenses in vitro. Exp Eye Res. 1983;36(5):645–53.
68. Libondi T, et al. Changes of some biochemical parameters of the lens in galactose-treated weaned rats with and without vitamin E therapy. Ophthalmic Res. 1985;17(1):42–8.
69. Creighton MO, et al. Modelling cortical cataractogenesis VII: effects of vitamin E treatment on galactose-induced cataracts. Exp Eye Res. 1985;40(2):213–22.
70. Ohta Y, et al. Preventive action of vitamin E-containing liposomes on cataractogenesis in young adult rats fed a 25% galactose diet. J Ocul Pharmacol Ther. 1997;13(6):537–50.
71. Ohta Y, et al. Preventive effect of topical vitamin E-containing liposome instillation on the progression of galactose cataract. Comparison between 5-week- and 12-week-old rats fed a 25% galactose diet. Exp Eye Res. 1999;68(6):747–55.

72. Yeum KJ, et al. Measurement of carotenoids, retinoids, and tocopherols in human lenses. Invest Ophthalmol Vis Sci. 1995;36(13):2756–61.
73. Leske MC, et al. Antioxidant vitamins and nuclear opacities: the longitudinal study of cataract. Ophthalmology. 1998;105(5):831–6.
74. Rouhiainen P, Rouhiainen H, Salonen JT. Association between low plasma vitamin E concentration and progression of early cortical lens opacities. Am J Epidemiol. 1996;144(5):496–500.
75. Lyle BJ, et al. Serum carotenoids and tocopherols and incidence of age-related nuclear cataract. Am J Clin Nutr. 1999;69(2):272–7.
76. Leske MC, et al. Biochemical factors in the lens opacities. Case-control study. The Lens Opacities Case-Control Study Group. Arch Ophthalmol. 1995;113(9):1113–9.
77. McNeil JJ, et al. Vitamin E supplementation and cataract: randomized controlled trial. Ophthalmology. 2004;111(1):75–84.
78. Stahl W, Sies H. Antioxidant activity of carotenoids. Mol Aspects Med. 2003;24(6):345–51.
79. Tapiero H, Townsend DM, Tew KD. The role of carotenoids in the prevention of human pathologies. Biomed Pharmacother. 2004;58(2):100–10.
80. Gale CR, et al. Plasma antioxidant vitamins and carotenoids and age-related cataract. Ophthalmology. 2001;108(11):1992–8.
81. Lyle BJ, et al. Antioxidant intake and risk of incident age-related nuclear cataracts in the Beaver Dam Eye Study. Am J Epidemiol. 1999;149(9):801–9.
82. Delcourt C, et al. Plasma lutein and zeaxanthin and other carotenoids as modifiable risk factors for age-related maculopathy and cataract: the POLA Study. Invest Ophthalmol Vis Sci. 2006;47(6):2329–35.
83. Moeller SM, et al. Associations between age-related nuclear cataract and lutein and zeaxanthin in the diet and serum in the carotenoids in the Age-related eye disease study, an ancillary study of the Women's health initiative. Arch Ophthalmol. 2008;126(3):354–64.
84. Trumbo PR, Ellwood KC. Lutein and zeaxanthin intakes and risk of age-related macular degeneration and cataracts: an evaluation using the Food and Drug Administration's evidence-based review system for health claims. Am J Clin Nutr. 2006;84(5):971–4.
85. Rajkumar S, et al. Activity of superoxide dismutase isoenzymes in lens epithelial cells derived from different types of age-related cataract. J Cataract Refract Surg. 2008;34(3):470–4.
86. Matsui H, et al. The effect of up- and downregulation of MnSOD enzyme on oxidative stress in human lens epithelial cells. Invest Ophthalmol Vis Sci. 2003;44(8):3467–75.
87. Reddy VN, et al. Effects of variation in superoxide dismutases (SOD) on oxidative stress and apoptosis in lens epithelium. Exp Eye Res. 2004;79(6):859–68.
88. Noor R, Mittal S, Iqbal J. Superoxide dismutase: applications and relevance to human diseases. Med Sci Monit. 2002;8(9):RA210–5.
89. Lin D, et al. Expression of superoxide dismutase in whole lens prevents cataract formation. Mol Vis. 2005;11:853–8.
90. Gottipati S, Cammarata PR. Mitochondrial superoxide dismutase activation with 17 beta-estradiol-treated human lens epithelial cells. Mol Vis. 2008;14:898–905.
91. Schriner SE, et al. Extension of murine life span by overexpression of catalase targeted to mitochondria. Science. 2005;308(5730):1909–11.
92. Zheng Y, et al. Resveratrol protects human lens epithelial cells against H_2O_2-induced oxidative stress by increasing catalase, SOD-1, and HO-1 expression. Mol Vis. 2010;16:1467–74.
93. Chamberlain CG, Mansfield KJ, Cerra A. Glutathione and catalase suppress TGFbeta-induced cataract-related changes in cultured rat lenses and lens epithelial explants. Mol Vis. 2009;15:895–905.
94. Ho YS, et al. Mice lacking catalase develop normally but show differential sensitivity to oxidant tissue injury. J Biol Chem. 2004;279(31):32804–12.
95. Cox AG, Winterbourn CC, Hampton MB. Mitochondrial peroxiredoxin involvement in antioxidant defence and redox signalling. Biochem J. 2010;425(2):313–25.

96. Wood ZA, et al. Structure, mechanism and regulation of peroxiredoxins. Trends Biochem Sci. 2003;28(1):32–40.
97. Lee W, Wells T, Kantorow M. Localization and H$_2$O$_2$-specific induction of PRDX3 in the eye lens. Mol Vis. 2007;13:1469–74.
98. Reddy VN, et al. Peroxide-induced damage in lenses of transgenic mice with deficient and elevated levels of glutathione peroxidase. Ophthalmologica. 1997;211(3):192–200.
99. Fridovich I. Superoxide anion radical (O2-.), superoxide dismutases, and related matters. J Biol Chem. 1997;272(30):18515–7.
100. Garner B, Davies MJ, Truscott RJ. Formation of hydroxyl radicals in the human lens is related to the severity of nuclear cataract. Exp Eye Res. 2000;70(1):81–8.
101. Garner B, et al. Redox availability of lens iron and copper: implications for HO* generation in cataract. Redox Rep. 1999;4(6):313–5.
102. Ruffett I, Ayres J, McBride D. Possible chemical pollution. Practitioner. 1992;236(1510):13–6.
103. Leske MC, Chylack Jr LT, Wu SY. The Lens Opacities Case-Control Study. Risk factors for cataract. Arch Ophthalmol. 1991;109(2):244–51.
104. Ramakrishnan S, et al. Smoking of beedies and cataract: cadmium and vitamin C in the lens and blood. Br J Ophthalmol. 1995;79(3):202–6.
105. Balaji M, Sasikala K, Ravindran T. Copper levels in human mixed, nuclear brunescence, and posterior subcapsular cataract. Br J Ophthalmol. 1992;76(11):668–9.
106. Dawczynski J, et al. Increased content of zinc and iron in human cataractous lenses. Biol Trace Elem Res. 2002;90(1–3):15–23.
107. Oppermann B, et al. Identification and spatial analysis of metallothioneins expressed by the adult human lens. Invest Ophthalmol Vis Sci. 2001;42(1):188–93.
108. Kantorow M, et al. Differential display detects altered gene expression between cataractous and normal human lenses. Invest Ophthalmol Vis Sci. 1998;39(12):2344–54.
109. Hawse JR, et al. Activation of metallothioneins and alpha-crystallin/sHSPs in human lens epithelial cells by specific metals and the metal content of aging clear human lenses. Invest Ophthalmol Vis Sci. 2003;44(2):672–9.
110. Hawse JR, et al. The role of metallothionein IIa in defending lens epithelial cells against cadmium and TBHP induced oxidative stress. Mol Vis. 2006;12:342–9.
111. Garner B, et al. Distribution of ferritin and redox-active transition metals in normal and cataractous human lenses. Exp Eye Res. 2000;71(6):599–607.
112. Levi S, et al. Analysis of ferritins in lymphoblastoid cell lines and in the lens of subjects with hereditary hyperferritinemia-cataract syndrome. Blood. 1998;91(11):4180–7.
113. Harrison PM, Arosio P. The ferritins: molecular properties, iron storage function and cellular regulation. Biochim Biophys Acta. 1996;1275(3):161–203.
114. Goralska M, et al. Iron metabolism in the eye: a review. Exp Eye Res. 2009;88(2):204–15.
115. McGahan MC, Fleisher LN. A micromethod for the determination of iron and total iron-binding capacity in intraocular fluids and plasma using electrothermal atomic absorption spectroscopy. Anal Biochem. 1986;156(2):397–402.
116. McGahan MC, Fleisher LN. Inflammation-induced changes in the iron concentration and total iron-binding capacity of the intraocular fluids of rabbits. Graefes Arch Clin Exp Ophthalmol. 1988;226(1):27–30.
117. McGahan MC. Does the lens serve as a "sink" for iron during ocular inflammation? Exp Eye Res. 1992;54(4):525–30.
118. Harned J, et al. Altered ferritin subunit composition: change in iron metabolism in lens epithelial cells and downstream effects on glutathione levels and VEGF secretion. Invest Ophthalmol Vis Sci. 2010;51(9):4437–46.
119. Vogt W. Oxidation of methionyl residues in proteins: tools, targets, and reversal. Free Radic Biol Med. 1995;18(1):93–105.
120. Stadtman ER, et al. Cyclic oxidation and reduction of protein methionine residues is an important antioxidant mechanism. Mol Cell Biochem. 2002;234–235(1–2):3–9.
121. Truscott RJ, Augusteyn RC. Oxidative changes in human lens proteins during senile nuclear cataract formation. Biochim Biophys Acta. 1977;492(1):43–52.

122. Garner MH, Spector A. Selective oxidation of cysteine and methionine in normal and senile cataractous lenses. Proc Natl Acad Sci U S A. 1980;77(3):1274–7.
123. Stadtman ER, Moskovitz J, Levine RL. Oxidation of methionine residues of proteins: biological consequences. Antioxid Redox Signal. 2003;5(5):577–82.
124. Kim HY, Gladyshev VN. Methionine sulfoxide reductases: selenoprotein forms and roles in antioxidant protein repair in mammals. Biochem J. 2007;407(3):321–9.
125. Marchetti MA, et al. Methionine sulfoxide reductases B1, B2, and B3 are present in the human lens and confer oxidative stress resistance to lens cells. Invest Ophthalmol Vis Sci. 2005;46(6):2107–12.
126. Kantorow M, et al. Methionine sulfoxide reductase A is important for lens cell viability and resistance to oxidative stress. Proc Natl Acad Sci U S A. 2004;101(26):9654–9.
127. Brennan LA, et al. Deletion of mouse MsrA results in HBO-induced cataract: MsrA repairs mitochondrial cytochrome c. Mol Vis. 2009;15:985–99.
128. Brennan LA, et al. Methionine sulfoxide reductase A (MsrA) restores alpha-crystallin chaperone activity lost upon methionine oxidation. Biochim Biophys Acta. 2009;1790(12):1665–72.
129. Ow YP, et al. Cytochrome c: functions beyond respiration. Nat Rev Mol Cell Biol. 2008;9(7):532–42.
130. Horwitz J. Alpha-crystallin can function as a molecular chaperone. Proc Natl Acad Sci U S A. 1992;89(21):10449–53.
131. Horwitz J. Alpha-crystallin. Exp Eye Res. 2003;76(2):145–53.
132. Jakob U, et al. Small heat shock proteins are molecular chaperones. J Biol Chem. 1993;268(3):1517–20.
133. Brady JP, et al. Targeted disruption of the mouse alpha A-crystallin gene induces cataract and cytoplasmic inclusion bodies containing the small heat shock protein alpha B-crystallin. Proc Natl Acad Sci U S A. 1997;94(3):884–9.
134. Kalinina EV, Chernov NN, Saprin AN. Involvement of thio-, peroxi-, and glutaredoxins in cellular redox-dependent processes. Biochemistry (Mosc). 2008;73(13):1493–510.
135. Raghavachari N, Lou MF. Evidence for the presence of thioltransferase in the lens. Exp Eye Res. 1996;63(4):433–41.
136. Raghavachari N, et al. Regulation of thioltransferase expression in human lens epithelial cells. Invest Ophthalmol Vis Sci. 2001;42(5):1002–8.
137. Moon S, Fernando MR, Lou MF. Induction of thioltransferase and thioredoxin/thioredoxin reductase systems in cultured porcine lenses under oxidative stress. Invest Ophthalmol Vis Sci. 2005;46(10):3783–9.
138. Xing K, Lou MF. The possible physiological function of thioltransferase in cells. FASEB J. 2003;17(14):2088–90.
139. Meyer LM, et al. Absence of glutaredoxin1 increases lens susceptibility to oxidative stress induced by UVR-B. Exp Eye Res. 2009;89(6):833–9.
140. Fernando MR, et al. Mitochondrial thioltransferase (glutaredoxin 2) has GSH-dependent and thioredoxin reductase-dependent peroxidase activities in vitro and in lens epithelial cells. FASEB J. 2006;20(14):2645–7.
141. Wu H, Xing K, Lou MF. Glutaredoxin 2 prevents H(2)O(2)-induced cell apoptosis by protecting complex I activity in the mitochondria. Biochim Biophys Acta. 2010;1797(10):1705–15.
142. Lillig CH, Holmgren A. Thioredoxin and related molecules: from biology to health and disease. Antioxid Redox Signal. 2007;9(1):25–47.
143. Matsui M, et al. Early embryonic lethality caused by targeted disruption of the mouse thioredoxin gene. Dev Biol. 1996;178(1):179–85.
144. Nonn L, et al. The absence of mitochondrial thioredoxin 2 causes massive apoptosis, exencephaly, and early embryonic lethality in homozygous mice. Mol Cell Biol. 2003;23(3):916–22.
145. Perez VI, et al. Thioredoxin 2 haploinsufficiency in mice results in impaired mitochondrial function and increased oxidative stress. Free Radic Biol Med. 2008;44(5):882–92.
146. Reddy PG, Bhuyan DK, Bhuyan KC. Lens-specific regulation of the thioredoxin-1 gene, but not thioredoxin-2, upon in vivo photochemical oxidative stress in the Emory mouse. Biochem Biophys Res Commun. 1999;265(2):345–9.

147. Yegorova S, Liu A, Lou MF. Human lens thioredoxin: molecular cloning and functional characterization. Invest Ophthalmol Vis Sci. 2003;44(8):3263–71.
148. Yegorova S, Yegorov O, Lou MF. Thioredoxin induced antioxidant gene expressions in human lens epithelial cells. Exp Eye Res. 2006;83(4):783–92.
149. Xing KY, Lou MF. Effect of age on the thioltransferase (glutaredoxin) and thioredoxin systems in the human lens. Invest Ophthalmol Vis Sci. 2010;51(12):6598–604.
150. Leveillard T, et al. Identification and characterization of rod-derived cone viability factor. Nat Genet. 2004;36(7):755–9.
151. Martin JL. Thioredoxin–a fold for all reasons. Structure. 1995;3(3):245–50.
152. Brennan LA, Lee W, Kantorow M. TXNL6 is a novel oxidative stress-induced reducing system for methionine sulfoxide reductase a repair of alpha-crystallin and cytochrome C in the eye lens. PLoS One. 2010;5(11):e15421.
153. Felberbaum-Corti M, et al. The redox sensor TXNL1 plays a regulatory role in fluid phase endocytosis. PLoS One. 2007;2(11):e1144.
154. Andersen KM, et al. Thioredoxin Txnl1/TRP32 is a redox-active cofactor of the 26 S proteasome. J Biol Chem. 2009;284(22):15246–54.
155. Haunhorst P, et al. Characterization of the human monothiol glutaredoxin 3 (PICOT) as iron-sulfur protein. Biochem Biophys Res Commun. 2010;394(2):372–6.
156. Zhang Y, et al. Evidence that dim1 associates with proteins involved in pre-mRNA splicing, and delineation of residues essential for dim1 interactions with hnRNP F and Npw38/PQBP-1. Gene. 2000;257(1):33–43.
157. Jeong W, et al. Identification and characterization of TRP14, a thioredoxin-related protein of 14 kDa. New insights into the specificity of thioredoxin function. J Biol Chem. 2004;279(5):3142–50.
158. Jeong W, et al. Roles of TRP14, a thioredoxin-related protein in tumor necrosis factor-alpha signaling pathways. J Biol Chem. 2004;279(5):3151–9.
159. Kaarniranta K, et al. Heat shock proteins as gatekeepers of proteolytic pathways-Implications for age-related macular degeneration (AMD). Ageing Res Rev. 2009;8(2):128–39.
160. Andley UP, et al. The molecular chaperone alphaA-crystallin enhances lens epithelial cell growth and resistance to UVA stress. J Biol Chem. 1998;273(47):31252–61.
161. Bagchi M, Katar M, Maisel H. Heat shock proteins of adult and embryonic human ocular lenses. J Cell Biochem. 2002;84(2):278–84.
162. Nagineni CN, Bhat SP. Lens fiber cell differentiation and expression of crystallins in co-cultures of human fetal lens epithelial cells and fibroblasts. Exp Eye Res. 1992;54(2):193–200.
163. Wang X, et al. Expression and regulation of alpha-, beta-, and gamma-crystallins in mammalian lens epithelial cells. Invest Ophthalmol Vis Sci. 2004;45(10):3608–19.
164. Wistow G, Piatigorsky J. Recruitment of enzymes as lens structural proteins. Science. 1987;236(4808):1554–6.
165. Litt M, et al. Autosomal dominant congenital cataract associated with a missense mutation in the human alpha crystallin gene CRYAA. Hum Mol Genet. 1998;7(3):471–4.
166. Bova MP, et al. Mutation R120G in alphaB-crystallin, which is linked to a desmin-related myopathy, results in an irregular structure and defective chaperone-like function. Proc Natl Acad Sci U S A. 1999;96(11):6137–42.
167. Cherian M, Abraham EC. Decreased molecular chaperone property of alpha-crystallins due to posttranslational modifications. Biochem Biophys Res Commun. 1995;208(2):675–9.
168. van Boekel MA, et al. The influence of some post-translational modifications on the chaperone-like activity of alpha-crystallin. Ophthalmic Res. 1996;28 Suppl 1:32–8.
169. Fujii N, et al. Comparison of post-translational modifications of alpha A-crystallin from normal and hereditary cataract rats. Amino Acids. 2004;26(2):147–52.
170. Rajan S, Horn C, Abraham EC. Effect of oxidation of alphaA- and alphaB-crystallins on their structure, oligomerization and chaperone function. Mol Cell Biochem. 2006;288(1–2):125–34.

Chapter 12
Protein Serine/Threonine Phosphatases-1 and -2A in Lens Development and Pathogenesis

Wen-Feng Hu, Xiao-Hui Hu, Weike Ji, Zhao-Xia Huang, Ling Wang, Zachary Woodward, Quan Dong Nguyen, and David Wan-Cheng Li

12.1 Introduction

The reversible phosphorylation and dephosphorylation of proteins play essential roles in regulating eukaryotic cellular events such as gene expression, cell cycle progression, cell proliferation, cell differentiation, homeostasis, transformation, and apoptosis [1–6]. Phosphoproteome studies have shown that about 33 % cellular proteins are regulated by phosphorylation and dephosphorylation [7]. Genome studies have demonstrated that while over 400 genes code for different serine/threonine kinases, only 40 genes encode the catalytic subunits for different serine/threonine phosphatases [3]. A much-reduced number of genes coding for the catalytic subunits of protein serine/threonine phosphatases are due to the fact that many regulatory subunits exist to modulate their specificity.

In eukaryotes, according to the differences of the substrates, the protein phosphatases can be divided into three major groups: protein serine/threonine phosphatases

W.-F. Hu • X.-H. Hu • Z.-X. Huang • L. Wang
Department of Ophthalmology and Visual Sciences, Truhlsen Eye Institute,
College of Medicine, University of Nebraska Medical Center, Omaha, NE, USA

Key Laboratory of Protein Chemistry and Developmental Biology of Education Ministry of China, College of Life Sciences, Hunan Normal University, Changsha, Hunan, China

W. Ji • Z. Woodward • Q.D. Nguyen
Department of Ophthalmology and Visual Sciences, Truhlsen Eye Institute,
College of Medicine, University of Nebraska Medical Center, Omaha, NE, USA

D.W.-C. Li (✉)
Department of Ophthalmology and Visual Sciences, Truhlsen Eye Institute,
College of Medicine, University of Nebraska Medical Center, Omaha, NE, USA

Key Laboratory of Protein Chemistry and Developmental Biology of Education Ministry of China, College of Life Sciences, Hunan Normal University, Changsha, Hunan, China
e-mail: davidli@unmc.edu

© Springer Science+Business Media New York 2015
M.A. Babizhayev et al. (eds.), *Studies on the Cornea and Lens*,
Oxidative Stress in Applied Basic Research and Clinical Practice,
DOI 10.1007/978-1-4939-1935-2_12

(PPPs), protein tyrosine phosphatases (PTPs), and dual-specific phosphatases (DSPs) [2–4].

Protein dephosphorylation at serine/threonine residues are executed by four major protein phosphatases based on their sensitivity to inhibitors and requirement for divalent cations, protein phosphatase-1 (PP-1), protein phosphatase-2A (PP-2A), protein phosphatase-2B (PP-2B), and protein phosphatase-2C (PP-2C) [2–4, 6], and a few minor phosphatases, including phosphatase-4 (PP-4), phosphatase-5 (PP-5), phosphatase-6 (PP-6), phosphatase-7 (PP-7) [8–11].

The PP-1 is composed of one catalytic subunit and one regulatory subunit. The catalytic subunits for PP-1 exist in four different isoforms, α, β, γ, and δ, which are highly homologous and only differ in the C-terminal domain [2–4]. The regulatory subunits are PP-1-interacting proteins either with inhibitory role on PP-1 activity, or with specific targeting function. Over 90 PP-1 interacting proteins or targeting proteins have been identified [3].

The PP-2A enzyme is a heterotrimer enzyme, which consists of a scaffold A subunit, a catalytic C subunit, and a variable regulatory B subunit. The C subunit of PP-2A interacts constitutively with the A subunit, which acts as a scaffolding protein for binding the C subunit and the variable regulatory B subunit. The A subunit exists in two isoforms, Aα and Aβ, which are derived from two different genes and are differentially expressed in most eukaryotic tissues [2–4]. The C subunit also exists in two isoforms, α and β, which are encoded by two different genes [2–4]. The B subunit has an important role in targeting PP-2A holoenzymes to specific subcellular proteins and is critical for regulating substrate specificity. Over 16 genes have been identified encoding for subfamilies of the regulatory subunits, including B, B', B", and B"' [2–4]. The PP-2B enzyme has two subunits and PP-2C is monomeric enzyme [2–4]. PP-2A, PP-2B, and PP-2C can be distinguished by their substrate specificity and subunit number. While PP-2B and PP-2C exhibit strict requirement for Ca^{2+} or Mg^{2+}, respectively, PP-2A does not. PP-2A can dephosphorylate phosphorylase-a, while PP-2B and PP-2C do not. Moreover, PP-1 and PP-2A are sensitive to okadaic acid and microcystin with different sensitivities, whereas PP-2B and PP-2C do not [2–4].

In eukaryotes, over 98 % dephosphorylation is performed by protein serine/threonine phosphatases, among which 90 % of intracellular protein serine/threonine phosphatase activity has been attributed to PP-1 and PP-2A [3]. Thus, our attention focuses on them. While the structures and general functions of PP-1 and PP-2A have well documented by different reviews [2–6], this chapter aims at summarizing their roles in regulating development and pathogenesis of the ocular lens.

12.2 Ocular Development

The mammalian eye is one of the most important organs, which is essentially a highly specialized extension of the brain. The eye development starts at the gastrulation when the eye field is in the early neural plate [12–21]. The optic pit appearances and the continued evagination of the optic pit induce the formation of the optic vesicles. The optic vesicle and the surface ectoderm move into close physical contact and interact with each other. As a result, the thickening of the surface ectoderm forms the lens placode. The lens placode and the optic vesicle coordinate invagination and then form the lens vesicle and optic cup. The optic cup has two layers, in which the inner layer forms the neural retina and the outer layer becomes the retinal pigment epithelium [12–15]. The posterior cells of the lens vesicle form the primary lens fiber cells. The lens epithelial cells (LECs) in the anterior continue to undergo proliferation and differentiation, generating secondary fiber cells. This process lasts the whole life. During this processing, crystallins are expressed at a high level and required for generating and maintaining lens transparency [12–15, 19–21]. The differentiation process includes cell elongation and the loss of cellular organelles including the nucleus, mitochondria, and endoplasmic reticulum besides high levels of expression of crystallins in the fiber cells [17, 18]. The surface ectoderm forms the cornea. The cornea epithelial cells secrete an extracellular matrix, which attracts neural crest-derived mesenchymal cells into the cornea and finally induces the dehydration and compaction of the posterior stroma and leads to the formation of mature cornea. The ciliary body and iris are both derived from the optic cup where the inner layer and the outer layer of the optic cup meet [21].

12.3 Roles of PP-1 in Lens Development and Cataractogenesis

PP-1 is expressed in all compartments of the mouse eye [22]. Among four eye tissues (retina, cornea, lens epithelium, and lens fiber), the catalytic subunits of PP-1 display a higher level of expression in retina and cornea, but a reduced level in lens epithelium and barely detectable in lens fiber [22]. In mouse, bovine, rat, and human lenses examined [22, 23], PP-1 displayed much higher expression levels than that of PP-2A [22, 23]. These results suggest that PP-1 have important functions in different eye tissues [22, 23].

Indeed, our recent studies have shown that PP-1 plays an essential role in regulating eye development of vertebrates. Using goldfish as a model system with morpholino oligo inhibition and shRNA silencing, we have shown that inhibition of PP-1 activity leads to the death of a majority of the treated embryos, and the surviving embryos displayed severe phenotype in the eye [24]. Knockdown of each catalytic subunit of PP-1 with morpholino oligomers leads to the death of some (PP-1α knockdown) or all (PP-1β or PP-1γ knockdown) the injected embryos. The surviving embryos from PP-1 knockdown displayed clear retardation in lens differentiation and abnormal retina differentiation [24]. Disturbance of normal PP-1 activity through overexpression of each subunit of PP-1 also caused death of majority of the injected embryos and led to abnormal development of goldfish eye. These results thus provide direct evidence that PP-1 plays a key role in governing normal eye formation during goldfish development.

Mechanistically, Pax-6 is one of the major downstream targets mediating the effects of PP-1 function for several reasons. First, we have previously shown that PP-1 is the major phosphatase modulating the phosphorylation status of Pax-6 [25]. Second, inhibition of PP-1 led to significantly changes in the phosphorylation status of Pax-6 [24]. Finally, the eye phenotype in Pax-6 knockdown fish was similar to that derived from overexpression of PP-1 [24]. In addition to Pax-6, PP-1 can also directly dephosphorylate the retinoblastoma protein (Rb) to modulate the functional status of the E2F transcription factor family members and thus control cell cycle progression. Ludlow et al. had previously demonstrated that PP-1 can directly dephosphorylate Rb to modulate its function in the late mitosis stage [26]. Furthermore, AKT pathway plays an important role to regulate signaling transduction pathways for cell proliferation, differentiation, and survival. We have recently shown that PP-1 is a major protein phosphatase to directly dephosphorylate AKT at Thr450 to modulate its activation status in human lens epithelial cells (HLECs) and retinal pigment epithelial cells (ARPE-19) [27].

PP-1 also plays an important role in maintaining survival of LECs. Inhibition of PP-1 activity by okadaic acid (100 nM) leads to the apoptosis of LECs in rabbit, rat, and human, through the up-regulated expression of p53 and the pro-apoptotic gene Bax [28–32]. While normal developmental apoptosis is necessary for the formation of lens vesicle and its subsequent development [33–38], stress-induced apoptosis leads to the development of cataracts. This conclusion is based on numerous lines of evidence. First, during development, apoptosis induced by gene mutations, expression of exogenous genes, and stress factors all cause microphthalmia or cataractogenesis [39–48]. One of the most striking examples came from the study of CREB-2 knockout. Normal lens development occurs between embryonic day (ED)12.5 and ED14.5 in CREB-2(−/−) mice. These mice displayed normal formation of the early lens vesicle, normal elongation of posterior primary fiber cells, and normal formation of the anterior LECs. However, the anterior LECs and their direct descendants at the equatorial poles of the lens underwent massive and synchronous p53-dependent apoptosis between ED14.5 and ED16.5, causing complete degeneration of the embryonic lens in these animals [39]. Second, we previously demonstrated that in the in vitro rat lens organ culture, various stress factors including oxidative stress, UB irradiation, and abnormal calcium insult-induced LEC apoptosis followed by the development of lens opacification [49–51]. In vitro organ culture study by another group demonstrated that inhibition of the Src kinase pathway lead to cataract formation, which is initiated by apoptosis of LECs in the treated lens [52]. Third, in the in vivo animal model studies with rat and rabbit, several laboratories have demonstrated that UV irradiation [53–55], diabetic condition [56, 57], selenite [58], and N-methyl-N-nitrosourea [59] all induced early apoptosis of LECs followed by cataractogenesis. It has also been shown that death of LECs was directly linked to age-related cataractogenesis [60]. Fourth, in human cataractous lenses, we have previously observed that varying levels of apoptotic cells were present in the capsular epithelia from cataractous patients [49]. These results have been confirmed by most other studies [61–64] but are in contrast with one report [65] in which the authors claimed there was no apoptosis. In a recent study [64], it was reported that

TUNEL staining was observed in 92.6 % specimens of cataractous lenses, whereas cells undergoing apoptosis were identified in only 8 % epithelia from non-cataractous lenses. This group concluded that the accumulation of small-scale epithelial losses during a lifetime may induce alterations in lens fiber formation and homeostasis and results in loss of lens transparency. In addition, apoptosis is also closely associated with human polar cataracts [66, 67]. Moreover, in his recent study, on ER stress-induced cataractogenesis, Shinohara's group found that apoptosis induced by ER stress plays an important role in ER stress-induced cataractogenesis [68].

In exploring molecular mechanisms by which PP-1 promotes survival, we have previously demonstrated that PP-1 can directly dephosphorylate p53 at Ser15 and Ser37 in HLECs to negatively regulate its pro-apoptotic activity [29].

Thus, as an abundant protein phosphatase, PP-1 is essential in governing vertebrate ocular development and pathogenesis.

12.4   Role of PP-2A in Ocular Development and Pathogenesis

As mentioned above, PP-2A is another major cellular phosphatase. It exists in both core enzyme (including the catalytic subunit and the scaffold subunit) and holoenzyme (including the core enzyme plus the regulatory subunit) in cells [2–4].

Both RT-PCR and western blot analysis have shown that all three subunits of PP-2A are expressed in the ocular tissues: retina, cornea, and lens [22]. In comparison to PP-1, PP-2A is less abundant in the ocular lens. Although PP-2A is less abundant in the ocular lens, its function is also important in both development and pathogenesis.

Using goldfish as a model, our recent work demonstrates that PP-2A directly regulates goldfish eye development. As PP-1, PP-2A is also essential to maintain embryonic survival. Inhibition of PP-2A by okadaic acid (20 nM) led to death of a majority of the embryos [69]. In the surviving embryos, it was found that over half of them had no eye structure. For those remaining embryos with eye structure, lens development stayed largely at the lens vesicle stage and no further lens cell differentiation was observed. In retina, cell proliferation was enhanced as reflected by more layers of cells. On the other hand, retinal differentiation (photoreceptor elongation) was not observed and the retinal cells remained round in shape.

The above results were further confirmed by PP-2A knockdown using morpholino oligomers against the catalytic subunit. Knockdown of PP-2A led to abnormal lens that stayed at the lens vesicle stage (the lens had one layer of epithelial cells but no differentiation). In addition, knockdown of PP-2A also blocks differentiation of both retina and lens tissues. To demonstrate that maintenance of a normal PP-2A level is necessary for eye development, the catalytic subunit of PP-2A was overexpressed. Overexpression changes the level of a particular catalytic subunit of PP-2A in the phosphatase. Interestingly, the phenotypes from the overexpression of two different catalytic subunits were similar, either generating an eye with thickened retina and larger lens, or lack of lens vesicle differentiation, suggesting that the two subunits have redundant functions. The former phenotype is a result of enhanced proliferation and deregulated differentiation, but the latter is due to a lack of differentiation. Together, these results point to the conclusion that the normal level of PP-2A activity is essential for normal eye development in goldfish, and likely the same in other vertebrates.

Indeed, in *Drosophila*, the catalytic subunit of PP-2A was found to be implicated in regulation of the R7 photoreceptor specification through both positive and negative regulation of MEK/ERK signaling pathway [70, 71]. In Xenopus, PP-2A was found to be involved in regulation of eye development (initial formation of optic vesicle, induced formation of lens vesicle, and subsequent differentiation of different compartments) [72]. Mechanistically, Rorick et al. have shown that one of the B family subunits for PP-2A is required for the IGF/PI3K/Akt pathway and that interfering with the PI3K/Akt pathway inhibits eye induction (formation) [72]. Moreover, during eye field separation, this subunit is also implicated in regulating the hedgehog-signaling pathway [72].

PP-2A is also implicated in cataractogenesis. As discussed in the section above, stress-induced apoptosis of LECs lead to cataractogenesis. Our recent studies have demonstrated that PP-2A can modulate p53-dependent apoptosis [73]. PP-2A complex has been implicated in cell cycle control as a mitotic suppressor through negative regulation of p34^{cdc} kinase activity and its substrates, and Kantorow et al. have shown that the mRNA level of the B regulatory subunit of PP-2A displays downregulation in the epithelial cells of the human cataract lens [74].

Our recent studies have shown that the two different scaffold subunits, Aα and Aβ, of PP-2A are regulated by numerous transcriptional factors, including Ets-1, SP1/SP3, CREB, and AP-2α [75, 76]. Abnormal expression level and activity of PP2A-Aα and PP2A-Aβ are known to induce carcinogenesis and likely ocular diseases.

The expression of PP-2A is strong in retina and cornea, which indicates that PP-2A has an important role in both retina and cornea. CaBP4 is a neuronal Ca^{2+}-binding protein, which is expressed in the retina, and mutations in the CaBP4 gene could induce autosomal recessive incomplete congenital stationary night blindness and cone-rod synaptic disorder. Francoise et al. reported that PP-2A is involved in CaBP4 dephosphorylation and modulate its function [77]. In rd mice, Palczewski et al. observed an increased level of PP-2A activity in the extract from rd/rd retinas compared to the PP-2A activity in the extract from rd/+retinas from postnatal days 5 to 10, which suggests that the elevated level of PP-2A activity may be implicated in the degeneration of the photoreceptor cells [78].

The corneal epithelium needs to renew all the time to maintain and protect the eye from exogenous noxious factors and bacterial infection, so cell proliferation and cell migration are required for corneal epithelial cell renewal and maintaining the cornea transparency [79, 80]. ERK and p38 pathways of the MAPK cascade are both important in cell proliferation and migration. Crosstalk interaction exists between different pathways of the MAPK cascade. Wang et al. demonstrated that PP-2A is involved in the signaling crosstalk between the ERK and p38 pathways to modulate EGF-induced control of cell proliferation in the cornea epithelium during corneal differentiation and corneal wound healing [79, 80].

Together, these studies show that PP-2A also plays an important role in vertebrate eye development and pathogenesis.

12.5 Conclusions

Protein phosphorylation modulates the functional status over 30 % of total cellular proteins and 98 % phosphorylation and dephosphorylation occur at the serine/threonine residue. PP-1 and PP-2A are the most abundant protein serine/threonine phosphatases and contribute to 90 % intracellular serine/threonine phosphatase activity. Both PP-1 and PP-2A are directly involved in control of eye development, especially lens differentiation. They do so by modulating the functions of major transcription factors such as Pax-6, and also major signaling pathways such as AKT pathway. PP-1 and PP-2A can also suppress stress-induced p53-dependent apoptosis, thus prevent stress-induced pathogenesis in the eye such as cataractogenesis and retina degeneration.

Acknowledgments This work was supported in part by the National Institutes of Health grants [EY015765 and EY018380], the Cooperative Innovation Center of Engineering and New Products for Developmental Biology of Hunan Province (20134486), the National Natural Science Foundation of China (81272228), and the Lotus Scholar Program, and the Chinese Scholarship Council (WKJ, XHH, WFH).

References

1. Hunter T. Protein kinases and phosphatases: the yin and yang of protein phosphorylation and signaling. Cell. 1995;80:225–36.
2. Gallego M, Virshup DM. Protein serine/threonine phosphatases: life, death, and sleeping. Curr Opin Cell Biol. 2005;17:197–202.
3. Moorhead GBG, Trinkle-Mulcahy L, Ulke-Lemée A. Emerging roles of nuclear protein phosphatases. Nat Rev Mol Cell Biol. 2007;8:234–44.
4. Mumby MC, Walter G. Protein serine/threonine phosphatases: structure, regulation, and functions in cell growth. Physiol Rev. 1993;73:673–99.
5. Xu Y, Xing Y, Chen Y, Chao Y, Lin Z, Fan E, Yu JW, Strack S, Jeffrey PD, Shi Y. Structure of the protein phosphatase 2A holoenzyme. Cell. 2006;127:1239–51.
6. Yan Q, Mao Y-W, Li DW. Protein serine/threonine phosphatases in the nervous system. Encycl Neurosci. 2009;4:3325–9.
7. Olsen JV, Blagoev B, Gnad F, Macek B, Kumar C, Mortensen P, Mann M. Global, in vivo, and site-specific phosphorylation dynamics in signaling networks. Cell. 2006;127:635–48.
8. Brewis ND, Street AJ, Prescott AR, Cohen PT. PPX, a novel protein serine/threonine phosphatase localized to centrosomes. EMBO J. 1993;12:987–96.
9. Chen MX, McPartlin AE, Brown L, Chen YH, Barker HM, Cohen PT. A novel human protein serine/threonine phosphatase, which possesses four tetratricopeptide repeat motifs and localizes to the nucleus. EMBO J. 1994;13:4278–90.

10. Bastians H, Ponstingl H. The novel human protein serine/threonine phosphatase 6 is a functional homologue of budding yeast Sit4p and fission yeast pp e1, which are involved in cell cycle regulation. J Cell Sci. 1996;109:2865–74.

11. Huang X, Honkanen RE. Molecular cloning, expression, and characterization of a novel human serine/threonine protein phosphatase, PP7, that is homologous to Drosophila retinal degeneration C gene product (rdgC). J Biol Chem. 1998;273:1462–8.

12. McAvoy JW. Induction of the eye lens. Differentiation. 1980;17:137–49.

13. Piatigorsky J. Lens differentiation in vertebrates. A review of cellular and molecular features. Differentiation. 1981;19:134–53.

14. Grainger RM. Embryonic lens induction: shedding light on vertebrate tissue determination. Trends Genet. 1992;8:349–55.

15. Wride MA. Cellular and molecular features of lens differentiation: a review of recent advances. Differentiation. 1996;61:77–93.

16. O'Rahilly R. The early development of the eye in staged human embryos. Contrib Embryol Carnegie Inst. 1966;38:1–42.

17. Kuwabara T, Imaizumi M. Denucleation process of the lens. Invest Ophthalmol Vis Sci. 1974;13:973–81.

18. Sanwal M, Muel AS, Chaudun E, Courtois Y, Counis MF. Chromatin condensation and terminal differentiation process in embryonic chicken lens in vivo and in vitro. Exp Cell Res. 1986;167:429–39.

19. McAvoy JW, Chamberlain CG. Growth factors in the lens. Prog Growth Factor Res. 1990;2:29–43.

20. Lang RA. Which factors stimulate lens fiber cell differentiation in vivo? Invest Ophthalmol Vis Sci. 1999;40:3075–8.

21. Chow RL. Early eye development in vertebrates. Annu Rev Cell Dev Biol. 2001;17:255–96.

22. Liu WB, Li Y, Zhang L, Chen HG, Sun S, Liu JP, Liu Y, Li DW. Differential expression of the catalytic subunits for PP-1 and PP-2A and the regulatory subunits for PP-2A in mouse eye. Mol Vis. 2008;14:762–73.

23. Li DW-C, Xiang H, Fass U, Zhang X-Y. Analysis of expression patterns of protein phosphatase-1 and phosphatase-2A in rat and bovine lenses. Invest Ophthalmol Vis Sci. 2001;42:2603–9.

24. Liu WB, Yan Q, Liu FY, Tang XC, Chen HG, Liu J, Nie L, Zhang XW, Ji WK, Hu XH, Hu WF, Woodward Z, Wu KL, Wu MX, Liu XL, Luo LX, Yu MB, Liu YZ, Liu SJ, Li DW. Protein serine/threonine phosphotase-1 is essential in governing normal development of vertebrate eye. Curr Mol Med. 2012;12:1361–71.

25. Yan Q, Liu WB, Qin J, Liu J, Chen HG, Huang X, Chen L, Sun S, Deng M, Gong L, Li Y, Zhang L, Liu Y, Feng H, Xiao Y, Liu Y, Li DW. Protein phosphatase-1 dephosphorylates Pax-6, a transcription factor controlling brain and eye development. J Biol Chem. 2007;282(19):13954–65.

26. Ludlow JW, Glendening CL, Livingston DM, DeCarprio JA. Specific enzymatic dephosphorylation of the retinoblastoma protein. Mol Cell Biol. 1993;13:367–72.

27. Xiao L, Gong LL, Yuan D, Deng M, Zeng XM, Chen LL, Zhang L, Yan Q, Liu JP, Hu XH, Sun SM, Liu J, Ma HL, Zheng CB, Fu H, Chen PC, Zhao JQ, Xie SS, Zou LJ, Xiao YM, Liu WB, Zhang J, Liu Y, Li DW. Protein phosphatase-1 regulates Akt1 signal transduction pathway to control gene expression, cell survival and differentiation. Cell Death Differ. 2010;17(9):1448–62.

28. Li DW, Fass U, Huizar I, Spector A. Okaidac acid-induced lens epithelial cell apoptosis requires inhibition of phosphatase-1 and is associated with induction of gene expression including p53 and bax. Eur J Biochem. 1998;257:351–61.

29. Li DW, Liu JP, Schmid PC, Schlosser R, Feng H, Liu WB, Yan Q, Gong L, Sun SM, Deng M, Liu Y. Protein serine/threonine phosphatase-1 dephosphorylates p53 at Ser-15 and Ser-37 to modulate its transcriptional and apoptotic activities. Oncogene. 2006;25:3006–22.

30. Yan Q, Liu J-P, Li DW-C. Apoptosis in the ocular lens: role in development and pathogenesis. Differentiation. 2006;74:195–211.

31. Zhang L, Yan Q, Liu JP, Zou LJ, Liu J, Sun S, Deng M, Gong L, Ji WK, Li DW. Apoptosis: its functions and control in the ocular lens. Curr Mol Med. 2010;10(9):864–75.
32. Hu WF, Gong L, Cao Z, Ma H, Ji W, Deng M, Liu M, Hu XH, Chen P, Yan Q, Chen HG, Liu J, Sun S, Zhang L, Liu JP, Wawrousek E, Li DW. αA- and αB-crystallins interact with caspase-3 and Bax to guard mouse lens development. Curr Mol Med. 2012;12(2):177–87.
33. Glucksmann A. Cell deaths in normal vertebrate ontogeny. Biol Rev. 1951;26:59–86.
34. Silver J, Hughes AFW. The role of cell death during morphogenesis of the mammalian eye. J Morphol. 1973;140:159–70.
35. Ishizaki Y, Voyvodic JT, Burne JF, Raff MC. Control of lens epithelial cell survival. J Cell Biol. 1993;121:899–908.
36. Ozeki H, Ogura Y, Hirabayashi Y, Shimada S. Suppression of lens stalk cell apoptosis by hyaluronic acid leads to faulty separation of the lens vesicle. Exp Eye Res. 2001;72:63–70.
37. Bozanic D, Tafra R, Saraga-Babic M. Role of apoptosis and mitosis during human eye development. Eur J Cell Biol. 2003;82:421–9.
38. Mohamed YH, Amemiya T. Apoptosis and lens vesicle development. Eur J Ophthalmol. 2003;13(1):1–10.
39. Hettmann T, Barton K, Leiden JM. Microphthalmia due to p53-mediated apoptosis of anterior lens epithelial cells in mice lacking the CREB-2 transcription factor. Dev Biol. 2000;222:110–23.
40. Blixt A, Mahlapuu M, Aitola M, Pelto-Huikko M, EneRback S, Carlsson P. A forkhead gene, FoxE3, is essential for lens epithelial proliferation and closure of the lens vesicle. Genes Dev. 2000;14:245–54.
41. Fromm L, Shawlot W, Gunning K, Butel JS, OveRbeek PA. The retinoblastoma protein-binding region of simian virus 40 large T antigen alters cell cycle regulation in lenses of transgenic mice. Mol Cell Biol. 1994;14:6743–54.
42. Pan H, Griep AE. Temporally distinct patterns of p53-dependent and p53-independent apoptosis during mouse lens development. Genes Dev. 1995;9:2157–69.
43. Morgenbesser SD, Schreiber-Agus N, Bidder M, Mahon KA, OveRbeek PA, Horner J, DePinho RA. Contrasting roles for c-Myc and L-Myc in the regulation of cellular growth and differentiation in vivo. EMBO J. 1995;14:743–56.
44. Nakamura T, Pichel JG, Williams-Simons L, Westphal H. An apoptotic defect in lens differentiation caused by human p53 is rescued by a mutant allele. Proc Natl Acad Sci U S A. 1995;92:6142–6.
45. Robinson ML, MacMillan-Crow LA, Thompson JA, OveRbeek PA. Expression of a truncated FGF receptor results in defective lens development in transgenic mice. Development. 1996;121:3959–67.
46. Gomez Lahoz E, Liegeois NJ, Zhang P, Engelman JA, Horner J, Silverman A, Burde R, Roussel MF, Sherr CJ, Elledge SJ, DePinho RA. Cyclin D- and E-dependent kinases and the p57(KIP2) inhibitor: cooperative interactions in vivo. Mol Cell Biol. 1997;19:353–63.
47. McCaffrey J, Yamasaki L, Dyson NJ, Harlow E, Griep AE. Disruption of retinoblastoma protein family function by human papillomavirus type 16 E7 oncoprotein inhibits lens development in part through E2F-1. Mol Cell Biol. 1999;19:6458–68.
48. de Iongh RU, Lovicu FJ, Overbeek PA, Schneider MD, Joya J, Hardeman ED, McAvoy JW. Requirement for TGFbeta receptor signaling during terminal lens fiber differentiation. Development. 2001;128:3995–4010.
49. Li WC, Kuszak JR, Dunn K, Wang RR, Ma W, Wang GM, Spector A, et al. Lens epithelial cell apoptosis appears to be a common cellular basis for non-congenital cataract development in humans and animals. J Cell Biol. 1995;130:169–81.
50. Li W-C, Kuszak JR, Wang G-M, Wu Z-Q, Spector A. Calcimycin-induced lens epithelial cell apoptosis contributes to cataract formation. Exp Eye Res. 1995;61:89–96.
51. Li W-C, Spector A. Lens epithelial cell apoptosis is an early event in the development of UVB-induced cataract. Free Radic Biol Med. 1996;20:301–11.
52. Zhou J, Leonard M, Van Bockstaele E, Menko AS. Mechanism of Src kinase induction of cortical cataract following exposure to stress: destabilization of cell-cell junctions. Mol Vis. 2007;13:1298–310.

53. Michael R, Vrensen GF, van Marle J, Gan L, SodeRberg PG. Apoptosis in the rat lens after *in vivo* threshold dose ultraviolet irradiation. Invest Ophthalmol Vis Sci. 1998;39:2681–7.
54. Michael R, Vrensen GF, van Marle J, Lofgren S, Soderberg PG. Repair in the rat lens after threshold ultraviolet radiation injury. Invest Ophthalmol Vis Sci. 2000;41:204–12.
55. Ayala M, Strid H, Jacobsson U, Söderberg PG. p53 Expression and apoptosis in the lens after ultraviolet radiation exposure. Invest Ophthalmol Vis Sci. 2007;48(9):4187–91.
56. Takamura Y, Kubo E, Tsuzuki S, Akagi Y. Apoptotic cell death in the lens epithelium of rat sugar cataract. Exp Eye Res. 2003;77:51–7.
57. Murata M, Ohta N, Sakurai S, Alam S, Tsai J, Kador PF, Sato S. The role of aldose reductase in sugar cataract formation: aldose reductase plays a key role in lens epithelial cell death (apoptosis). Chem Biol Interact. 2001;130–132(1–3):617–25.
58. Tamada Y, Fukiage C, Nakamura Y, Azuma M, Kim YH, Shearer TR. Evidence for apoptosis in the selenite rat model of cataract. Biochem Biophys Res Commun. 2000;275:300–6.
59. Yoshizawa K, Oishi Y, Nambu H, Yamamoto D, Yang J, Senzaki H, Miki H, Tsubura A. Cataractogenesis in neonatal Sprague-Dawley rats by N-methyl-N-nitrosourea. Toxicol Pathol. 2000;28:555–64.
60. Wolf N, Penn P, Pendergrass W, Van Remmen H, Bartke A, Rabinovitch P, Martin GM. Age-related cataract progression in five mouse models for anti-oxidant protection or hormonal influence. Exp Eye Res. 2005;81:276–85.
61. Nishi O, Nishi K. Apoptosis in lens epithelial cells of human cataracts. J Eye. 1998;15:1309–13 (In Japanese).
62. Takamura Y, Sugimoto Y, Kubo E, Takahashi Y, Akagi Y. Immunohistochemical study of apoptosis of lens epithelial cells in human and diabetic rat cataracts. Jpn J Ophthalmol. 2001;45:559–63.
63. Okamura N, Ito Y, Shibata MA, Ikeda T, Otsuki Y. Fas-mediated apoptosis in human lens epithelial cells of cataracts associated with diabetic retinopathy. Med Electron Microsc. 2002;35:234–41.
64. Charakidas A, Kalogeraki A, Tsilimbaris M, Koukoulomatis P, Brouzas D, Delides G. Lens epithelial apoptosis and cell proliferation in human age-related cortical cataract. Eur J Ophthalmol. 2005;15(2):213–20.
65. Harocopos GJ, Alvares KM, Kolker AE, Beebe DC. Human age-related cataract and lens epithelial cell death. Invest Ophthalmol Vis Sci. 1998;39:2696–706.
66. Mihara E, Miyata H, Nagata M, Ohama E. Lens epithelial cell damage and apoptosis in atopic cataract-histopathological and immunohistochemical studies. Jpn J Ophthalmol. 2000;44:695–6.
67. Lee EH, Wan XH, Song J, Kang JJ, Cho JW, Seo KY, Lee JH. Lens epithelial cell death and reduction of anti-apoptotic protein Bcl-2 in human anterior polar cataracts. Mol Vis. 2002;11:235–40.
68. Mulhern ML, Madson CJ, Danford A, Ikesugi K, Kador PF, Shinohara T. The unfolded protein response in lens epithelial cells from galactosemic rat lenses. Invest Ophthalmol Vis Sci. 2006;47(9):3951–9.
69. Liu WB, Hu XH, Zhang XW, Deng MX, Nie L, Hui SS, Duan W, Tao M, Zhang C, Liu J, Hu WF, Huang ZX, Li L, Yi M, Li TT, Wang L, Liu Y, Liu SJ, Li DW. Protein serine/threonine phosphotase-2A are differentially expressed and regulates eye development in vertebrates. Curr Mol Med. 2013;13(8):1376–84.
70. Wassarman DA, Solomon NM, Chang HC, Karim FD, Therrien M, Rubin GM. Protein phosphatase 2A positively and negatively regulates Ras1-mediated photoreceptor development in Drosophila. Genes Dev. 1996;10:272–8.
71. Edwards SC, O'Day PM, Herrera DC. Characterization of protein phosphatases type 1 and type 2A in Limulus nervous tissue: their light regulation in the lateral eye and evidence of involvement in the photoresponse. Vis Neurosci. 1996;13:73–85.
72. Rorick AM, Mei W, Liette NL, Phiel C, El-Hodiri HM, Yang J. PP2A:B56epsilon is required for eye induction and eye field separation. Dev Biol. 2007;302:477–93.

73. Qin J, Chen HG, Yan Q, Deng M, Liu J, Doerge S, Ma W, Dong Z, Li DW. Protein phosphatase-2a is a target of epigallocatechin-3-gallate and modulates p53-Bak apoptotic pathway. Cancer Res. 2008;68:4150–62.
74. Kantorow M, Kays T, Horwitz J, Huang Q, Sun J, Piatigorsky J, Carper D. Differential display detects altered gene expression between cataractous and normal human lenses. Invest Ophthalmol Vis Sci. 1998;39:2344–54.
75. Liu J, Ji W, Sun S, Zhang L, Chen HG, Mao Y, Liu L, Zhang X, Gong L, Deng M, Chen L, Han WJ, Chen PC, Hu WF, Hu X, Woodward Z, Liu WB, Xiao YM, Liang SP, Liu Y, Liu SJ, Li DW. The PP2A-Aβ gene is regulated by multiple transcriptional factors including Ets-1, SP1/SP3, and RXRα/β. Curr Mol Med. 2012;12(8):982–94.
76. Chen HG, Han WJ, Deng M, Qin J, Yuan D, Liu JP, Xiao L, Gong L, Liang S, Zhang J, Liu Y, Li DW. Transcriptional regulation of PP2A-A alpha is mediated by multiple factors including AP-2alpha, CREB, ETS-1, and SP-1. PLoS One. 2009;4(9):e7019.
77. Haeseleer F, Sokal I, Gregory FD, Lee A. Protein phosphatase 2A dephosphorylates CaBP4 and regulates CaBP4 function. Invest Ophthalmol Vis Sci. 2013;54(2):1214–26.
78. Palczewski K, Farber DB, Hargrave PA. Elevated level of protein phosphatase 2A activity in retinas of rd mice. Exp Eye Res. 1991;53:101–5.
79. Wang Z, Yang H, Tachado SD, Capó-Aponte JE, Bildin VN, Koziel H, Reinach PS. Phosphatase-mediated crosstalk control of ERK and p38 MAPK signaling in corneal epithelial cells. Invest Ophthalmol Vis Sci. 2006;47:5267–75.
80. Tanifuji-Terai N, Terai K, Hayashi Y, Chikama T, Kao WW. Expression of keratin 12 and maturation of corneal epithelium during development and postnatal growth. Invest Ophthalmol Vis Sci. 2006;47:545–51.

Chapter 13
Proteases in Lens and Cataract

Puttur Santhoshkumar, Rama Kannan, and K. Krishna Sharma

Abbreviations

APH   Acylpeptide hydrolase
LC-MS   Liquid chromatography mass spectrometry
LMW   Low molecular weight
WIS   Water-insoluble
WS   Water-soluble

13.1   Lens Structure and Function

Although the lens has a limited number of cells and cell types, its development is complex [1, 2]. The eye lens is composed of a single layer of epithelial cells on its anterior portion, an interior mass of concentrically arranged, elongated fiber cells,

This review chapter is dedicated to the memory of Dr. Beryl J. Ortwerth

P. Santhoshkumar
Department of Ophthalmology, University of Missouri—Columbia School of Medicine, Columbia, MO, USA

R. Kannan
Departments of Ophthalmology and Biochemistry, University of Missouri—Columbia School of Medicine, Columbia, MO, USA

K.K. Sharma (✉)
Departments of Ophthalmology and Biochemistry, University of Missouri—Columbia School of Medicine, Columbia, MO, USA

Department of Ophthalmology, University of Missouri School of Medicine, Columbia, MO 65212, USA
e-mail: sharmak@health.missouri.edu

© Springer Science+Business Media New York 2015   221
M.A. Babizhayev et al. (eds.), *Studies on the Cornea and Lens*,
Oxidative Stress in Applied Basic Research and Clinical Practice,
DOI 10.1007/978-1-4939-1935-2_13

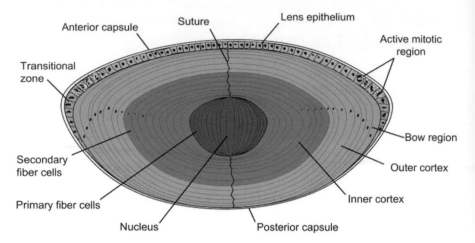

Fig. 13.1 Schematic of cross-sectional view of a human lens showing various layers. The lens epithelium is a single layer of cells covering the anterior section. At the bow region, the epithelial cells elongate, lose their organelles, and differentiate to form fiber cells. The central nucleus is made up of primary fiber cells. The cortex comprises secondary fiber cells and is divided into inner and outer cortex. The outermost layer covering the entire lens is called the capsule

and a collagenous capsule that surrounds the lens (Fig. 13.1) [3]. Lens development begins around embryonic day 33 in humans with invagination of the lens placode to form the lens vesicle. The differentiating epithelial cells then start filling the vesicle until they have obliterated the whole cavity by the end of the seventh week. The "first formed" fiber cells that occupy the center of the lens become the embryonic nucleus. To accomplish this at the equatorial region, the epithelial cells differentiate, elongate, and become fiber cells and synthesize large quantities of the structural proteins called crystallins. At the same time all potential light-scattering organelles are removed from the fiber cells in a programmed process involving the proteases [4]. New fiber cells continue to be formed at the equatorial zone throughout life.

The lens derives all of its nutrients and oxygen from the aqueous humor and vitreous body. Because the lens is composed of fibers representing a range of ages, it is an attractive tissue for studying the effects of aging on protein structure and function [5]. Lens fiber cells are arranged in layers, with little extracellular space to disrupt the regular array [6]. Transparency is achieved by the tightly ordered packing and assembly of crystallins, the soluble proteins in lens fibers [7]. Protein concentrations in the lens are very high (450 mg/mL), yet there is no colloidal osmotic pressure (and hence no ability to draw in water). Both the packing and hydration properties of the proteins confer the refractive power to the lens. Although the lens has protective mechanisms that preserve its function during most of the human life span, these mechanisms begin to deteriorate with aging, leading to the gradual accumulation and aggregation of modified proteins, and a concomitant loss in the optical quality of the lens, and, ultimately, cataract formation [3, 8–11].

Crystallins make up about 90 % of the lens proteins and their properties account for the high refractive index of the lens [3, 7, 12]. Other proteins in the lens include cytoskeletal and membrane proteins, such as actin, filensin, and spectrin; transporters and channel proteins; proteins that constitute cell–cell junction and communication; and enzymes involved in metabolism, protein synthesis, and degradation [3, 7]. Crystallins produced within the lens during the formation of lens fiber cells have exceptional longevity. There are three classes of crystallins—α-, β-, and γ-crystallins—encoded by a diverse groups of genes. α-Crystallin constitutes 40 % of the total lens protein mass and is composed of two types of 20-kDa subunits, αA and αB. The αA- and αB-crystallins coexist, in roughly a 3:1 ratio, as polydisperse hetero-oligomers [12]. α-Crystallin belongs to the small heat shock protein (sHsp) family and functions as both a molecular chaperone and a structural protein [12, 13]. These properties are particularly relevant to the preservation of β- and γ-crystallins in the non-aggregating form to maintain lens transparency. β-Crystallins are a family of basic (βB1, βB2, βB3) and acidic (βA1, βA2, βA3, and βA4) polypeptides of 22–28 kDa. They exist as hetero-oligomers. γ-Crystallins exist only as monomers of 20 kDa. Both β- and γ-crystallins serve as structural proteins in lens [7].

13.2 Lens Proteases and Their Effect on Lens Proteins

Table 13.1 lists the various proteases and peptidases that have been identified or isolated from the lens and lens epithelial cells, underscoring the critical role the lens proteolytic system plays during lens development and morphogenesis [2, 4, 14, 15]. Most of the proteases expressed during lens development complete their major roles

Table 13.1 Proteases and peptidases in lens and lens epithelial cells

Name	References
Acylpeptide hydrolase	[26]
ADAM-12, 19, ADAMTS7	[27]
Aminopeptidase III	[28, 29]
Calpains	[22, 23, 30–34]
Caspases 3, 6, 7	[35]
Cathepsin B, D	[36]
Dipeptidase	[37]
Leucine aminopeptidase	[38]
Matrix metalloproteases	[27, 39, 40]
Membrane proteinase	[21]
Prolyl oligopeptidase	[41]
Proteasome	[16, 42–45]
Secretases	[46]
Serine-type protease	[19, 20, 47]
Thimet oligopeptidase	[18]
Trypsin-like protease	[17, 47]

by the time the fiber cells attain maturity, but they remain in the lens, together with peptidases, and exert a low level of activity. With time, their cumulative effects contribute to the age-related truncation of crystallins, the cleavage of cytoskeletal proteins, membrane proteins such as aquaporins and connexins, and the generation of crystallin fragments. A number of proteases have been implicated in various aspects of lens development, organelle degradation, and lens physiology. Several endopeptidase activities have been demonstrated in mature lens fiber cells [16–25].

To highlight some of the other evidence linking proteolytic activity to protein aggregation preceding cataractogenesis, proteases that cause truncation of α-crystallin missing five residues from its C-terminus have been identified as calpains [48]. In addition, in studies of cataract formation in rat lenses, calpains have been shown to induce accelerated truncation in α- and β-crystallins [49]. Proteolytic activity that cleaves αA-, αB-, and βA3/A1-crystallin to generate peptide fragments has been demonstrated in human lenses [19, 50]. Matrix-assisted laser desorption ionization imaging mass spectrometry (MALDI-MS imaging) studies have shown age-related changes in crystallin and other proteins in various species, including bovine, rabbit, guinea pig, rat and human lenses [51–55]. MALDI imaging studies have shown progressive truncation of αA- and αB-crystallins and aquaporin in concentric layers of fiber cell, with the highest cleavage in the oldest region of the lens [53, 55]. Further, the abundance of cleaved products increase with age in a given lens reinforcing the likelihood that an age-related increase in crystallin degradation occurs during aging [56]. The major bonds cleaved in vivo are on the carboxyl side of Asn101, Asp151, Ser168, and Ser169 of the A chain and on the carboxyl side of Thr170 of the B chain of bovine α-crystallin [11, 57]. The other truncation observed in αA-crystallin is due to cleavage at the carboxyl side of Ser172 [58], and the amount of cleaved αA- and αB-crystallin is high in the water-insoluble (WIS) fraction than in water-soluble (WS) fraction [59–61]. One study found that in aged human lenses, αA-crystallin is truncated by three residues and αB is truncated by one or six residues at the N-terminus [62]. Mass spectrometric analysis of WS and WIS fractions has identified several of the previously known sites of truncation as well as a few additional ones [60, 63–68]. But the protease(s) responsible for α-crystallin truncation at multiple sites has not been identified. Truncation of βB1-, βA3/A1-, and βB2-crystallins in vivo has been reported, based on the identification of several truncated and cleaved β- and γ-crystallins in both WS and WIS fractions of human lenses [59, 60]. Since truncation alters the stability of βB2-crystallin in vitro [49], it is hypothesized that in vivo βB2-crystallin truncation may be responsible for its aggregation with aging.

Proteasome is the main intracellular proteolytic system responsible for clearing the damaged proteins. Lens fiber cells and lens epithelial cells have a completely functional ubiquitin proteasome pathway (UPP), although the protein levels and activities of the protein components are lower in the inner region of the lens than in the outer region [42, 69]. The age-dependent decrease in proteasome activity in human lens may partly reflect a decreased rate of clearance of damaged proteins and peptides [70]. The ubiquitin-dependent proteolytic pathway in the lens is involved in the degradation of oxidatively modified proteins. However, due to an age-related decrease in ubiquitin conjugation activity, the oxidatively modified crystallins accumulate in the lens, primarily in the nuclear region. The accumulation of modified crystallins is believed to be responsible, at least in part, for the development of age-related protein aggregation and light scattering [9, 69]. Recent studies have shown that inhibition of the proteasome activity involved in the differentiation of lens fiber cells prevents normal maturation of fiber cells [15]. While calpains have been implicated in the maturation of rodent lens crystallins [49] and to some extent in the development of cataract in sheep [34, 63], the role of calpains in human cataractogenesis is not yet established because human lens also contains calpain inhibitor [30, 71]. The presence of lens proteins cleaved at specific sites during aging [57] provides credence to the idea that, although protease activity begins to decline in lens fibers as they mature, the aging lens nevertheless has appreciable levels of protease and peptidase activity [14, 29, 72]. Such activity may be exemplified in posterior capsular opacification (PCO) after cataract surgery. Proteases, such as matrix metalloproteases (MMPs), have been implicated in the development of PCO. In addition, the increased expression of TIMP3, ADAM-12, ADAM-19, and ADAMTS7 genes in response to transforming growth factor β (TGFβ) treatment suggests that these proteases might play a role in the development of PCO as well [27].

13.3 Effect of Proteolysis on α-Crystallin Subunits and Chaperone Activity

The lens WIS fraction (which is mostly composed of α-crystallin aggregates) contains greater amounts of C-terminally truncated αA- and αB-crystallins than does the WS fraction [58, 60–62, 67]. Additionally, C-terminally truncated α-crystallin has reduced chaperone activity [73, 74], suggesting that C-terminal truncation affects oligomerization of α-crystallin and its chaperone activity. Studies of the oligomeric size, structure, hydrodynamic properties, and chaperone function of the C-terminally truncated human αA-crystallin mutants suggest that truncation of crystallins plays a role in high-molecular-weight aggregate formation and loss of chaperone activity in the aging lens [73–75]. A portion of αA-crystallin is in its deamidated form in the aged lens. Surprisingly, a study of truncated αA-crystallin in its deamidated form revealed that the C-terminally truncated (αA-1-140) and deamidated (αA-N123D) form possesses better chaperone activity than the truncated form [76], raising the possibility that deamidation in a truncated αA-crystallin might be beneficial. In vitro studies have shown that the C-terminally truncated αA-crystallin (αA-1-162) is susceptible to degradation by the UPP. However, oligomerization of the truncated αA-crystallin with wild-type αA- or αB-crystallin was found to render it more resistant to UPP, perhaps one of the reasons for the accumulation of truncated crystallins in vivo [77].

The effect of α-crystallin truncation on oligomerization is apparent from our studies with recombinant αA151-crystallin. The addition of increasing amounts of αA151 to wild-type αA-crystallin causes a concentration-dependent increase in the oligomeric mass of αA-crystallin. When wild-type αA-crystallin and αA151 were mixed at a 3:1 ratio, the mass of the hetero-oligomers was 14.0 ± 0.04^{e5}, whereas the oligomers formed when αA- and αA151 were mixed at 2:1 ratio was 20.0 ± 0.01^{e5} and both these masses were significantly higher than the mass of oligomers formed by wild-type αA-crystallin. There was also an increase in the hydrodynamic radius of αA-crystallin + αA151 oligomers (11.8 ± 0.3 nm), as compared to the hydrodynamic radius of oligomers of wild-type αA-crystallin (8.0 ± 0.2 nm). While the reasons for the increase in the size of oligomers in a dynamic α-crystallin molecule is not known, it might be due to the change in the subunit exchange rate following truncation. When labeled full-length αA + truncated αA151 hetero-oligomers, prepared at 2:1 ratio, were examined by fluorescence energy transfer assay, the subunit exchange rate was 0.0486/min, significantly lower than the 0.0572/min in the control experiment, suggesting that the C-terminal truncation affects the dynamic state of α-crystallin. Thus, the slower exchange rate may result in the retention of a larger number of subunits bound to the oligomer, leading to larger aggregate formation.

13.4   Analysis of Human Lenses and Identification of Crystallin Fragments

The concentration of crystallin fragments (<18 kDa) is higher in aged and cataract lenses than in normal young lenses [29, 49, 52, 59, 60, 66, 78, 79]. Similarly, we found that the concentration of low-molecular-weight crystallin fragments (<3 kDa) is greater in aged and cataractous lenses than in young lenses [56, 80]. In our studies of peptide levels in young lenses (20 years old; $n=5$), aged lenses (70 years old; $n=5$), and cataract lenses (74 years old; $n=5$), the total peptide level in young lenses was 0.29 mg/lens, whereas in old and cataract (with light brown nucleus) lenses peptide levels were only marginally different (0.9 mg/lens and 1.1 mg/lens, respectively). A study of peptide distribution within the lens by MALDI tissue imaging showed that the inner most region of the lens, which represents the oldest lens fiber cells, has the greatest amount of peptides, suggesting an age-dependent accumulation of the peptides in vivo [9, 52, 65].

Our liquid chromatography–mass spectrometry (LC-MS) studies clearly point to age-dependent peptide levels in the lens, with aged lenses having a higher concentration than young and middle-aged lenses. The total ion current (TIC) elution profile of WS peptides in young, middle-aged, and aged lenses demonstrated more peptides in the aged lens than in the middle-aged lens. When the peptide ion populations from LC-MS data sets of lens samples from young and middle-aged lenses were analyzed by Agilent Mass Profiler software and the peptides showing statistically significant abundance changes across samples were plotted in a log/log plot, a total of 1,261 peptides was found in WS lens fractions and majority of them at a higher concentration in the WS fraction of middle-aged lenses. When the data were filtered to exclude peaks whose abundance was <1 % of the most abundant, there were 128 peptides, of which 18 were unique to young lenses and 102 were unique to middle-aged lenses, and only 8 peptides were shared (Fig. 13.2a). Thus, there was a fivefold increase in the number of distinct peptides in middle-aged lenses when compared to young lenses. In the comparison between middle-aged and aged lenses, 1,804 peptides were detected and they were clearly more abundant in aged lenses than middle-aged lenses. Further analysis showed that 78 peptides were unique to middle-aged lenses, 199 unique to aged lenses, and only 74 peptides were shared (Fig. 13.2b). Thus, there is a clear progression of peptide appearance from young to aged lenses. Interesting results also emerged from our LC-MS

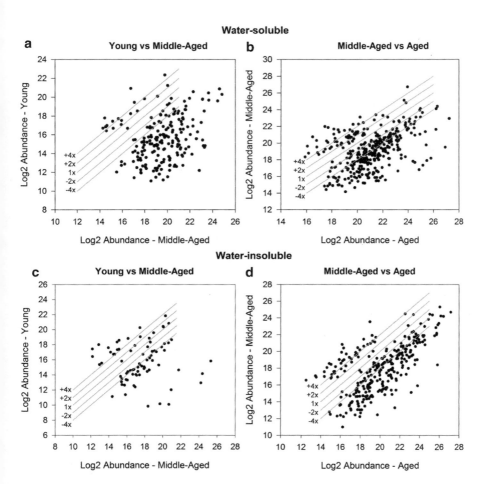

Fig. 13.2 A log/log plot displaying fold changes in water-soluble (WS) (**a, b**) and water-insoluble (WIS) (**c, d**) peptides in young vs. middle-aged lenses (**a, c**) and middle-aged vs. aged human lenses (**b, d**). The WS and WIS fractions from the lenses were treated with 6 M urea and filtered using 10 kDa filter to isolate the LMW peptides. The peptides were analyzed by LC-MS on an Agilent 6520 Accurate Mass QTOF mass spectrometer at University of Missouri Proteomic Core facility

analyses of peptides in the urea-soluble fraction of the lens proteins from different age groups. In the young vs. middle-aged comparison of peptides, 718 peptides were detected. When data were filtered to exclude peaks whose abundance is <1 % of the most abundant, only 79 peptides differentially abundant in both samples were detected and most of the peptides appear in middle-aged lenses (Fig. 13.2c). This finding indicates the prominent presence of WIS, urea-soluble peptides in middle-aged lenses. In the middle-aged vs. aged comparison, 1,502 peptides were detected with a bias towards middle-aged lens. However, after applying the 1 % threshold to the data, there were 195 peptides, of which 13 were unique to middle-aged lenses, 157 unique to aged lenses, and 25 were shared (Fig. 13.2d). Thus, the age-related increase in peptide abundance in both WS and urea-soluble fractions of the lens attests to an age-dependent increase in the cleavage of crystallins.

A number of factors could be involved in the accumulation of low-molecular-weight peptides (<10 kDa) with age, such as the activation of proteases, decreased clearing of the peptides due to diminished peptidase activities, the binding of the peptides to crystallins, and making them resistant to further degradation by peptidases, and increased nonenzymatic cleavage of crystallins in aging lenses due to free radicals generated in vivo. The presence of latent protease and nonenzymatic cleavage of lens crystallins is well documented [20, 64, 68]. Our studies with lens acylpeptide hydrolase (APH) demonstrate that a truncated form of this enzyme has the potential to cleave lens proteins and it is present in aged lenses [26, 81, 82]. Analysis of different regions of the bovine lens revealed the least amount of peptidase activity in older lens fiber cells with the highest amount of crystallin fragments (Fig. 13.3). Further, on the basis of the activity distribution of major peptidases in the lens, it is evident that the activity of leucine aminopeptidease, aminopeptidase III, prolyloligopeptidase, and APH, whose action together would result in complete hydrolysis of any peptide, is significantly less in the nuclear region as compared to the outer and inner cortical region of the same lens (Fig. 13.3).

We have identified a large number of the peptides in human lenses of different age groups [56, 80]. Several of the peptides were from the N-terminal domain of αA- or αB-crystallins or the chaperone site [56, 83]. There were also representative peptides from β- and γ-crystallins. However, it has not been possible from the peptides identified to reconstruct the full sequence of crystallins, suggesting that some of the peptides are fully hydrolyzed into amino acids. How all of these peptides are formed and why some are not completely hydrolyzed into amino acids in vivo are not known. Specific proteases are likely involved in the generation of some peptides, whereas other peptides are likely generated by free radical-mediated cleavage of peptide bonds. Recently, we synthesized peptide substrates with αA-62-70 and αA-76-85 sequence and demonstrated that the lens has proteases that cleave these synthesized substrates at the site(s) corresponding to the in vivo cleavage that results in the release of αA-66-80 peptide from αA-crystallin [25]. Human lens has αA-66-80 and βA3-188-215 peptides [56] and their truncated forms, likely generated by the action of aminopeptidases. A separate study using synthetic peptides showed that spontaneous cleavage of peptide bonds might be occurring to generate peptides with Ser-terminal residues [64].

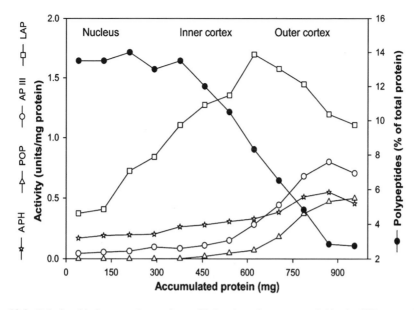

Fig. 13.3 Relationship between lens polypeptide levels and protease activities in different regions of the bovine lens. Decapsulated lenses were agitated in 3 mL portions of water and, after certain amount of fiber cells were dissolved, the undissolved portion of the lens was transferred to another container and the process was repeated 12 times to completely dissolve the lens. The 12 fractions were further homogenized in a dounce homogenizer. Leucine aminopeptidase (LAP) activity was assayed at pH 8.5 using Leu-pNA after activation with MgCl$_2$. Aminopeptidase III (APIII), acyl-peptide hydrolase (APH), and prolyl oligopeptidase (POP) activities were assayed at pH 7.5 using Arg-pNA, *N*-Acetyl-Ala-pNA, and tBOC-Ala-Ala-pNA as substrates, respectively. The polypeptide (<18 kDa) content in each fraction was measured following SDS-PAGE by image analysis. The polypeptide levels decrease with increasing protease activities and vice versa suggesting a correlation between accumulation of lens polypeptide and proteolytic activity of lens

13.5   Crystallin-Derived Peptides Induce Protein Aggregation

Proteolysis has been implicated as one of the cataract-inducing events [14, 84]. However, until recently, the molecular mechanism of crystallin fragment-induced protein aggregation in the lens has not been explained, although the intrinsic aggregation propensity of amyloidogenic peptides has been shown to correlate with protein aggregation disease [85, 86]. We were the first to demonstrate that peptides derived from oxidized crystallins induce the aggregation of crystallins by interaction at specific regions [87]. Subsequently we found that αB-1-18 (MDIAIHHPWIRRPFFPFH) and βA3/A1-102-117 (SD(N)AYHIERLMSFRPIC) peptides, present in aged and cataract lenses, induce aggregation of αB-crystallin [87, 88]. The αB-crystallin complexed with βA3/A1-102-117 peptide showed diminished chaperone activity, thereby preventing its ability to suppress protein aggregation. Peptide cross-linking and mass spectrometric studies showed that the βA3/A1 102-117 peptide interacted with αB-crystallin regions 70-74, 83-107,

121-123, 150-157, and 164-175, which are part of the α-crystallin domain and C-terminal extension. These data suggest that the βA3/A1-102-117 peptide acts like an anti-chaperone peptide because of its ability to interact with αB-crystallin at the chaperone site [88].

Our extensive studies with αA66-80 peptide and its truncated forms αA-66-75, αA-67-75, and αA-67-80, which are present in aged lenses and in increasing amounts in cataract lenses [56, 80, 83] show that these peptides interact with purified α-crystallin and cause aggregation and precipitation (Fig. 13.4). Of all the crystallin-derived peptides, αA-66-80 was found to be the most effective in precipitating α-crystallin (~70 %). Among the three truncated αA-66-80 peptides, αA-67-80 precipitated and aggregated the α-crystallin the most (49 %), followed by αA-67-75 (30 %) and αA-66-75 (20 %). Under similar experimental conditions, a peptide from a different region of αA-crystallin (αA-43-56) caused only 10 % protein precipitation. The interaction of the αA-66-80 peptide with α-crystallin also diminished α-crystallin chaperone activity pointing to one of the peptide interaction site(s) in α-crystallin subunits as a chaperone site, which was indeed subsequently confirmed by isotope-labeled cross-linking and mass spectroscopic studies with αB-crystallin and αA-66-80 [83]. The αB-70-107 and 164-174 regions were identified as αA-66-80 binding sites, and αB-72-92 region is now known to be one of the chaperone sites in αB-crystallin [89].

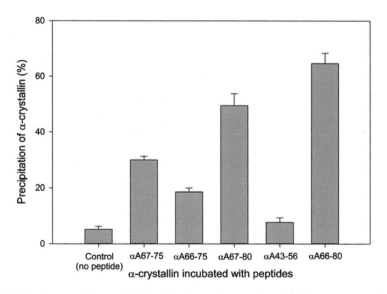

Fig. 13.4 Certain crystallin peptides found in the aged human lens precipitate α-crystallin in vitro. α-Crystallin and peptides in the ratio of 2:1 (w/w) were incubated in 500 μL of 50 mM phosphate buffer (in duplicates) at 37 °C for 24 h. After incubation, samples are centrifuged at 8,000 rpm for 30 min. Pellets were re-dissolved in 20 μL of 6 M urea and made up to 200 μL with buffer. Protein concentration of the re-dissolved pellets was estimated by Bio-Rad protein assay reagent

13.6 Mechanism of Proteolysis-Mediated Crystallin Aggregation and Development of Cataract

Data from in vitro and in vivo experiments show that lens opacity resulting from crystallin protein aggregation bears the hallmarks of increased proteolysis and decreased α-crystallin chaperone activity, features observed in aging cataract-prone lenses [9, 11, 33–35, 49, 55, 60, 65, 73, 74]. Proteins from aging lenses show increased posttranslational modifications. As depicted in the Fig. 13.5, it can be argued that the crystallin fragments accumulating in vivo due to proteolysis (enzymatic + nonenzymatic) occupy the central position in the pathway for lens opacification because of both their ability to induce aggregation and precipitation of α-crystallin and their anti-chaperone activity [56, 80, 83, 87, 88, 90]. The interaction of crystallin-derived peptides with α-crystallin can bring about structural changes that encompass hydrophobicity enhancement and the propensity to aggregation with additional crystallin molecules. These mechanisms are supported by our

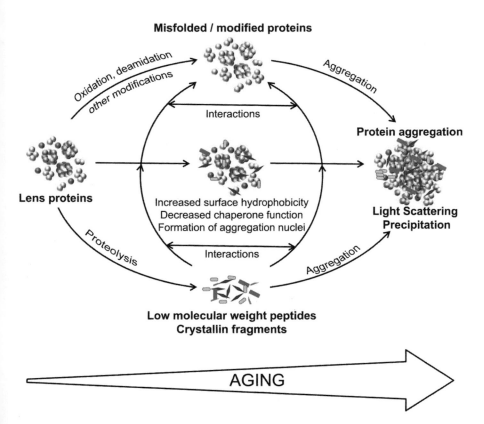

Fig. 13.5 Schematic representation of the proposed role of proteases and crystallin fragments in lens aging and cataract formation

observation that after peptide interaction with α-crystallin, there is increased binding of hydrophobic site-specific probe bis-ANS [83]. Further, time-lapse recording has shown that α-crystallin–peptide complexes, once formed, attract additional protein molecules and the aggregates begin to grow [80]. The importance of the interactions between α-crystallin and crystallin-derived peptides and the accumulation of crystallin fragments as a result of proteolysis garners additional support from the observation of higher amounts of crystallin fragments in transgenic lenses expressing the protease APH prior to the development of lens opacity [91].

13.7 Animal Models for Studying Lens Proteases in Cataractogenesis

Various attempts have been made to develop animal models to study the role of proteases in cataract formation [33, 34, 71, 91–93]. Hyperbaric oxygen-treated Guinea pig lens is a good model for investigating the mechanisms of age-related cataract formation. Analysis of hyperbaric oxygen-treated lenses shows glutathione loss, increased protein degradation, protein aggregation and insolubilization, protein cross-linking, membrane abnormalities, etc. all features like the changes observed in the aging human lens [94–97]. The hyperbaric oxygen-treated Guinea pig lens shows specific cleavages of crystallins and the presence of some of peptides seen in human lenses. Yet to be determined is to what extent does proteolysis contribute to cataract formation in this attractive model for human age-related cataract. A sheep model with an autosomal dominant gene for cortical cataract has been developed to test the role of calpain in cataract formation [34]. The cataract lenses from this model show cleavage of cytoskeletal proteins, similar to that occurring after calpain activation. Although calpain inhibitor was found to be very effective in controlling cataract formation under lens culture conditions, application of calpain inhibitor topically to the animal eyes was less effective, with only 27 % of the eyes showing decreased opacity in the initial 3 months compared to opacity in the controls that did not receive treatment [33]. It is unclear whether the ineffective delivery of the inhibitor was the reason for the limited success in this study. Nevertheless, this model offers an opportunity to test the involvement of calpain in particular and proteases in general in the development of cataract.

A fully functional ubiquitin proteasome system is required for lens development. Transgenic mice expressing K6W-ubiquitin (K6W-Ub) have been utilized to study the importance of K6W in ubiquitin in particular and UPP system in general. The transgenic lines of K6W-Ub exhibited defective lens development and alteration in protein expression [98]. The defect in lens development and cataract formation in mice that express K6W-Ub in the lens may be related to the changes in protein expression and intact K6 is required for progress of a proper differentiation program. We recently created transgenic mice that overexpress APH specifically in the lens to test the role of protease in the generation and accumulation of peptides [91]. Cataract development was seen at various postnatal days in the majority of mice expressing active APH (wt-APH). Cataract onset and severity correlated with the total APH protein and activity in the lens. Transgenic mice carrying inactive APH (mt-APH) did not develop cataract. Cataract development also correlated with N-terminal cleavage of the APH to generate a 57-kDa protein, along with an increased accumulation of low-molecular-weight (LMW) peptides, similar to those found in aging human and cataract lenses. Figure 13.6 shows the profile of crystallin fragments found in transgenic lens 1 month postnatally, prior to the development of cataract. In this animal model with mature cataract, the crystallin modifications and cleavage were quite dramatic. Affected lenses showed capsule rupture at the posterior pole, with expulsion of the lens nucleus and degenerating fiber cells. This study suggests that the cleaved APH fragment might be responsible for the accumulation of distinct LMW peptides that promote protein aggregation in lenses expressing wt-APH since earlier studies showed that 57 kDa form of APH is catalytically active [81]. The animal models such as this would be useful to test the use of protease inhibitors to prevent cataracts induced by proteolysis.

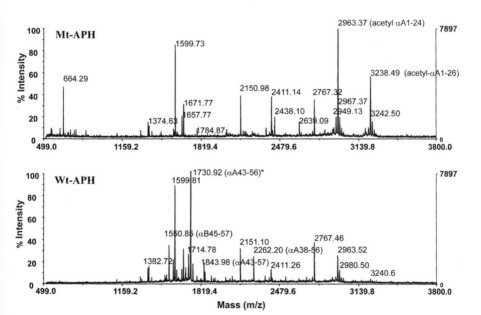

Fig. 13.6 MALDI TOF MS analysis of low-molecular-weight (LMW) peptides in 30-day postnatal lenses of transgenic mice expressing inactive acylpeptide hydrolase (mt-APH) and active APH (wt-APH). The lenses were homogenized in 6 M urea buffer containing protease inhibitors and filtered through 10 kDa filter to collect the LMW peptides. The lenses expressing wt-APH had significantly more peptides compared to the control group (mt-APH). The sequences identified by Nanospray QqTOF MS/MS analysis are given in *parenthesis*. Some of the peptides present distinctly in wt-APH group are also found in aging human lenses. (*Note*: Asterisk Intensity was truncated for scaling purpose). This research was originally published in J. Biol. Chem. (P. Santhoshkumar, L. Xie, M. Raju, K. Krishna Sharma. Lens crystallin modifications and cataract in transgenic mice overexpressing acylpeptide hydrolase. J. Biol. Chem. 2014, 289, 9039-9052. © the American Society for Biochemistry and Molecular Biology)

13.8 Future Directions of Lens Protease Research

Optimal function of proteases is clearly required for lens development and maturation. An imbalance between proteolysis and removal of crystallin fragments can result in the accumulation of protein breakdown products that are detrimental to the lens. The interaction of crystallin degradation products with α-crystallin may compromise the lens function. The available animal models, assay methods, new protease inhibitors, and delivery methods offer greater opportunities to understand the molecular mechanisms involved in proteolysis-mediated cataract formation and development of therapeutic molecules in coming years.

Acknowledgments We thank Brian Mooney, Ph.D. from the University of Missouri Proteomic Core for the help and discussions in the mass spectrometric analysis of the peptides and Sharon Morey for help in the preparation of the manuscript. Work in author's laboratory is supported in part by National Institutes of Health grants EY023219 and EY019878.

References

1. Lovicu FJ, Robinson ML, editors. Development of the ocular lens. New York: Cambridge University Press; 2004.
2. Wride MA. Lens fibre cell differentiation and organelle loss: many paths lead to clarity. Philos Trans R Soc Lond B Biol Sci. 2011;366:1219–33.
3. Jaffe NS, Horwitz J. Lens and cataract. In: Podos SM, Yanoff M, editors. Text book of ophthalmology, vol. 3. New York: Gower Medical Publishing; 1991.
4. Bassnett S. On the mechanism of organelle degradation in the vertebrate lens. Exp Eye Res. 2009;88:133–9.
5. Duncan G, Wormstone IM, Davies PD. The aging human lens: structure, growth, and physiological behaviour. Br J Ophthalmol. 1997;81:818–23.
6. Kuszak JR, Zoltoski RK, Sivertson C. Fibre cell organization in crystalline lenses. Exp Eye Res. 2004;78:673–87.
7. Bloemendal H, de Jong W, Jaenicke R, Lubsen NH, Slingsby C, Tardieu A. Ageing and vision: structure, stability and function of lens crystallins. Prog Biophys Mol Biol. 2004;86:407–85.
8. Benedek GB. Cataract as a protein condensation disease: the Proctor Lecture. Invest Ophthalmol Vis Sci. 1997;38:1911–21.
9. Sharma KK, Santhoshkumar P. Lens aging: effects of crystallins. Biochim Biophys Acta. 2009;1790:1095–108.
10. Roy D, Spector A. Absence of low molecular weight alpha crystallin in nuclear region of old human lenses. Proc Natl Acad Sci U S A. 1976;73:3484–7.
11. Miesbauer LR, Zhou X, Yang Z, Yang Z, Sun Y, Smith DL, Smith JB. Post-translational modifications of water-soluble human lens crystallins from young adults. J Biol Chem. 1994;269:12494–502.
12. Horwitz J. Alpha-crystallin. Exp Eye Res. 2003;76:145–53.
13. Horwitz J. α-Crystallin can function as a molecular chaperone. Proc Natl Acad Sci U S A. 1992;89:10449–53.
14. Wride MA, Geatrell J, Guggenheim JA. Proteases in eye development and disease. Birth Defects Res C Embryo Today. 2006;78:90–105.
15. Zandy AJ, Bassnett S. Proteolytic mechanisms underlying mitochondrial degradation in the ocular lens. Invest Ophthalmol Vis Sci. 2007;48:293–302.

16. Imai F, Yoshizawa A, Fujimori-Tonou N, Kawakami K, Masai I. The ubiquitin proteasome system is required for cell proliferation of the lens epithelium and for differentiation of lens fiber cells in zebrafish. Development. 2010;137:3257–68.

17. Tse SS, Ortwerth BJ. Activation and release of a trypsin-like proteinase from bovine lens alpha-crystallin. Exp Eye Res. 1982;34:659–74.

18. Chaerkady R, Sharma KK. Characterization of a bradykinin-hydrolyzing protease from the bovine lens. Invest Ophthalmol Vis Sci. 2004;45:1214–23.

19. Gupta R, Chen J, Srivastava OP. A serine-type protease activity of human lens betaA3-crystallin is responsible for its autodegradation. Mol Vis. 2010;16:2242–52.

20. Srivastava OP, Srivastava K. Characterization of a sodium deoxycholate-activatable proteinase activity associated with betaA3/A1-crystallin of human lenses. Biochim Biophys Acta. 1999;1434:331–46.

21. Srivastava OP. Characterization of a highly purified membrane proteinase from bovine lens. Exp Eye Res. 1988;46:269–83.

22. David LL, Shearer TR. Purification of calpain II from rat lens and determination of endogenous substrates. Exp Eye Res. 1986;42:227–38.

23. Ueda Y, McCormack AL, Shearer TR, David LL. Purification and characterization of lens specific calpain (Lp82) from bovine lens. Exp Eye Res. 2001;73:625–37.

24. Swanson AA, Davis RM, Meinhardt NC. Proteases in human lenses and their possible significance. Curr Eye Res. 1985;4:43–8.

25. Hariharapura R, Santhoshkumar P, Krishna Sharma K. Profiling of lens protease involved in generation of αA-66-80 crystallin peptide using an internally quenched protease substrate. Exp Eye Res. 2013;109:51–9.

26. Sharma KK, Ortwerth BJ. Bovine lens acylpeptide hydrolase. Purification and characterization of a tetrameric enzyme resistant to urea denaturation and proteolytic inactivation. Eur J Biochem. 1993;216:631–7.

27. Dawes LJ, Elliott RM, Reddan JR, Wormstone YM, Wormstone IM. Oligonucleotide microarray analysis of human lens epithelial cells: TGFbeta regulated gene expression. Mol Vis. 2007;13:1181–97.

28. Sharma KK, Ortwerth BJ. Isolation and characterization of a new aminopeptidase from bovine lens. J Biol Chem. 1986;261:4295–301.

29. Sharma KK, Ortwerth BJ. Aminopeptidase III activity in normal and cataractous lenses. Curr Eye Res. 1986;5:373–80.

30. David LL, Varnum MD, Lampi KJ, Shearer TR. Calpain II in human lens. Invest Ophthalmol Vis Sci. 1989;30:269–75.

31. De Maria A, Shi Y, Kumar NM, Bassnett S. Calpain expression and activity during lens fiber cell differentiation. J Biol Chem. 2009;284:13542–50.

32. Fukiage C, Nakajima E, Ma H, Azuma M, Shearer TR. Characterization and regulation of lens-specific calpain Lp82. J Biol Chem. 2002;277:20678–85.

33. Morton JD, Lee HY, McDermott JD, Robertson LJ, Bickerstaffe R, Jones MA, Coxon JM, Abell AD. A macrocyclic calpain inhibitor slows the development of inherited cortical cataracts in a sheep model. Invest Ophthalmol Vis Sci. 2013;54:389–95.

34. Robertson LJ, Morton JD, Yamaguchi M, Bickerstaffe R, Shearer TR, Azuma M. Calpain may contribute to hereditary cataract formation in sheep. Invest Ophthalmol Vis Sci. 2005;46:4634–40.

35. Morozov V, Wawrousek EF. Caspase-dependent secondary lens fiber cell disintegration in alphaA-/alphaB-crystallin double-knockout mice. Development. 2006;133:813–21.

36. Eisenhauer DA, Berger JJ, Peltier CZ, Taylor A. Protease activities in cultured beef lens epithelial cells peak and then decline upon progressive passage. Exp Eye Res. 1988;46:579–90.

37. Sulochana KN, Ramakrishnan S, Arunagiri K. Purification and characterization of a new enzyme dipeptidase from human lens. Exp Eye Res. 1996;62:221–9.

38. Taylor A. Aminopeptidases: structure and function. FASEB J. 1993;7:290–8.

39. Sachdev NH, Di Girolamo N, Nolan TM, McCluskey PJ, Wakefield D, Coroneo MT. Matrix metalloproteinases and tissue inhibitors of matrix metalloproteinases in the human lens: implications for cortical cataract formation. Invest Ophthalmol Vis Sci. 2004;45:4075–82.
40. Descamps FJ, Martens E, Proost P, Starckx S, Van den Steen PE, Van Damme J, Opdenakker G. Gelatinase B/matrix metalloproteinase-9 provokes cataract by cleaving lens betaB1 crystallin. FASEB J. 2005;19:29–35.
41. Sharma KK, Ortwerth BJ. Purification and characterization of prolyl oligopeptidase from bovine lens. Exp Eye Res. 1994;59:107–15.
42. Pereira P, Shang F, Hobbs M, Girao H, Taylor A. Lens fibers have a fully functional ubiquitin-proteasome pathway. Exp Eye Res. 2003;76:623–31.
43. Ray K, Harris H. Purification of neutral lens endopeptidase: close similarity to a neutral proteinase in pituitary. Proc Natl Acad Sci U S A. 1985;82:7545–9.
44. Trayhurn P, van Heyningen R. Neutral proteinase activity in the human lens. Exp Eye Res. 1976;22:251–7.
45. Wagner BJ, Margolis JW, Singh I. Bovine lens multicatalytic proteinase complex. Enzyme Protein. 1993;47:202–9.
46. Li G, Percontino L, Sun Q, Qazi AS, Frederikse PH. Beta-amyloid secretases and beta-amloid degrading enzyme expression in lens. Mol Vis. 2003;9:179–83.
47. Srivastava OP, Ortwerth BJ. Isolation and characterization of a 25K serine proteinase from bovine lens cortex. Exp Eye Res. 1983;37:597–612.
48. Yoshida H, Murachi T, Tsukahara I. Limited proteolysis of bovine lens alpha-crystallin by calpain, a Ca2+-dependent cysteine proteinase, isolated from the same tissue. Biochim Biophys Acta. 1984;798:252–9.
49. David LL, Azuma M, Shearer TR. Cataract and the acceleration of calpain-induced beta-crystallin insolubilization occurring during normal maturation of rat lens. Invest Ophthalmol Vis Sci. 1994;35:785–93.
50. Sharma KK, Kester K, Elser N. Identification of new lens protease(s) using peptide substrates having in vivo cleavage sites. Biochem Biophys Res Commun. 1996;218:365–70.
51. Grey AC, Schey KL. Distribution of bovine and rabbit lens alpha-crystallin products by MALDI imaging mass spectrometry. Mol Vis. 2008;14:171–9.
52. Grey AC, Schey KL. Age-related changes in the spatial distribution of human lens alpha-crystallin products by MALDI imaging mass spectrometry. Invest Ophthalmol Vis Sci. 2009;50:4319–29.
53. Han J, Schey KL. MALDI tissue imaging of ocular lens alpha-crystallin. Invest Ophthalmol Vis Sci. 2006;47:2990–6.
54. Schey KL, Wang ZJLW, Qi Y. Aquaporins in the eye: expression, function, and roles in ocular disease. Biochim Biophys Acta. 2014;1840:1513–23.
55. Stella DR, Floyd KA, Grey AC, Renfrow MB, Schey KL, Barnes S. Tissue localization and solubilities of alphaA-crystallin and its numerous C-terminal truncation products in pre- and postcataractous ICR/f rat lenses. Invest Ophthalmol Vis Sci. 2010;51:5153–61.
56. Santhoshkumar P, Udupa P, Murugesan R, Sharma KK. Significance of interactions of low molecular weight crystallin fragments in lens aging and cataract formation. J Biol Chem. 2008;283:8477–85.
57. Van Kleef SM, Willems-Thijssen W, Hoenders HJ. Intracellular degradation and deamidation of alpha-crystallin subunits. Eur J Biochem. 1976;66:477–83.
58. Hanson SR, Hasan A, Smith DL, Smith JB. The major in vivo modifications of the human water-insoluble lens crystallins are disulfide bonds, deamidation, methionine oxidation and backbone cleavage. Exp Eye Res. 2000;71:195–207.
59. Harrington V, McCall S, Huynh S, Srivastava K, Srivastava OP. Crystallins in water soluble-high molecular weight protein fractions and water insoluble protein fractions in aging and cataractous human lenses. Mol Vis. 2004;10:476–89.
60. Harrington V, Srivastava OP, Kirk M. Proteomic analysis of water insoluble proteins from normal and cataractous human lenses. Mol Vis. 2007;13:1680–94.

61. Takemoto L. Increased cleavage of the c-terminal serine from alpha-A crystallin present in the high molecular weight aggregate fraction from human and bovine lenses. Curr Eye Res. 1999;19:450–5.

62. Kamei A, Iwase H, Masuda K. Cleavage of amino acid residue(s) from the N-terminal region of alpha A- and alpha B-crystallins in human crystalline lens during aging. Biochem Biophys Res Commun. 1997;231:373–8.

63. Robertson LJ, David LL, Riviere MA, Wilmarth PA, Muir MS, Morton JD. Susceptibility of ovine lens crystallins to proteolytic cleavage during formation of hereditary cataract. Invest Ophthalmol Vis Sci. 2008;49:1016–22.

64. Su SP, Lyons B, Friedrich M, McArthur JD, Song X, Xavier D, Truscott RJ, Aquilina JA. Molecular signatures of long-lived proteins: autolytic cleavage adjacent to serine residues. Aging Cell. 2012;11:1125–7.

65. Su SP, McArthur JD, Aquilina JA. Localization of low molecular weight crystallin peptides in the aging human lens using a MALDI mass spectrometry imaging approach. Exp Eye Res. 2010;91:97–103.

66. Su SP, McArthur JD, Truscott RJW, Aquilina JA. Truncation, cross-linking and interaction of crystallins and intermediate filament proteins in the aging human lens. Biochim Biophys Acta. 2011;1814:647–56.

67. Thampi P, Hassan A, Smith JB, Abraham EC. Enhanced C-terminal truncation of alphaA- and alphaB-crystallins in diabetic lenses. Invest Ophthalmol Vis Sci. 2002;43:3265–72.

68. Voorter CE, de Haard-Hoekman WA, van den Oetelaar PJ, Bloemendal H, de Jong WW. Spontaneous peptide bond cleavage in aging alpha-crystallin through a succinimide intermediate. J Biol Chem. 1988;263:19020–3.

69. Shang F, Gong X, Palmer HJ, Nowell Jr TR, Taylor A. Age-related decline in ubiquitin conjugation in response to oxidative stress in the lens. Exp Eye Res. 1997;64:21–30.

70. Viteri G, Carrard G, Birlouez-Aragon I, Silva E, Friguet B. Age-dependent protein modifications and declining proteasome activity in the human lens. Arch Biochem Biophys. 2004;427:197–203.

71. Fukiage C, Azuma M, Nakamura Y, Tamada Y, Shearer TR. Calpain-induced light scattering by crystallins from three rodent species. Exp Eye Res. 1997;65:757–70.

72. Srivastava OP, Srivastava K. Human lens membrane proteinase: purification and age-related distributional changes in the water-soluble and insoluble protein fractions. Exp Eye Res. 1989;48:161–75.

73. Takemoto L, Emmons T, Horwitz J. The C-terminal region of alpha-crystallin: involvement in protection against heat-induced denaturation. Biochem J. 1993;294(Pt 2):435–8.

74. Andley UP, Mathur S, Griest TA, Petrash JM. Cloning, expression, and chaperone-like activity of human alphaA-crystallin. J Biol Chem. 1996;271:31973–80.

75. Kallur LS, Aziz A, Abraham EC. C-Terminal truncation affects subunit exchange of human αA-crystallin with αB-crystallin. Mol Cell Biochem. 2008;308:85–91.

76. Chaves JM, Srivastava K, Gupta R, Srivastava OP. Structural and functional roles of deamidation and/or truncation of N- or C-termini in human alpha A-crystallin. Biochemistry. 2008;47:10069–83.

77. Wu M, Zhang X, Bian Q, Taylor A, Liang JJ, Ding L, Horwitz J, Shang F. Oligomerization with wt alphaA- and alphaB-crystallins reduces proteasome-mediated degradation of C-terminally truncated alphaA-crystallin. Invest Ophthalmol Vis Sci. 2012;53:2541–50.

78. Sharma KK, Kester K. Peptide hydrolysis in lens: role of leucine aminopeptidase, aminopeptidase III, prolyloligopeptidase and acylpeptidehydrolase. Curr Eye Res. 1996;15:363–9.

79. Srivastava OP. Age-related increase in concentration and aggregation of degraded polypeptides in human lenses. Exp Eye Res. 1988;47:525–43.

80. Santhoshkumar P, Raju M, Sharma KK. αA-crystallin peptide 66SDRDKFVIFLDVKHF80 accumulating in aging lens impairs the function of α-crystallin and induces lens protein aggregation. PLoS One. 2011;6:e19291.

81. Chongcharoen K, Sharma KK. Characterization of trypsin-modified bovine lens acylpeptide hydrolase. Biochem Biophys Res Commun. 1998;247:136–41.
82. Senthilkumar R, Reddy PN, Sharma KK. Studies on trypsin-modified bovine and human lens acylpeptide hydrolase. Exp Eye Res. 2001;72:301–10.
83. Kannan R, Santhoshkumar P, Mooney BP, Sharma KK. The alphaA66-80 peptide interacts with soluble alpha-crystallin and induces its aggregation and precipitation: a contribution to age-related cataract formation. Biochemistry. 2013;52:3638–50.
84. David LL, Shearer TR. Role of proteolysis in lenses: a review. Lens Eye Toxic Res. 1989;6:725–47.
85. Stefani M. Structural features and cytotoxicity of amyloid oligomers: implications in Alzheimer's disease and other diseases with amyloid deposits. Prog Neurobiol. 2012;99:226–45.
86. Villar-Pique A, de Groot NS, Sabate R, Acebron SP, Celaya G, Fernandez-Busquets X, Muga A, Ventura S. The effect of amyloidogenic peptides on bacterial aging correlates with their intrinsic aggregation propensity. J Mol Biol. 2012;421:270–81.
87. Senthilkumar R, Chaerkady R, Sharma KK. Identification and properties of anti-chaperone-like peptides derived from oxidized bovine lens betaL-crystallins. J Biol Chem. 2002;277:39136–43.
88. Rao G, Santhoshkumar P, Sharma KK. Anti-chaperone βA3/A1102-117 peptide interacting sites in human αB-crystallin. Mol Vis. 2008;14:666–74.
89. Bhattacharyya J, Padmanabha Udupa EG, Wang J, Sharma KK. Mini-alphaB-crystallin: a functional element of alphaB-crystallin with chaperone-like activity. Biochemistry. 2006;45:3069–76.
90. Udupa PE, Sharma KK. Effect of oxidized betaB3-crystallin peptide (152-166) on thermal aggregation of bovine lens gamma-crystallins: identification of peptide interacting sites. Exp Eye Res. 2005;80:185–96.
91. Santhoshkumar P, Xie L, Raju M, Reneker L, Sharma KK. Lens crystallin modifications and cataract in transgenic mice overexpressing acylpeptide hydrolase. J Biol Chem. 2014;289:9039–52.
92. Swanson AA, Davis RM, Meinhardt NC, Kuck KD, Kuck Jr JF. Proteases in the Emory mouse cataract. Invest Ophthalmol Vis Sci. 1985;26:1035–7.
93. Mitton KP, Kamiya T, Tumminia SJ, Russell P. Cysteine protease activated by expression of HIV-1 protease in transgenic mice. MIP26 (aquaporin-0) cleavage and cataract formation in vivo and ex vivo. J Biol Chem. 1996;271:31803–6.
94. Bantseev V, Oriowo OM, Giblin FJ, Leverenz VR, Trevithick JR, Sivak JG. Effect of hyperbaric oxygen on guinea pig lens optical quality and on the refractive state of the eye. Exp Eye Res. 2004;78:925–31.
95. Freel CD, Gilliland KO, Mekeel HE, Giblin FJ, Costello MJ. Ultrastructural characterization and Fourier analysis of fiber cell cytoplasm in the hyperbaric oxygen treated guinea pig lens opacification model. Exp Eye Res. 2003;76:405–15.
96. Padgaonkar VA, Lin LR, Leverenz VR, Rinke Λ, Reddy VN, Giblin FJ. Hyperbaric oxygen in vivo accelerates the loss of cytoskeletal proteins and MIP26 in guinea pig lens nucleus. Exp Eye Res. 1999;68:493–504.
97. Simpanya MF, Ansari RR, Suh KI, Leverenz VR, Giblin FJ. Aggregation of lens crystallins in an in vivo hyperbaric oxygen guinea pig model of nuclear cataract: dynamic light-scattering and HPLC analysis. Invest Ophthalmol Vis Sci. 2005;46:4641–51.
98. Shang F, Wilmarth PA, Chang ML, Liu K, David LL, Caceres MA, Wawrousek E, Taylor A. Newborn mouse lens proteome and its alteration by lysine 6 mutant ubiquitin. J Proteome Res. 2014;13:1177–89.

Chapter 14
Photosensitized Oxidation of Lens Proteins Exposed to UVA-Visible Light at Low Oxygen Concentration: Its Effect on the Proteasome System

Eduardo Silva, Felipe Ávila, and Bertrand Friguet

Abbreviations

AGEs	Advanced glycation end products
ASC	Ascorbate
BLPs	Bovine lens proteins
CML	Carboxymethyllysine
GDC	Glucose decomposition chromophore
Glc	Glucose
LP	Lens protein
RH	Reduced substrate
RH$^{\cdot+}$	R$^{\cdot}$ and R$_{Oxid}$ represent the intermediate radical cation, the neutral radical and the oxidized form of the substrate, respectively
ROS	Reactive oxygen species
S	Sensitizer
Threo	Threose

E. Silva (✉)
Faculty of Chemistry, Department of Physical Chemistry, Pontifical Catholic University of Chile, Casilla 306, Correo 22, Santiago, Chile
e-mail: esilva@uc.cl

F. Ávila
Laboratory of Cellular Biology of Aging, UR4-IFR83, Pierre-and-Marie-Curie University, Paris, France

Interdisciplinary Research Program of Excellence in Healthy Aging (PIEI-ES), Universidad de Talca, Talca, Chile

B. Friguet
Laboratory of Cellular Biology of Aging, UR4-IFR83, Pierre-and-Marie-Curie University, Paris, France

Adaptation Biologique et Vieillissement, Institute of Biology Paris-Seine, INSERM U1164, UPMC University, Paris, France

© Springer Science+Business Media New York 2015
M.A. Babizhayev et al. (eds.), *Studies on the Cornea and Lens*,
Oxidative Stress in Applied Basic Research and Clinical Practice,
DOI 10.1007/978-1-4939-1935-2_14

14.1   Introduction

The eye lens is a tissue composed mainly by proteins that are characterized by the absence or reduction of turnover; consequently, they are sensitive to age-related and environmental damage. The accumulation of modified proteins during aging is associated with lens opacification or senile cataract. Throughout life, the eye lens suffers constant exposure to light that is essential for vision. In addition this light has also been postulated to play an important role in systemic health through the interaction with photosensitive retinal ganglion cells that set the circadian clock [1, 2]. However, there is evidence of the occurrence of photochemical modifications in the eye lens proteins during aging, which could contribute to the development of cataracts [3, 4]. To understand and explain the photochemical processes that can occur in the eye lens, it is necessary to consider the principles governing these reactions. The first law of photochemistry, now known as the Grotthuss–Draper law, establishes that "Only the light that is absorbed by a system can cause chemical change." Consequently, the absorption properties of the chemical constituents of the lens determine the feasibility of the occurrence of photochemical processes in this tissue. The main photochemical activity of the lens proteins, constituting approximately 35 % of the wet weight and nearly all the dry weight of the lens, should thus be due to the absorption capacity of the aromatic chromophores of the amino acids Phe $(\varepsilon_{259nm}=2\times10^2/M\ cm)$, Trp $(\varepsilon_{279nm}=5.2\times10^3/M\ cm)$, and Tyr $(\varepsilon_{278nm}=1.1\times10^3/M\ cm)$ present in the lens proteins, and only radiation in the UVB region can provoke direct photochemical modifications. Indeed, several authors have concluded that UVB radiation induces anterior lenticular opacities [5–7] in lenses exposed to this type of UV light. Under normal conditions, ozone and molecular oxygen block electromagnetic radiation below 280 and 240 nm, respectively, and no UVC reaches the Earth's surface. UVB is received in varying amounts, depending on the diurnal variation of ocular UV exposure [8] and the location on the Earth [9]. The cornea (100 % absorption below 280 nm, 92 % at 300 nm, and 45 % at 320 nm) and aqueous humor (6 % absorption at 300 nm and 16 % at 320 nm) filter most of the UVB that reaches the eye [10–12]. The deleterious mechanisms, associated with photo-induced damage in the eye lens, mediated by UVA (315–400 nm) and even visible light radiation (400–700 nm), are subjects of increasing interest [4, 13, 14]. The occurrence of UVA-visible-mediated photochemical processes requires the presence of chromophores, with photosensitizing properties that absorb in this region of the electromagnetic spectrum. Numerous chromophores have been identified in human cataractous eye lenses [15–20], and despite their diverse chemical origin, many of them can act as photosensitizers [21–26], which can be precursors of damage to structural and functional components of the eye lens fibers, including crystallin proteins, ascorbic acid, glutathione, and the proteasome, among others. These types of reactions have been proposed to play a relevant role in the onset of human age-related nuclear cataract [24, 26]. A photosensitizer (S) can be defined as a compound that is electronically excited through the direct absorption of photons, and this energy is transferred to another molecule that can result in a direct modification or be an intermediary of further modifications. The photosensitizing properties can be manifested through Type I, Type II, or mixed Type I-Type II mechanisms [27, 28], where the Type I mechanism is favored at low oxygen concentrations.

Type I mechanism

$$S + h\upsilon \rightarrow {}^1S \rightarrow {}^3S \tag{14.1}$$

$${}^3S + RH \rightarrow S^{\bullet -} + RH^{\bullet +} \rightarrow SH^{\bullet} + R^{\bullet} \tag{14.2}$$

$$S^{\bullet -} + O_2 \rightarrow S + O_2^{\bullet -} \tag{14.3}$$

$$2SH^{\bullet} \rightarrow S + SH_2 \tag{14.4}$$

$$SH_2 + O_2 \rightarrow S + H_2O_2 \tag{14.5}$$

$$H_2O_2 + O_2^{\bullet -} \rightarrow {}^-OH + {}^{\bullet}OH + O_2 \tag{14.6}$$

$$2R^{\bullet} \rightarrow R-R \tag{14.7}$$

$$R \text{ and}/\text{or } R^{\bullet} + H_2O_2/O_2^{\bullet -}/{}^{\bullet}OH/O_2 \rightarrow R_{Oxid} \tag{14.8}$$

Type II mechanism

$$S + h\upsilon \rightarrow {}^1S \rightarrow {}^3S \tag{14.9}$$

$${}^3S + O_2 \rightarrow S + {}^1O_2 \tag{14.10}$$

$$RH + {}^1O_2 \rightarrow R_{Oxid} \tag{14.11}$$

In these mechanistic formulations, S, ^{1}S, and ^{3}S represent the sensitizer (S) in the ground state and in the excited singlet and triplet states, respectively; $S^{\bullet -}$, $SH^{\bullet}$, and SH_2 are the radical anion, the neutral radical, and the reduced form of S, respectively; RH corresponds to the reduced substrate; and $RH^{\bullet +}$, $R^{\bullet}$, and R_{Oxid} represent the intermediate radical cation, the neutral radical, and the oxidized form of the substrate, respectively. It can be appreciated in Eqs. 14.1–14.11 that the oxygen concentration modulates the prevalence and efficiency of Type I o Type II photosensitized mechanisms [29, 30]. At low oxygen concentrations, the Type I mechanism that implies a direct electron transfer between the substrate and the triplet excited state of the sensitizer (Eq. 14.2), followed by a sequence of reactions (Eqs. 14.3–14.8), is favored. It is important to consider that processes represented by Eqs. 14.1, 14.2, 14.4, and 14.7 do not require the presence of oxygen in the reaction medium. The oxygen pressure in the lens epithelium reaches values between 17 and 23 mmHg (approximately 2.2–3.0 % O_2) [31], and this pressure drops steadily toward the center, leading to an oxygen pressure <2 mmHg in the core [32]. The trace amounts of

O_2 present in the lens nucleus produce significant levels of oxidized products in the nucleus of the guinea pig lens exposed to UVA light [33] and contribute to the regeneration of the sensitizer (Eqs. 14.3, 14.5, and 14.10). The hypoxia in the deep cortical layers of the normal lens serves as a trigger for the organelle loss process that ensures the transparency of the tissue [34]. Nevertheless, with age, an extent of vitreous liquefaction, concomitant with a modification in the oxygen distribution that increases the susceptibility of the lens to oxidative injury, has been observed [35]. These processes, together with the impediment to glutathione diffusion in older normal human lenses [36], could promote the decomposition of ascorbic acid and, therefore, the generation of colored advanced glycation end products (AGEs).

14.2 Dye-Sensitized Capacity of Advanced Glycation End Products

In the literature, there is an extensive list of low molecular weight compounds that changes their relative concentration in the human eye lens during aging and in non-diabetic cataracts [26, 37–40]. These compounds have been assigned different possible roles based on whether they act as sensitizers and prooxidants, antioxidants or benign filters that protect the posterior parts of the eye from light damage; some compounds still have an uncertain role [25, 41].

In this section, we discuss the photochemical mechanisms that take place during the irradiation of amino acid models of advanced glycation end products and their effects on key molecules that are targets of oxidative stress in the eye lens.

The generation of advanced glycation end products (AGEs) in lens proteins has been proposed as a major protein modification contributing to the formation of age-onset and diabetic cataracts [20, 26, 42–46]. In the eye lens, the formation of AGEs can occur with either glucose (important in diabetic patients with high sugar levels) or ascorbic acid oxidation products [45–48]. The degradation of glucose and ascorbic acid can produce intermediaries, such as threosone and 3-deoxythreosone, that possesses carbonyl groups that can react with proteins, giving rise to AGEs [49, 50]. Evidence of the presence of tetroses has been reported in the aqueous humor of cataractous subjects [51] as well in the water-soluble and -insoluble fractions of lens proteins from cataractous patients [52], indicating the relevance of the nonenzymatic degradation of glucose and ascorbic acid.

Most of the known AGEs can be formed from nonenzymatic degradation intermediates, as another route to the well-known cellular metabolic pathways. These intermediates generate electrophilic compounds, such as methylglyoxal, which can also give rise to AGEs [53–55]. The generation of AGEs via nonenzymatic reactions is especially important in mature eye lens fibers, where the contribution to AGEs generated through metabolic pathways is expected to be negligible due to the absence of organelles and a reduced or absent glucose metabolic activity [56].

In proteins, glycation process is initiated by a nucleophilic attack of sulfhydryl or amino groups to carbonyl groups of carbohydrates or their degradation products, which leads to the formation of a Schiff's base, followed by an Amadori rearrangement (aldoses) or Heyns rearrangement (ketoses). The generation of intermediary compounds containing new carbonyl groups allows the reaction with another nucleophilic group, initiating multistep transformations that ultimately lead to a variety of structures known generically as AGEs. The generation of these compounds is accompanied by the simultaneous appearance of a brown coloration with absorption bands in the UVA and visible regions of the spectra. Some of these chromophores, which also have fluorophore properties ($\lambda_{em} = 450$ nm), have been probed as sensitizers when they are irradiated in the UVA region [23, 57]. However, there are no comparative studies concerning their photosensitizing activities when the irradiations are performed at low oxygen pressure and using different wavelengths to investigate the probable simultaneous occurrence of two or more sensitizing processes in the reaction mixture. These conditions are very important because they simulate the physiology of a tissue, such as the eye lens, that has very low or no protein turnover, is not vascularized, is exposed to the direct effect of light and whose proteins are exposed to glycation for a long period of time, depending on the age of the subject. Research concerning UVA-absorbing compounds in the eye lens and their photochemical properties has been a subject of increasing interest [12, 16, 26, 58, 59]. Under aerobic conditions, excited states of AGEs are deactivated, generating singlet oxygen as the predominant specie [60] and overcoming the generation of superoxide radical anion [61] and hydrogen peroxide [62]. Nonetheless, assessments of the actual photosensitizing potential of AGEs did not take into account the low oxygen concentration in this tissue, which can have a direct influence on the prevalence of the Type I photosensitizing mechanism. Consequently, these studies overlooked the reactive species produced through this pathway. In recent years, new studies have been performed in vitro, using models of AGE as photosensitizers and taking into consideration the low oxygen pressure of the lens. The purpose of these studies, whose main results are shown below, was the comparative evaluation of the photosensitizing activity of AGEs obtained in vitro through the incubation of ascorbate (ASC), glucose (Glc), or threose (Threo) with lysine (Lys), employing the characteristic low oxygen pressure present in the eye lens [63]. The essential amino acid tryptophan (Trp) and vitamin C were used as targets to evaluate the efficiencies of the photo-processes which were performed with poly- and monochromatic light at 5 % O_2.

Table 14.1 shows the quantum yields for Trp decomposition when this amino acid was irradiated with monochromatic light (344 or 367 nm) in the presence of the different sensitizers. It was observed that the quantum yields are from 2.8- to 5.7-times lower at 367 nm than at 344 nm; this finding confirms the existence of at least two species with different sensitizing capacities in the samples. When the emission properties of the chromophores and their photosensitizing activities during Trp decomposition were compared, it was observed that at 344 nm (Fig. 14.1a) and 367 nm (Fig. 14.1b), Trp decomposition quantum yields increased when the fluorescence emission yields of the sensitizers decreased. This inverse correlation can be

Table 14.1 Initial quantum
yields for Trp sensitized
decomposition at 344 and
367 nm in the presence of the
different chromophores
(absorbance of 0.2 at 365 nm)

Sensitizers	$\phi_{344} \times 10^3$	$\phi_{367} \times 10^3$
ASC	5.41 ± 0.98	1.23 ± 0.01
Lys/Glc	1.72 ± 0.31	0.61 ± 0.06
Lys/ASC	1.38 ± 0.28	0.55 ± 0.11
Lys/Threo	1.31 ± 0.66	0.23 ± 0.18

Reproduced from Photochem. Photobiol.
2007, 83: 563-569 with permission from
Wiley

The irradiations were performed under low
oxygen pressure (5 % O_2)

Data are the mean $\pm$ S.D. of three independent experiments

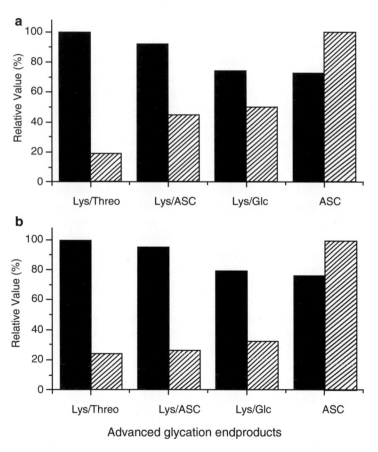

Fig. 14.1 Relative Trp photodecomposition quantum yields (*striped columns*) and emission fluorescence efficiencies (*black columns*) at 344 (**a**) and 367 nm (**b**). One Hundred percent was
assigned to the highest value of each set of data. Fluorescence emission efficiencies were expressed
as relative emission intensities at the corresponding wavelengths divided by the fraction of
absorbed energy $(1 - T_\lambda)$. Reproduced from Photochem. Photobiol. 2007, 83: 563-569 with permission from Wiley

mainly explained by considering that photosensitization is a triplet process and its quantum yield is inversely related to the fluorescence yield.

Taking into account that physiological ASC concentrations in the eye lens can be extremely high (between 1 and 2 mM) [64] and that its decomposition promotes the generation of colored compounds and also glycating agents, this vitamin was used as a target of the photo-processes mediated by amino acid models of AGEs studied in this work. The obtained ASC decomposition quantum yields mediated by AGE are significantly higher ($\phi_{367} \times 10^3$: 256 ± 10 for ASC, 110 ± 8 for Lys/Glc, 67.9 ± 3.9 for Lys/ASC and 26.2 ± 2.8 for Lys/Threo) than those obtained for Trp decomposition (see Table 14.1), and the efficiency on the photodegradation was also dependent on the type of carbohydrate used as a precursor of AGE synthesis. The presence of copper ions in the reaction media can contribute to an active propagation of ASC oxidation. To avoid this effect, the experiments were carried out in the presence of the metal chelator DTPA.

Importantly, the efficient photosensitizing activity of the degradation products of ASC can autosensitize its own decomposition resulting in the complete depletion of this antioxidant. Trp in free form or as a constituent of proteins is normally one of the main targets of photosensitizing processes mediated by AGEs [29]. It has been demonstrated that the AGE-photosensitized Trp oxidation rate increases with AGE concentration and is optimal at 5 % oxygen [29]. By contrast, for ASC, the photo-oxidation rate increases with oxygen concentration. As ASC is more rapidly photo-degraded than Trp, the antioxidant competitively protects Trp from oxidation up to 1 mM, although not absolutely [29]. There is no exact information in reference to the chemical nature of the compounds responsible for this coloration, but the main products of ASC degradation are well documented [49, 52, 65–68].

The mechanistic aspects of the photo-processes mediated by a model of AGEs, obtained by the incubation of Nα-acetyl-L-lysine, Nα-acetyl-L-arginine and D-GLUCOSE, have also been analyzed [69]. To investigate the photosensitizing mechanism of AGEs, Trp was used as a model target and the photodecomposition quantum yields at 367 nm of this amino acid were determined at different oxygen concentrations. FeCy, a known electron-scavenger [70], was used with the aim of interfering with the electron-transfer mechanism, whereas deuterium oxide buffer was used to enhance singlet oxygen-mediated photodamage.

The results shown in Table 14.2 support a combined Type I–Type II mechanism for all oxygen concentrations, with Type I and Type II favored at low and high oxygen concentrations, respectively. The fraction of Type I ($f_{Type\ I}$) and Type II ($f_{Type\ II}$) mechanisms that contributes to Trp photodecomposition depends on both Trp and oxygen concentration, and the rate constants of the reactions between triplet AGEs and Trp or O_2. This relationship is expressed by the equations below:

$$f_{TypeI} = k_{et}\left[Trp\right] / \left(k_{ic} + k_P + k_{et}\left[Trp\right] + k_q\left[O_2\right]\right), \quad (14.12)$$

$$f_{TypeII} = k_q\left[O_2\right] / \left(k_{ic} + k_P + k_{et}\left[Trp\right] + k_q\left[O_2\right]\right), \quad (14.13)$$

Table 14.2 Quantum yields for AGE-sensitized Trp decomposition at 367 nm under different oxygen pressures

Trp photodecomposition quantum yields ($\phi_{367} \times 10^3$)			
O_2 (%)	Control	Added FeCy[a]	D_2O buffer[b]
5	0.55	0.28	1.44
20	0.35	0.29	1.74
100	0.49	0.38	2.42

Reproduced from Photochem. Photobiol. 2009, 85: 185-194 with permission from Wiley
Initial concentration of Trp was 50 µM and AGEs were set at an absorbance of 0.2 at 365 nm
[a]FeCy was added at a final concentration of 0.1 mM
[b]Phosphate buffer 100 mM was prepared in D_2O (pD=7.4)

where k_{et}, k_{ic}, k_P, and k_q correspond to the rate constants of the electron transfer between triplet AGE and Trp; internal conversion; phosphorescence emission; and the quenching of triplet AGE by oxygen, respectively.

Given that the Trp concentration was the same in all the experiments, the contribution of Type I and Type II photo-processes depends solely on the oxygen concentration and the values for the rate constants of the processes.

At 5 % O_2, the Type I photosensitizing mechanism is predominant because the presence of FeCy considerably drops the quantum yield to 51 % with respect to the control. A similar result was reported previously for the predominantly Type I sensitizer, riboflavin [28]. In the same conditions, D_2O increases the damage 2.6 times, which is low if we consider that deuterium oxide prolongs the lifetime of singlet oxygen approximately 10–17 times compared to water [71]. This result indicates a small contribution of singlet oxygen at low oxygen concentration.

At 20 % O_2, the quantum yield drops to 64 % with respect to 5 % O_2. This result is consistent with the finding that AGEs acts preferentially as Type I sensitizers. Therefore, the fourfold increase in the oxygen concentration disfavors the interaction between the triplet sensitizer and tryptophan, which is reflected in a lower damage to the target. The presence of FeCy does not inhibit the process to a great extent (down to 80 % of the control), which agrees with the fact that at 20 % oxygen, the Type I mechanism is disfavored. A high oxygen concentration favors the quenching of the triplet state of the sensitizer by ground state oxygen in a Type II mechanism that generates singlet oxygen, which is observed as a fivefold increase in the quantum yield when the reaction is performed in deuterium oxide buffer.

At 100 % O_2, the quantum yield rises again to 89 % of the value observed at 5 % O_2. This result was surprising because for a Type I sensitizer, an even greater decrease in the quantum yield was expected compared to the value at 20 % O_2. A plausible explanation for this behavior is that at 100 % O_2, the increased amount of singlet oxygen generated in this condition, compensates for the loss in a Type I mechanism, thereby balancing the total efficiency of the process. The influence of

FeCy and D_2O in the quantum yield shows the same behavior as that observed at 20 % O_2 and is consistent with the predominance of the Type II mechanism at high oxygen concentrations. Singlet oxygen concentrations from 1.2 to 2.7 μmol/mL have been reported for purified threose AGE compounds irradiated with UVA [23].

Notwithstanding the fact that an enhancement of the photodamage in the presence of deuterium oxide was observed for all oxygen concentrations, supporting the participation of the Type II mechanism, the increase was small compared to the expected enhancement of the lifetime of singlet oxygen. The ratio between the quantum yields in D_2O and in H_2O is 2.62 at 5 % O_2, which is low compared to that for known Type II sensitizers such as methylene blue, where a ratio of 8.29 in oxygen-saturated solutions was observed [72]. In the same study, a ratio of 4.99 was observed for riboflavin in aerated solutions, which agrees well with the values of 4.97 and 4.94 observed for 20 % and 100 % oxygen concentrations, respectively, that were found for AGE in our conditions. These data indicate that Type I photo-processes are the principal mechanism by which AGEs induce photodamage under low oxygen conditions, which corresponds with the in vivo condition in the eye lens.

14.3 Autosensitized Oxidation of Colored and Glycated Bovine Lens Proteins Exposed to UVA-Visible Light at Low Oxygen Concentrations [73]

Photooxidative damage produced by AGEs generated during aging has been proposed as a possible mechanism of cataract formation [3, 23, 25, 26, 57]. In this part of the chapter, the photosensitizing ability of the chromophores generated by the incubation of bovine lens proteins (BLPs) with ASC will be analyzed, taking into consideration the actual low oxygen concentration present in the eye lens [74]. As lenticular AGEs are believed to mainly derive from the reaction between crystallins and the oxidation products of ASC [16], ASC was chosen to induce the glycation reaction and the concomitant production of the colored compounds required to perform photochemical studies of AGEs bound to proteins.

Glycated samples of lens proteins (10 mg/mL) were incubated for 3 and 6 days in the presence of 30 mM ASC and 5 μM $CuSO_4$, respectively, resulting in an increase in the absorbance in the UVA-visible region that is associated with the generation of AGEs (Fig. 14.2a). These compounds originate from the initial reaction between amino groups of basic amino acid residues (e.g., the ε-amino groups of Lys residues) and carbonyl groups arising from the decomposition products of ASC. This reaction leads to the formation of a Schiff's base, followed by an Amadori rearrangement. The generation of intermediary compounds containing new carbonyl groups allows the reaction with another amino group, initiating the multistep transformation and ultimately leading to a variety of structures known as advanced glycation end products (AGEs). The ASC-glycated proteins exhibited a 57 % loss in their amino group content at the end of 6 days (Fig. 14.2c), an amount that is

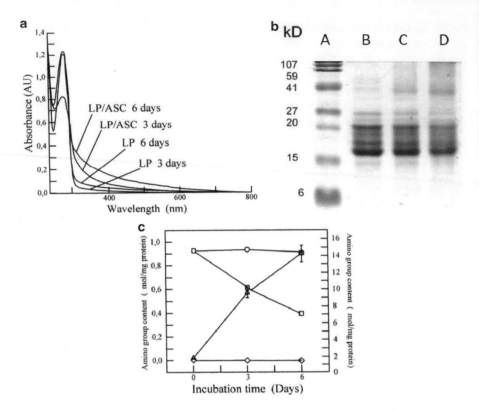

Fig. 14.2 Glycation of bovine lens proteins (LP) with ascorbic acid (ASC). (**a**) Shows the UV-visible spectra and time course production of advanced glycation end products. Samples were dialyzed and the spectra were recorded at a concentration of 1 mg/mL. (**b**) Shows the changes in the SDS-PAGE patterns of the samples. *Lanes a–d* correspond to the MW standard, 1, 3 and 6 days of incubation, respectively. (**c**) Shows the loss in amino group content (*open square*) and the rise in carbonyl groups (*filled triangle*) due to the glycoxidation processes. Controls for amino group loss (*open circle*) and carbonyl group formation (*filled diamond*) were performed by incubating the protein in the absence of ASC. Reproduced from Photochem. Photobiol Sci. 2008, 7: 718-724 by permission of The Royal Society of Chemistry (RSC) for the European Society for Photobiology, the European Photochemistry Association, and the RSC

more than sufficient to explain the presence of 27 % cross-linked proteins (Fig. 14.2b). The majority of the generated cross-links corresponds to dimeric forms, which derive from the formation of these AGE-chromophores in which two amino groups from different protein subunits participate. This fact will determine the protein residues that will be exposed to the reactive species generated during the photosensitization processes mediated by these covalently bonded chromophores.

Protein carbonyl groups were also detected in parallel with the formation of AGEs and the decrease in amino group content (Fig. 14.2c). Under the experimental conditions employed in this work, carbonyl groups could be introduced into proteins by a variety of oxidative processes. One of these processes is related to the prooxidative capacity of ASC in the presence of trace amounts of Cu^{2+}, especially when the antioxidant reaches low concentrations and is not able to trap the reactive oxygen species (ROS) generated by the reoxidation of the reduced transition metal [75]. It is possible that this process occurs due to oxidative Maillard reactions catalyzed by the copper present in the incubation medium. In this case, the Amadori products undergo a tautomerization process, which gives rise to a protein enediol. This reaction marks the start of glycoxidation, in which the enediol is oxidized to a protein dicarbonyl in a reaction catalyzed by the transition metal and with the participation of molecular oxygen, which is reduced to a superoxide radical anion and finally is transformed to the hydroxyl radical via hydrogen peroxide in a new step that requires the presence of metals [76, 77]. The different ROS generated during the glycoxidation process are responsible for the oxidation that affects the side chains of reactive amino acids, such as His, Trp, or Tyr [78, 79]. This phenomenon was experimentally observed as a decrease in the fluorescence intensity of the Trp residues in the glycated proteins. Notwithstanding the fact that the precursors of AGEs include different intermediates that contain carbonyl groups, such as Amadori or Heyns products, it has been reported previously that neither of these products contributes to the 2,4-dinitrophenylhydrazine DNPH reaction because these groups are rapidly lost in a first-order reaction [80]. Consequently, the observed increase in the carbonyl content found during the incubation of lens proteins with ASC must be mostly attributed to oxidative processes mediated by ROS.

It has been demonstrated that low level CL is emitted after human serum albumin oxidation mediated by AGEs and ROS formation [63]. This CL lasts for several minutes after the end of the photolysis and is due to the formation of excited states generated in the decomposition of peroxides and/or hydroperoxides formed during the oxidation of Trp residues inside the proteins [81]. The CL observed after the exposure of ASC-glycated lens proteins to visible light at low oxygen pressure indicates the occurrence of autosensitized photo-processes inside the proteins mediated by the covalently bonded AGE-chromophores (Fig. 14.3). As a consequence of the process, some amino acid residues become targets of this oxidation. The high sensitivity of this method allows the observation of CL even when the samples are illuminated with visible light, notwithstanding the fact that the absorption properties of the irradiated samples are very low in this region. This result is especially relevant because it indicates that even visible light can contribute to the enhanced photo-processes found in aged lenses. Native proteins incubated in the absence of ASC were insensitive to the action of visible light.

Previous experiments performed with small molecular weight AGEs prepared with amino acids instead of the protein showed that the efficiency of these compounds at low oxygen pressure during Trp oxidation is higher than that found in an aerobic atmosphere [29], indicating the prevalence of Type I sensitized photo-processes. When lens proteins were irradiated with UVA-visible light in the presence of low molecular weight AGEs, the formation of cross-linking was observed (Fig. 14.4), indicating that Type I processes involving an electron transfer between the reactants took place, giving rise to radical intermediaries that are responsible for the aggregation processes. The absence of modifications in the electrophoretic

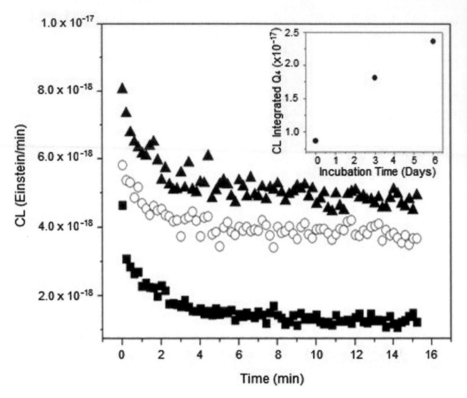

Fig. 14.3 Time profile of the chemiluminescence (CL) observed after the 5 min irradiation of bovine lens proteins glycated with ascorbic acid under 5 % O_2 pressure. CL measurement was performed after 30 s at 25 °C. The *insert* shows the correlation between the CL integrated along the first 20 min of decay and the incubation time. Symbols: (*filled square*) 6 days incubated native proteins in the absence of ASC, (*open circle*) 3 days glycation and (*filled triangle*) 6 days glycation. Reproduced from Photochem. Photobiol Sci. 2008, 7: 718-724 by permission of The Royal Society of Chemistry (RSC) for the European Society for Photobiology, the European Photochemistry Association, and the RSC

patterns of the glycated and subsequently irradiated proteins (Fig. 14.5) indicates that radical species generated during the electron transfer of a Type I mechanism are not able to give rise to the intermolecular cross-linking observed when the same experiments were performed using amino acid models of AGE-chromophores as sensitizers (Fig. 14.4). In the case of the protein-bound AGEs, the sensitizing damage is restricted to the area near the sensitizer. The fact that the main increase in the carbonyl content after the irradiation of the glycated lens proteins was found in the cross-linked fractions (Fig. 14.6), which are those that contain most of the AGE-chromophores, suggests that the damage is located in the vicinity where the sensitizers are positioned. Nonetheless, considering that the protein concentration can be assumed to be as high as 300 mg/mL in the lens of the eye [67], the possibility of intermolecular reactions sensitized by the AGEs cannot be discarded.

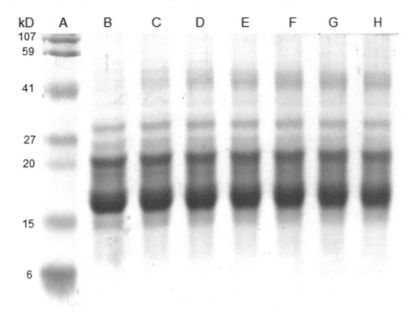

Fig. 14.4 SDS-PAGE of bovine lens proteins previously irradiated with UVA-visible light in the presence of small molecular weight AGEs generated from the incubation of ascorbic acid with *N*-acetyl-lysine and *N*-acetyl-arginine. *Lanes a–h* correspond to MW standard, 0, 10, 20, 30, 40, 50 and 60 min irradiated samples. Reproduced from Photochem. Photobiol Sci. 2008, 7: 718-724 by permission of The Royal Society of Chemistry (RSC) for the European Society for Photobiology, the European Photochemistry Association, and the RSC

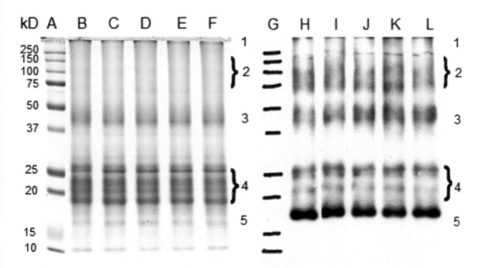

Fig. 14.5 Irradiation of 6-days glycated lens proteins with UVA-visible light at 5 % O$_2$ pressure. *Lanes a–f* show the SDS-PAGE analysis of the irradiated samples corresponding to the MW standard, 0, 15, 30, 45 and 60 min of irradiation, respectively. *Lanes g–l* show the Oxyblot analysis of the same samples. Reproduced from Photochem. Photobiol Sci. 2008, 7: 718-724 by permission of The Royal Society of Chemistry (RSC) for the European Society for Photobiology, the European Photochemistry Association, and the RSC

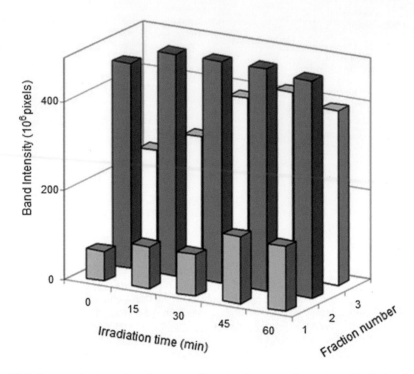

Fig. 14.6 Increase in the carbonyl content of previously glycated proteins with the irradiation time. The intensity of the bands was measured with Image Master 1D Software. No modifications were observed on bands 4 and 5 of Fig. 14.5 and, therefore, they were not included in this figure. Reproduced from Photochem. Photobiol Sci. 2008, 7: 718-724 by permission of The Royal Society of Chemistry (RSC) for the European Society for Photobiology, the European Photochemistry Association, and the RSC

The production of singlet oxygen when protein preparations from aged lenses were irradiated with UVA light has also been reported [57, 82]. The actual oxygen content in the lens is a topic that is still under discussion. Values of 28 mmHg in the cortex of the rabbit lens [83] and 2.5 mmHg for the bovine nucleus [33], and even the possibility of an essentially anaerobic lens [84] have been reported. In this context, the Type I sensitizing mechanism, which in its first step does not require the presence of molecular oxygen and involves the direct interaction between the sensitizer in triplet state and an adequate quencher, could play a very important role in the photosensitized events that occur in vivo in the eye lens. In addition to this finding, Ortwerth et al. reported the generation of hydrogen peroxide [62] and superoxide anion [61] as a result of the UVA irradiation of human lens proteins. This reactive species can derive from the Type I mechanism. Sensitized photochemical reactions mediated by AGEs are characterized by low quantum efficiency, but this fact is contrasted by the important contribution of these chromophores to the absorption of the incident radiation impinging on the lens [30], including the radiation of the visible region.

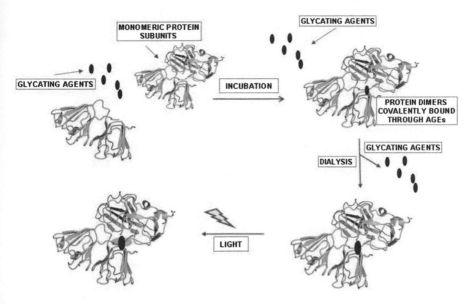

Scheme 14.1 Effect of UVA-visible light on bovine eye lens proteins, previously cross-linked in the absence of light, after incubation in the presence of ascorbate (ASC). The photochemical damage is circumscribed to the close vicinity of the chromophore

As a final remark, these results have demonstrated that the covalently bonded chromophores located in the cross-linked and glycated lens proteins can induce photooxidative damage when the samples are exposed to UVA-visible light under the low oxygen pressure present in the eye lens. Nonetheless, this damage is restricted to the area close to the chromophores and cannot explain the protein cross-linking observed during aging and in patients with cataracts (see Scheme 14.1).

14.4 Simultaneous Chemical and Photochemical Protein Cross-Linking of Eye Lens Proteins Incubated in the Presence of Ascorbate and Exposed UVA-Visible Light [85]

The in vitro incubation of lens proteins with ascorbic acid, which is present at high concentrations in the eye lens, results in glycation and protein cross-linking [16, 52]. Additionally, the glycated proteins are characterized by their sensitizer activity, which is very similar to that of the water-insoluble fraction from aged human lenses [48]. Colored ascorbic acid oxidation products obtained at 60 °C in oxidative conditions generate tryptophan photooxidation and glucose 6-phosphate dehydrogenase deactivation when these products are irradiated with UVA-visible light [63].

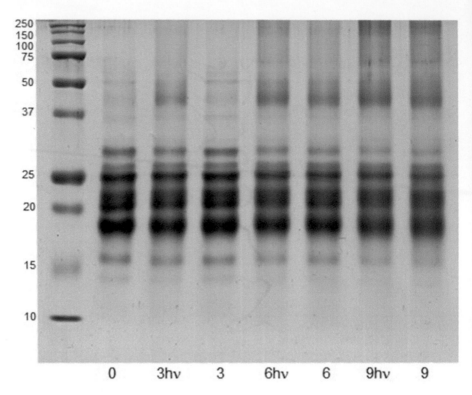

Fig. 14.7 SDS-PAGE profiles of bovine eye lens proteins (10 mg/mL) incubated during 0, 3 and 9 days at low oxygen concentration with ASC (3 mM). The samples were incubated in the absence and presence of UVA-visible light (3, 6, 9 and 3hʋ, 6hʋ, 9hʋ, respectively). *Lane 1* corresponds to a molecular weight marker. Reproduced from Photochem. Photobiol Sci. 2010, 9: 1351-1358 by permission of The Royal Society of Chemistry (RSC) for the European Society for Photobiology, the European Photochemistry Association, and the RSC

The glycating and photochemical properties of the decomposition products of ASC make them particularly useful in experiments in which eye lens proteins are incubated in the presence of this vitamin, when the solutions are in the dark and also exposed to light, using a low concentration of oxygen, simulating the physiological conditions of the eye lens. The results observed in Fig. 14.7 show that the incubations performed in the presence of UVA-visible light generate an increase in the amount of protein cross-linking compared to the samples maintained in the dark.

Using densitometry, it was possible to assess the increased cross-linking in the samples exposed to UVA-visible light during the incubation. The cross-linking was estimated by considering all the fractions above 37 kDa. The following quotients were obtained: $(\text{cross-linking})_{3D h\upsilon}/(\text{cross-linking})_{3D} = 1.04$; $(\text{cross-linking})_{6D h\upsilon}/(\text{cross-linking})_{6D} = 1.16$ and $(\text{cross-linking})_{9D h\upsilon}/(\text{cross-linking})_{9D} = 1.35$. When the protein samples were incubated in the dark or exposed to light in the absence of ASA, no significant modification of the electrophoretic patterns was observed. The fact that both irradiated and nonirradiated samples were characterized by the same decrease in amino group content and a similar rate of ASA consumption indicates that the additional cross-linking cannot be associated with an extended process of ascorbylation accompanied by protein aggregation. The nonoccurrence at 9 days of additional protein cross-linking when the samples were incubated under exposure to UVA-visible light in the absence of ASA indicates that colored AGEs bound to protein, which could be present in the high molecular weight aggregates, do not have photosensitizing capacity to induce these types of photo-processes (Fig. 14.7, lane 2). This result is in agreement with previously reported results [73]. Notwithstanding the fact that AGEs are efficient sensitizers of ASA photoconversion [29, 63], in this work, an increase in the decomposition of ASA was not observed as a consequence of the light because the rate of AGE formation was significantly slower than that of the ASA decomposition rate, which mainly occurs within the first 24 h of incubation. The photosensitizing capacity of other chromophores in the reaction medium, unrelated to AGEs and arising from the decomposition of ASA in secondary reactions, can explain the initially small but continuously increasing protein cross-linking over time, when the incubations were performed under UVA-visible light. In concordance with this assumption, protein cross-linking and oxidation were observed when colored solutions of ASA, obtained after 3, 6, and 9 days incubation of at 37 °C, were used as photosensitizers [85].

Scheme 14.2 summarizes the main chemical and photochemical processes that occur when bovine eye lens proteins are incubated in the presence of ASC at low oxygen concentration and maintained in the dark or exposed to UVA-visible light.

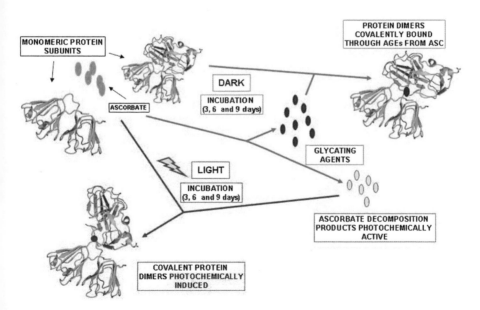

Scheme 14.2 *Blue arrows* show the chemical processes that occur during the incubation of bovine eye lens proteins with ASC in the dark. When the same experiment was repeated by exposing the incubation mixture to UVA-visible light, the occurrence of photochemical processes was observed (*red arrows*) in addition to the reactions that occurred in the absence of light (*blue arrows*). The experiments were performed at 37 °C, in phosphate buffer (pH 7.4), under low oxygen pressure (5 % O$_2$)

14.5   Age-Dependent Protein Modifications and Declining Proteasome Activity in the Human Lens

The ocular lens is made up of approximately 60 % water and 38 % proteins [22, 86], which constitute almost 35 % of the wet weight of the organ and 100 % of the total dry weight. These proteins are called lens crystallins. The lens is characterized by a low molecular oxygen concentration because it lacks vascularization, and it is also characterized by a reduced cellular exchange [56, 86]. This type of environment leads to the accumulation of the modifications suffered by its structural components throughout the lifetime, which eventually leads to the aging of the organ and, in certain cases, the formation of cataracts [22].

In the aged lenses, higher molecular weight proteins tend to accumulate. The amount of these products increases as the individual gets older [87, 88].

Once synthesized, all cellular proteins of an organism have a certain lifetime and, therefore, are continuously degraded to their constituent amino acids. The lack of protein turnover in the central region of the lens causes an increase in the concentration of oxidation-modified crystallins because the greater lifetime of proteins in this zone increases the possibility that they undergo chemical modifications and oxidative stress [88–90]. The external part of the lens is protected by a large number of antioxidant enzymes, but these mechanisms are reduced in the core of the lens [91].

The damaged proteins must be repaired or degraded because the accumulation of such material is associated with malfunctioning of cells, tissues, and organs. In addition, oxidative stress, to which the eye lens is permanently exposed, is one of the causes for the aging of cellular components; in addition, several diseases [92], including cataract [93] formation [94], have been related to oxidative stress [95]. The identification and degradation of these proteins are fundamental mechanisms of cellular defense, and the proteasome is the main proteolytic complex that performs this cleaning process [96]. Proteasomes are enzymatic complexes of high molecular weight that are widely distributed in the cytosol as well as in the nucleus of the cell [97]. The proteins that are prone to be degraded by the proteasome must present one or more of the following characteristics [98]: (1) are oxidized and, consequently, are toxic for the cell; (2) are incorrectly assembled and, therefore, incapable of fulfilling their functions properly, and (3) have a short lifetime, which applies to the great majority of intracellular proteins [99]. This latter fact implies that the proteasome is involved in the regulation of the cellular cycle in plants and animals because it directly modulates the activities of other proteins within the cell [100]. Therefore, the proteasome is involved in functions that are absolutely necessary for cellular growth and viability. In mammalian cells, the 20S proteasome is constituted by four stacked heptameric rings that form a hollow cylinder. Each ring consists of seven different subunits that belong to two families, α and β, that are located in the outer and inner part of the cylinder, respectively [97]. Two activators, named 19S and 11S, can bind the 20S proteasome on each side of the cylinder and give rise to different complexes [96]. The 26S proteasome, which consists of the 19S and 20S particles, degrade proteins in a ubiquitin-dependent and ATP-dependent manner

[101]. Most of the substrates that are degraded by proteasomes, especially proteins with short half-lives, must be tagged for their degradation by the addition of a poly-ubiquitin chain [102]. Ubiquitin is a 76-residue peptide that is present in eukaryotic cells [103]. Ubiquitin binds covalently to proteins through a highly regulated energy-dependent process [104]. The number of ubiquitin units that bind to the protein as a chain is not constant, and it is thought that the length of the polyubiqui-tin chain helps to fix the proteins to the proteasome entrance for the time necessary to denature it completely and, thus, prepare it for complete degradation.

The number of pathologies derived from alterations of proteasome activity and the consecutive failure in the intracellular protein degradation is considerable. In fact, the accumulation of damaged proteins that would be degraded under normal conditions is a common characteristic of several neurodegenerative diseases [105], such as Parkinson's [106, 107] and Alzheimer's disease [108, 109]. In addition, aging is accompanied by an alteration of the cellular metabolism that leads to oxi-dized protein accumulation and cross-linking and a reduction in some enzymatic activities [110, 111]. The molecular basis of the less efficient protein degradation related to aging is still uncertain, but it is suspected that the proteasome impairment that occurs with age is somehow related to a less efficient control of the intracellular protein turnover.

Several studies have concluded that proteins are susceptible to oxidative damage, which alters their proteolytic susceptibility. These findings demonstrate that mild oxidative damage increases proteolytic degradation, whereas extensive oxidative damage causes decreased proteolysis due to cross-linking, aggregation, and decreased solubility [112, 113]. It has been demonstrated that the inhibition of the proteolytic activity of the proteasome causes an increase in the amount of oxidized proteins in several cell types, which is also found in cataractous lenses. Several studies aimed at determining a proteasome activity status in aged individuals have suggested a cor-relation of the two phenomena, which would explain the accumulation of oxidized proteins in the aged [114–117]. More recent studies of proteasome activity in kerati-nocytes showed that the reduction of the proteolytic activity could be related to a decrease in the cellular proteasome content [118]. It has also been demonstrated that the proteolytic activity of the proteasome in a cataractous eye lens epithelium is lower compared to that of cells obtained from healthy lens epithelium [116].

A systematic comparative study of the modifications that occur in eye lens pro-teins, as a function of AGEs, and the effect that they have in the amount, integrity, and functionality of the proteasome have been performed [119]. For this purpose, soluble and insoluble protein fractions of human lenses corresponding to various age groups were characterized in terms of their levels of glyco-oxidative damage (Figs. 14.8a, b and 14.9a, b). The data obtained on water-insoluble (previously solubilized using a hydrolytic treatment) and -soluble proteins (Figs. 14.9 and 14.8, respectively) showed that an increase in the anti-AGE and anti-carboxymethyllysine (CML) immunoreactivity occurred as a function of age, which is consistent with previous results [120]. In the lens fibers, the generation of CML arises as an end product of the reaction resulting from either the oxidative degradation of the Amadori product or the reaction between oxidative degradation products of glucose

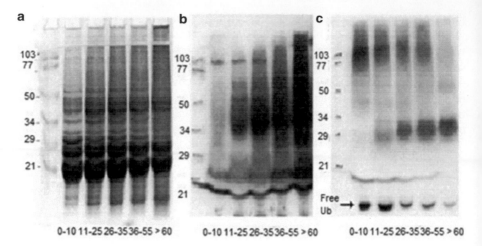

0-10 11-25 26-35 36-55 >60   0-10 11-25 26-35 36-55 >60   0-10 11-25 26-35 36-55 >60

Fig. 14.8 SDS-PAGE and Western blot analyses of the lens proteins from the water soluble fraction. *Lane 1* corresponds to pre-stained molecular-weight markers. The different age groups are marked for each of the other lanes. (**a**) SDS-PAGE of the total soluble proteins of human eye lenses grouped by age. (**b**) Western blot of the total soluble proteins of human eye lenses grouped by age using anti-*N*-epsilon-carboxymethyllysine (CML) polyclonal antibodies. (**c**) Western blot of the soluble proteins of human eye lenses grouped by age using anti-ubiquitin (Ub) polyclonal antibodies. Reproduced from Arch. Biochem. Biophys. 2004, 427: 197-203 with permission from Elsevier Inc

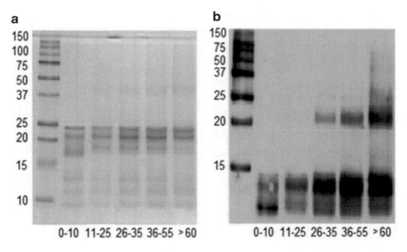

0-10 11-25 26-35 36-55 >60   0-10 11-25 26-35 36-55 >60

Fig. 14.9 SDS-PAGE and Western blot analyses of peptides from the water insoluble lens fraction. *Lane 1* corresponds to pre-stained molecular-weight markers. The various age groups are marked on each of the other lanes. (**a**) SDS-PAGE data on the insoluble proteins from human eye lenses of individuals from various age-groups. (**b**) Western blot of the insoluble proteins of human eye lenses corresponding to various age-groups, obtained using anti-CML polyclonal antibodies. Reproduced from Arch. Biochem. Biophys. 2004, 427: 197-203 with permission from Elsevier Inc

or ASC. CML has been identified as a good molecular indicator of the aging process in many tissues and has been reported to accumulate during degenerative processes, such as those involved in the development of cataracts, arthritis, atherosclerosis, and Alzheimer's disease. The protein hydrolysates of the water-insoluble protein fraction showed that increasingly high molecular weight peptides were present as a function of age (Fig. 14.9a). This finding could indicate the presence of increasing covalent cross-links in the proteins. In the soluble fraction, it was found that anti-CML reactivity is mainly associated with high molecular weight proteins. These findings suggest that glyco-oxidation reactions may contribute directly to the aggregation process during aging.

Lens fibers and lens epithelial cells have a completely functional ubiquitin-proteasome pathway, although the protein levels and the activities of the protein components decrease in the inner region of the lens [121]. Because most of the substrates that are degraded by 20S proteasomes must be tagged for degradation by a polyubiquitin chain [122], it was of interest to determine the presence of and monitor the fate of ubiquitinated proteins in the eye lens as a function of age. A decrease in the free ubiquitin and high molecular weight ubiquitin-protein conjugate content was found to occur with age, as well as an increase in the 30 kDa fraction, possibly corresponding to the crystalline subunits (Fig. 14.8c). Ubiquitinated proteins were present in the lenses of the youngest age group(s), but only the high molecular weight proteins were found in relatively small proportions; these proteins were simultaneously reactive to anti-CML. Taken together, these findings suggest that these proteins cannot be degraded by the proteasome. In addition, they may undergo further cross-linking over time, resulting in their insolubilization, and they may therefore gradually disappear from the soluble fraction with age. However, the proteasome seems to efficiently degrade low molecular weight crystallins, as indicated by the absence of anti-CML reactivity (Fig. 14.8c, lane 2). The increased levels of low molecular weight ubiquitinated crystallins with age (Fig. 14.8c) constitute an efficient functional response of the ubiquitin system to the increasing crystalline glyco-oxidation (Fig. 14.2b), despite the decrease in the rate of proteolysis of the proteasomes present in the human lens soluble fraction (Fig. 14.10). This loss of activity can be expected in view of the decreasing proteasome levels observed in the lenses of the various age groups. In a previous report, no correlations between proteasome activity and age were observed, but these results were based mainly on lenses with cataracts and included only three clear lenses of elderly individuals [117] and, therefore, cannot be compared with the results reported in this chapter. The decrease in the proteasome concentration may partly reflect a decrease in the rate of damaged protein elimination. In addition, the efficiency of the proteasome activity also seems to be affected by age, at least as far as the peptidylglutamyl-peptide hydrolase activity is concerned.

Using an anti-CML immunoaffinity column and an anti-20S antibody, the presence of the 20S proteasome was detected in the bound fraction, indicating that the proteasome itself had undergone glyco-oxidation. This observation might, therefore, explain the alteration of its proteolytic activities. Interestingly, minimal modification of the proteasome was observed in the sample from the youngest age group (Fig. 14.11).

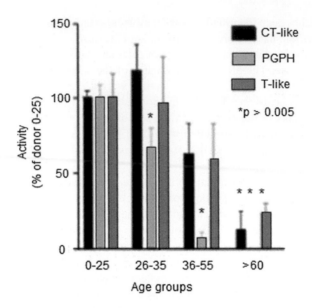

Fig. 14.10 Decrease of chymotrypsin-like, peptidylglutamyl-peptide hydrolase, and trypsin-like activities of the proteasome present in human eye lens soluble fraction. The activities are expressed as the percentage of the value obtained for the youngest sample. For each activity, measurements were performed in triplicate in two separate experiments. Error bars indicate the standard error obtained using the Student t test and a significant difference ($p < 0.05$) with the activity of the youngest sample is marked with an *asterisk*. Reproduced from Arch. Biochem. Biophys. 2004, 427: 197-203 with permission from Elsevier Inc.

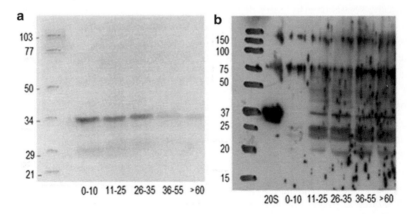

Fig. 14.11 Western blot detection of the proteasome in lens soluble fraction before and after separation of the CML-modified proteins. (**a**) Western blot of the total soluble fraction of human eye lens of individuals from different age groups using anti-20S polyclonal antibodies. *Lane 1* corresponds to pre-stained molecular-weight markers. The different age groups are marked for each of the other lanes. (**b**) Western blot of the total CML-modified soluble proteins of human eye lens of individuals from different age groups using anti-20S. The CML-modified proteins have been previously isolated using an immuno-affinity column prepared with CML-antibodies. *Lane 1* corresponds to 20S proteasome standard. Both Western blot experiments reported in (**a**) and (**b**) were performed twice. Reproduced from Arch. Biochem. Biophys. 2004, 427: 197-203 with permission from Elsevier Inc

This finding suggests that proteasomes, as well as the other proteins in the eye lens, may undergo glycation and/or glyco-oxidation reactions that result in both structural and functional changes. Together with the decrease in the proteasome content, the drop in the proteolytic activity of this enzymatic complex may therefore contribute to the age-related accumulation of modified proteins in the lens. Furthermore, oxidated proteins are known to be suitable substrates for the proteasome, but they are also able to act as proteasome inhibitors when they are greatly modified [123]. Cross-linked proteins resulting either from prolonged oxidation or from reactions with lipid peroxidation products, such as 4-hydroxy-2-nonenal, have previously been reported to inhibit proteasomes [113, 124–126]. Therefore, these inhibitory proteins, as well as glycated cross-linked proteins, may also participate in the age-dependent decrease in proteasome activity observed in the human lens.

Wagner and Margolis have previously reported that the 20S proteasome partially purified from bovine lens exhibited an age-related decrease in both the trypsin-like and peptidylglutamyl-peptide hydrolase activities, whereas the chymotrypsin-like activity was only slightly affected [127]. The reason we observed a more pronounced age-related decrease in the peptidylglutamyl-peptide hydrolase activity in the human lens may be because it is more susceptible to inactivation by oxidative processes [128]. In addition, this peptidase activity was also found to be more sensitive to inactivation in the treatment of human dermal fibroblasts with the glycating agent, glyoxal, although this was not found to be the case when glyoxal treatment was applied to the 20S proteasome in vitro [129].

14.6 Photosensitized Reactions Mediated by the Major Chromophore Arising from Glucose Decomposition Result in the Oxidation and Cross-Linking of Lens Proteins and Activation of the Proteasome

Although numerous colored products derived from the Maillard reaction have been identified in the lens during aging and in diabetic patients [130] the photochemical contribution of the degradation products of glucose has not been addressed.

In this chapter, a novel glucose-derived chromophore found after in vitro glucose decomposition but also in the water-soluble fractions of the human eye lens from cataractous patients is discussed. This chromophore, named GDC, possesses strong photosensitizing properties leading to increased protein oxidation, protein cross-linking, and the activation of the proteasome activity upon UVA irradiation at low oxygen concentrations.

GDC can be detected in vitro after long periods (10 days or more) of glucose incubation in phosphate buffer in the presence of trace amounts of metal [131].

When solutions incubated with glucose were analyzed by HPLC-MS, using detection by total ion current (TIC), four peaks with different retention times were observed. The colored compound ($\lambda_{Max} = 365$ nm), which was purified several times

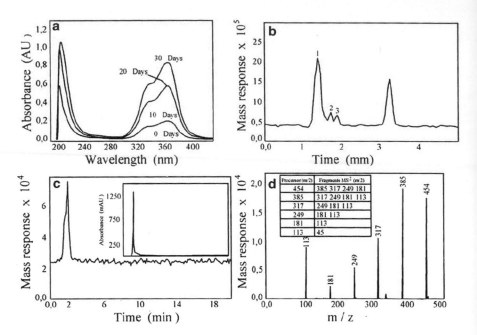

Fig. 14.12 Chemical characterization of a chromophore derived from glucose decomposition. (**a**) Absorption spectra of glucose solutions after 10, 20 or 30 days of incubation. Glucose solutions (30 mM) in buffer phosphate (100 mM, pH 7.4) were bubbled with oxygen at 5 % during 30 min and then sealed in order to keep the low oxygen concentration. The incubations were performed at 37 °C, in sterile conditions. (**b**) HPLC-MS elution profiles of glucose solutions (30 mM) after 30 days of incubation at low oxygen concentration and pH 7.4. (**c**) HPLC profiles of the fraction previously purified through a Sephadex G-15 column and analyzed by MS detection using detection by total ion current (TIC). The *inset* shows the chromatogram analyzed by absorption spectroscopy (365 nm). (**d**) Negative electrospray ionization mass spectra of the glucose-derived chromophore. The *inset* table shows the fragmentation of each one of the peaks observed in the spectra. Reproduced from Biochim. Biophys. Acta 2012, 1822: 564-572 with permission from Elsevier B.V.

by size-exclusion chromatography, showed representative mass spectra peaks at 453, 385, 317, 249, 181, and 113 m/z (Fig. 14.12). The mass spectra and the fragmentation of this chromophore correspond to a molecular ion of 68 m/z, which is in agreement with the loss of one furan unit. The same loss of the molecular ion of 68 m/z was observed five times along the consecutive fragmentation pattern, suggesting the presence of 5 furan units in the structure of GDC. These results are in agreement with previously published data on a glucose-derived degradation product that contains furan units [132, 133]. There is a wide variety of furan precursors because this product can be generated by the degradation and/or recombination of smaller fragments [134].

Although the process of GDC formation is rather slow, it is relatively faster compared to the reaction between glucose and the ε-amino groups of Lys or the guanidinium groups of Arg residues in eye lens proteins. In fact, when eye lens proteins at a concentration of 10 mg/mL were incubated at 37 °C in the presence of glucose

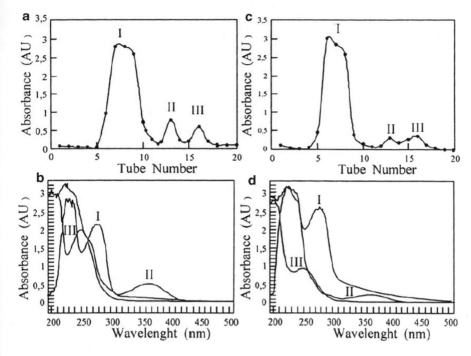

Fig. 14.13 Size exclusion chromatogram for normal and cataractous human eye lens. Elution profiles analyzed at 280 nm, of the water-soluble human eye lens proteins from normal (**a**) and cataractous (**c**) individuals, fractioned on a Sephadex G-15 column. The absorption spectra of the fractions I, II and III shown in (**a**) and (**c**) are indicated in (**b**) and (**d**), respectively. Reproduced from Biochim. Biophys. Acta 2012, 1822: 564-572 with permission from Elsevier B.V.

(30 mM) at a low oxygen concentration (5 %), the first chemical modifications (in terms of loss of amino groups) appear at times significantly longer than 10 days. The slow rate observed for the Schiff base production has been directly linked with the low amount of the open-chain form of glucose because this monosaccharide is found mainly as a cyclic structure formed by an intramolecular hemiacetal in aqueous solutions [135].

The generation of a colored compound during the decomposition of glucose in vitro has motivated research to assess the presence of this derivative in the water-soluble fractions derived from human cataractous lenses. The size-exclusion chromatography (Sephadex G-15) analysis of these fractions, obtained from normal and cataractous human eye lenses, indicated the presence of three components with absorption peaks at 280 nm (Fig. 14.13a, c). In both cases, the first fraction (I) contained water-soluble eye lens proteins, and the presence of coloration in the UVA-visible region was observed in cataractous lenses (Fig. 14.13d). Fractions II and III do not contain proteins (Fig. 14.13b, d); however, a defined band centered at approximately 365 nm in fraction II was found to be decreased in cataractous lenses (Fig. 14.13d). Fraction III presents a maximum near 260 nm and does not absorb in the UVA region, which is similar to the spectral properties of ASC. The analysis of the elution profile after reversed-phase chromatography of fraction II obtained from cataractous lenses (Fig. 14.14a) and from the glucose solution incubated for 30 days at 37 °C in the dark, under a 5 % oxygen atmosphere, indicates that one of the components found in the cataractous lenses has the same spectrum as GDC (Fig. 14.14b, inset B-1). These results suggest that autoxidative pathways contribute to lens fiber aging, not only through protein glycation but also by the generation of a chromophore that could exacerbate in vivo photosensitized reactions. Recently, it has been found that the ASC oxidation, in the presence or absence of proteins, generates a chromophore with the same spectral and chromatographic properties as GDC [85].

To study the photosensitizing capacity of GDC, BLPs were used as a photochemical target. The irradiation of BLPs in the presence of GDC at a low oxygen concentration resulted in increased protein cross-linking and extensive protein oxidation, which is in agreement with the behavior that we have previously reported in aged lenses (Fig. 14.15a, b) [119]. No changes were observed in samples incubated in the dark (lane 1) or in those exposed to UVA-visible light (lane 3), indicating that BLPs are not sensitive to the irradiation conditions used in this experiment. However, the occurrence of protein cross-linking (Fig. 14.15a, lane 2), increased protein carbonylation (Fig. 14.15b, lane 2), and the decreased protein levels of the beta 5 proteasome catalytic subunit present in BLP extracts (Fig. 14.4d, lane 2) were observed in samples containing the GDC. No changes in the protein levels of the 20S proteasome alpha 2 subunit were observed (Fig. 14.15c, lane 2).

The effect of GDC-mediated photosensitization reactions on the 20S proteasome present in the BLP mixture has also been addressed. An increased chymotrypsin-like peptidase activity in the proteasome was observed in the presence of GDC upon irradiation. Proteasome activity was increased more than eight and fourfold at 5 % and 20 % oxygen, respectively (Table 14.3). To explain the differences between both irradiated systems, follow-up experiments were carried out using D_2O as a solvent, which is known to increase the lifetime of singlet oxygen [136]. Samples irradiated under 5 and 20 % atmospheres were characterized by a significant decrease of the proteasomal chymotrypsin-like activity, indicating a deleterious effect of singlet oxygen on proteasome activity. Previous studies on skin fibroblasts treated with UVA showed a decrease in proteasome activity, which was correlated to $^1O_2$ formation because NaN_3, a potent quencher of $^1O_2$, prevented the decline of the proteasome activity [137]. A continuous increase of the ratio $(enzymatic\ activity)_{H_2O} / (enzymatic\ activity)_{D_2O}$ was observed over time for both oxygen concentrations. These results suggest that the species generated by Type I and Type II mechanisms have opposite effects on proteasome activity, with Type I functioning as a positive activator, whereas Type II mechanisms lead to the impairment of proteasome function. The observed activation of the chymotrypsin-like activity in the proteasome may be due to the conversion of latent 20S proteasomes into their active form, as previously described [138]. In addition, an increased susceptibility of the irradiation-activated proteasome was observed in the presence of GDC to thermal inactivation, in agreement with results that had been previously reported [139]. Taken together, these data suggest that Type I photo-processes medi-

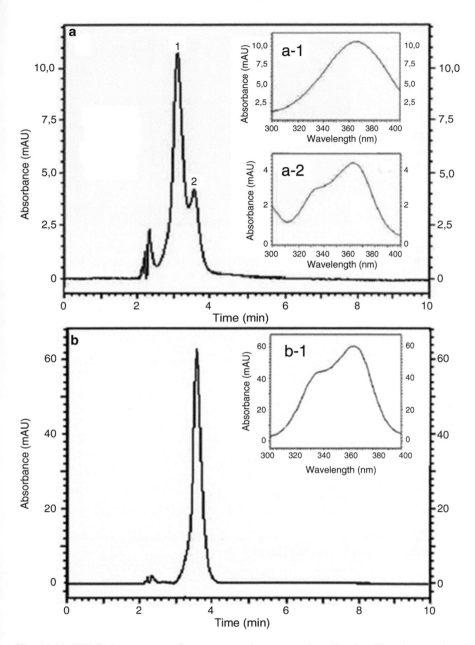

Fig. 14.14 HPLC chromatogram for cataractous human eye lens (fraction II) and comparison with the glucose-derived chromophore. Reverse-phase HPLC chromatogram at 365 nm absorbance for the fraction II of Fig. 14.13c (**a**). (**b**) Represents the reversed-phase HPLC chromatogram at 365 nm absorbance for a glucose solution previously incubated during 30 days at 37 °C in the dark, under 5 % oxygen atmosphere, in buffer phosphate 100 mM, pH 7.4. The *insets* show the absorption spectra of the fractions shown in the chromatogram. Reproduced from Biochim. Biophys. Acta 2012, 1822: 564-572 with permission from Elsevier B.V.

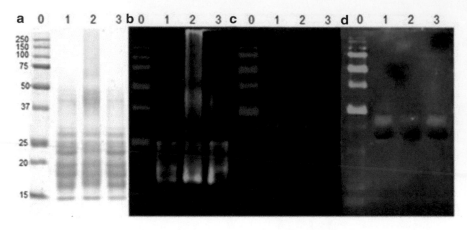

Fig. 14.15 Protein oxidation and proteasome levels within BLP irradiated in the presence or absence of the glucose-derived chromophore. SDS-PAGE (**a**), Oxy-Blot (**b**) and Western-Blots corresponding to 20S proteasome subunits α-2 (**c**) and β-5 (**d**), of native BLP (*lanes 1*), BLP previously irradiated with UVA-visible light during 1 h, in the presence (*lanes 2*) or absence (*lanes 3*) of colored glucose decomposition product. Reproduced from Biochim. Biophys. Acta 2012, 1822: 564–572 with permission from Elsevier B.V.

Table 14.3 Proteasome chymotrypsin-like peptidase activity after photosensitized reactions mediated by glucose-derived chromophore

	Irradiation at 5 %O_2		Irradiation at 20 %O_2	
Irradiation time (min)	Activity H_2O (rfu/min)	Activity D_2O (rfu/min)	Activity H_2O (rfu/min)	Activity D_2O (rfu/min)
0	67 ± 14	46 ± 6	67 ± 14	46 ± 6
20	320 ± 8	37 ± 16	201 ± 14	24 ± 4
40	402 ± 21	35 ± 11	218 ± 54	16 ± 5
60	452 ± 34	15 ± 15	199 ± 12	9 ± 10

Reproduced from Biochim. Biophys. Acta 2012, 1822: 564-572 with permission from Elsevier B.V.
rfu/min relative fluorescence units/min. The experiments were performed at two different oxygen concentrations (5 and 20 %) and using H_2O and D_2O

ated by GDC induce conformational changes in the proteasome, thus explaining the enhanced peptidase activity observed. Interestingly, an increase in low molecular weight crystallin peptides has been reported in human eye lenses during aging and cataractous disease [140, 141], this fact is especially important in the nuclear region [142, 143]. The origin of these peptides in vivo is still uncertain. However, it has been postulated that they could arise as a result of the unregulated proteolysis of crystallins and nonenzymatic mechanisms [144]. This fact highlights the relevance of studying the proteolytic mechanisms involved in human nuclear cataracts and their regulation by photosensitized mechanisms.

14.7 Concluding Remarks

The eye lens is a tissue permanently exposed to UVA-visible light and is therefore prone to be affected by the occurrence of photochemical modifications involving the crystalline proteins, which are its main constituents. Because the native crystalline is transparent, photosensitizers are required to induce photochemical processes. With age different chemical modifications occur, one of which involves reactions between the proteins and different glycating compounds, which gives rise to colored advanced glycation end products (AGE), which have the properties of sensitizers in the UVA-visible region. These compounds can act through a Type I radical mediated and/or a Type II oxygen-mediated photosensitizing mechanism. The Type I mechanism in particular is very important in the eye lens considering the low oxygen concentration that is present in this tissue. AGE models generated by the incubation of basic amino acids and sugars induce oxidative modifications and cross-linking of the eye lens proteins when irradiated with UVA-visible light under conditions of low oxygen concentration. When AGEs are generated using the lens proteins as the target of the glycation and the colored proteins are then exposed to UVA-visible light, oxidative processes are observed in the immediate vicinity of the sensitizer, but no cross-linking occurs under these conditions. Colored compounds produced during the decomposition of ASC and glucose have been isolated and used as sensitizers. They have the property of inducing an increase in the amount of carbonyl groups and of cross-linking the eye lens proteins when they are irradiated with UVA-visible light at low oxygen concentrations. A similar chromophore is found in cataractous human eye lenses.

An increase in the UVA-visible absorption of the eye lens proteins of humans is observed with increasing age and this coloration is extremely high in patients with cataracts.

The soluble and insoluble protein fractions of these human lenses from the various age groups show an increased glycol-oxidative damage with age, as measured on the basis of anti-carboxymethyl-lysine immune reactivity. Concomitantly, decreases in the proteasome contents and the peptidase activities are observed in the water-soluble fraction. An increased proteasome peptidase activity is observed when lens proteins are exposed to UVA-visible light under low oxygen concentration in the presence of a colored glucose decomposition product. However, the use of D_2O results in decreased proteasome activity, suggesting that singlet oxygen promotes the impairment of proteasome activity.

Taken together, the data presented in this chapter suggest that a colored, low molecular weight compound formed via the decomposition of glucose and ASC produces both oxidation and protein cross-linking in a manner analogous to that which occurs in vivo in the eye lens.

Acknowledgments These studies were supported by CONICYT (Chile), bi-national projects ECOS (France)/CONICYT (Chile), and the European Union.

References

1. Turner PL, Mainster MA. Circadian photoreception: ageing and the eye's important role in systemic health. Br J Ophthalmol. 2008;92:1439–44.
2. Turner PL, Van Someren EJW, Mainster MA. The role of environmental light in sleep and health: effects of ocular aging and cataract surgery. Sleep Med Rev. 2010;14:269–80.
3. Ortwerth BJ, Bhattacharyya J, Shipova E. Tryptophan metabolites from human lenses and the photooxidation of ascorbic acid by UVA light. Invest Ophthalmol Vis Sci. 2009;50:3311–9.
4. Kessel L, Eskildsen L, Lundeman JH, Jensen OB, Larsen M. Optical effects of exposing intact human lenses to ultraviolet radiation and visible light. BMC Ophthalmol. 2011;11:41–7.
5. Jose JG, Pitts DG. Wavelength dependence of cataracts in albino mice following chronic exposure. Exp Eye Res. 1985;41:545–63.
6. Andley UP, Malone JP, Townsend RR. Inhibition of lens photodamage by UV-absorbing contact lenses. Invest Ophthalmol Vis Sci. 2011;52:8330–41.
7. Giblin FJ, Lin LR, Leverenz VR, Dang L. A class I (Senofilcon A) soft contact lens prevents UVB-induced ocular effects, including cataract, in the rabbit in vivo. Invest Ophthalmol Vis Sci. 2011;52:3667–75.
8. Gao N, Hu L-W, Gao Q, Ge T-T, Wang F, Chu C, Yang H, Liu Y. Diurnal variation of ocular exposure to solar ultraviolet radiation based on data from a manikin head. Photochem Photobiol. 2012;88:736–43.
9. McCarty CA, Taylor HR. Recent developments in vision research: light damage in cataract. Invest Ophthalmol Vis Sci. 1996;37:1720–3.
10. Boettner EA, Wolter JR. Transmission of the ocular media. Invest Ophthalmol Vis Sci. 1962;1:776–83.
11. Sliney DH. Physical factors in cataractogenesis: ambient ultraviolet radiation and temperature. Invest Ophthalmol Vis Sci. 1986;27:781–90.
12. Tsentalovich YP, Sherin PS, Kopylova LV, Cherepanov IV, Grilj J, Vauthey E. Photochemical properties of UV filter molecules of the human eye. Invest Ophthalmol Vis Sci. 2011;52:7687–96.
13. Giblin FJ, Lin LR, Simpanya MF, Leverenz VR, Fick CE. A class I UV-blocking (senofilcon A) soft contact lenses prevents UVA-induced yellow fluorescence and NADH loss in the rabbit lens nucleus in vivo. Exp Eye Res. 2012;102:17–27.
14. Hiramoto K, Yamate Y, Kobayashi H, Ishii M. Long-term ultraviolet A irradiation of the eye induces photoaging of the skin in mice. Arch Dermatol Res. 2012;304:39–45.
15. Borkman RF, Tassin JD, Lerman S. Fluorescence lifetimes of chromophores in intact human lenses and lens proteins. Exp Eye Res. 1981;32:313–22.
16. Cheng R, Lin B, Lee KW, Ortwerth BJ. Similarity of the yellow chromophores isolated from human cataracts with those from ascorbic acid-modified calf lens proteins: evidence for ascorbic acid glycation during cataract formation. Biochim Biophys Acta. 2001;1537:14–26.
17. Thiagarajan G, Shirao E, Ando K, Inoue A, Balasubramanian D. Role of xanthurenic acid 8-O-beta-glucoside, a novel fluorophore that accumulates in the brunescent human eye lens. Photochem Photobiol. 2002;76:368–72.
18. Cheng RZ, Lin B, Ortwerth BJ. Separation of the yellow chromophores in individual brunescent cataracts. Exp Eye Res. 2003;77:313–25.
19. Cheng R, Feng Q, Argirov OK, Ortwerth BJ. Structure elucidation of a novel yellow chromophore from human lens protein. J Biol Chem. 2004;279:45441–9.
20. Argirov OK, Lin B, Ortwerth BJ. 2-ammonio-6-(3-oxidopyridinium-1-yl)hexanoate (OP-lysine) is a newly identified advanced glycation end product in cataractous and aged human lenses. J Biol Chem. 2004;279:6487–95.
21. Dillon J. New trends in photobiology: the photophysics and photobiology of the eye. J Photochem Photobiol B. 1991;10:23–40.
22. Krishna CM, Uppuluri S, Riesz P, Zigler Jr JS, Balasubramanian D. A study of the photodynamic efficiencies of some eye lens constituents. Photochem Photobiol. 1991;54:51–8.

23. Ortwerth BJ, Prabhakaram M, Nagaraj RH, Linetsky M. The relative UV sensitizer activity of purified advanced glycation end products. Photochem Photobiol. 1997;65:666–72.
24. Argirova MD, Breipohl W. Glycated proteins can enhance photooxidative stress in aged and diabetic lenses. Free Radic Med. 2000;36:1251–9.
25. Zigman S. Lens UVA photobiology. J Ocul Pharmacol Ther. 2000;16:161–5.
26. Balasubramanian D. Photodynamics of cataract: an update on endogenous chromophores and antioxidants. Photochem Photobiol. 2005;81:498–501.
27. Yoshimura A, Ohno T. Lumiflavin-sensitized photooxygenation of indole. Photochem Photobiol. 1988;48:561–5.
28. Silva E, Ugarte R, Andrade A, Edwards AM. Riboflavin-sensitized photoprocesses of tryptophan. J Photochem Photobiol B. 1994;23:43–8.
29. De la Rochette A, Birlouez-Aragon I, Silva E, Morliere P. Advanced glycation endproducts as UVA photosensitizers of tryptophan and ascorbic acid: consequence for the lens. Biochim Biophys Acta. 2003;1621:235–41.
30. Silva E, Quina FH. Photoinduced processes in the eye lens: do flavins really play a role? In: Silva E, Edwards AM, editors. Flavins. Photochemistry and photobiology. Cambridge: RSC Publishing; 2006.
31. Shui YB, Fu JJ, García C, Dattilo LK, Rsjagopal R, Mc Millan S, Mak G, Holekamp NM, Lewis A, Beebe DC. Oxygen distribution in the rabbit eye and oxygen consumption by the lens. Invest Ophthalmol Vis Sci. 2006;47:1571–80.
32. McNulty R, Wang H, Mathias RT, Ortwerth BJ, Truscott RJW, Bassnett S. Regulation of tissue oxygen levels in the mammalian lens. J Physiol. 2004;559:883–98.
33. Giblin FJ, Leverenz VR, Padgaonkar VA, Unakar NJ, Dang L, Lin LR, Lou MF, Reddy VN, Borchman D, Dillon JP. UVA light in vivo reaches the nucleus of the guinea pig lens and produces deleterious, oxidative effects. Exp Eye Res. 2002;75:445–58.
34. Bassnett S, McNulty R. The effect of elevated intraocular oxygen on organelle degradation in the embryonic Chicken lens. J Exp Biol. 2003;206:4353–61.
35. Haracopos GP, Shui YB, McKinnon M, Holekamp NM, Gordon MO, Beebe DC. Importance of vitreous liquefaction in age-related cataract. Invest Ophthalmol Vis Sci. 2004;45:77–85.
36. Sweeney M, Truscott RJW. An impediment to glutathione diffusion in older human lenses: a possible precondition for nuclear cataract. Exp Eye Res. 1998;67:587–95.
37. Roberts JE, Finley EL, Patat SA, Schey KL. Photooxidation of lens proteins with xanthurenic acid: a putative chromophore for cataractogenesis. Photochem Photobiol. 2001;74:740–4.
38. Davies MJ, Truscott RJW. Photo-oxidation of proteins and its role in cataractogenesis. J Photochem Photobiol B. 2002;63:114–25.
39. Zigman S. Environmental near-UV radiation and cataracts. Optom Vis Sci. 1995;72:899–907.
40. Shirao Y, Shirao E, Iwase T, Inoue A, Matsukawa S. Comparison of non-tryptophan fluorophores in protein-free extract brunescent and non-brunescent human cataract. Jpn J Ophthalmol. 2000;44:198–204.
41. Dillon J, Atherton SJ. Time resolved spectroscopic studies on the intact human lens. Photochem Photobiol. 1990;51:465–8.
42. Sell DR, Monnier VM. Conversion of arginine into ornithine by advanced glycation in senescent human collagen and lens crystallins. J Biol Chem. 2004;279:54173–84.
43. Nagaraj RH, Oya-Ito T, Padayatti PS, Kumar R, Mehta S, West K, Levison B, Sun J, Crabb JW, Padival AK. Enhancement of chaperone function of alpha-crystallin by methyl-glyoxal modification. Biochemistry. 2003;42:10746–55.
44. Cheng R, Lin B, Ortwerth BJ. Rate of formation of AGEs during ascorbate glycation and during aging in human lens tissue. Biochim Biophys Acta. 2002;1587:65–74.
45. Stevens VJ, Rouzer CA, Monnier VM, Cerami A. Diabetic cataract formation: potential role of glycosylation of lens crystallins. Proc Natl Acad Sci U S A. 1978;75:2918–22.
46. Oimomi M, Maeda Y, Hata F, Kitamura Y, Matsumoto S, Baba S, Iga T, Yamamoto M. Glycation of cataractous lens in non-diabetic senile subjects and in diabetic patients. Exp Eye Res. 1988;46:415–20.

47. Ortwerth BJ, Olesen PR. Ascorbic acid-induced crosslinking of lens proteins: evidence supporting a Maillard reaction. Biochim Biophys Acta. 1988;956:10–22.
48. Ortwerth BJ, Linetzky M, Olesen P. Ascorbic acid glycation of lens proteins produces UVA sensitizers similar to those in human lens. Photochem Photobiol. 1995;62:454–62.
49. Simpson G, Ortwerth BJ. The non-oxidative degradation of ascorbic acid at physiological conditions. Biochim Biophys Acta. 2000;1501:12–24.
50. Gobert J, Glomb MA. Degradation of glucose: reinvestigation of reactive α-dicarbonyl compounds. J Agric Food Chem. 2009;57:8591–7.
51. Harding JJ, Chasset P, Rixon KC, Bron AJ, Harvey DJ. Sugars including erythronic and threonic acids in human aqueous humor. Curr Eye Res. 1999;19:131–6.
52. Nemet I, Monnier VM. Vitamin C degradation products and pathways in the human lens. J Biol Chem. 2011;286:37128–36.
53. Lederer MO, Klaiber RG. Cross-linking of proteins by Maillard processes: characterization and detection of lysine-arginine cross-links derived from glyoxal and methylglyoxal. Bioorg Med Chem. 1999;7:2499–507.
54. Nakamura K, Nakazawa Y, Ienaga K. Acid-stable fluorescent advanced glycation end products: vesperlysines A, B, and C are formed as crosslinked products in the Maillard reaction between lysine or proteins with glucose. Biochem Biophys Res Commun. 1997;232: 227–30.
55. Padayatti PS, Ng AS, Uchida K, Glomb MA, Nagaraj RH. Argpyrimidine, a blue fluorophore in human lens proteins: high levels in brunescent cataractous lenses. Invest Ophthalmol Vis Sci. 2001;42:1299–304.
56. Beebe DC, Truscott RJW. Counterpoint: the lens fluid circulation model-A critical appraisal. Invest Ophthalmol Vis Sci. 2010;51:2306–10.
57. Ortwerth BJ, Chemoganskiy V, Olesen PR. Studies on singlet oxygen formation and UVA light-mediated photobleaching of the yellow chromophores in human lenses. Exp Eye Res. 2002;74:217–29.
58. Brondsted AE, Lundeman JH, Kessel L. Short wavelength light filtering by the natural human lens and IOLs-implications for entrainment of circadian rhythm. Acta Ophthalmol. 2013; 91:52–7.
59. Mizdrak J, Hains PG, Truscott RJ, Jamie JF, Davies MJ. Tryptophan-derived ultraviolet filter compounds covalently bound to lens proteins are photosensitizers of oxidative damage. Free Radic Biol Med. 2008;44:1108–19.
60. Ortwerth BJ, Casserly TA, Olesen PR. Singlet oxygen production correlates with His and Trp destruction in brunescent cataract water-insoluble proteins. Exp Eye Res. 1998;67:377–80.
61. Linetsky M, James H-L, Ortwerth BJ. The generation of superoxide anion by the UVA irradiation. Exp Eye Res. 1996;63:67–74.
62. Linetsky M, Ortwerth BJ. The generation of hydrogen peroxide by UVA irradiation of human lens proteins. Photochem Photobiol. 1995;62:87–93.
63. Fuentealba D, Galvez M, Alarcón E, Lissi E, Silva E. Photosensitized activity of advanced glycation endproducts on tryptophan, glucose 6-phosphate dehydrogenase, human serum albumin and ascorbic acid evaluated al low oxygen pressure. Photochem Photobiol. 2007;83:563–9.
64. De La Rochette A, Silva E, Birlouez-Aragon I, Manzini M, Edwards AM, Morliere P. Riboflavin photodegradation and photosensitizing effects are highly dependent on oxygen and ascorbate concentrations. Photochem Photobiol. 2000;72:815–20.
65. Nishikawa Y, Toyoshima Y, Kurata T. Identification of 3,4-dihydroxy-2-oxo-butanal (L-threosone) as an intermediate compound in oxidative degradation of dehydro-L-ascorbic acid and 2,3-diketo-L-gulonic acid in a deuterium oxide phosphate buffer. Biosci Biotechnol Biochem. 2001;65:1707–12.
66. Kimoto E, Hideiko T, Ohmoto T, Choami M. Analysis of the transformation products of dehydro-L-ascorbic acid by ion-pairing high-performance liquid chromatography. Anal Biochem. 1993;214:38–44.

67. Ortwerth BJ, Speaker JA, Prabakharam M, Lopez MG, Yinan E, Feather MS. Ascorbic acid glycation: the reactions of L-threose in lens tissue. Exp Eye Res. 1994;58:665–74.
68. Fayle SE, Gerrard JA, Simmons L, Meade SJ, Reid EA, Johnston AC. Crosslinkage of proteins by dehydroascorbic acid and its degradation products. Food Chem. 2000;70:193–8.
69. Fuentealba D, Friguet B, Silva E. Advanced glycation endproducts induce photocrosslinking and oxidation of bovine lens proteins through type-I mechanism. Photochem Photobiol. 2009;85:185–94.
70. Govindjee, Xu C, Schansker G, Van Rensen JJS. Chloroacetates as inhibitors of photosystem II: effects on electron acceptor site. J Photochem Photobiol B. 1997;37:107–17.
71. Vileno B, Lekka M, Sienkiewicz A, Marcoux P, Kulik AJ, Kasas S, Catsicas S, Graczyk A, Forró L. Singlet oxygen (1Δg)-mediated oxidation of cellular and subcellular components: ESR and AFM assays. J Phys Condens Matter. 2005;17:1471–82.
72. Silva E, Gaule J. Light-induced binding of riboflavin to lysozyme. Radiat Environ Biophys. 1977;14:303–10.
73. Ávila F, Matus A, Fuentealba D, Lissi E, Friguet B, Silva E. Autosensitized oxidation of glycated bovine lens proteins irradiated with UVA-visible light at low oxygen concentration. Photochem Photobiol Sci. 2008;7:718–24.
74. Huang M, Ellozy AR, Zhang L, Merriam J, Dillon J. The diffusion of oxygen in the mammalian lens. Invest Ophthalmol Vis Sci. 2001;42:S284.
75. Kanner J, Mendel H, Budowski P. Prooxidant and antioxidant effects of ascorbic acid and metal salts in a β-carotene-linoleate model system. J Food Sci. 1977;42:60–4.
76. Hunt JV, Bottoms MA, Mitchinson MJ. Oxidative alterations in the experimental glycation model of diabetes mellitus are due to protein-glucose adduct oxidation. Biochem J. 1993; 291:529–35.
77. Ahmed N. Advanced glycation endproducts-role in pathology of diabetic complications. Diab Res Clin Pract. 2005;67:3–21.
78. Silva E. Sensitized photo-oxidation of amino acids in proteins. In: Eyzaguirre J, editor. Chemical modifications of enzymes. Active site studies. Chichester: Ellis Horwood; 1987.
79. Shacter E. Quantification and significance of protein oxidation in biological simples. Drug Metab Rev. 2000;32:307–26.
80. Liggins J, Furth AJ. Role of protein-bond carbonyl groups in the formation of advanced glycation end products. Biochim Biophys Acta. 1997;1361:123–30.
81. Aspée A, Lissi EA. Kinetics and mechanism of the chemiluminescence associated with the free radical-mediated oxidation of amino acids. Luminescence. 2000;15:273–82.
82. Linetsky M, Ortwerth BJ. Quantitation of the singlet oxygen produced by UVA irradiation of human lens proteins. Photochem Photobiol. 1997;65:522–9.
83. Kwan M, Niinikoshi J, Hunt TK. In vivo measurements of oxygen. Invest Ophthalmol. 1972;11:108–14.
84. Eaton JW. Is the lens canned? Free Radic Biol Med. 1991;11:207–13.
85. Ávila F, Friguet B, Silva E. Simultaneous chemical and photochemical protein crosslinking induced by irradiation of eye lens proteins in the presence of ascorbate: the photosensitizing role of an UVA-visible-absorbing decomposition product of vitamin C. Photochem Photobiol Sci. 2010;9:1351–8.
86. Wannemacher CF, Spector A. Protein synthesis in the core of calf lens. Exp Eye Res. 1968;7:623–5.
87. Harding JJ. Aggregation of proteins in human cataract. Ophthalmic Res. 1979;11:429–32.
88. Srivastava OP. Age-related increase in concentration and aggregation of degraded polypeptides in human lenses. Exp Eye Res. 1988;47:525–43.
89. Satoh K. Age related changes in the structural proteins of human lens. Exp Eye Res. 1972;14:53–7.
90. Truscott RJW, Augusteyn RC. Changes in human lens proteins during nuclear cataract formation. Exp Eye Res. 1977;24:159–70.
91. Berman E. Biochemistry of the eye. New York: Plenum Press; 1991. p. 203–74.

92. Carrard G, Bulteau A, Petropolous I, Friguet B. Impairment of proteasome structure and function in aging. Int J Biochem Cell Biol. 2002;34:1461–74.
93. Hosler M, Wang-Su S, Wagner B. Targeted disruption of specific steps of the ubiquitin-proteasome pathway by oxidation in lens epithelial cells. Int J Biochem Cell Biol. 2003;35:685–97.
94. Shang F, Nowell T, Taylor A. Removal of oxidatively damaged proteins from lens cells by the ubiquitin-proteasome pathway. Exp Eye Res. 2001;73:229–38.
95. Spector A, Ma W, Wang R, Kleiman N. Microperoxidases catalytically degrade reactive oxygen species and may be anti-cataract agents. Exp Eye Res. 1997;65:457–70.
96. Grune T. Oxidative stress, aging and the proteasomal system. Biogerontology. 2000; 1:31–40.
97. Gaczynska M, Osmulski P, Ward W. Caretaker or undertaker? The role of proteasome in aging. Mech Ageing Dev. 2001;122:235–54.
98. Orlowski M, Wilk S. Catalytic activities of the 20S proteasome, a multicatalytic proteinase complex. Arch Biochem Biophys. 2000;383:1–16.
99. Wójcik C. Proteasomes in apoptosis: villains or guardians? Cell Mol Life Sci. 1999; 56:908–17.
100. Sorimachi H, Ishiura S, Suzuki K. Structure and physiological function of calpains. Biochem J. 1997;328:721–32.
101. Coux O, Tanaka K, Goldberg A. Structure and functions of the 20S and 26S proteasomes. Annu Rev Biochem. 1996;65:801–47.
102. Hochstrasser M. Ubiquitin-dependent protein degradation. Annu Rev Genet. 1996; 30:405–39.
103. Shang F, Gong X, Palmer H, Nowell T, Taylor A. Age-related decline in ubiquitin conjugation in response to oxidative stress in the lens. Exp Eye Res. 1997;64:21–30.
104. Shaojun Y, Wang S, Cai H, Wagner B. Changes in three types of ubiquitin mRNA and Ubiquitin-protein conjugate levels during lens development. Exp Eye Res. 2002;75:271–84.
105. Chung-Kenny K, Dawson-Valina L, Dawson-Ted M. The role of the ubiquitin-proteasomal pathway in Parkinson's disease and other neurodegenerative disorders. Trends Neurosci. 2001;24:s7–14.
106. Vu P, Sakamoto K. Ubiquitin-mediated proteolysis and human disease. Mol Genet Metab. 2000;71:261–6.
107. Leroy E, Boyer R, Auburger G, Leube B, Ulm G, Mezey E, Harta G, Brownstein MJ, Jonnalagada S, Chernova T, Dehejia A, Lavedan C, Gasser T, Steinbach PJ, Wilkinson KD, Polymeropoulos MH. The ubiquitin pathway in Parkinson's disease. Nature. 1998;395: 451–2.
108. Checler F, Alves da Costa C, Ancolio K, Chevallier N, Lopez-Perez E, Marambaud P. Role of the proteasome in Alzheimer's disease. Biochim Biophys Acta. 2000;1502:133–8.
109. Muller S, Schwartz L. Ubiquitin in homeostasis, developments and disease. Bioessays. 1995;17:677–84.
110. Friguet B, Bulteau A, Chondrogianni N, Conconi M, Petropoulos I. Protein degradation by the proteasome and its implications in aging. Ann N Y Acad Sci. 2000;908:143–54.
111. Merker K, Stolzing A, Grune T. Proteolysis, caloric restriction and aging. Mech Ageing Dev. 2001;122:595–615.
112. Reinheckel T, Sitte N, Ullrich O, Kuckelkorn U, Davies K, Grune T. Comparative resistance of the 20S and 26S proteasome to oxidative stress. Biochem J. 1998;335:637–42.
113. Friguet B, Stadtman E, Sweda L. Modification of glucose-6-phosphate dehydrogenase by 4-hydroxynonenal. Formation of cross-linked protein than inhibits the multicatalytic protease. J Biol Chem. 1994;269:21639–43.
114. Taylor A, Davies KJA. Protein oxidation and loss of protease activity may lead to cataract formation in the aged lens. Free Radic Biol Med. 1987;3:371–7.
115. Andersson M, Sjöstrand J, Karlsson JO. Proteolytic cleavage of N-Succ-Leu-Leu-Val-Tyr-AMC by the proteasome in lens epithelium from clear and cataractous human lenses. Exp Eye Res. 1998;67:231–6.

116. Andersson M, Sjöstrand J, Karlsson JO. Differential inhibition of three peptidase activities of the proteasome in human lens epithelium by health and oxidation. Exp Eye Res. 1999;69: 129–38.
117. Zetterberg M, Petersen A, Sjostrand J, Karlsson JO. Proteasome activity in human lens nuclei and correlation with age, gender and severity of cataract. Curr Eye Res. 2003;27:45–53.
118. Petropoulos I, Conconi M, Wang X, Hoenel B, Bregegere F, Milner Y, Friguet B. Increase of oxidatively modified protein is associated with a decrease of proteasome activity and content in aging epidermal cells. J Gerontol A Biol Sci Med Sci. 2000;55:B220–7.
119. Viteri G, Carrard G, Birlouez-Aragón I, Silva E, Friguet B. Age-dependent protein modifications and declining proteasome activity in the human lens. Arch Biochem Biophys. 2004; 427:197–203.
120. Ikeda K, Nagai R, Sakamoto T, Sano H, Araki T, Sakata N, Nakayama H, Yoshida M, Ueda S, Horiuchi S. Immunochemical approaches to AGE-structures: characterization of anti-AGE antibodies. J Immunol Methods. 1998;215:95–104.
121. Pereira P, Shang F, Hobbs M, Girao H, Taylor A. Lens fibers have a fully functional ubiquitin-proteasome pathway. Exp Eye Res. 2003;76:623–31.
122. Louie JL, Kapphahn RJ, Ferrington DA. Proteasome function and protein oxidation in the aged retina. Exp Eye Res. 2002;75:271–84.
123. Bulteau AL, Szweda L, Friguet B. Age-dependent declines in proteasome activity in the heart. Arch Biochem Biophys. 2002;397:298–304.
124. Davies KJ. Degradation of oxidized proteins by the 20S proteasome. Biochimie. 2001; 83:301–10.
125. Friguet B, Szweda LI. Inhibition of the multicatalytic proteinase (proteasome) by 4-hydroxy-2-nonenal cross-linked protein. FEBS Lett. 1997;405:21–5.
126. Sitte N, Huber M, Grune T, Ladhoff A, Doecke WD, von Zglinicki T, Davies KJ. Proteasome inhibition by lipofucsin/ceroid during postmitotic aging of fibroblasts. FASEB J. 2000;14: 1490–8.
127. Wagner B, Margolis J. Age-dependent association of isolated bovine lens multicatalytic proteinase complex (proteasome) with heat-shock protein 90, an endogenous inhibitor. Arch Biochem Biophys. 1995;323:455–62.
128. Conconi M, Szweda LI, Levine RL, Stadman E, Friguet B. Age-related decline of rat liver multicatalytic proteinase activity and protection from oxidative inactivation by heat-shock protein 90. Arch Biochem Biophys. 1996;331:232–40.
129. Bulteau A, Verbeke P, Petropoulos I, Chafotte A, Friguet B. Proteasome inhibition in glyoxal-treated fibroblasts and resistance of glycated glucose-6-phosphate dehydrogenase to 20 S proteasome degradation in vitro. J Biol Chem. 2001;276:45662–8.
130. Reddy V, Beyaz A. Inhibitors of the Maillard reaction and the AGE breakers as therapeutics for multiple diseases. Drug Discov Today. 2006;11:646–54.
131. Ávila F, Trejo S, Baraibar MA, Friguet B, Silva E. Photosensitized reactions mediated by the major chromophore arising from glucose decomposition, result in oxidation and cross-linking of lens proteins and activation of the proteasome. Biochim Biophys Acta. 2012;1822: 564–72.
132. Frank O, Hofmann T. Characterization of key chromophores formed by nonenzimatic browning of hexoses and L-alanine by using the color activity concept. J Agric Food Chem. 2000;48:6303–11.
133. Frank O, Heuberger S, Hofman T. Structure determination of a novel 3 (6H)-Pyranone chromophore and clarification of its formation from carbohydrates and primary amino acids. J Agric Food Chem. 2001;49:1595–600.
134. Limacher A, Kerler J, Davidek T, Schmalzried F, Blank I. Formation of furan and methylfuran by Maillard-type reactions in model systems and food. J Agric Food Chem. 2008; 56:3639–47.
135. Pischetsrieder M. Chemistry of glucose and biochemical pathways of biological interest. Perit Dial Int. 2000;20:S26–230.

136. Merkel PB, Kearns DR. Remarkable solvent effects on the lifetime of 1Δg oxygen. J Am Chem Soc. 1972;94:1029–30.
137. Catalgol B, Ziaja I, Breusing N, Jung T, Höhn A, Alpertunga B, Schroeder P, Chondrogianni T, Gonos ES, Petropoulos I, Friguet B, Klotz LO, Krutmann J, Grune T. The proteasome is an integral part of solar ultraviolet A radiation-induced gene expression. J Biol Chem. 2009;284:30076–86.
138. Lu C, Liu Y. Electron transfer oxidation of tryptophan and tyrosine by triplet states and oxidized radicals of flavin sensitizers: a laser flash photolysis study. Biochim Biophys Acta. 2002;1571:71–6.
139. Conconi M, Petropoulos I, Emod I, Turlin E, Biville F, Friguet B. Protection from oxidative inactivation of the 20S proteasome by heat-shock proteína 90. Biochem J. 1998; 333:407–15.
140. Sharman KK, Ortwerth BJ. Aminopeptidase III activity in normal and cataractous lenses. Curr Eye Res. 1986;5:373–80.
141. Santhoshkumar P, Udupa P, Murugesan R, Sharma KK. Significance of interactions of low molecular weight crystallin fragments in lens aging and cataract formation. J Biol Chem. 2008;283:8477–85.
142. Han J, Schey KL. MALDI tissue imaging of ocular lens alpha-crystallin. Invest Ophthalmol Vis Sci. 2006;47:2990–6.
143. Su SP, MacArthur JD, Aquilina JA. Localization of low molecular weight crystallin peptides in the aging human lens using a MALDI mass spectrometry imaging approach. Exp Eye Res. 2010;91:97–103.
144. David LL, Shearer TR. Role of proteolysis in lenses: a review. Lens Eye Toxic Res. 1989;6:725–47.

Chapter 15
p53 Regulates Developmental Apoptosis and Gene Expression to Modulate Lens Differentiation

Xiang-Cheng Tang, Fang-Yuan Liu, Zhong-Wen Luo, Weike Ji, Xiao-Hui Hu, Wen-Feng Hu, Zhao-Xia Huang, Zhengfeng Wang, Zachary Woodward, Yi-Zhi Liu, Quan Dong Nguyen, and David Wan-Cheng Li

15.1   Introduction

Since Spemann initiated the study on lens development in the beginning of the twentieth century [1], the eye lens has been one of the best models for the investigation of tissues induction and exploration of the molecular mechanisms mediating development under control by different transcription factors [2–4]. One of such transcription factors is p53, which was initially discovered by the British scientist, David Lane [5], and later was proved to be a tumor suppressor [5–8]. As a tumor

X.-C. Tang
State Key Laboratory of Ophthalmology, Zhongshan Ophthalmic Center,
Sun Yat-Sen University, #54 Xianlie Road, Guangzhou, Guangdong 510060, China

Department of Ophthalmology and Visual Sciences, Truhlsen Eye Institute,
College of Medicine, University of Nebraska Medical Center, Omaha, NE 68198-5540, USA

F.-Y. Liu • Z.-W. Luo • Y.-Z. Liu • D.W.-C. Li (✉)
State Key Laboratory of Ophthalmology, Zhongshan Ophthalmic Center,
Sun Yat-Sen University, #54 Xianlie Road, Guangzhou, Guangdong 510060, China
e-mail: davidli@unmc.edu

W. Ji • Z. Wang • Z. Woodward • Q.D. Nguyen
Department of Ophthalmology and Visual Sciences, Truhlsen Eye Institute,
College of Medicine, University of Nebraska Medical Center, Omaha, NE 68198-5540, USA

X.-H. Hu • W.-F. Hu • Z.-X. Huang • D.W.-C. Li (✉)
Department of Ophthalmology and Visual Sciences, Truhlsen Eye Institute,
College of Medicine, University of Nebraska Medical Center, Omaha, NE 68198-5540, USA

Key Laboratory of Protein Chemistry and Developmental
Biology of Education Ministry of China, College of Life Sciences,
Hunan Normal University, Changsha, Hunan, China
e-mail: davidli@unmc.edu

© Springer Science+Business Media New York 2015
M.A. Babizhayev et al. (eds.), *Studies on the Cornea and Lens*,
Oxidative Stress in Applied Basic Research and Clinical Practice,
DOI 10.1007/978-1-4939-1935-2_15

suppressor, p53 has been one of the most extensively studied proteins in the past three decades [9–19]. P53 has been proven to be "the guardian of the genome" [8], acts as a master regulator of apoptosis [20–23] and can control cell cycle progression [24, 25]. In addition, numerous studies have shown that the p53 plays an essential role in regulating differentiation of various types of cells including muscle cells [26, 27], B-cell [28], and erythrocytes [29–31]. More recently, our laboratory and several others have shown that p53 can regulate lens differentiation via control of developmental apoptosis and also lens-specific gene expression. This chapter summarizes these recent studies.

15.2 Development of the Vertebrate Eye Lens

In vertebrates, the eye lens is derived from the ectoderm in the brain region. Under the induction of the optic vesicle (forming future retina), the ectoderm above the optic vesicle thickens to form the lens placode and then invaginate into a lens pit together with the optic vesicle. The lens pit is further deepened and then detached from the ectoderm to become a lens vesicle [32–36], which contains a monolayer of lens epithelial cells. Once the lens vesicle is formed, the epithelial cells in the posterior end undergo differentiation marked by cell elongation, and synthesis of different families of lens crystallins. These cells eventually occupy the hollow lens vesicle and become primary lens fiber cells. As development proceeds, lens growth continues. The epithelial cells at the subequatorial region divide, migrate to the lens equatorial region, and undergo differentiation, which is marked by cell elongation, loss of cellular organelles, and eventual denucleation, forming an organelle-free zone (OFZ) [32–36]. During this process, p53 plays key roles, which are described below.

15.3 P53 Knockout Leads to Cataractogenesis

Donehower et al. have demonstrated that homozygous p53-deficient mice appear to develop normally but are highly susceptible to spontaneous tumors [37]. This result somehow brought in a wrong prevalent view that p53 has no function in eye development. In the later studies, Ikeda compared two different strains of p53-deficient mice and found that the eyes are abnormal in adult B6 mice though the eyes in 129/Sv mice seem to be normal [38]. After a careful examination of adult mouse eyes from p53-deficient mice with ophthalmic scope, Reichel et al. revealed a high frequency of cataracts [39]. Thus, it is clear that p53 plays important functions during eye lens development.

15.4   p53 Is Expressed During Lens Development

In our previous study, we examined the expression level of the p53 during mouse eye development [40]. The expression of p53 was clearly detected in retina, lens, and cornea at stage 14.5 ED. The expression level becomes gradually increased in the corresponding eye tissues as development moves from 14.5 ED to 17.5 ED. From 17.5 ED to newborn mouse, p53 became decreased in the lens fiber cells but remains relatively stable in the epithelial cells. These results are consistent with those reported by others [41, 42] where it was found that p53 is expressed in the lens epithelial cells of the central and pre-equatorial zones and in the bow region of lens fibers. In a more detailed study by Geatrell et al. on spatiotemporal expression patterns of Mdm2, p53, and Mdm4/X during various stages of lens development and on the comparison of the expression patterns of p53 in both mouse and chick, they concluded that p53 plays a role during eye lens development [43].

15.5   p53-Dependent Apoptosis Plays an Important Role During Lens Development

Cell death in lens was described as "mitotic metabolites" in 1900s [44]. Later, Glucksmann (1951) systematically summarized cell death at various stages of the lens formation: before and during lens ectoderm invagination, during detachment of the lens vesicle from the ectoderm, and in the process of regression of peridemal plug [45]. In 1973, Silver and Hughes described this cell death as cell degeneration and use the term "necrotic center" [46], which is quite similar to the concept of apoptosis described by Kerr, Wyllie, and Currie [47]. In 1990s, with the development of modern cell biology technologies "in situ terminal deoxynucleotidyl transferase (TdT)-mediated dUTP nick-end labeling (TUNEL)" [48], the existence of apoptosis and its possible role during lens development were unequivocally confirmed [49], whether this apoptosis is p53 dependent or not remains to be elusive.

In their transgenic mouse study, DePinho's group found that *Rb*-deficient can cause unchecked proliferation, impaired expression of differentiation markers, and eventual inappropriate apoptosis of these lens cells [50]. While in embryos with dual null in Rb and p53, the observed apoptosis was completely suppressed. This result clearly showed that apoptosis derived from Rb-deficient lenses is p53 dependent. About the same time, through overexpression of the viral protein E6, Griep's group also confirmed that inactivation of p53 lead to cataractogenesis [51] and moreover, they also confirmed that inactivation of Rb in lens by overexpression of viral protein E7 caused apoptosis of lens epithelial cells which was largely p53 dependent [51, 52]. Nakamura et al. generated mouse models expressing exogenous wild-type or mutant p53 [53, 54] and found that overexpression of wild-type p53 mice develops microphthalmia as a result of the defect in fiber formation due to p53-dependent apoptosis of the differentiating lens cells. On the other hand, the mutant p53 transgenic mice display a faint central cataract caused by the presence of remnants of fiber cell nuclei (incomplete differentiation). This group also generated a double transgenic mice expressing exogenous wild-type and mutant p53 at the same time. Interestingly, the transgenic mice expressing both the wild-type and the mutant human p53 gene product display normal lens phenotypes. Together these results suggest that p53-dependent apoptosis is crucial for crafting the lens, and moreover, the tumor suppressor plays an important role during lens denucleation. In addition, these results demonstrated that the mutant p53 allele can interfere with the apoptotic function of the wild-type product but not with its ability to activate p21 and MDM2, this result is consistent with later observations that p53-dependent apoptosis does not require transcriptional activation of a number of downstream genes [55, 56]. The most direct evidence that p53-dependent apoptosis causes lens pathology is derived from the study by Hettmann et al. [57]. In their CREB-2 knockout study, they found lens development occurs between ED12.5 and ED14.5 in CREB-2(−/−) mice. These mice also displayed normal formation of the early lens vesicle, normal elongation of posterior primary fiber cells, and normal formation of the anterior lens epithelial cells. However, the anterior lens epithelial cells and their direct descendants at the equatorial poles of the lens underwent massive and synchronous p53-dependent apoptosis between ED14.5 and ED16.5, causing complete degeneration of the embryonic lens, and eventual microphthalmia in these animals [57]. The nuclear receptor coactivator 6 (NCOA6) seems to support lens differentiation by suppressing apoptosis. Cvekl's group recently showed that overexpression of the N-terminal fragment of NCOA6, an 81-amino acid LXXLL motif, caused both p53-dependent and p53-independent apoptosis and eventual cataract formation [58]. In a more recent p53 deletion study [58], it was found that p53 knockout alone increased the number of fiber cells that failed to withdraw from the cell cycle, these cells improperly proliferate and migrate to the posterior end of the lens and caused phenotype resembling posterior subcapsular cataracts. Thus, p53-dependent apoptosis can protect the lens against posterior subcapsular cataract formation by suppressing the proliferation of fiber cells [59].

15.6   The Signaling Pathway Components Mediating p53-Dependent Apoptosis in Lens

As a master regulator of apoptosis in different cells, p53 regulates apoptosis through two different mechanisms. First, it acts as a transcriptional factor to regulate expression of many genes involved in apoptosis [60–70]. Second, it can be localized in mitochondria where it antagonizes the anti-apoptotic ability of Bcl-2 and Bcl-XL [71].

In the adult lens, we have previously demonstrated that stress-induced apoptosis seems to act as a common cellular mechanism by which various stress factors induce cataractogenesis [72–76]. To explore the apoptotic control machinery, we have shown that during apoptosis of lens epithelial cells induced through inhibition of protein serine/threonine phosphatases by okadaic acid, p53 is clearly upregulated and its downstream target gene Bax is also upregulated [77, 78]. Inhibition of protein synthesis can suppress okadaic acid-induced upregulation of p53 and Bax, and also apoptosis. The p53 target gene, Bax, is also actively involved in Staurosporine-, UV-, and abnormal calcium-induced apoptosis of lens epithelial cells [79, 80]. During lens development, alpha-crystallins modulate Bax and caspase-3 activity to control developmental apoptosis [81, 82].

Our recent studies have demonstrated that in lens, p53 can directly regulate another important proapoptotic gene, Bak [40, 83]. Moreover, Bak is highly expressed in embryonic lens and the p53-Bax axis regulates not only apoptosis but also lens differentiation (see more discussion below) [40]. Bak is also implicated in staurosporine-induced apoptosis of lens epithelial cells [79]. Bcl-2, another p53 downstream target gene [60], also plays an important role in regulating lens differentiation [84, 85]. Overexpression of Bcl-2 causes microphthalmia [85].

15.7 p53 Directly Regulates Other Transcription Factors to Control Lens Differentiation

As a major transcription factor, p53 has been shown to regulate over 130 target genes and this list is growing year after year [70, 86]. In our recent studies, we have demonstrated that p53 can directly regulate two key lens transcription factors, c-Maf and Prox1 [87].

The c-Maf genes from human, mouse, and rat all contain two p53 binding sites: one in the promoter region, the other one in the first intron. Electrophoretic mobility shifting assay (EMSA) shows that p53 can bind to both sites. However, reporter gene activity analysis demonstrates that while the p53 binding site in the promoter region confers strong positive regulation on c-Maf expression, the second p53 site in intron may have a negative effect [87]. In the p53 knockout mice, expression of c-Maf was clearly downregulated in the embryonic lens from ED11.5 to ED13.5. This is also true in the lens placode stage [87]. These results reveal that p53 plays an important role in regulating c-Maf expression during the early embryonic developmental stages. By ED17.5, the p53 regulation of c-Maf expression has gradually decreased. Prior to our work, Xie and Cvekl [88] demonstrated that Pax-6 also plays a major role in regulating c-Maf expression and the downstream lens crystallin genes. Together, it is clear that during the early lens development, Pax-6 and p53 are two major players in regulating c-Maf expression. Whether Pax-6 and p53 interact with each other to control c-Maf remains to be explored.

The c-Maf gene is actively involved in control of lens differentiation. Loss of c-Maf function causes severe defects in lens differentiation including lack of primary fiber cell elongation and absence of lens crystallin gene expression induction in the differentiating lens fiber cells [88–91].

Prox1 also plays an essential role during lens development. It is one of the essential genes controlling lens fiber elongation. Lack of Prox1 function leads to absence of the lens fiber elongation [92, 93]. Prox1 also regulates differentiation of other tissues [94–96]. At the molecular level, deficiency of Prox1 downregulates expression of p27 and p57, leading to enhanced cell proliferation but loss of cell polarity [92, 93]. In our recent study, we have demonstrated that p53 also directly controls Prox1 [87]. The Prox1 gene from humans contains a conserved p53 binding site in the distal promoter region (–915 to –941). In contrast, the Prox1 gene from mice contains two p53 binding sites, one in similar position as found in humans, the other one near the proximal promoter (–356 to –382). EMSA and dose-dependent response analysis reveal that these p53 sites are active in mediating Prox1 control by p53 [87]. During mouse lens development, p53 exerts important control on Prox1 expression as reflected by the fact that knockout of p53 leads to significant downregulation of Prox1 from ED12.5 to ED17.5. This regulation becomes gradually less important in the late embryonic stages and adult lens [87].

Together, p53 can regulate lens differentiation through its control on other transcription factors.

15.8   p53 Directly Regulates Lens-Crystallin Genes to Control Lens Differentiation

Besides its control on other transcription factors, our recent studies have also demonstrated that p53 can regulate expression of differentiation markers, the lens crystallin genes [97–98]. The αA-crystallin gene contains one p53 binding site in the promoter region and two p53 binding sites in the first intron, all three sites can bind to p53 [97], mediating its positive regulation by wild-type p53 but negative regulation by p53-dominant negative mutant. In the developing lens, ChIP assay revealed that p53 is directly bound to the αA-crystallin gene promoter. In the p53 knockout mice, expression of the αA-crystallin gene was downregulated. The βA3/A1-crystallin gene also has three p53 binding sites in the first and second introns. Again, EMSA revealed that these sites are active in p53 bind and mediate positive and negative regulation by wild-type and dominant negative p53, respectively. In the developing lens, these sites also bind to p53 as demonstrated by ChIP assays. The p53 (–/–) mice have clearly downregulated βA3/A1-crystallin expression at ED11 although this downregulation becomes less clear at the later developmental stages [97].

15.9   Conclusions

In summary, p53 plays an important role in the ocular lens. It is necessary to mediate normal developmental apoptosis so that the unwanted cells during lens vesicle formation and the lens stalk cells during lens differentiation can be removed without disturbing normal lens physiology. On the other hand, enhanced apoptosis due to gene mutation or knockout also leads to microphthalmia or absence of eye development. This process is executed through regulation of Bak, Bax, and other components. p53 is also actively involved in regulation of lens differentiation. It does so either by regulating the major transcription factors including c-Maf and Prox1, or directly controls differentiation markers, the lens crystallin genes. The future study will certainly elucidate the signaling network components mediating p53 control of both developmental apoptosis and differentiation-specific genes, and moreover the synergistic effects between p53 and other transcription factors.

Acknowledgements This work was supported in part by the National Institutes of Health grants, the National Natural Science Foundation of China (81272228), the collaborative grant from Zhongshan Ophthalmic Center (8282012-XCT, FYL, ZWL, YZL, and DWL), the Cooperative Innovation Center of Engineering and New Products for Developmental Biology of Hunan Province (20134486) and the Chinese Scholarship Council (WKJ, XHH, WFH).

References

1. Spemann H. Über Correlationen in der Entwicklung des Auges. Verhand Anat Ges. 1901;15:61–79.
2. Lovicu FJ, McAvoy JW. Growth factor regulation of lens development. Dev Biol. 2005;280(1):1–14.
3. Donner AL, Lachke SA, Maas RL. Lens induction in vertebrates: variations on a conserved theme of signaling events. Semin Cell Dev Biol. 2006;17:676–85.
4. Ogino H, Ochi H, Reza HM, Yasuda K. Transcription factors involved in lens development from the preplacodal ectoderm. Dev Biol. 2012;363(2):333–47.
5. DeLeo AB, Jay G, Appella E, Dubois GC, Law LW, Old LJ. Detection of a transformation-related antigen in chemically induced sarcomas and other transformed cells of the mouse. Proc Natl Acad Sci USA. 1979;76:2420–4.
6. Lane DP, Crawford LV. T antigen is bound to a host protein in SV40-transformed cells. Nature. 1979;278:261–3.
7. Linzer DI, Levine AJ. Characterization of a 54K dalton cellular SV40 tumor antigen present in SV40-transformed cells and uninfected embryonal carcinoma cells. Cell. 1979;17(1):43–52.
8. Lane DP. Cancer. p53, guardian of the genome. Nature. 1992;358:15–6.
9. Panier S, Boulton SJ. Double-strand break repair: 53BP1 comes into focus. Nat Rev Mol Cell Biol. 2014;15(1):7–18.
10. Muller PA, Vousden KH. Mutant p53 in cancer: new functions and therapeutic opportunities. Cancer Cell. 2014;25(3):304–17.
11. Shiloh Y, Ziv Y. The ATM protein kinase: regulating the cellular response to genotoxic stress, and more. Nat Rev Mol Cell Biol. 2013;14(4):197–210.
12. Kilpivaara O, Aaltonen LA. Diagnostic cancer genome sequencing and the contribution of germline variants. Science. 2013;339(6127):1559–62.
13. Purvis JE, Lahav G. Encoding and decoding cellular information through signaling dynamics. Cell. 2013;152(5):945–56.
14. Wade M, Li YC, Wahl GM. MDM2, MDMX and p53 in oncogenesis and cancer therapy. Nat Rev Cancer. 2013;13(2):83–96.
15. Muller PA, Vousden KH. p53 Mutations in cancer. Nat Cell Biol. 2013;15(1):2–8.

16. Sperka T, Wang J, Rudolph KL. DNA damage checkpoints in stem cells, ageing and cancer. Nat Rev Mol Cell Biol. 2012;13(9):579–90.

17. Hermeking H. MicroRNAs in the p53 network: micromanagement of tumour suppression. Nat Rev Cancer. 2012;12(9):613–26.

18. Freed-Pastor WA, Prives C. Mutant p53: one name, many proteins. Genes Dev. 2012;26(12):1268–86.

19. Sahin E, DePinho RA. Axis of ageing: telomeres, p53 and mitochondria. Nat Rev Mol Cell Biol. 2012;13(6):397–404.

20. Shen Y, White E. P53-dependent apoptosis pathways. Adv Cancer Res. 2005;82:55–84.

21. Vousden KH, Lu X. Live or let die: the cell's response to p53. Nat Rev Cancer. 2002;2(8):594–604.

22. Oren M. Decision making by p53: life, death and cancer. Cell Death Differ. 2003;10(4):431–42.

23. Manfredi JJ. p53 and apoptosis: it's not just in the nucleus anymore. Mol Cell. 2003;11(3):552–4.

24. Haffner R, Oren M. Biochemical properties and biological effects of p53. Curr Opin Genet Dev. 1995;5:84–90.

25. Waga S, Hannon GJ, Beach D, Stillman B. The p21 inhibitor of cyclin-dependent kinases controls DNA replication by interaction with PCNA. Nature. 1994;369:574–8.

26. Parker SB, Eichele G, Zhang P, Rawls A, Sands AT, Bradley A, Olson EN, Harper JW, Elledge SJ. p53-Independent expression of p21Cip1 in muscle and other terminally differentiating cells. Science. 1995;267:1024–7.

27. Halevy O, Novitch BG, Spicer DB, Skapek SX, Rhee J, Hannon GJ, Beach D, Lassar AB. Correlation of terminal cell cycle arrest of skeletal muscle with induction of p21 by MyoD. Science. 1995;267(5200):1018–21.

28. Shaulsky G, Goldfinger N, Peled A, Rotter V. Involvement of wild-type p53 in pre-B-cell differentiation in vitro. Proc Natl Acad Sci USA. 1991;88(20):8982–6.

29. Quang CT, Wessely O, Pironin M, Beug H, Ghysdael J. Cooperation of Spi-1/PU.1 with an activated erythropoietin receptor inhibits apoptosis and Epo-dependent differentiation in primary erythroblasts and induces their Kit ligand-dependent proliferation. EMBO J. 1997;16(18):5639–53.

30. Metz T, Harris AW, Adams JM. Absence of p53 allows direct immortalization of hematopoietic cells by the myc and raf oncogenes. Cell. 1995;82(1):29–36.

31. Weiss MJ, Orkin SH. Transcription factor GATA-1 permits survival and maturation of erythroid precursors by preventing apoptosis. Proc Natl Acad Sci USA. 1995;92(21):9623–7.

32. McAvoy JW. Induction of the eye lens. Differentiation. 1980;17:137–49.

33. Bloemendal H. Molecular and cellular biology of the eye lens. New York: Wiley; 1981.

34. Piatigorsky J. Lens differentiation in vertebrates: a review of cellular and molecular features. Differentiation. 1981;19:134–53.

35. Zhang L, Yan Q, Liu JP, Zou LJ, Liu J, Sun S, Deng M, Gong L, Ji WK, Li DW. Apoptosis: its functions and control in the ocular lens. Curr Mol Med. 2010;10(9):864–75.

36. Cvekl A, Duncan MA. Genetic and epigenetic mechanisms of gene regulation during lens development. Progress Retinal Eye Res. 2007;26:555–97.

37. Donehower LA, Harvey M, Slagle BL, McArthur MJ, Montgomery Jr CA, Butel JS, Bradley A. Mice deficient for p53 are developmentally normal but susceptible to spontaneous tumours. Nature. 1992;356:215–21.

38. Ikeda S, Hawes NL, Chang B, Avery CS, Smith RS, Nishina PM. Severe ocular abnormalities in C57BL/6 but not in 129/Sv p53-deficient mice. Invest Ophthalmol Vis Sci. 1999;40:1874–8.

39. Reichel MB, Ali RR, D'Esposito F, Clarke AR, Luther PJ, Bhattacharya SS, Hunt DM. High frequency of persistent hyperplastic primary vitreous and cataracts in p53-deficient mice. Cell Death Differ. 1998;5:156–62.

40. Deng M, Chen P, Liu F, Fu S, Tang H, Fu Y, Xiong Z, Hui S, Ji W, Zhang X, Zhang L, Gong L, Hu X, Hu W, Sun S, Liu J, Xiao L, Liu WB, Xiao YM, Liu SJ, Liu Y, Li DWC. The p53-Bak

apoptotic signaling axis plays an essential role in regulating differentiation of the ocular lens. Curr Mol Med. 2012;12:901–16.

41. Pokroy R, Tendler Y, Pollack A, Zinder O, Weisinger G. p53 Expression in the normal murine eye. Invest Ophthalmol Vis Sci. 2002;43:1736–41.

42. Ayala M, Strid H, Jacobsson U, Söderberg PG. p53 Expression and apoptosis in the lens after ultraviolet radiation exposure. Invest Ophthalmol Vis Sci. 2007;48:4187–91.

43. Geatrell JC, Gan PM, Mansergh FC. Apoptosis gene profiling reveals spatio-temporal regulated expression of the p53/Mdm2 pathway during lens development. Exp Eye Res. 2009;88:1137–51.

44. Rabl C. Uber den Bau und die Entwicklung der lines. III. Die lines der Saugethiere: Ruckblick und Schluss. Z Wiss Zool. 1900;67:1–138.

45. Glucksmann A. Cell death in normal vertebrate ontogeny. Biol Rev. 1951;26:59–86.

46. Silver J, Hughes AFW. The role of cell death during morphogenesis of the mammalian eye. J Morphol. 1973;140:159–70.

47. Kerr JF, Wyllie AH, Currie AR. Apoptosis: a basic biological phenomenon with wide-ranging implications in tissue kinetics. Br J Cancer. 1972;26:239–57.

48. Gavrieli Y, Sherman Y, Ben-Sasson SA. Identification of programmed cell death in situ via specific labeling of nuclear DNA fragmentation. J Cell Biol. 1992;119:493–501.

49. Ishizaki Y, Voyvodic JT, Burne JF, Raff MC. Control of lens epithelial cell survival. J Cell Biol. 1993;121(4):899–908.

50. Morgenbesser SD, Williams BO, Jacks T, DePinho RA. p53-Dependent apoptosis by Rb-deficiency in the developing mouse lens. Nature. 1994;371:72–4.

51. Pan H, Griep AE. Altered cell cycle regulation in the lens of HPV-16 E6 or E7 transgenic mice: implications for tumor suppressor gene function in development. Genes Dev. 1994;8:1285–99.

52. Pan H, Griep AE. Temporally distinct patterns of p53-dependent and p53-independent apoptosis during mouse lens development. Genes Dev. 1995;9:2157–69.

53. Nakamura T, Pichel JG, Williams-Simons L, Westphal H. An apoptotic defect in lens differentiation caused by human p53 is rescued by a mutant allele. Proc Natl Acad Sci USA. 1995;92(13):6142–6.

54. Nakajima Y, NakamuraT ET, Murata Y. Loss of one allele of the p53 gene in the lens epithelial tumor in transgenic mice suppresses apoptosis induced by a topoisomerase I inhibitor (CPT-11). Cancer Lett. 2002;179(2):165–73.

55. Caelles C, Helmberg A, Karin M. p53-Dependent apoptosis in the absence of transcriptional activation of p53-target genes. Nature. 1994;370:220–3.

56. Wagner AJ, Kokontis JM, Hay N. Myc-mediated apoptosis requires wild-type p53 in a manner independent of cell cycle arrest and the ability of p53 to induce p21waf1/cip1. Genes Dev. 1994;8:2817–30.

57. Hettmann T, Barton K, Leiden JM. Microphthalmia due to p53-mediated apoptosis of anterior lens epithelial cells in mice lacking the CREB-2 transcription factor. Dev Biol. 2000;222:110–23.

58. Wang WL, Li Q, Xu J, Cvekl A. Lens fiber cell differentiation and denucleation are disrupted nuclear receptor box of NCOA6 and result in p53-dependent. Mol Biol Cell. 2010;21(14):2453–68.

59. Wiley LA, Rajagopal R, Dattilo LK, Beebe DC. The tumor suppressor gene Trp53 protects the mouse lens against posterior subcapsular cataracts and the BMP receptor Acvr1 acts as a tumor suppressor in the lens. Dis Model Mech. 2011;4:484–95.

60. Miyashita T, Krajewski S, et al. Tumor suppressor p53 is a regulator of bcl-2 and baxgene expression in vitro and in vivo. Oncogene. 1994;9:1799–805.

61. Miyashita T, Reed JC. Tumor suppressor p53 is a direct transcriptional activator of the human bax gene. Cell. 1995;80:293–9.

62. Wu GS, Burns TF, et al. KILLER/DR5 is a DNA damage-inducible p53-regulated death receptor gene. Nat Genet. 1997;17:141–3.

63. Wen-Schaub LB, Zhang W, et al. Wild-type human p53 and a temperature-sensitive mutant induce Fas/APO-1 expression. Mol Cell Biol. 1995;15:3032–40.
64. Buckbinder L, Talbott R, et al. Induction of the growth inhibitor IGF-binding protein 3 by p53. Nature. 1995;377:646–9.
65. Israeli D, Tessler E, et al. A novel p53-inducible gene, PAG608, encodes a nuclear zinc finger protein whose overexpression promotes apoptosis. EMBO J. 1997;16(14):4384–92.
66. Polyak K, Xia Y, Zweier JL, Kinzler KW, Vogelstein B. A model for p53-induced apoptosis. Nature. 1997;389:300–5.
67. El-Deiry WS. Regulation of p53 downstream genes. Seminar Cancer Biol. 1998;8:345–57.
68. Brown L, Boswell S, Raj L, Lee SW. Transcriptional targets of p53 that regulate cellular proliferation. Crit Rev Eukaryot Gene Expr. 2007;17(1):73–85.
69. Weisz L, Oren M, Rotter V. Transcription regulation by mutant p53. Oncogene. 2007;26(15):2202–11.
70. Riley T, Sontag E, Chen P, Levine A. Transcriptional control of human p53-regulated genes. Nat Rev Mol Cell Biol. 2008;9:402–12.
71. Mihara M, Erster S, et al. p53 has a direct apoptogenic role at the mitochondria. Mol Cell. 2003;11:577–90.
72. Li WC, Kuszak JR, Dunn K, Wang RR, Ma W, Wang GM, Spector A, et al. Lens epithelial cell apoptosis appears to be a common cellular basis for non-congenital cataract development in humans and animals. J Cell Biol. 1995;130:169–81.
73. Li W-C, Kuszak JR, Wang G-M, Wu Z-Q, Spector A. Calcimycin-induced lens epithelial cell apoptosis contributes to cataract formation. Exp Eye Res. 1995;61:89–96.
74. Li W-C, Spector A. Lens epithelial cell apoptosis is an early event in the development of UVB-induced cataract. Free Radic Biol Med. 1996;20:301–11.
75. Yan Q, Liu J-P, Li DW-C. Apoptosis in the ocular lens: role in development and pathogenesis. Differentiation. 2006;74:195–211.
76. Chen P, Ji W-K, Liu F-y, Tang H-J, Shujun F, Zhang X, Liu M, Gong L, Deng M, Wen-Feng H, Xiao-Hui H, Chen X-W, Li Z-l, Li X, Liu J-P, Li DW-C. Alpha-Crystallin and Carcinogenesis. Curr Mol Med. 2012;12(9):1164–73.
77. Li DW, Fass U, Huizar I, Spector A. Okaidac acid-induced lens epithelial cell apoptosis requires inhibition of phosphatase-1 and is associated with induction of gene expression including p53 and bax. Eur J Biochem. 1998;257:351–61.
78. Li DW-C, Xiang H, Mao Y-W, Wang J, Fass U, Zhang X-Y, Xu C. Caspase-3 is actively involved in okadaic acid-induced lens epithelial cell apoptosis. Exp Cell Res. 2001;266:279–91.
79. Li DW-C, Liu J-P, Mao YW, et al. Calcium-activated RAF/MEK/ERK signaling pathway mediates p53-dependent apoptosis and is abrogated by alphaB-crystallin through inhibition of RAS activation. Mol Biol Cell. 2005;16:4437–53.
80. Mao Y-W, Liu J, Xiang H, Li DW-C. Human αA and αB-crystallins bind to Bax and Bcl-X$_S$ to sequester their translocation during staurosporine-induced apoptosis. Cell Death Differ. 2004;11:512–26.
81. Hu WF, Gong L, Cao Z, Ma H, Ji W, Deng M, Liu M, Hu XH, Chen P, Yan Q, Chen HG, Liu J, Sun S, Zhang L, Liu JP, Wawrousek E, Li DW. αA- and αB-crystallins interact with caspase-3 and Bax to guard mouse lens development. Curr Mol Med. 2012;12(2):177–87.
82. Moorozov V, Wawrousek EF. Caspase-dependent secondary lens fiber cell disintegration in αA/αB-crystallin double knockout mice. Development. 2006;133:813–21.
83. Qin J, Chen HG, Yan Q, Deng M, Liu J, Doerge S, Ma W, Dong Z, Li DW. Protein phosphatase-2A is a target of epigallocatechin-3-gallate and modulates p53-Bak apoptotic pathway. Cancer Res. 2008;68:4150–62.
84. Wride MA, Parker E, Sanders EJ. Members of the bcl-2 and caspase families regulate nuclear degeneration during chick lens fibre differentiation. Dev Biol. 1999;213(1):142–56.
85. Fromm L, Overbeek PA. Inhibition of cell death by lens-specific overexpression of bcl-2 in transgenic mice. Dev Genet. 1997;20(3):276–87.

86. Lane D, Levine A. p53 Research: the past thirty years and the next thirty years. Cold Spring Harb Perspect Biol. 2010;2(12):a000893.
87. Liu FY, Tang XC, Deng M, Chen P, Ji W, Zhang X, Gong L, Woodward Z, Liu J, Zhang L, Sun S, Liu JP, Wu K, Wu MX, Liu XL, Yu MB, Liu Y, Li DWC. The tumor suppressor p53regulates c-Maf and Prox-1 to control lens differentiation. Curr Mol Med. 2012;12:917–28.
88. Xie Q, Cvekl A. The orchestration of mammalian tissue morphogenesis through a series of coherent feed-forward loops. J Biol Chem. 2011;286(50):43259–71.
89. Kawauchi S, Takahashi S, Nakajima O, et al. Regulation of lens fiber cell differentiation by transcription factor c-Maf. J Biol Chem. 1999;274(27):19254–60.
90. Czerny T, Busslinger M. DNA-binding and transactivation properties of Pax-6: three amino acids in the paired domain are responsible for the different sequence recognition of Pax-6 and BSAP (Pax-5). Mol Cell Biol. 1995;15:2858–71.
91. Ring BZ, Cordes SP, Overbeek PA, Barsh GS. Regulation of mouse lens fiber cell development and differentiation by the Maf gene. Development. 2000;127:307–17.
92. Tomarev SI, Zinovieva RD, Chang B, Hawes NL. Characterization of the mouse Prox1 gene. Biochem Biophys Res Commun. 1998;248(3):684–9.
93. Wigle JT, Chowdhury K, Gruss P, Oliver G. Prox1 function is crucial for mouse lens-fibre elongation. Nat Genet. 1999;21(3):318–22.
94. Wilting J, Papoutsi M, Christ B, et al. The transcription factor Prox1 is a marker for lymphatic endothelial cells in normal and diseased human tissues. FASEB J. 2002;16(10):1271–3.
95. Mishima K, Watabe T, Saito A, et al. Prox1 induces lymphatic endothelial differentiation via integrin alpha9 and other signaling cascades. Mol Biol Cell. 2007;18(4):1421–9.
96. Westmoreland JJ, Kilic G, Sartain C, et al. Pancreas-specific deletion of Prox1 affects development and disrupts homeostasis of the exocrine pancreas. Gastroenterology. 2012;142(4):999–1009.
97. Ji WK, Tang XC, Yi M, Chen PQ, Liu FY, Hu XH, Hu WF, Fu SJ, Liu JF, Wu KL, Wu MX, Liu XL, Luo LX, Huang S, Liu ZZ, Yu MB, Liu YZ, Li DW-C. p53 directly regulates αA- and βA3/A1-crystallin genes to modulate lens differentiation. Curr Mol Med. 2013;13(6):968–78.
98. Hu X-H, Nie Q, Yi M, Li T-T, Wang Z, Huang Z-X, Gong X-D, Zhou L, Ji W, Hu W, Liu J, Wang L, Zhu J, Liu W-B, Nguyen QD, Li DW-C. The p53 regulates gammaA-Crystallin gene during mouse lens development. Curr Mol Med. 2014;14(9):1197–1204.

Chapter 16
Etiology and Prevention of Cataract

Rasiah Pratheepa Kumari, Srinivasagan Ramkumar, Bency Thankappan, and Kumarasamy Anbarasu

Abbreviations

ALCAR	Acetyl-L-carnitine
BFSP1	Beaded filament structural protein-1
BFSP2	Beaded filament structural protein 2
CAT	Catalase
CD1	Curcumin salicylidenecurcumin
CD2	Benzalidenecurcumin
CHMP4B	Chromatin modifying protein-4B
CRYAA	Alpha-crystallin A chain
CRYAB	Alpha-crystallin B chain
CRYBA1	Beta-crystallin A1
CRYBA4	Beta-crystallin A4
CRYBB1	Beta-crystallin B1
CRYBB2	Beta-crystallin B2
CRYBB3	Beta-crystallin B3
CRYGC	Gamma-crystallin C
CRYGD	Gamma-crystallin D
CRYGS	Gamma-crystallin S
ECCE	Extracapsular cataract extraction
FVN	Flavanoid fraction *Vitex negundo*
GCNT2	Glucosaminyl (*N*-acetyl) transferase 2
GJA3	Gap junction protein alpha 3

R.P. Kumari • S. Ramkumar • B. Thankappan • K. Anbarasu (✉)
Department of Marine Biotechnology, Bharathidasan University,
Tiruchirappalli, Tamil Nadu, India
e-mail: pratheepakumari@gmail.com; ramm.bio@gmail.com;
benzbt11@gmail.com; anbubdu@gmail.com

© Springer Science+Business Media New York 2015
M.A. Babizhayev et al. (eds.), *Studies on the Cornea and Lens*,
Oxidative Stress in Applied Basic Research and Clinical Practice,
DOI 10.1007/978-1-4939-1935-2_16

GJA8	Gap junction protein alpha 8
Gpx	Glutathione peroxidase
GR	Glutathione reductase
GSH	Reduced glutathione
GST	Glutathione-S-transferase
HSF4	Heat shock transcription factor 4
ICCE	Intracapsular cataract extraction
iNOS	Inducible nitric oxide synthase
IR3G	Isorhamnetin-3-glucoside
LIM2	Lens intrinsic membrane protein 2
M	Molarity
MAF	Musculoaponeurotic fibrosarcoma protein family
MIP	Major intrinsic protein
NAC	N-acetylcarnosine
NADPH	Nicotinamide adenine dinucleotide phosphate
NCX	Sodium calcium exchanger
NO	Nitric oxide
PBA	Prevent Blindness America
PITX3	Pituitary homeobox 3
PMCA	Plasma membrane-associated calcium ATPase
ROS	Reactive oxygen species
SOD	Superoxide dismutase
TMEM114	Transmembrane protein 114
UV-B	Ultraviolet-B
WI	Urea-soluble fraction/water-insoluble fraction
WS	Water-soluble fraction

16.1   Introduction

Cataract represents one of the major causes of blindness worldwide. In India, cataract accounts for about 50–80 % of the blindness and in United States 22 million people encounter cataract and are expected to reach 30 million by 2020 [1]. According to Prevent Blindness America (PBA), cataract cases heads (>50 %) among all the other causes of blindness such as glaucoma, macular degeneration, and diabetic retinopathy. Cataract may develop as a result of both genetic (congenital mutation) and non-genetic factors like aging. During cataractogenesis, the lens of the eye becomes cloudy which is transparent otherwise. The vision may be blurred or misty; shadow and color vision or less vivid characterized by myopia and sensitivity to light.

Cataract is treated by removal of the cloudy lens by phacoemulsification system aided by ultrasound energy that fragments the lens which is then aspirated from the capsular bag and replaced by a synthetic intraocular lens. Clinical studies have markedly reported profound aftermath effects of cataract surgery. Approximately,

25 % of the cases encounter posterior capsular opacification during 5 years of post-operative period [2], the most common cause of loss of vision after cataract surgery worldwide. Another significant and compelling component of surgery is the expensive surgical cost. Therefore, developing strategies to thwart the occurrence and identification of potential risk factors should be considered as a prime epitome in prevention and management of cataract. Pharmacological interventions are aimed at reducing the damage rendered to lenticular epithelium and fiber cells by oxidation, glycation, posttranslational modifications and other risk factors that are being in trial for effective management of cataract. Anticataractogenic agents of natural and synthetic origin may pave as another milestone in the intervention of cataractogenesis. Further research is warranted to account the effectiveness of antioxidant supplementation and the stage of administration to counteract the relative incidence of cataract.

16.2   Etiology of Cataractogenesis

Cataract is a multifactorial disease characterized by opacification and clouding of the eye lens due to crystallin protein unfolding and aggregation. The mechanism of cataractogenesis is not merely a gene mutation resulting in the phenotype, preferably caused by either genetic and/or many other environmental and metabolic factors contributing to the altered gene expression or protein alterations. Intriguingly, epidemiological and family history has revealed a positive correlation of the genetic component and cataractogenic process (Table 16.1). There appears to be a central role for posttranslational modification of the lenticular proteins in human cataract. Unambiguous perturbations in light scattering, protein tertiary structural changes, conformational changes, redox components, phase transitions, and aggregation state play a pivotal role in the cataractogenic process [3]. Protein truncations and deletions during posttranslational modifications in crystallin proteins are witnessed as prelude of human cataract [4]. Point mutations of α-crystallin genes, structural anomalies caused by racemization, isomerization, truncations, aggregation, and insolubilization are reported to cause human cataract [5]. Isomerization of

Table 16.1 Risk factors of cataract

Unmodifiable risk factors	Modifiable risk factors
1. Race	1. Oxidative stress-related factors
2. Sex	2. Maternal malnutrition
3. Age	3. Medication and physical exercise
4. Genetics	4. Smoking and alcoholism
5. Family history	5. Exposure to UV and ionizing radiations
6. Body mass index	6. Nutrition deficiency, low antioxidant intake
7. Diabetes	7. Environmental pollution
	8. Eye injury

αA-crystallin at the Asp58 and Asp151 has been reported recently [6] as a major issue regarding the conformational changes rendered to the crystallin protein. D-isomerization and β-linkage of aspartyl residues affect the quaternary structure of crystallin proteins because of conformational changes in the main chain that remains the prime reason for aggregation and insolubilization of lens proteins. Abnormal aggregations and heterogeneous assemblies of the lens proteins end up with loss in chaperone activity and such a decrease avails light scattering and opacification [7]. However, it is often foolhardy to draw a common pathway depicting the series of biochemical and biophysical changes in the lens.

Steady state fall in the free radical scavenging capacity of the redox components contributes to the subtle clinical progression of nuclear opacification. Prospective cohort study, patient, and case control study demonstrate a strong positive correlation between enzymatic, nonenzymatic antioxidants and relative incidence of cataract [8]. Profound liquefaction of the nucleus and cortex observed in hyper mature cataract is the generic reason behind light scattering. Autosomal dominant mutations are responsible for early childhood congenital cataract causing bilateral vision loss. Majority of the mutations reported in genes, disrupt the tertiary structure culminating in aggregation and precipitation of crystallin. Apart from structural gene mutations, point mutations in the cytoskeletal components [Major intrinsic protein (MIP)], aquaporin0, membrane proteins, gap junctional proteins (Connexin 46 & 50), other heat shock components ends up in loss of lenticular homeostasis [9], and membrane disintegrations in fiber cells [10] can also cause/increase the risk of congenital and age-related cataract.

16.2.1 Crystallin Lens Proteins: The Key Components of the Eye Lens

The eye lens fiber cells are composed of high concentration of (>90 %) specific soluble proteins called crystallin. Crystallin protein is classified into three major groups such as α-, β-, and γ-crystallin that act as components of the lens and are vital for lens transparency, viscosity, and high refractive index. The oligomeric α-crystallin molecule exists in dynamic state continuously involved in rapid exchange and dissociation of subunits. The homologous and heterologous interactions of the crystallin proteins within/with other crystallin and membrane proteins are yet another vital factor of lens transparency [11]. They also physically and functionally interact with both the cell membrane and cytoskeleton. Functional changes in α-crystallin have been shown to modify cell–cell interactions and lead to pathology in vivo [12], the well-known examples are various neurodegenerative diseases and cancer. Crystallin proteins are arranged in regular mode with a shorter range order lesser than the wavelength of light. The fiber cells maintain a unique signature of crystallin expression contributing to higher concentration leading to molecular crowding thence enhancing α-/β-, α-/α-, β-/γ-, α-/γ-, and γ-/γ-interactions and associations. It is envisaged that minor encumbrance in the protein–protein interaction phenomenon directly contemplates in lens transparency [4, 13–19]. Number of in vitro and in vivo assays proved that the homogeneous and heterogeneous interaction of α-crystallin prevents the cluster form of aggregates in native as well as different physiological conditions [20]. The highly ordered array in the lens is accountable for refractive index and transparency; this is achieved by the crystallin proteins and characterized for being able to reach in higher concentration without aggregating and scattering light. This highly conserved, small heat shock protein prevents the aberrant physiological changes taking place in the eye lens protein during stress conditions.

Transparency and proper light refraction of the lens depend on a unique arrangement of tightly packed fiber cells, which in turn rely on a defined protein structure. The human lens has a protein concentration of 33 % of its wet weight, which is twice that of most other tissues such as brain = 10 % and muscle = 18 % [21]. The crystallins are intracellular proteins contained within the epithelium and plasma membrane of the lens fiber cells. α-Crystallin constitutes subunits αA and αB; each subunit polypeptide has a molecular weight of about 20 kDa and possesses the ability to form oligomers of 200–800 kDa. The subunits are held together by hydrogen bonds and hydrophobic interactions. Crystallins appear to be specifically involved in the transformation of epithelial cells in the lens fiber cells. At the time of human birth 1.6 million fiber cells are found that increases to 3 million at the age of 20 and 3.5 million at 80 years of age [22]. The rate of synthesis of α-crystallins is seven times higher in epithelial cells than in the cortical fibers, indicating a significant decrease in rate of synthesis after the transformation. β-crystallins account for 55 % (by weight) of the water-soluble proteins in the lens and γ-crystallins are the smallest form of the crystallins, with a molecular weight in the range of 20 kDa. The conversion of water-soluble lens protein to insoluble protein is the indication of cataract during aging process, the acceleration of insoluble protein leads to form protein aggregation. As there is negligible protein turnover in mature fiber cells [23], most of these proteins are surprisingly stable and remain in the lens for the duration of an individual's life span [24]. Methods have been developed to isolate lens protein, which normally involves sequential buffer extraction from decapsulated lens tissues [24]. Lens proteins extracted by diluted aqueous buffer are termed as water-soluble protein, which accounts for up to 80–90 % of total proteins in normal lenses and consists of almost entirely structural protein known as crystallins [25]. Lens proteins that are solubilized in 7–8 M urea are termed as water-insoluble (WI)/inclusion proteins, which consist of denatured crystallin and cytoskeletal proteins [24, 26, 27]. The insoluble protein fraction possesses high-molecular weight disulphide (S-S)-linked protein aggregates. The urea-soluble fraction (WI) contains cytoskeleton proteins that provide the structural framework of the lens cells and the fiber plasma membranes that resemble erythrocyte plasma membranes in many aspects. As the fiber cells begin to elongate, the MIP can be detected in membranes and throughout the mass of the lens. It is not found in the epithelial cell and seems to be associated with the differentiation of epithelial cells in to fiber cells. The MIP is concentrated in the gap junctions and is the predominant protein of the junction-enriched membrane proteins. It is an inherent part of the membrane, where it can be localized by immunofluorescence.

16.2.2 Cataract: A Progressive Deterioration of Vision

During stress conditions, over expression of αB-crystallin has been reported in various non-lenticular tissues especially in cardiac and skeletal muscles related to myopathy, carcinoma, and neurodegenerative diseases. αB-crystallin exists as molecular chaperone prevent the misfolding of proteins leading to aggregation and amyloid fiber formation in lens and other organs [28]. Moreover, elevated levels of αB-crystallin confers anti-apoptotic effects in many cells including retinal pigment epithelial cells, and always detected as biomarker of oxidative stress-induced apoptosis [29] in various muscular and neurodegenerative diseases. Researchers have also reported that temperature above 30 °C could enhance the chaperone-like activity of αA-crystallin to several-fold [30] and the protein structural stability persist at 100 °C with little unfolding condition, even though, it will revert back to normal when cooled to 21 °C [31]. Failure of this chaperone activity is mostly due to mutation or alterations during posttranslation modifications in α-, β-, and γ-crystallin and leading to aggregation of misfolded proteins resulting in diseased state [32, 33]. There are lots of ongoing epidemiological studies to figure out risk factors; however, there are only a few factors recognized and investigated in detail like UV-B exposure, low antioxidant intake, certain medications, cigarette smoking, diabetes, and gout as well as family history [34]. In contrast to these age-related forms of cataract, congenital cataracts or cataracts in early childhood are rather rare but avoidable causes of blindness reported in both developed and developing countries with a frequency of 30 cases among 100,000 births; with a further 10 cases being diagnosed by the age of 15 years (mainly as dominant forms). Rates are likely to be higher in developing countries because of Rubella infections and consanguinity for the recessive forms [35].

Congenital cataract is detectable at birth or during the first decade of life due to different causes, including metabolic disorders (galactosemia), infections during embryogenesis [36], gene defects, and chromosomal abnormalities [37]. Cataract may be an anomaly, observed in association with other ocular developmental abnormality, or part of a multisystem syndrome, such as Down's syndrome, Wilson's disease, and myotonic dystrophy [38]. Inherited cataracts correspond to 8–25 % of congenital cataract [39] and the commonest mode of inheritance is the autosomal dominant form. Appearance of the lens opacities seen in families with inherited cataract is classified into five groups: lamellar, coralliform, stellate, anterior, posterior polar, and finally an "undefined" group. At least 34 loci in the human genome have been reported to be associated with various forms of pediatric cataract. Autosomal dominant and recessive forms of cataracts have been caused by mutations in 22 different genes encoding crystallins CRYAA [40], CRYAB [41], CRYBA1 [42], CRYBA4 [43], CRYBB1 [44], CRYBB2 [45], CRYBB3 [46], CRYGC, CRYGD [47], and CRYGS [48], cytoskeletal proteins BFSP1 [49] and BFSP2 [50], membrane proteins GJA3 [51] and GJA8 [52], MIP [53] and LIM2 [54], transcription factors HSF4 [55], PITX3 [56], and MAF [57], glucosaminyl (N-acetyl) transferase 2 (GCNT2) [58], chromatin modifying protein-4B CHMP4B [59] and TMEM114 [60] (Table 16.2). On the basis of current studies, mutations in

Table 16.2 List of reported mutants and the congenital cataract cases worldwide

S.No	Gene	Mutation	Type	Cataract	Family	Reference
1	CRYAA	W9X	Nonsense	Autosomal recessive	Jewish Persian	[61]
		R12C	Novel	Zonular	Danish	[62]
		R21L	Novel	Congenital	German	[63]
		R21W	Novel	Zonular	Danish	[62]
		R49C	Novel missense	Autosomal dominant	Caucasian	[34].
		R54C	Novel nonsense	Congenital	Saudi	[64]
		F71L	Novel	Age related	Indian	[65]
		G98R	Novel	Putative	Indian	[66]
		R116C	Novel	Congenital	Danish	[62]
		R116H	Novel	Congenital	Danish	[62]
2	CRYAB	D140N	Novel Missense	Lamellar	Chinese	[67, 68]
		R11H	Novel	Nuclear	Chinese	[69]
		R56W	Novel Missense	Juvenile	Saudi	[70]
		P20S	Novel	Posterior polar	Chinese	[67, 68]
		R56W	Novel	Congenital	Saudi Arabia	[64]
		A171T	Missense	Pediatric	South India	[71]
		R120G	Point	Posterior polar	France	[72]
3	CRYBB	G220X	Nonsense	Autosomal dominant	Chinese	[44]
4	CRYBB2	D128V	Novel	Congenital	German	[73]
		V187M	Missense	Congenital	Basotho	[74]
		Q155X	Nonsense	Congenital coronary	Chinese	[75]
5	CRYBB1	S228P	Novel missense	Autosomal dominant	Chinese	[76]
		X253R	Novel	Congenital	UK	[77]
		Q223X	Novel nonsense	Autosomal dominant	Chinese	[78]
6	CRYBB3	G165R	Point	Autosomal recessive	Pakistani	[46]
7	CRYBA1/A3	G91 DEL	Deletion	Autosomal dominant congenital	Chinese	[79]
	CRYBA1/A3	G91DEL	Deletion	Congenital nuclear lactescent	Swiss	[80]
8	CRYGC	C109X	Nonsense	Autosomal dominant	Chinese	[81]
		T5P	Novel	Coppock	Swiss	[47]

(continued)

Table 16.2 (continued)

S.No	Gene	Mutation	Type	Cataract	Family	Reference
9	CRYGD	G61C	Novel missense	Congenital coralli form	Chinese	[75]
		R15S	Novel missense	Congenital coralli form	Chinese	[82]
		P24T	Novel missense	Congenital coralli form	Chinese	[82]
		P23T	Missense	Coral like	Chinese	[83]
		R14C	Missense	Congenital	Chinese	[84]
		R58H	Missense	Aculeiform	Mexican	[85]
		494DEL G	Deletion	Congenital nuclear	Chinese	[82]
		R14C	Missense	Coralli form	Chinese	[84]
		R36S	Missense	Crystal	Czech	[86]
		R58H	Missense	Aculeiform	Swiss	[47]
		E107A	Missense	Nuclear	Mexico	[39]
		Y134X	Novel	Congenital	Danish	[62]
10	CRYGS	G18V	Missense	Dominant progressive cortical	Chinese	[48]

about half of the affected families occurs in crystallin gene, a quarter in connexins and the remaining is evenly split between membrane proteins, intermediate filament proteins, and transcription factors. However, the relative contribution of these classes of genes to pediatric cataracts is still unclear.

Age-related/senile cataract is a progressive disorder of the lens affecting transparency accompanied with marked light scattering. Oxidative stress-related factors, exposure to radiations, smoking, low antioxidant status, and exposure to irradiations are the key factors that trigger cataractogenic process above the age of 60. Free radicals and glycation are the perpetrator causing cross-linking of the lenticular proteins, and the aggregated proteins scatter light. With aging phenomena, the antioxidant defense machinery gets defoliated and the system is overridden by oxidative stress [87], specifically the level of GSH and its precursor amino acids are significantly reduced. The reason for the decline in GSH is possibly the alterations in the function of GSH transporters [88]. In the case of secondary cataract like diabetic cataract, the protein breakdown process is accelerated during hyperglycemia. Aldose reductase facilitates the transfer of glucose into sorbitol, which is impermeable to the membrane; accumulation of sorbitol ends with glycative and osmotic stress [89]. Any injury pertaining in the lens tissue may lead to hydration of the protein causing dense cortical cataract. The extent of opacification depends on the type and depth of injury.

16.2.2.1   Stratification of Opacified Lens

According to the degree of maturity, typical classification is made to describe morphological classification of cataract.

Immature	Asymptomatic cataract with demarcations between opaque areas
Intumescent	Swollen with water, probably due to osmotic stress
Mature	The entire cortex is white and opacified
Hyper mature	Pronounced liquefaction of the nucleus and cortex

Generic terms such as total/diffuse, anterior polar, lamellar, nuclear, posterior polar, posterior lentiglobus, and posterior subcapsular are applied to different forms of cataract. In general surgeons label the shape and structure of cataract like punctuate, pulcerent, coroliform, coronary, floriform, retrodot, sunflower, blue dot, and sutural. Retrospective of the form, shape, and incidence, cataract surgery is one of the most cost-effective interventions in the field of medicine, resulting in almost immediate visual rehabilitation [90]. Nowadays, phacoemulsification is applied in the management of cataract because of its earlier refractive stabilization, reduced induced astigmatism, and milder postoperative inflammation, all resulting in faster visual rehabilitation. It has been shown that improvement in visual acuity following cataract surgery is accompanied by considerable gains in real-life activities, emotional and social life components [91].

16.3   Avalanche of Biochemical Reactions

Significant inroads are being made to elucidate the series of changes taking place during the cataractogenic process. The sequel begins with the biochemical or physical insults carried on by phase separation of crystallins into soluble and insoluble aggregates, distortion of antioxidant defense, reduction in GSH level, loss in protein secondary and tertiary structure; any of which may result in light scattering. However, when the key factor reckon to be a mutation in the crystallin gene or maternal malnutrition, this results in congenital cataract. If the factor is an environmental insult such as radiation, hyperglycemia, or oxidation, this may contribute to age-related cataract.

16.3.1   *Mutation and Its Cardinal Role in Congenital Cataract*

Based on the underlying gene functions, cataract is caused by mutations in crystallin, membrane/cytoskeleton proteins, and transcription factors [92]. Both αA- and αB-crystallin are encoded by two dissimilar genes: CRYAA and CRYAB genes, respectively. The homologous and heterologous interaction between the different crystallin proteins confers a dynamic state which is essential for lens transparency [15]. Mutations in CRYAA, CRYAB, CRYBB, CRYGC, and CRYGD and truncations in the N- and C-terminal regions of crystallin proteins (CRYAA and CRYAB) have been reported to develop opacities. During the past decade, a large body of evidences has been filed on the protein–protein interaction; chaperone activity and subunit exchange of crystallin proteins play crucial role in the prevention of cataract [93, 94]. Literature evidences that recombinantly expressed and purified mutants of αA-, αB-, β-, and γ-crystallins and their truncations have altered structural conformation, solubility, stability in different pH, temperature and functional characteristics, specifically loss of chaperone activity, altered hydrophobicity, and subunits protein–protein interactions leading to cause cataract. Nevertheless, the degree and pattern of interaction varies for each mutation and truncation. The mutants and truncated forms of crystallins are prone to aggregation as sign of loss in native structures.

Gap junctional proteins such as connexins 43, 46, and 50 are localized on the cell membranes spanning the intracellular communications between adjacent cells. These gap junctional proteins facilitate the transportation of ions (K^+, Ca^{2+}) and small molecules including metabolites (e.g., glucose) and second messengers [95]. Localized point mutations such as GJA8/Cx50 and GJA3/Cx46 in the connexin gene elicits poor interaction of the protein product with the neighboring protein suggesting that mutations in connexin genes can lead to dominant and recessive forms of cataract.

Cytoskeletal proteins are inherent compositions of the lens bestowing to the structural, cell motility, maintenance of cell volume and shape. R278W and delE233 are reported mutations in the Beaded Filament Structural Proteins (BFSP) also termed Filensin, a highly divergent intermediate filament. This mutation has been reported as the important causative factor for severe congenital cataract characterized by nuclear, sutural, and cortical cataract. Certain mutations in the transcriptional regulators like PITX3, MAF, and HSF4 peculiarly S13N, 10q24-q25, R288P, K297R, A19D, R73H, I86V, L114P, R119C, and R175P have been implicated in the cataractogenic process [9]. The structural confirmation and interaction between crystallin, membrane, and cytoskeleton proteins are the key factor for determining the protein complex, molecular assembly, and maintaining lens transparency [19].

16.3.2 Oxidation the Perpetrator of Senile Cataract

Oxidative stress is the key factor in senile and secondary form of cataractogenesis. Environmental insult to lenticular proteins, photochemical damage, and oxidative assaults by hydrogen peroxide, superoxide, hydroxyl, and reactive nitrogen species induces damage to the lens epithelial cells [96]. The continuous exposure of above photochemical factors and oxidative stress induces free radical formation in the eye lens even though it is counteracted by number of antioxidant defense molecule in this avascular organ. The organ is inherently aided with endogenic antioxidant defense machinery that comprises enzymatic: SOD, CAT, Gpx, GR, thioredoxin, and nonenzymatic antioxidants: GSH, Vitamin C, etc. [97]. SOD, a chain-breaking antioxidant catalyzes the dismutation of the superoxide radical into molecular oxygen and hydrogen peroxide. The enzyme exists in two forms, one containing Mn^{2+}, restricted to the mitochondria, and a cytosolic form containing Zn^{2+} and Cu^{2+}. Subsequently CAT, a hemoprotein that requires NADPH for regeneration to its active form catalyzes the reduction of H_2O_2 to water and molecular oxygen. GSH in the lenticular tissue is the major factor involved in maintaining protein sulfhydryl groups by directly scavenging the reactive oxygen species. The glutathione redox cycle is vital for maintaining lenticular transparency by detoxifying the generated reactive oxygen species. Gpx is a selenoprotein, which catalyzes the reduction of hydroperoxides with the assistance of its reducing substrate GSH [98]. With the advent of risk factors such as aging the antioxidant system is challenged detrimentally, where the reactive free radicals direct the epithelial cells to apoptosis resulting in severe damage of the eye lens. Withal, the protein content undergoes certain irreversible posttranslational modifications in particular oxidation, deamidation, racemization, and truncations. However, with age, it appears that these protective mechanisms decrease in activity, resulting in elevated H_2O_2 levels, ultimately leading to opacification. Irrevocable ionic imbalance stimulated by the continuous efflux of calcium, sodium ions by the calcium ATPases, sodium potassium ATPases, NCX, PMCA [99–101]. Evidently, studies from human cataracteous lens also depicted reduced activity of sodium potassium ATPase, PMCA [102] where the ion exporting mechanisms are incapable of balancing the passive leakage of calcium, sodium, and potassium ions. Pioneering studies have suggested the biochemical alteration in the lens is reflected in the ratio of soluble and insoluble protein. Mutations and proteolysis of lenticular crystallin, breakdown of cytoskeletal contents specifically in actin, vimentin, and spectrin are strongly associated with the increase insoluble content that favors opacification and cataract development [15, 19]. These cascade of events trigger the deterioration of the lenticular milieu culminating in sever mature cataract.

16.3.3 Glycation, a Rudimentary Factor of Diabetic Cataract

Glycation is often considered as the hallmark of diabetes mellitus, a metabolic syndrome characterized by hyperglycemia and insulin resistance. Diabetic individuals pose an increased risk in the development of posterior subcapsular cataract due to nonenzymatic glycation of eye lens proteins, oxidative stress, and activated polyol pathway in glucose disposition. Aldose reductase facilitates the conversion of glucose to sorbitol using NADPH as the cofactor. Accumulation of sorbitol results in osmotic swelling, osmotic gradient, and tissue damage [89]. The patho-mechanism is convicted by the formation of advanced glycation end products, the end products of millard reaction culminates in lens protein alteration. The millard process adds glucose carbonyl group to the free amino group of protein or amino acid forming Schiff base adducts, which in turn forms stable amadori products [103] that can disrupt the potential arrangements of lens protein that lead to progressive cataractogenesis.

16.4   Prevention and Management

The nature of day-to-day life of individuals and occupational demands plays a vital role to create awareness of visual impairment and management. Cataract (other than congenital form) is one of the leading causes of avertable blindness worldwide. The burden of visual impairment is directly correlated with the loss of productivity. Accumulation of damage from the environment, deterioration of defense, repair mechanisms, and genetic predisposition [104] are the major contributing factors. Limited access to health care, lack of awareness to obtain healthy balanced food, medication, and UV-radiation are considered as a potential ground for visual impairment and increased rates of cataract surgery.

16.4.1   Strategies for Cataract Prevention

16.4.1.1   Natural Antioxidants and Prophylaxis

A great body of epidemiological data in animal model suggests a direct positive correlation between higher level of antioxidant intake and decreased incidence of cataract. Oxidative stress is inextricably related to cataractogenesis process, explaining the underlying role of free radicals. Antioxidants are small biomolecules with electronegative centers posing the capability to scavenge free radicals, modulating the antioxidant enzymes and chelate metal ions. The innate defense mechanism is challenged by the onset of cataractogenesis; however, these antioxidants may reduce the stress burden on the redox components and thereby averting the deleterious cascade of cellular damage [105]. Research in the recent years have been focused to identify genetic variations and gene regulation by dietary factors that could serve as diagnostic tools for individualized intervention and novel therapeutic strategies for cataractogenesis [106]. Oxidative stress is counter acted by innate antioxidant enzyme such as SOD, CAT, Gpx, GR, and GST. Antioxidant substances are small biological molecules that resist the damage to the structural and functional biomolecules at relatively lower concentrations. Antioxidants are versatile components that serve by scavenging the free radicals, modulating the antioxidant enzymes and chelating metal ions. A massive body of evidences by epidemiological and intervention studies state that the incidence of cataract is directly proportional to the dietary intake of antioxidant supplements [107].

Extensive studies on animal and random placebo control trial presumes that oxidation precedes opacification. Vitamin C is a routine neutralizer of free radicals, a nonenzymatic antioxidant element present in higher concentration both in the eye lens and aqueous humor [108]. Vitamin C and E are implicated in preventing experimental cataract by reducing the levels of lipid peroxidation in the aminothiazole-induced cataract. It should be noted that external administration of antioxidants can only be beneficial when the endogenous system is jeopardized. These antioxidants can directly involve in the detoxification process or trigger the activation of redox components at the level of both transcription and translation. In selenite-induced rat model, Ca^{2+} homeostasis is abruptly disturbed which triggers the calcium-dependent protease calpain and Lp82. Elanchezhian et al. [109] have reported that a decline in Ca^{2+} was precluded by the antioxidant property of ALCAR. Presumably, the autolytic process of calpain after activation was evidenced by a lowered expression levels of Lp82 protein and decreased expression levels of m-calpain mRNA transcripts. According to Muralidharan et al. [110], the gap junctional proteins such as connexin 46, connexin 50, and PMCA 1 are the key regulators of calcium homeostasis in the lens. Moreover, in silico study was accomplished by homology modeling of the functional domains of the connexin protein and a concomitant docking analysis with ALCAR was executed. The results suggest the formation of strong hydrogen bond between ALCAR and the functional domains that explains the interaction between the antioxidant and the protein moiety at the atomic level.

16.4.1.2 Lead Compounds of Plant Origin: A Probable Development of Anticataractogenic Agent

During the past decade, enormous data has emerged on the action of plant compounds and extracts in experimental cataractogenesis (Table 16.3). Investigations on the potent pharmacological action of the bioactive extracts such as Astaxanthin, C-Phycocyanin, caffeine, curcumin, elagic acid, lycopene, flavonoid fraction of *Vitex negundo*, Isorhamnetin-3-glucoside, Drevogenin D, proanthocyanidin, *Emilia sonchifolia*, *Crataegus pinnatifida*, *Cineraria maritime*, *Pleurotus ostreatus*, *Trigonella foenum-graecum*, *Embelica officinalis*, *Camellia sinensis*, *Brassica oleracea*, *Ginkgo biloba*, garlic, onion, rutin, tetramethylpyrazine, *Citrus aurantium*, etc. possibly prevented cataractogenesis by thwarting oxidative stress. However, the side effects and mode of action is not well characterized so far. The prominent evidence on the anticataractogenic potential of plant sources lies by protecting the antioxidant defense machinery. So far, no plant or natural products have been tested for clinical trials that warrant extensive study on the modality of pharmacological action.

Rutin is an abundant flavone glycoside present in herbs that exhibit pronounced antioxidant and free radical scavenging capacity. Extensive studies have been carried out on the anticataractogenic potential of rutin against selenite-induced model. The flavones have been recognized to protect the enzymatic and nonenzymatic antioxidants such as SOD, CAT, Gpx, GR, GST [143], and ascorbate. The major function

Table 16.3 Natural compounds and the possible mode of action as potent anticataractogenic agent in animal model

S. No.	Component	Source	Mode of action	Reference
1	Aqueous	Apricots	Conserves the lens enzymatic antioxidants and scavenges the reactive species	[111]
2	Aqueous extract	*Embelica officinalis*	Maintains the mean activities of antioxidant enzymes	[112]
3	Aqueous extract	Garlic	Protects the antioxidant balance	[113]
4	Astaxanthin	Plants, algae, and marine animals	Interaction of astaxanthin with selenium through conjugated polyene	[114]
5	Caffeine	Plant origin	Influence the lens metabolism by inhibiting cyclic adenosyl monophosphate phosphodiesterase	[115]
6	C-Phycocyanin	*Spirulina platensis*	Modulating the antioxidant enzyme status	[116]
7	Curcumin	*Cucurma longa*	Antioxidant property	[117, 118]
8	Danshensu	*Salvia miltiorrhiza*	Conserve the mean activities of antioxidant enzymes	[119]
9	Drevogenin D	*Dregea volubilis*	Protects against calpain activation	[120]
10	Elagic acid	Raspberries, pomegranate, walnuts, grapes, and blackcurrants	Maintaining of antioxidant enzyme activities and decreased malondialdehyde levels	[121]
11	*Emilia sonchifolia* Flavonoids	*Emilia sonchifolia*	The flavonoids modulate lens opacification and oxidative stress	[122]
12	Ethanolic extract	*Camellia sinensis*	Acts primarily by preserving the antioxidant defense system	[123]
13	Ethanolic extract	*Cineraria maritime*	Conserves the enzymatic antioxidant system	[124]
14	Ethanolic extract	*Pleurotus ostreatus*	Protects against oxidative stress	[125]

(continued)

Table 16.3 (continued)

S. No.	Component	Source	Mode of action	Reference
15	Extract	*Ocimum sanctum*	Restoration of the antioxidant defense system and inhibition of protein insolubilization	[126]
16	Extract	*Ginkgo biloba*	Significantly retards the progression of lens opacification	[127]
17	Extract	*Origanum vulgare*	Based on direct or indirect antioxidant mechanisms	[128]
18	Extract	Onion	Maintains the inherent antioxidant levels	[113]
19	Extract	*Trigonella foenum-graecum*	Protects against experimental cataract by virtue of its antioxidant properties	[129]
20	Flavonoid fraction of leaves	*Moringa oleifera*	Maintains the activities of antioxidant enzymes and sulfhydryl content and prevented reactive oxygen species generation and lipid peroxidation	[130]
21	Flower buds	*Cleistocalyx operculatus*	Significantly reduces the glucose, sorbitol, and fructose levels in diabetic rat lenses	[131]
22	Galloyl glucoses	*Cornus officinalis*	Inhibitory activity against the formation of advanced glycation end products	[132]
23	Hydroethanolic extract	*Pleurotus florida*	By maintenance of higher levels of protective antioxidant enzymes as well as water-soluble protein in the lens	[133]
24	Isoquinoline alkaloids	*Tinospora cordifolia*	Inhibit lens aldose reductase	[134]
25	Isorhamnetin-3-glucoside	*Cochlospermum religiosum*	Maintaining the Ca^{2+} ATPase activity, preventing oxidative stress damage, accumulation of calcium, and prevention of lipid peroxidation	[135]

(continued)

Table 16.3 (continued)

S. No.	Component	Source	Mode of action	Reference
26	Itone™	*Azadirachta indica*—5 %, *Moringa pterygosperma*—5 %, *Eclipta alba*—5 %, *Boerhaavia diffusa*—7.5 %, *Carum copticum*—2 %, *Terminalia chebula*—5 %, Terminalia belerica—5 %, *Emblica officinalis*—5 %, *Santalum album*—5 %, Mukta—1 %, *Ocimum sanctum*—5 %, *Vitex negundo*—5 %, *Curcuma longa*—5 %, *Mentha piperata*—2 %, *Cinnamomum camphora*—3 %, *Amomum subulatum*—5 %, *Rosa centifolia*—7.5 %, Saindhava laban—2 % and honey—10 %.	Antioxidant property	[133]
27	Leaves extract	*Hydrocotyl bonariensis,*	Moderately increased percentage soluble protein in the lens but did not increase the antioxidant status	[136]
28	Leaves extract	*Crataegus pinnatifida*	Conserves the activities of antioxidant enzyme SOD, CAT, and GSH	[137]
29	Luteolin	*Vitex negundo*	Inhibits calpain activation, maintains the antioxidant status of lens tissue	[138]
30	Lycopene	Tomatoes, other red fruits and vegetables	Antioxidant property	[139]
31	Melatonin	Secretory product of melatonin	Increases antioxidant activity	[140]

(continued)

Table 16.3 (continued)

S. No.	Component	Source	Mode of action	Reference
32	Methanolic extract	*Brassica oleracea*	Maintains antioxidant status and ionic balance through Ca^{2+} ATPase pump, inhibition of lipid peroxidation, calpain activation, and protein insolubilization	[141]
33	Proanthocyanidin	Grape seeds	Protect the antioxidant enzyme modules, inhibits the generation of NO and iNOS	[142]
34	Rutin	Plant origin	Protecting the antioxidant defense mechanism	[143]
35	Saffron	Saffron	Prevents lipid peroxidation, protein oxidation, proteolysis, and insolubilization of the lens water-soluble fraction	[144]
36	Scopoletin	*Magnolia fargesii*	Inhibits aldose reductase activity	[145]
37	Seed coat polyphenols	*Eleusine coracana*	Acts by inhibiting Aldose reductase reversibly by noncompetitive inhibition	[146]
38	Tetramethylpyrazine	*Ligusticum wallichii*	Preserves lens transparency by maintain normal levels of activity of SOD, Gpx and CAT	[147]
39	Triphala	*Emblica officinalis, Treminalia chebula* and *Terminalia belerica*	Protects antioxidant enzyme activity	[148]
40	Tuber	*Zingiber officinalis*	Acts by its antiglycating potential and to a lesser extent by inhibition of the polyol pathway	[149]
41	Water extract	*Aralia elata*	Inhibits aldose reductase and acts in vitro as an antioxidant	[150]
42	Water extract	*Morus alba*	Antioxidative potential of flavonoids	[151]

of the lens crystallin has been reported to be the chaperone-like activity that mediates the proper folding of other proteins, a function similar to the small heat shock proteins [16]. During the onset of cataractogenesis, the loss of chaperone activity of crystallin protein is considered as the trigger for opacification. The key factor towards the deprivation of chaperone activity is the elevated oxidative stress affirmed by sodium selenite. However, treatment with plant-derived antioxidants reverts back the chaperone activity and retains the structure and function of α-, β-, and γ-crystallins. This corollary was authenticated by in vivo experiment in rat pups. After induction of cataract, the eye lenses depicted a diminished chaperone activity which was screened in terms of heat-induced aggregation, DTT aggregation assay, and refolding attributes of β-crystallin [152].

Curcumin is a bright yellow colored oil-soluble principal component of curcuminods; it is considered as the major element of food ingredient and Indian Ayurvedic medicine. Many preclinical studies have proven that curcumin elicit pronounced antitumor, antioxidant, anti-inflammatory, antidiabetic, and antimicrobial properties commonly used by rural peoples in the prevention and treatment of several diseases [153]. Extensive studies have been recorded on the anticataractogenic activity of curcumin on sodium selenite-induced rat model. Manikandan et al. [117] reported that the generation of free radicals such as O_2^-, OH^-, and NO are the major causative agents that turn off the antioxidant defense mechanism. Hence the deterioration of respiratory inhibitors indicates a jeopardized normal physiology of the lens metabolism. Pretreatment with curcumin efficiently suppressed the ROS generated when induced with selenium when compared to simultaneous and posttreatment strategies. Administration with curcumin also conserved the antioxidant status of the liver, kidney, and serum [118]. Sodium selenite induction impairs Ca^{2+}, Na^+/K^+ ATPases that contributes to the elevated calcium levels, activation of calpain proteases, and imbalance of electrolytes. However, the treatment with curcumin likely protects the active sites of ATPases.

The accretion of peroxynitrite validates the fact that sodium selenite not only kindles oxidative stress but evokes nitrosative stress as well. In reality, the inducible iNOS are more critical reactive nitrogen species since they are generated in large quantities and files massive assault on the thiol groups of cytosolic proteins. The iNOS generated can directly jeopardize the vulnerable thiol groups of lenticular protein. The administration of curcumin circumvents the nitrosative stress and protects the lens tissues from nitrosative stress [154]. However, the solubility and relative bioavailability recounts curcumin to be a poor candidate for therapeutic needs.

A group of researchers have derivatized curcumin with aldehydes like salicyladehyde and benzaldehyde by Knovenegal condensation to yield two derivatives of CD1 and CD2, respectively. These derivatives are effectual in subduing the shifts induced by sodium selenite and also sustained their activity of the antioxidant defense system. Besides the attenuation of cataractogenesis, both the derivatives displayed the fundamental antiradical scavenging property also. The presence of an additional phenolic OH group in CD1 is considered as a free radical quenching active center and accounts for the appreciable antioxidant property of the derivative [155].

Vitex negundo is a naturally occurring herb that exhibited anti-inflammatory, analgesic, and antihistamine properties. The flavonoid fraction incurred from *V. negundo* exhibits a wide range of lenticular protective mechanisms. Flavonoids are excellent scavengers of free radicals and can perform as a better antioxidant; this facet bestows the anticataractogenic potential of the FVN [156]. Reactive oxygen species developed in the lenticular tissue was probed by 2′, 7′-dichlorofluorescein diacetate (oxidant-sensitive probe). The measure of ROS generated was accelerated in the sodium selenite-induced lens tissue that reflected in the fluorescent intensity of oxidant probe. However, the flavonoid fraction of *V. negundo* actively scavenged the ROS generated as that of the standard antioxidant quercetin [138]. Free radical-mediated lipid peroxidation was inhibited by the FVN and further progression of damage to the vital lenticular proteins was suppressed. The loss of protein activity is in accordance to the fact that the calpains undergo rapid auto degradation in response to calcium overload [157]. *V. negundo* extract has the ability to positively modulate the antioxidant enzyme activities such as SOD, CAT, Gpx, and GR. GSH is the prime target of sodium selenite, which is almost depleted on induction of cataract. Luteolin, an active constituent present in *V. negundo* is a key factor that defended against ROS generation and conserved the GSH at near normal levels [138].

IR3G is identified as a bioactive component isolated from *Cochlospermum religiosum*. IR3G by the virtue of its inherent antioxidant acquit protected the lens protein from selenite cataractogenesis by maintaining the Ca^{2+}ATPase activity, preventing oxidative stress, accumulation of calcium, and prevention of lipid peroxidation [135].

Drevogenin D, an aglycone of volubiloside A has been isolated from *Dregea volubilis*. This compound testified potent activity in maintaining the calcium homeostasis and related calpain activation. In spite of selenite cataract induction, the Ca^{2+} ATPase were maintained at near normal levels. Substrate gels testified and attested the casenolytic function of calpain and Lp82 were conserved on treating the lenses with appropriate dosage of Drevogenin D [158]. The triterpenoid acts as an equipotent antioxidant by critically conserving the sulfhydryl oxidation of Ca^{2+} ATPase thereby averting the dysfunction of the calcium homeostatic machinery [120].

Proanthocyanidin extracted from grape seed exhibit a broad range of pharmacological and therapeutic activities against free radicals both in in vitro and in vivo models. In corroboration with the earlier findings, the proanthocyanidin extracted from grape seeds inhibit the calpain activation caused due to selenite and have shown to protect the antioxidant enzyme modules, which has the ability to protect against selenite-induced cataractogenesis proven by histomorphology study [142]. According to Zhang and Hu [142], the proanthocyanidin inhibits the incidence of cataract in dose-dependent manner by the reduction in the generation of NO and iNOS.

Apricots contain carotene which protects the cellular components from oxidative damage. The other utilitarian components of apricot are Vitamin E, Vitamin C, isocloregonic acid, caffeic acid, 4-methyl catechol, chlorogenic acid, catechin, epicatechin, catechol, flavonols, and p-coumaric acid [159]. Investigations on the anticataractogenic effect of apricots in a selenite model revealed its positive modulatory effect on the lens morphology, antioxidant enzymes, and total nitrite levels [111]. The data substantiates the antioxidant potential of the components present in apricots which synergistically promotes its anti-cataract potency.

16.4.1.3 Synthetic Drugs

Chemically synthesized compounds pose on inherent ability to block the active sites of key enzymes and proteins. Recent evidences have suggested the potent role of NAC has excellent singlet oxygen scavenging, anti-peroxidase activity leading to delay the onset of age-related cataract. NAC was established to penetrate the cornea and maintain a longer therapeutic concentration after application as ophthalmic suspension [160, 161]. Clinical interventions on cataracteous patients cleared the visual opacities and conserved visual functions [162, 163]. Taking the intracellular buffering, sustained release of carnosine of NAC into account, it can be contemplated that L-carnosine released from NAC is an antiglycating antioxidant that potentially evades lipid peroxides, singlet oxygen, hydroxyl radicals thereby preventing opacities at the initial event of cataractogenesis.

As mentioned earlier aldolase reductase is the rate limiting enzyme in the polyol pathway demonstrated to lead the accumulation of sorbitol. Researchers have focused on in silico-based drug designing complementary to the active site of this particular enzyme. Aldose reductase inhibitors restrain glycation-related protein modification that can substantially delay or prevent the onset and development of diabetic complications. Calpain, a calcium-dependent cysteine protease is activated on loss in calcium homeostasis, whose over activation is entailed in cleavage of preferred calpain substrates such as cytoskeletal protein and regulatory protein. Exogenous administration of calpain inhibitors are being investigated for their potent part in averting cataract. Macrocyclic aldehydes, CAT811 [164], CAT505, and CAT0059 [165], have been demonstrated to elicit efficient activity against ovine-inherited cataract model. SJA6017, another candidate of the calpain inhibitor cadre, has been proven as a potent inhibitor by in silico approach and animal experimentation [166]. These inhibitors can be applied as alternative drug to prevent/therapy for cataract without surgery.

16.4.2   Management of Cataract

Comprehensive assessment of prevalence studies, blind registries, and records on cataract surgical rate has proven that cataract remains the prime cause of blindness worldwide. However, as of now there is no drug treatment for cataract, but sight can be restored by the surgical removal of the opacified lens and implantation of an intraocular lens. Promising therapy, that can prevent or delay the formation of cataract, could probably unleash the burden on surgical eye care. Anecdotal evidence and randomized clinical trials have pondered on the channelization of modifiable risk factors especially smoking, alcoholism, diabetes, administration of steroids, and UV exposure. Age-related cataract can be addressed by medical, environmental, and nutritional approaches. Numerous scientific evidences suggest that daily intake of balanced food constituting low calorific value, fruits, vegetables, whole grains, nuts, and legumes; polysaturated fats and regular walking practices may delay the formation of cataract during life span [167].

16.4.2.1 Cataract Surgery

Predominantly surgeons suggest surgical removal of the cataracteous lens as the appropriate and the only possible treatment of cataract. Cataract surgery is indicated/recommended for patients when the visual ability reduces the overall function and quality of life. Early detection and surgery, optical rehabilitation, and close follow-up are essential for good outcome. Cataract extraction surgery is performed in two modes, ECCE and ICCE/phacoemulsification. Following cataract extraction visual acuity of the patient mainly relies on the type and method of intraocular lens implanted. The propensity for postoperative inflammation, fibrin reaction, infections, capsular opacification, endophthalmitis, glaucoma, retinal detachment, and vitreous hemorrhage is well pronounced in surgery. Pristine extraction of cataract is wholly bestowed to the skill of the surgeon. Prevention of cataract is an ideal approach in developing and third world countries, due to inefficient cataract surgery and additional issues such as medical accessibility, lack of technical experts and sophisticated instrumentation, high surgical cost, postsurgical problems, and indeterminate length of waiting time. Hence, a basic understanding of the etiology of cataractogenesis, effective health-related studies, public awareness, suitable alternative preventive/therapeutic drug development to prevent cataract or delay the cataract development is most essential.

Acknowledgements The authors kindly acknowledge the financial assistance from DST PURSE and INSPIRE in the form of fellowship to first and third authors (DST/INSPIRE Fellowship/2010(129)), respectively. All the authors also gratefully acknowledge the financial support granted by the funding agencies, DST-YS (SR/FT/LS-153/2008), DST-JSPS (DST/INT/JSPS/P-119/2011).

References

1. Murthy G, Gupta SK, John N, Vashist P. Current status of cataract blindness and Vision 2020: the right to sight initiative in India. Indian J Ophthalmol. 2008;56:489–94.
2. Brar GS, Ram J, Pandav SS, Reddy GS, Singh U, Gupta A. Postoperative complications and visual results in uniocular pediatric traumatic cataract. Ophthalmic Surg Lasers. 2001;32:233–8.
3. Spector A. Oxidative stress-induced cataract: mechanism of action. FASEB J. 1995;9: 1173–82.
4. Takemoto L, Sorensen CM. Protein-protein interactions and lens transparency. Exp Eye Res. 2008;87:496–501.
5. Numoto N, Kita A, Fujii N, Miki K. A P39R mutation at the N-terminal domain of human αB-crystallin regulates its oligomeric state and chaperone-like activity. Biochem Biophys Res Commun. 2012;425:601–6.
6. Shimizu K, Kita A, Fujii N, Miki K. Structural features of isomerizable aspartyl residues in human alpha-crystallins. Mol Vis. 2012;18:1823–7.
7. Fujii N, Kawaguchi T, Sasaki H. Simultaneous stereoinversion and isomerization at the Asp-4 residue in betaB2-crystallin from the aged human eye lenses. Biochemistry. 2011;50: 8628–35.
8. Michael R, Bron AJ. The ageing lens and cataract: a model of normal and pathological ageing. Philos Trans R Soc Lond B Biol Sci. 2011;366:1278–92.

9. Santana A, Waiswo M. The genetic and molecular basis of congenital cataract. Arq Bras Oftalmol. 2011;74:136–42.
10. Garner MH, Roy D, Rosenfeld L, Garner WH, Spector A. Biochemical evidence for membrane disintegration in human cataracts. Proc Natl Acad Sci. 1981;78:1892–5.
11. Slingsby C, Wistow GJ, Clark AR. Evolution of crystallins for a role in the vertebrate eye lens. Protein Sci. 2013;22:367–80.
12. Andley UP. Effects of alpha-crystallin on lens cell function and cataract pathology. Curr Mol Med. 2009;9:887–92.
13. Abraham EC, Huaqian J, Aziz A, Kumarasamy A, Datta P. Role of the specifically targeted lysine residues in the glycation dependent loss of chaperone activity of alpha A- and alpha B-crystallins. Mol Cell Biochem. 2008;310:235–9.
14. Fu L, Liang JJ. Detection of protein-protein interactions among lens crystallins in a mammalian two-hybrid system assay. J Biol Chem. 2002;277:4255–60.
15. Liu BF, Anbarasu K, Liang JJ. Confocal fluorescence resonance energy transfer microscopy study of protein-protein interactions of lens crystallins in living cells. Mol Vis. 2007;13:854–61.
16. Kumarasamy A, Abraham EC. Interaction of C-terminal truncated human alphaA-crystallins with target proteins. PLoS One. 2008;3:3175.
17. Ponce A, Sorensen C, Takemoto L. Role of short-range protein interactions in lens opacifications. Mol Vis. 2006;12:879–84.
18. Raju I, Kumarasamy A, Abraham EC. Multiple aggregates and aggresomes of C-terminal truncated human alpha A-crystallins in mammalian cells and protection by alpha B-crystallin. PLoS One. 2011;6:e19876.
19. Song S, Hanson MJ, et al. Protein-protein interactions between lens vimentin and alphaB-crystallin using FRET acceptor photobleaching. Mol Vis. 2008;14:1282–7.
20. Mohr BG, Dobson CM, Garman SC, Muthukumar M. Electrostatic origin of in vitro aggregation of human gamma-crystallin. J Chem Phys. 2013;139:121914.
21. Cotlier E, Sharma YR. Aspirin and senile cataracts in rheumatoid arthritis. Lancet. 1981;1:338–9.
22. Kusak JR, Brown HG. Embryology and anatomy of the lens. In: Albert DM, Jakobiec FA, editors. Principles and practice of ophthalmology: basic sciences. Philadelphia: WB Saunders; 1994.
23. Lynnerup N, Kjeldsen H, Heegaard S, Jacobsen C, Heinemeier J. Radiocarbon dating of the human eye lens crystallines reveal proteins without carbon turnover throughout life. PLoS One. 2008;3:e1529.
24. Berman ER. Biochemistry of the eye. New York: Springer; 1991.
25. Bloemendal H. The vertebrate eye lens. Science. 1977;197:127–38.
26. Harrington V, Srivastava OP, Kirk M. Proteomic analysis of water-insoluble proteins from normal and cataractous human lenses. Mol Vis. 2007;13:1680–94.
27. Srivastava OP, Kirk MC, Srivastava K. Characterization of covalent multimers of crystallins in aging human lenses. J Biol Chem 2004;279:10901–9.
28. Horwitz J. Alpha-crystallin can function as a molecular chaperone. Proc Natl Acad Sci. 1992;89:10449–53.
29. Xu F, Yu H, Liu J, Cheng L. Alphab-crystallin regulates oxidative stress-induced apoptosis in cardiac H9c2 cells via the PI3K/AKT pathway. Mol Biol Rep. 2013;40:2517–26.
30. Rao CM, Raman B, Ramakrishna T, Rajaraman K, Ghosh D, Datta S, Trivedi VD, Sukhaswami MB. Structural perturbation of alpha-crystallin and its chaperone-like activity. Int J Biol Macromol. 1998;22:271–81.
31. Lin SY, Ho CJ, Li MJ. Thermal stability and reversibility of secondary conformation of alpha-crystallin membrane during repeated heating processes. Biophys Chem. 1998;74:1–10.
32. Fujii N, Kaji Y, Nakamura T, Motoie R, Mori Y, Kinouchi T. Collapse of homochirality of amino acids in proteins from various tissues during aging. Chem Biodivers. 2010;7:1389–97.

33. Fujii N, Satoh K, Harada K, Ishibashi Y. Simultaneous stereoinversion and isomerization at specific aspartic acid residues in alpha A-crystallin from human lens. J Biochem. 1994;116:663–9.
34. Mackay DS, Andley UP, Shiels A. Cell death triggered by a novel mutation in the alphaA-crystallin gene underlies autosomal dominant cataract linked to chromosome 21q. Eur J Hum Genet. 2003;11:784–93.
35. Foster A, Gilbert C, Rahi J. Epidemiology of cataract in childhood: a global perspective. J Cataract Refract Surg. 1997;1:601–4.
36. Rahi JS, Sripathi S, Gilbert CE, Foster A. Childhood blindness in India: causes in 1318 blind school students in nine states. Eye (Lond). 1995;9:545–50.
37. Eckstein M, Vijayalakshmi P, Killedar M, Gilbert C, Foster A. Aetiology of childhood cataract in south India. Br J Ophthalmol. 1996;80:628–32.
38. He W, Li S. Congenital cataracts: gene mapping. Hum Genet. 2000;106:1–13.
39. Messina-Baas OM, Gonzalez-Huerta LM, Cuevas-Covarrubias SA. Two affected siblings with nuclear cataract associated with a novel missense mutation in the CRYGD gene. Mol Vis. 2006;12:995–1000.
40. Litt M, Kramer P, LaMorticella DM, Murphey W, Lovrien EW, Weleber RG. Autosomal dominant congenital cataract associated with a missense mutation in the human alpha crystallin gene CRYAA. Hum Mol Genet. 1998;7:471–4.
41. Berry V, Francis P, Reddy MA, Collyer D, Vithana E, MacKay I, Dawson G, Carey AH, Moore A, Bhattacharya SS, Quinlan RA. Alpha-B crystallin gene (CRYAB) mutation causes dominant congenital posterior polar cataract in humans. Am J Hum Genet. 2001;69:1141–5.
42. Kannabiran C, Rogan PK, Olmos L, Basti S, Rao GN, Kaiser-Kupfer M, Hejtmancik JF. Autosomal dominant zonular cataract with sutural opacities is associated with a splice mutation in the betaA3/A1-crystallin gene. Mol Vis. 1998;4:21.
43. Billingsley G, Santhiya ST, Paterson AD, Ogata K, Wodak S, Hosseini SM, Manisastry SM, Vijayalakshmi P, Gopinath PM, Graw J, Heon E. CRYBA4, a novel human cataract gene, is also involved in microphthalmia. Am J Hum Genet. 2006;79:702–9.
44. Mackay DS, Boskovska OB, Knopf HL, Lampi KJ, Shiels A. A nonsense mutation in CRYBB1 associated with autosomal dominant cataract linked to human chromosome 22q. Am J Hum Genet. 2002;71:1216–21.
45. Litt M, Carrero-Valenzuela R, LaMorticella DM, Schultz DW, Mitchell TN, Kramer P, Maumenee IH. Autosomal dominant cerulean cataract is associated with a chain termination mutation in the human beta-crystallin gene CRYBB2. Hum Mol Genet. 1997;6:665–8.
46. Riazuddin SA, Yasmeen A, Yao W, Sergeev YV, Zhang Q, Zulfiqar F, Riaz A, Riazuddin S, Hejtmancik JF. Mutations in betaB3-crystallin associated with autosomal recessive cataract in two Pakistani families. Invest Ophthalmol Vis Sci. 2005;46:2100–6.
47. Heon E, Priston M, Schorderet DF, Billingsley GD, Girard PO, Lubsen N, Munier FL. The gamma-crystallins and human cataracts: a puzzle made clearer. Am J Hum Genet. 1999;65:1261–7.
48. Sun H, Ma Z, Li Y, Liu B, Li Z, Ding X, Gao Y, Ma W, Tang X, Li X, Shen Y. Gamma-S crystallin gene (CRYGS) mutation causes dominant progressive cortical cataract in humans. J Med Genet. 2005;42:706–10.
49. Ramachandran RD, Perumalsamy V, Hejtmancik JF. Autosomal recessive juvenile onset cataract associated with mutation in BFSP1. Hum Genet. 2007;121:475–82.
50. Conley YP, Erturk D, Keverline A, Mah TS, Keravala A, Barnes LR, Bruchis A, Hess JF, FitzGerald PG, Weeks DE, Ferrell RE, Gorin MB. A juvenile-onset, progressive cataract locus on chromosome 3q21-q22 is associated with a missense mutation in the beaded filament structural protein-2. Am J Hum Genet. 2000;66:1426–31.
51. Rees MI, Watts P, Fenton I, Clarke A, Snell RG, Owen MJ, Gray J. Further evidence of autosomal dominant congenital zonular pulverulent cataracts linked to 13q11 (CZP3) and a novel mutation in connexin 46 (GJA3). Hum Genet. 2000;106:206–9.

52. Shiels A, Mackay D, Ionides A, Berry V, Moore A, Bhattacharya S. A missense mutation in the human connexin50 gene (GJA8) underlies autosomal dominant "zonular pulverulent" cataract, on chromosome 1q. Am J Hum Genet. 1998;62:526–32.

53. Berry V, Francis P, Kaushal S, Moore A, Bhattacharya S. Missense mutations in MIP underlie autosomal dominant 'polymorphic' and lamellar cataracts linked to 12q. Nat Genet. 2000;25:15–7.

54. Pras E, Levy-Nissenbaum E, Bakhan T, Lahat H, Assia E, Geffen-Carmi N, Frydman M, Goldman B. A missense mutation in the LIM2 gene is associated with autosomal recessive presenile cataract in an inbred Iraqi Jewish family. Am J Hum Genet. 2002;70:1363–7.

55. Bu L, Jin Y, Shi Y, Chu R, Ban A, Eiberg H, Andres L, Jiang H, Zheng G, Qian M, Cui B, Xia Y, Liu J, Hu L, Zhao G, Hayden MR, Kong X. Mutant DNA-binding domain of HSF4 is associated with autosomal dominant lamellar and Marner cataract. Nat Genet. 2002;31:276–8.

56. Semina EV, Ferrell RE, Mintz-Hittner HA, Bitoun P, Alward WL, Reiter RS, Funkhauser C, Daack-Hirsch S, Murray JC. A novel homeobox gene PITX3 is mutated in families with autosomal-dominant cataracts and ASMD. Nat Genet. 1998;19:167–70.

57. Vanita V, Singh D, Robinson PN, Sperling K, Singh JR. A novel mutation in the DNA-binding domain of MAF at 16q23.1 associated with autosomal dominant "cerulean cataract" in an Indian family. Am J Med Genet A. 2006;140:558–66.

58. Pras E, Raz J, Yahalom V, Frydman M, Garzozi HJ, Hejtmancik JF. A nonsense mutation in the glucosaminyl (N-acetyl) transferase 2 gene (GCNT2): association with autosomal recessive congenital cataracts. Invest Ophthalmol Vis Sci. 2004;45:1940–5.

59. Shiels A, Bennett TM, Knopf HL, Yamada K, Yoshiura K, Niikawa N, Shim S, Hanson PI. CHMP4B, a novel gene for autosomal dominant cataracts linked to chromosome 20q. Am J Hum Genet. 2007;81:596–606.

60. Jamieson RV, Farrar N, Stewart K, Perveen R, Mihelec M, Carette M, Grigg JR, McAvoy JW, Lovicu FJ, Tam PP, Scambler P, Lloyd IC, Donnai D, Black GC. Characterization of a familial t(16;22) balanced translocation associated with congenital cataract leads to identification of a novel gene, TMEM114, expressed in the lens and disrupted by the translocation. Hum Mutat. 2007;28:968–77.

61. Pras E, Frydman M, Levy-Nissenbaum E, Bakhan T, Raz J, Assia EI, Goldman B. A nonsense mutation (W9X) in CRYAA causes autosomal recessive cataract in an inbred Jewish Persian family. Invest Ophthalmol Vis Sci. 2000;41:3511–5.

62. Hansen L, Yao W, Eiberg H, Kjaer KW, Baggesen K, Hejtmancik JF, Rosenberg T. Genetic heterogeneity in microcornea-cataract: five novel mutations in CRYAA, CRYGD, and GJA8. Invest Ophthalmol Vis Sci. 2007;48:3937–44.

63. Graw J, Klopp N, Illig T, Preising MN, Lorenz B. Congenital cataract and macular hypoplasia in humans associated with a de novo mutation in CRYAA and compound heterozygous mutations in P. Graefes Arch Clin Exp Ophthalmol. 2006;244:912–9.

64. Khan AO, Abu Safieh L, Alkuraya FS. Later retinal degeneration following childhood surgical aphakia in a family with recessive CRYAB mutation (p.R56W). Ophthalmic Genet. 2010;31:30–6.

65. Bhagyalaxmi SG, Srinivas P, Barton KA, Kumar KR, Vidyavathi M, Petrash JM, Bhanuprakash Reddy G, Padma T. A novel mutation (F71L) in alphaA-crystallin with defective chaperone-like function associated with age-related cataract. Biochim Biophys Acta. 2009;1792:974–81.

66. Santhiya ST, Soker T, Klopp N, Illig T, Prakash MV, Selvaraj B, Gopinath PM, Graw J. Identification of a novel, putative cataract-causing allele in CRYAA (G98R) in an Indian family. Mol Vis. 2006;12:768–73.

67. Liu M, Ke T, Wang Z, Yang Q, Chang W, Jiang F, Tang Z, Li H, Ren X, Wang X, Wang T, Li Q, Yang J, Liu J, Wang QK. Identification of a CRYAB mutation associated with autosomal dominant posterior polar cataract in a Chinese family. Invest Ophthalmol Vis Sci. 2006;47:3461–6.

68. Liu Y, Zhang X, Luo L, Wu M, Zeng R, Cheng G, Hu B, Liu B, Liang JJ, Shang F. A novel alpha B-crystallin mutation associated with autosomal dominant congenital lamellar cataract. Invest Ophthalmol Vis Sci. 2006;47:1069–75.
69. Chen Q, Ma J, Yan M, Mothobi ME, Liu Y, Zheng F. A novel mutation in CRYAB associated with autosomal dominant congenital nuclear cataract in a Chinese family. Mol Vis. 2009;15:1359–65.
70. Safieh LA, Khan AO, Alkuraya FS. Identification of a novel CRYAB mutation associated with autosomal recessive juvenile cataract in a Saudi family. Mol Vis. 2009;15:980–4.
71. Devi RR, Yao W, Vijayalakshmi P, Sergeev YV, Sundaresan P, Hejtmancik JF. Crystallin gene mutations in Indian families with inherited pediatric cataract. Mol Vis. 2008; 14:1157–70.
72. Sacconi S, Feasson L, Antoine JC, Pecheux C, Bernard R, Cobo AM, Casarin A, Salviati L, Desnuelle C, Urtizberea A. A novel CRYAB mutation resulting in multisystemic disease. Neuromuscul Disord. 2012;22:66–72.
73. Pauli S, Soker T, Klopp N, Illig T, Engel W, Graw J. Mutation analysis in a German family identified a new cataract-causing allele in the CRYBB2 gene. Mol Vis. 2007;13:962–7.
74. Mothobi ME, Guo S, Liu Y, Chen Q, Yussuf AS, Zhu X, Fang Z. Mutation analysis of congenital cataract in a Basotho family identified a new missense allele in CRYBB2. Mol Vis. 2009;15:1470–5.
75. Li FF, Zhu SQ, Wang SZ, Gao C, Huang SZ, Zhang M, Ma X. Nonsense mutation in the CRYBB2 gene causing autosomal dominant progressive polymorphic congenital coronary cataracts. Mol Vis. 2008;14:750–5.
76. Wang J, Ma X, Gu F, Liu NP, Hao XL, Wang KJ, Wang NL, Zhu SQ. A missense mutation S228P in the CRYBB1 gene causes autosomal dominant congenital cataract. Chin Med J (Engl). 2007;120:820–4.
77. Willoughby CE, Shafiq A, Ferrini W, Chan LL, Billingsley G, Priston M, Mok C, Chandna A, Kaye S, Heon E. CRYBB1 mutation associated with congenital cataract and microcornea. Mol Vis. 2005;11:587–93.
78. Yang J, Zhu Y, Gu F, He X, Cao Z, Li X, Tong Y, Ma X. A novel nonsense mutation in CRYBB1 associated with autosomal dominant congenital cataract. Mol Vis. 2008;14: 727–31.
79. Qi Y, Jia H, Huang S, Lin H, Gu J, Su H, Zhang T, Gao Y, Qu L, Li D, Li Y. A deletion mutation in the betaA1/A3 crystallin gene (CRYBA1/A3) is associated with autosomal dominant congenital nuclear cataract in a Chinese family. Hum Genet. 2004;114:192–7.
80. Ferrini W, Schorderet DF, Othenin-Girard P, Uffer S, Heon E, Munier FL. CRYBA3/A1 gene mutation associated with suture-sparing autosomal dominant congenital nuclear cataract: a novel phenotype. Invest Ophthalmol Vis Sci. 2004;45:1436–41.
81. Yao K, Jin C, Zhu N, Wang W, Wu R, Jiang J, Shentu X. A nonsense mutation in CRYGC associated with autosomal dominant congenital nuclear cataract in a Chinese family. Mol Vis. 2008;14:1272–6.
82. Zhang LY, Gong B, Tong JP, Fan DS, Chiang SW, Lou D, Lam DS, Yam GH, Pang CP. A novel gammaD-crystallin mutation causes mild changes in protein properties but leads to congenital coralliform cataract. Mol Vis. 2009;15:1521–9.
83. Mackay DS, Andley UP, Shiels A. A missense mutation in the gammaD crystallin gene (CRYGD) associated with autosomal dominant "coral-like" cataract linked to chromosome 2q. Mol Vis. 2004;10:155–62.
84. Gu F, Li R, Ma XX, Shi LS, Huang SZ, Ma X. A missense mutation in the gamma D-crystallin gene CRYGD associated with autosomal dominant congenital cataract in a Chinese family. Mol Vis. 2006;12:26–31.
85. Zenteno JC, Morales ME, Moran-Barroso V, Sanchez-Navarro A. CRYGD gene analysis in a family with autosomal dominant congenital cataract: evidence for molecular homogeneity and intrafamilial clinical heterogeneity in aculeiform cataract. Mol Vis. 2005;11:438–42.

86. Kmoch S, Brynda J, Asfaw B, Bezouska K, Novak P, Rezacova P, Ondrova L, Filipec M, Sedlacek J, Elleder M. Link between a novel human gammaD-crystallin allele and a unique cataract phenotype explained by protein crystallography. Hum Mol Genet. 2000;9:1779–86.
87. Kisic B, Miric D, Zoric L, Ilic A, Dragojevic I. Antioxidant capacity of lenses with age-related cataract. Oxid Med Cell Longev. 2012;2012:467130.
88. Tombran-Tink J, Barnstable CJ. Ocular transporters in ophthalmic diseases and drug delivery. Totowa: Humana; 2008.
89. Muthenna P, Akileshwari C, Saraswat M, Bhanuprakash Reddy G. Inhibition of advanced glycation end-product formation on eye lens protein by rutin. Br J Nutr. 2012;107:941–9.
90. Lansingh VC, Carter MJ, Martens M. Global cost-effectiveness of cataract surgery. Ophthalmology. 2007;114:1670–8.
91. Lamoureux EL, Fenwick E, Pesudovs K, Tan D. The impact of cataract surgery on quality of life. Curr Opin Ophthalmol. 2011;22:19–27.
92. Ponnam SP, Ramesha K, Matalia J, Tejwani S, Ramamurthy B, Kannabiran C. Mutational screening of Indian families with hereditary congenital cataract. Mol Vis. 2013;19:1141–8.
93. Anbarasu K, Ramkumar S, Thankappan B, Abraham EC. S1–3 Truncations in αA-crystallins: molecular basis for senile cataract (Symposium 1, The 37th Annual Meeting of the SSOEL-Japan). Viva Origino. 2012;40:9.
94. Anbarasu K, Ramkumar S, Thankappan B, Abraham EC. Towards to clarify the etiology of cataract by crystallin protein-protein interaction studies. J Jpn Soc Cataract Res. 2013;25:65–74.
95. Jiang JX, Goodenough DA. Heteromeric connexons in lens gap junction channels. Proc Natl Acad Sci U S A. 1996;93:1287–91.
96. Halliwell B. Biochemistry of oxidative stress. Biochem Soc Trans. 2007;35:1147–50.
97. Varma SD, Chand D, Sharma YR, Kuck JF, Richards RD. Oxidative stress on lens and cataract formation: role of light and oxygen. Curr Eye Res. 1984;3:35–57.
98. Hayes JD, McLellan LI. Glutathione and glutathione-dependent enzymes represent a co-ordinately regulated defence against oxidative stress. Free Radic Res. 1999;31:273–300.
99. Hightower K, McCready J. Selenite-induced damage to lens membranes. Exp Eye Res. 1994;58:225–9.
100. Wang Z, Bunce GE, Hess JL. Selenite and Ca^{2+} homeostasis in the rat lens: effect on Ca-ATPase and passive Ca^{2+} transport. Curr Eye Res. 1993;12:213–8.
101. Wang Z, Hess JL, Bunce GE. Calcium efflux in rat lens: Na/Ca-exchange related to cataract induced by selenite. Curr Eye Res. 1992;11:625–32.
102. Paterson CA, Delamere NA. ATPases and lens ion balance. Exp Eye Res. 2004;78: 699–703.
103. Hollenbach S, Thampi P, Viswanathan T, Abraham EC. Cleavage of in vitro and in vivo formed lens protein cross-links by a novel cross-link breaker. Mol Cell Biochem. 2003;243: 73–80.
104. Rodriguez-Rodero S, Fernandez-Morera JL, Menendez-Torre E, Calvanese V, Fernandez AF, Fraga MF. Aging genetics and aging. Aging Dis. 2011;2:186–95.
105. Kovacic P, Somanathan R. Mechanism of anesthetic toxicity: metabolism, reactive oxygen species, oxidative stress, and electron transfer. ISRN Anesthesiol. 2011;2011:10.
106. Delcourt C. Application of nutrigenomics in eye health. Forum Nutr. 2007;60:168–75.
107. Shichi H. Cataract formation and prevention. Expert Opin Investig Drugs. 2004;13: 691–701.
108. Hosseini HRJ, Aminlari M, Khalili MR. Prevention of selenite-induced cataract by L-cysteine and vitamin C in rats. Iran Red Crescent Med J. 2008;10:281–7.
109. Elanchezhian R, Sakthivel M, Isai M, Geraldine P, Thomas PA. Evaluation of lenticular antioxidant and redox system components in the lenses of acetyl-L-carnitine treatment in BSO-induced glutathione deprivation. Mol Vis. 2009;15:1485–91.
110. Muralidharan AR, Leema G, Annadurai T, Anitha TS, Thomas PA, Geraldine P. Deciphering the potential efficacy of acetyl-L-carnitine (ALCAR) in maintaining connexin-mediated lenticular homeostasis. Mol Vis. 2012;18:2076–86.

111. Doganay S, Duz C, Firat PG, Cankaya C, Kutukde D, Cigremis Y. The effect of apricots on the experimental cataract model formed by sodium selenite. Food Chem Toxicol. 2013;55:371–7.

112. Nair NK, Patel K, Gandhi T. Effect of aqueous extract of *Embelica officinalis* on selenite induced cataract in rats. Iran J Pharm Res. 2010;9:147–52.

113. Javadzadeh A, Ghorbanihaghjo A, Bonyadi S, Rashidi MR, Mesgari M, Rashtchizadeh N, Argani H. Preventive effect of onion juice on selenite-induced experimental cataract. Indian J Ophthalmol. 2009;57:185–9.

114. Liao JH, Chen CS, Maher TJ, Liu CY, Lin MH, Wu TH, Wu SH. Astaxanthin interacts with selenite and attenuates selenite-induced cataractogenesis. Chem Res Toxicol. 2009;22:518–25.

115. Varma SD, Hegde KR, Kovtun S. Inhibition of selenite-induced cataract by caffeine. Acta Ophthalmol. 2010;88:e245–9.

116. Kumari RP, Sivakumar J, Thankappan B, Anbarasu K. C-phycocyanin modulates selenite-induced cataractogenesis in rats. Biol Trace Elem Res. 2013;151:59–67.

117. Manikandan R, Thiagarajan R, Beulaja S, Sudhandiran G, Arumugam M. Curcumin prevents free radical-mediated cataractogenesis through modulations in lens calcium. Free Radic Biol Med. 2010;48:483–92.

118. Manikandan R, Thiagarajan R, Beulaja S, Sudhandiran G, Arumugam M. Effect of curcumin on selenite-induced cataractogenesis in wistar rat pups. Curr Eye Res. 2010;35:122–9.

119. Qi HP, Wei SQ, Zhang LQ, Gao XC, Yu NN, Bi S, Cui H. Preventive effect of danshensu on selenite-induced cataractogenesis in cultured rat lens. Clin Experiment Ophthalmol. 2013;41:172–9.

120. Biju PG, Rooban BN, Lija Y, Devi VG, Sahasranamam V, Abraham A. Drevogenin D prevents selenite-induced oxidative stress and calpain activation in cultured rat lens. Mol Vis. 2007;13:1121–9.

121. Sakthivel M, Elanchezhian R, Ramesh E, Isai M, Jesudasan CN, Thomas PA, Geraldine P. Prevention of selenite-induced cataractogenesis in Wistar rats by the polyphenol, ellagic acid. Exp Eye Res. 2008;86:251–9.

122. Lija Y, Biju PG, Reeni A, Cibin TR, Sahasranamam V, Abraham A. Modulation of selenite cataract by the flavonoid fraction of *Emilia sonchifolia* in experimental animal models. Phytother Res. 2006;20:1091–5.

123. Gupta SK, Halder N, Srivastava S, Trivedi D, Joshi S, Varma SD. Green tea (*Camellia sinensis*) protects against selenite-induced oxidative stress in experimental cataractogenesis. Ophthalmic Res. 2002;34:258–63.

124. Anitha TS, Annadurai T, Thomas PA, Geraldine P. Prevention of selenite-induced cataractogenesis by an ethanolic extract of *Cineraria maritima*: an experimental evaluation of the traditional eye medication. Biol Trace Elem Res. 2011;143:425–36.

125. Isai M, Elanchezhian R, Sakthivel M, Chinnakkaruppan A, Rajamohan M, Jesudasan CN, Thomas PA, Geraldine P. Anticataractogenic effect of an extract of the Oyster mushroom, *Pleurotus ostreatus*, in an experimental animal model. Curr Eye Res. 2009;34:264–73.

126. Gupta SK, Srivastava S, Trivedi D, Joshi S, Halder N. *Ocimum sanctum* modulates selenite-induced cataractogenic changes and prevents rat lens opacification. Curr Eye Res. 2005;30:583–91.

127. Thiagarajan G, Chandani S, Rao SH, Samuni AM, Chandrasekaran K, Balasubramanian D. Molecular and cellular assessment of *Ginkgo biloba* extract as a possible ophthalmic drug. Exp Eye Res. 2002;75:421–30.

128. Dailami KN, Azadbakht M, Pharm ZR, Lashgari M. Prevention of selenite-induced cataractogenesis by *Origanum vulgare* extract. Pak J Biol Sci. 2010;13:743–7.

129. Gupta SK, Kalaiselvan V, Srivastava S, Saxena R, Agrawal SS. *Trigonella foenum-graecum* (Fenugreek) protects against selenite-induced oxidative stress in experimental cataractogenesis. Biol Trace Elem Res. 2010;136:258–68.

130. Sasikala V, Rooban BN, Priya SG, Sahasranamam V, Abraham A. *Moringa oleifera* prevents selenite-induced cataractogenesis in rat pups. J Ocul Pharmacol Ther. 2010;26:441–7.

131. Mai TT, Yamaguchi K, Yamanaka M, Lam NT, Otsuka Y, Chuyen NV. Protective and anti-cataract effects of the aqueous extract of *Cleistocalyx operculatus* flower buds on beta-cells of streptozotocin-diabetic rats. J Agric Food Chem. 2010;58:4162–8.

132. Lee J, Jang DS, Kim NH, Lee YM, Kim J, Kim JS. Galloyl glucoses from the seeds of *Cornus officinalis* with inhibitory activity against protein glycation, aldose reductase, and cataracto-genesis ex vivo. Biol Pharm Bull. 2011;34:443–6.

133. Ganeshpurkar A, Bhadoriya SS, Pardhi P, Jain AP, Rai G. In vitro prevention of cataract by Oyster Mushroom *Pleurotus florida* extract on isolated goat eye lens. Indian J Pharmacol. 2011;43:667–70.

134. Patel MB, Mishra S. Isoquinoline alkaloids from *Tinospora cordifolia* inhibit rat lens aldose reductase. Phytother Res. 2012;26:1342–7.

135. Devi VG, Rooban BN, Sasikala V, Sahasranamam V, Abraham A. Isorhamnetin-3-glucoside alleviates oxidative stress and opacification in selenite cataract in vitro. Toxicol In Vitro. 2010;24:1662–9.

136. Ajani EO, Salako AA, Sharlie PD, Akinleye WA, Adeoye AO, Salau BA, Adebawo OO. Chemopreventive and remediation effect of Hydrocotyl bonariensis Comm. Ex Lam (Apiaceae) leave extract in galactose-induced cataract. J Ethnopharmacol. 2009;123: 134–42.

137. Wang T, Zhang P, Zhao C, Zhang Y, Liu H, Hu L, Gao X, Zhang D. Prevention effect in selenite-induced cataract in vivo and antioxidative effects in vitro of *Crataegus pinnatifida* leaves. Biol Trace Elem Res. 2011;142:106–16.

138. Rooban BN, Sasikala V, Devi VG, Sahasranamam V, Abraham A. Prevention of selenite induced oxidative stress and cataractogenesis by luteolin isolated from *Vitex negundo*. Chem Biol Interact. 2012;196:30–8.

139. Gupta SK, Trivedi D, Srivastava S, Joshi S, Halder N, Verma SD. Lycopene attenuates oxida-tive stress induced experimental cataract development: an in vitro and in vivo study. Nutrition. 2003;19:794–9.

140. Kilic A, Selek S, Erel O, Aksoy N. Protective effects of melatonin on oxidative-antioxidative balance and cataract formation in rats. Ann Ophthalmol (Skokie). 2008;40:22–7.

141. Vibin M, Siva Priya SG, Sasikala BNR, Sahasranamam V, Abraham A. Broccoli regulates protein alterations and cataractogenesis in selenite models. Curr Eye Res. 2010;35:99–107.

142. Zhang X, Hu Y. Inhibitory effects of grape seed proanthocyanidin extract on selenite-induced cataract formation and possible mechanism. J Huazhong Univ Sci Technolog Med Sci. 2012;32:613–9.

143. Isai M, Sakthivel M, Ramesh E, Thomas PA, Geraldine P. Prevention of selenite-induced cataractogenesis by rutin in Wistar rats. Mol Vis. 2009;15:2570–7.

144. Makri OE, Ferlemi AV, Lamari FN, Georgakopoulos CD. Saffron administration prevents selenite-induced cataractogenesis. Mol Vis. 2013;19:1188–97.

145. Kim J, Kim CS, Lee YM, Sohn E, Jo K, Shin SD, Kim JS. Scopoletin inhibits rat aldose reductase activity and cataractogenesis in galactose-fed rats. Evid Based Complement Alternat Med. 2013;2013:8.

146. Chethan S, Dharmesh SM, Malleshi NG. Inhibition of aldose reductase from cataracted eye lenses by finger millet (*Eleusine coracana*) polyphenols. Bioorg Med Chem. 2008;16:10085–90.

147. Li N, Zhu YX, Deng XG, Gao Y, Zhu YG, He MF. Protective effects and mechanism of tetra-methylpyrazine against lens opacification induced by sodium selenite in rats. Exp Eye Res. 2011;93:98–102.

148. Gupta SK, Kalaiselvan V, Srivastava S, Agrawal SS, Saxena R. Evaluation of anticataract potential of Triphala in selenite-induced cataract: in vitro and in vivo studies. J Ayurveda Integr Med. 2010;1:280–6.

149. Saraswat M, Suryanarayana P, Reddy PY, Patil MA, Balakrishna N, Reddy GB. Antiglycating potential of Zingiber officinalis and delay of diabetic cataract in rats. Mol Vis. 2010;10: 1525–37.

150. Chung YS, Choi YH, Lee SJ, Choi S, Lee J, Kim H, Hong EK. Water extract of *Aralia elata* prevents cataractogenesis *in vitro* and *in vivo*. J Ethnopharmacol. 2005;101:49–54.

151. El-Sayyad HI, El-Sherbiny MA, Sobh MA, Abou-El-Naga AM, Ibrahim MA, Mousa SA. Protective effects of *Morus alba* leaves extract on ocular functions of pups from diabetic and hypercholesterolemic mother rats. Int J Biol Sci. 2011;7:715–28.
152. Sasikala V, Rooban BN, Sahasranamam V, Abraham A. Rutin ameliorates free radical mediated cataract by enhancing the chaperone activity of alpha-crystallin. Graefes Arch Clin Exp Ophthalmol. 2013;251:1747–55.
153. Ammon HP, Wahl MA. Pharmacology of Curcuma longa. Planta Med. 1991;57:1–7.
154. Manikandan R, Thiagarajan R, Beulaja S, Chindhu S, Mariammal K, Sudhandiran G, Arumugam M. Anti-cataractogenic effect of curcumin and aminoguanidine against selenium-induced oxidative stress in the eye lens of Wistar rat pups: an in vitro study using isolated lens. Chem Biol Interact. 2009;181:202–9.
155. Radha A, Rukhmini SD, Vilasini S, Sakunthala PR, Sreedharan B, Velayudhan MP, Abraham A. Bioactive derivatives of curcumin attenuate cataract formation in vitro. Chem Biol Drug Design. 2012;80:887–92.
156. Rooban BN, Lija Y, Biju PG, Sasikala V, Sahasranamam V, Abraham A. *Vitex negundo* attenuates calpain activation and cataractogenesis in selenite models. Exp Eye Res. 2009;88:575–82.
157. David LL, Shearer TR. Calcium-activated proteolysis in the lens nucleus during selenite cataractogenesis. Invest Ophthalmol Vis Sci. 1984;25:1275–83.
158. Biju PG, Devi VG, Lija Y, Abraham A. Protection against selenite cataract in rat lens by Drevogenin D, a triterpenoid aglycone from *Dregea volubilis*. J Med Food. 2007;10:308–15.
159. Korekar G, Stobdan T, Arora R, Yadav A, Singh SB. Antioxidant capacity and phenolics content of apricot (*Prunus armeniaca* L.) kernel as a function of genotype. Plant Foods Hum Nutr. 2011;66:376–83.
160. Babizhayev MA, Yermakova VN, Sakina NL, Evstigneeva RP, Rozhkova EA, Zheltukhina GA. *N* alpha-acetylcarnosine is a prodrug of L-carnosine in ophthalmic application as antioxidant. Clin Chim Acta. 1996;254:1–21.
161. Babizhayev MA, Yermakova VN, Semiletov YA, Deyev AI. The natural histidine-containing dipeptide *N*-alpha-acetylcarnosine as an antioxidant for ophthalmic use. Biochemistry (Mosc). 2000;65:588–98.
162. Babizhayev MA, Deyev AI, Yermakova VN, Semiletov YA, Davydova NG, Doroshenko VS, Zhukotskii AV, Goldman IM. Efficacy of *N*-acetylcarnosine in the treatment of cataracts. Drugs R D. 2002;3:87–103.
163. Babizhayev MA, Deyev AI, Yermakova VN, Semiletov YA, Davydova NG, Kurysheva NI, Zhukotskii AV, Goldman IM. *N*-acetylcarnosine, a natural histidine-containing dipeptide, as a potent ophthalmic drug in treatment of human cataracts. Peptides. 2001;22:979–94.
164. Morton JD, Lee HY, McDermott JD, Robertson LJ, Bickerstaffe R, Jones MA, Coxon JM, Abell AD. A macrocyclic calpain inhibitor slows the development of inherited cortical cataracts in a sheep model. Invest Ophthalmol Vis Sci. 2013;54:389–95.
165. Lee HY, Morton JD, Robertson LJ, McDermott JD, Bickerstaffe R, Abell AD, Jones MA, Mehrtens JM, Coxon JM. Evaluation of a novel calpain inhibitor as a treatment for cataract. Clin Experiment Ophthalmol. 2008;36:852–60.
166. Tamada Y, Fukiage C, Mizutani K, Yamaguchi M, Nakamura Y, Azuma M, Shearer TR. Calpain inhibitor, SJA6017, reduces the rate of formation of selenite cataract in rats. Curr Eye Res. 2001;22:280–5.
167. Everitt AV, Hilmer SN, Brand-Miller JC, Jamieson HA, Truswell AS, Sharma AP, Mason RS, Morris BJ, Le Couteur DG. Dietary approaches that delay age-related diseases. Clin Interv Aging. 2006;1:11–31.

Chapter 17
The Effects of Lutein in Preventing Cataract Progression

Rijo Hayashi

17.1 Introduction

Oxidation has been reported as one of the important causes of cataract formation. The decreases of antioxidative substances, such as Cu,Zn-superoxide dismutase (SOD), reduced glutathione (GSH), and L-ascorbic acid, have been reported in the cataractous lenses [1–3]. Antioxidants are presumed to be effective in the preventing cataracts. Several studies report the decreased prevalence of cataracts with dietary antioxidant nutrients, such as vitamin C [4–6] and vitamin E (Vit E, [7]).

17.2 General Information of Lutein

Among antioxidant nutrients, lutein is a noteworthy antioxidant that is proven to be effective in preventing age-related macular degeneration [8]. Also, lutein has been reported the decreased risk of cataract formation [9, 10]. Lutein and zeaxanthin are the only carotenoids that have been proven to exist in the lens [11] and they are not evenly distributed. The concentrations decrease from the epithelium to the nucleus [12]. These two xanthophylls are characterized by the presence of a hydroxyl group attached to each of the two terminal β-ionone rings in the molecule thus making them more hydrophilic than other carotenoids, such as α-carotene, β-carotene, and lycopene. The hydrophilic properties allow lutein and zeaxanthin to react with

R. Hayashi (✉)
Department of Ophthalmology, Koshigaya Hospital, Dokkyo Medical University,
Koshigaya, Saitama, Japan
e-mail: lhayashi@dokkyomed.ac.jp

© Springer Science+Business Media New York 2015
M.A. Babizhayev et al. (eds.), *Studies on the Cornea and Lens,*
Oxidative Stress in Applied Basic Research and Clinical Practice,
DOI 10.1007/978-1-4939-1935-2_17

singlet oxygen that is generated in the aqueous phase more efficiently than other nonpolar carotenoids [13].

Dietary lutein and zeaxanthin are absorbed by the intestinal mucosa, reach the circulating blood, and subsequently taken up by hepatocytes to be incorporated into other lipoproteins, such as low density lipoprotein, then transported to various tissues [14]. It has been reported that a single dietary dose of lutein reaches peak concentration in the serum at approximately 16-h post-ingestion [15] and the concentration in the serum reaches a consistent level after 1–3 months of dosage [16].

It has been reported that macular lutein concentration increases with dietary lutein supplementation [17] and serum lutein [18], which suggests the definite transmission for dietary lutein into ocular tissue. The uptake of carotenoids into the intestine and transfer to the eye is known to be highly variable [19]. There are several factors affecting the absorption of lutein, including age, body composition, and gender [20, 21].

Several in vitro studies have investigated and support the antioxidant effects of lutein. Lutein has been reported to decrease intracellular H_2O_2 accumulation by scavenging superoxide and H_2O_2 [22]. It has also been reported that supplementation with lutein can effectively block H_2O_2-induced protein oxidation, lipid peroxidation, and DNA damage in lens epithelial cells (LECs) [23]. It is reported that carotenoids are effective antioxidants, especially at a low partial pressure of oxygen such as in the lens [24, 25].

17.3   Epidemiological Studies on Relationship Between Cataract and Lutein

Several epidemiological studies have indicated that a decreased risk of cataracts is associated with a high dietary intake of lutein [7, 10, 26, 27]. Studies have also indicated the decrease of cataracts in patients with higher serum lutein [10, 27]. However, others studies did not reveal a statistically significant association between lutein and cataracts [28–31].

There are several factors that may influence the results of these epidemiological studies. First, some studies were cross-sectional [29, 30]. These studies investigated serum lutein levels and cataract severity simultaneously, which may mean that the cause–effect relationship is not completely comprehensive. Second, most of the previous epidemiological studies measured the amount of antioxidant nutrients contained in food in which the dosage of lutein varies. For example, the amount of lutein intake in the study by Vu et al. was less than half of that in other studies [9, 26]. Third, some studies investigated the frequency of cataract surgery [28] instead of the grades of cataract when surveying the effects of antioxidant nutrients. The performed cataract surgery depends on the individual demand, and the surgery may not accurately represent the severity of the cataract. Overall, most of the previous studies investigated the correlation between the grades of cataract and the intake amount or serum levels of antioxidant nutrients from different individuals. Confounding factors, which are also important causes of cataract formation, such as genetic factors, could interfere with the results when comparing the effects of nutrition among different individuals.

17.4 Our Studies on Lutein and Cataract

To control the possible confounding factors that may interfere the correlation between lutein and cataract described above, we conducted studies that investigated the anterior capsule from the same patients before and after intake of antioxidant supplement, Ocuvite + lutein® [32]. By measuring the anterior capsule of lens sampled from both eyes of the same patient, the individual factors, such as gene, ultraviolet irradiation, and nutrition, can be controlled. Also, the amount of nutrients can be controlled by taking from supplement. Most of all, the oxidative status in the anterior capsule of lens are good indicators for oxidation in lens. In our studies, we measured the mRNA expression of antioxidative enzymes in anterior capsule before and after intake of antioxidant supplement.

Measuring changes in the synthesis of antioxidative enzymes after taking a supplement may be important for verifying the effectiveness of antioxidant supplementation in cataract prevention. Because the first stage of protein synthesis is initiated by mRNA transcription, measuring changes in mRNA expression may indicate changes in the synthesis of certain proteins, such as antioxidative enzymes. Therefore, measuring changes in the mRNA expression levels of antioxidative enzymes in the lenticular anterior capsule, which mostly occur in LECs, may be important for verifying the effects of antioxidant supplementation on the oxidation in the lens, thus, on the prevention of cataracts.

In one of our previous studies, we measured the mRNA expression levels of glucose-6-phosphate dehydrogenase (G6PDH) and 18S rRNA (18S) in the anterior capsules of lenses from cataract patients before and after taking the antioxidant supplement Ocuvite + lutein®.

17.4.1 Methodology

The composition of the antioxidant supplement used in this study, Ocuvite + lutein®, is described in Table 17.1.

Thirteen patients, seven male patients (mean ± SD age, 69.9 ± 4.3 years) and six female patients (mean ± SD age, 79.8 ± 9.6 years) were included in the study. The informed consents were given according to the tenets of the Declaration of Helsinki. Approval from the institutional human experimentation committee was also granted. All of the patients had the same WHO classification and the same type of lens opacity

Table 17.1 The composition
of Ocuvite + lutein®

Substance	Amount
Lutein	6.0 mg
Vitamin C	300.0 mg
Vitamin E	60.0 mg
Vitamin B_2	3.0 mg
β-carotene	1,200.0 µg
Niacin	12.0 mg
Zinc	9.0 mg
Selenium	45.0 µg
Copper	0.6 mg
Manganese	1.5 mg

in both eyes. Patients with ocular complications, such as uveitis and retinopathy, were excluded. Patients with systemic diseases, such as diabetes, were also excluded.

Central pieces of the anterior capsule measuring 5 mm in diameter were collected before and after supplementation as pre- and post-intake samples continuous curvilinear capsulorhexis during cataract surgery. Three tablets (the recommended daily dosage) of Ocuvite + lutein®, the antioxidant supplement, were administered orally every day beginning the day after the first surgery for 6 weeks. Both the pre- and post-intake samples were frozen immediately after collection and stored at −80 °C until measured.

The anterior capsule pieces were homogenized and proteolyzed using MagNa Lyser Green Beads (Roche). The RNA was purified with the MagNa Pure Compact RNA Isolation Kit (Roche). Using a Transcriptor High Fidelity cDNA Synthesis Kit (Roche), cDNA was reverse transcribed by the Gene Amp PCR system 9700 (AB). The cDNA of the control RNA (both positive and negative control RNA included with the kit) was also reverse transcribed at the same time. Using the cDNA, Universal® Probes (Roche) and primers (Nihon Gene Research, Inc.), which are shown in Table 17.2, the mRNAs of G6PDH and 18S were quantified using the Light Cycler® LC480 (Roche) for RT-PCR.

The expression level of each mRNA was represented as a Cp value, where Cp was defined as the threshold cycle of RT-PCR at which the amplified product was detected. The Cp values were calculated based on a standard curve that was determined from an internal standard and the control RNA for each substance. The mRNA levels of each substance were indicated relative to internal standard and control RNA.

17.4.2 Results

In post-intake samples, the expression levels of G6PDH and 18S were significantly increased (Fig. 17.1). Both genders had the same tendencies. However, a significant difference only existed among the female patients. There were no differences

Table 17.2   Probes and primer of the mRNAs of G6PDH and 18S

ProbeFinder has designed an optimal real-time PCR assay for:
NM_004285.3 Homo sapiens hexose-6-phosphate dehydrogenase (glucose 1-dehydrogenase) (H6PD), mRNA

Assay rank 1
Use probe **#89** (cat. no. 04689143001)

Primer	Length	Position	Tm	%GC	Sequence
Left	23	983 - 1005	60	43	tggagatcatcatgaaagagacc
Right	20	1037 - 1056	60	55	gcgaatgacaccgtactcct
Amplicon (74 nt)					

tggagatcatcatgaaagagaccgtggatgctgaaggccgcaccagcttctatgaggagtacggtgtcattcgc

X03205.1| X03205:EMBL| X03205:SILVA-SSU Human 18S ribosomal RNA

Assay rank 1
Use probe **#48** (cat. no. 04688082001)

Primer	Length	Position	Tm	%GC	Sequence
Left	20	1617 - 1636	60	45	gcaattattccccatgaacg
Right	20	1665 - 1684	59	50	gggacttaatcaacgcaagc
Amplicon (68 nt)					

gcaattattccccatgaacgaggaattcccagtaagtgcgggtcataagcttgcgttgattaagtccc

Fig. 17.1   In post-intake samples, the expression levels of G6PDH and 18S were significantly increased. Both genders had the same tendencies. However, a significant difference only existed among the female patients. There were no differences between genders in pre- or post-intake samples

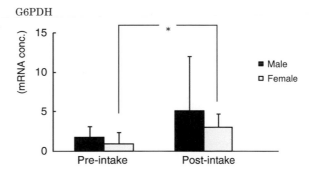

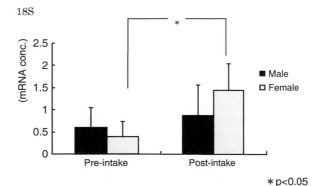

$* p < 0.05$

Fig. 17.2 There were no significant correlations between the expression levels in the pre- and post-intake samples. However, the tendencies toward lower expression of 18S in the pre-intake samples and higher expression of 18S in the post-intake samples were shown in both genders

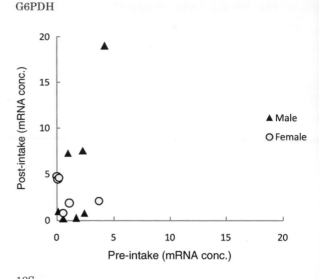

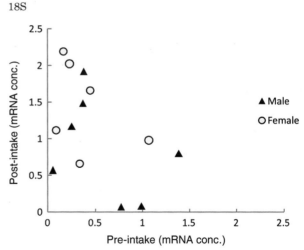

between genders in pre- or post-intake samples. There were no significant correlations between the expression levels of 18S in the pre- and post-intake samples (Fig. 17.2). However, the tendencies toward lower expression of 18S in the pre-intake samples and higher expression of 18S in the post-intake samples were shown in both genders.

There were significant positive correlations between G6PDH and 18S expression levels in the post-intake samples among male patients (Fig. 17.3). G6PDH expression increased by more than four times and 18S expression almost doubled in the post-intake samples compared with the expression levels in the pre-intake samples.

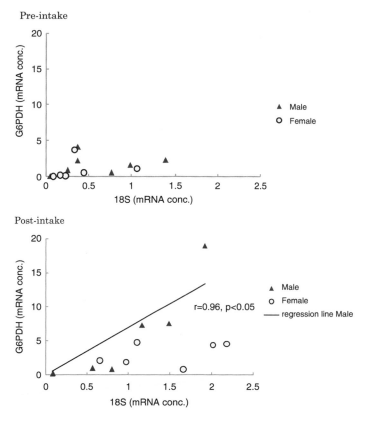

Fig. 17.3 There were significant positive correlations between G6PDH and 18S expression levels in the post-intake samples among male patients. G6PDH expression increased by more than four times and 18S expression almost doubled in the post-intake samples compared with the expression levels in the pre-intake samples

17.4.3   The Effects of Increases in Expression of G6PDH

As the lens is under the anaerobic status, the pentose phosphate cycle is more important than TCA cycles for energy production from glucose in the lens. G6PDH is known as the rate-determining enzyme of the pentose phosphate cycle, thus the increase in gene expression of G6PDH in the post-intake samples may indicate the activation of the pentose phosphate cycle followed by an increase of NAD(P)H in the LECs. Glutathione reductase (GR) is activated due to an increase in the level of coenzyme NAD(P)H, leading to an increase in the level of GSH. Thus, the increasing GSH and GR in turn increase the scavenging of peroxides after supplement intake.

G6PDH showed significantly higher expression levels in the post-intake samples from the female patients more significantly than those from the male patients. Innate antioxidative abilities decrease in elderly female patients due to hormonal imbalance.

Estrogens are known to have antioxidant effects [33, 34] due to direct scavenging of free radicals [35] and upregulation of antioxidative enzymes [36]. In postmenopausal female patients, peroxidation increases because of dramatically decreased levels of estrogen. Moreover, NAD(P)H from the pentose phosphate cycle is a coenzyme in the estrogen antioxidation cycle which continuously regenerates the antioxidative potential of estrogens [37]. After supplement intake, an increase in G6PDH is thought to increase NAD(P)H to compensate for the decrease in estrogen among the postmenopausal female patients in this study.

17.4.4 The Effects of Increases in Expression of 18S

The 18S is the active center of protein synthesis in the 40S ribosomal subunit. The increased expression levels of 18S in post-intake samples suggest an increase in the transcription and synthesis of proteins and enzymes, including enzymes involved in the pentose phosphate cycle. Furthermore, the amount a product of the pentose phosphate cycle, ribose-5-phosphate (R5P), may also increase with G6PDH after supplement intake. R5P is necessary for nucleic acid synthesis and is related to the production of the ribosome. Therefore, gene transcription is possibly activated and increased with the increase in G6PDH. This is another possible mechanism for the proportional increase in 18S expression and G6PDH expression after taking the supplement.

17.5 Conclusion

The mRNA expression of G6PDH increased with 18S proportionally in cataractous lenses after intake of the supplement including lutein. It suggests the antioxidative effect of lutein in the lens and that may be effective in preventing the progression of cataract formation.

References

1. Lin J. The association between copper ions and peroxidative reaction in diabetic cataract. J Japanese Ophthalmol Soc. 1996;100:672–9.
2. Obara Y. The oxidative stress in the cataract formation. J Japanese Ophthalmol Soc. 1995;99:1303–41.
3. Truscott RJ. Age-related nuclear cataract-oxidation is the key. Exp Eye Res. 2005;80: 701–25.
4. Tan AG, Mitchell P, Flood VM, et al. Antioxidant nutrient intake and the long-term incidence of age-related cataract: the Blue Mountain Eye Study. Am J Clin Nutr. 2008;87:1899–905.
5. Taylor A, Jacques PF, Nadler D, et al. Relationship in humans between ascorbic acid consumption and levels of total and reduced ascorbic acid in lens, aqueous humor, and plasma. Curr Eye Res. 1991;10:751–9.

6. Yoshida M, Takashima Y, Inoue M, JPHC Study Group, et al. Prospective study showing that dietary vitamin C reduced the risk of age-related cataracts in a middle-aged Japanese population. Eur J Nutr. 2007;46:118–24.
7. Jacques PF, Taylor A, Moeller S, et al. Long-term nutrient intake and 5-year change in nuclear lens opacities. Arch Ophthalmol. 2005;123:517–26.
8. Age-related Eye Disease Study Group. A randomized, placebo-controlled, clinical trial of high dose supplementation with vitamins C and E, beta carotene and zinc for age-related macular degeneration and vision loss. AREDS report no 8. Arch Ophthalmol. 2001;119:1417–36.
9. Chasan-Taber L, Willett WC, Seddon JM, et al. A prospective study of carotenoid and vitamin A intakes and risk of cataract extraction in US women. Am J Clin Nutr. 1999;70:509–16.
10. Moeller S, Voland R, Tinker L, et al. Associations between age-related nuclear cataract and lutein and zeaxanthin in the diet and serum in the carotenoids in the age-related eye disease study (CAREDS), an ancillary study of the women's health initiative. Arch Ophthalmol. 2008;126:354–64.
11. Yeum KJ, Taylor A, Tang G, et al. Measurement of carotenoids, reninoids and tocopherols in human lenses. Invest Ophthalmol Vis Sci. 1995;36:2756–62.
12. Yeum KJ, Shang FM, Schalch WM, et al. Fat-soluble nutrient concentrations in different layers of human cataractous lenses. Curr Eye Res. 1999;19:502–5.
13. Ojima F, Sakamoto H, Ishigura Y, et al. Consumption of carotenoids in photosensitized oxidation of human plasma and plasma low-density lipoprotein. Free Rad Bio Med. 1993;15:377–84.
14. Yeum KJ, Russell RM. Carotenoid biovailability and bioconversion. Annu Rev Nutr. 2002;22:483–504.
15. Kostic D, White WS, Olson JA. Intestinal absorption, serum clearance, and interactions between lutein and beta-carotene when administered to human adults in separate or combined oral doses. Am J Clin Nutr. 1995;62:604–10.
16. Rosenthal JM, Kim J, Monastario F, et al. Dose-ranging study of lutein supplementation in person aged 60 years or older. Invest Ophthalmol Vis Sci. 2006;47:5227–33.
17. Bone RA, Landerum JT. Dose-dependent response of serum lutein and macular pigment optical density to supplementation with lutein esters. Arch Biochem Biophys. 2010;504:50–5.
18. Renzi LM, Hammond BR, Dengler M, et al. The relation between serum lipids and lutein and zeaxanthin in the serum and retina: results from cross-sectional, case-control and case study designs. Lipids Health Dis. 2012;11:33–42.
19. Mares JA, LaRowe TL, Snodderly DM, CAREDS Macular Pigment Study Group and Investigators, et al. Predictors of optical density of lutein and zeaxanthin in retinas of older women in the carotenoids in age-related eye disease study, an ancillary study of the women health initiative. Am J Clin Nutr. 2006;84:1107–22.
20. Brady WE, Mares-Perlman JA, Bowen P, et al. Human serum caroteoid concentrations are related to physiologic and lifestyle factors. J Nutr. 1996;126:129–37.
21. Williams AW, Boileau TW, Erdman Jr JW. Factors influencing the uptake and absorption of carotenoids. Proc Soc Exp Biol Med. 1998;218:106–8.
22. Kim JH, Na HJ, Kim CK, et al. The non-provitamin A carotenoid, lutein, inhibits HF-kappaB-dependent gene expression through redox-based regulation of the phosphatidylinositol 3-kinase/PTEN/Akt and NF-kappaB-inducing kinase pathway: role of H2O2 in NF-kappaB activation. Free Radic Biol Med. 2008;45:885–96.
23. Gao S, Qin T, Liu Z, et al. Lutein and zeaxanthin supplementation reduces H_2O_2—induce oxidative damage in human lens epithelial cells. Mol Vis. 2011;17:3180–90.
24. Spector A, Wang GM, Wang RR, et al. The prevention of cataract caused by oxidative stress in culture rat lenses. I H_2O_2 and photochemically induced cataract. Curr Eye Res. 1993;12:163–79.
25. Woodall AA, Lee SWM, Weesie RJ, et al. Oxidation of carotenoids by free radicals: relationship between structure and reactivity. Biochem Biophys Acta. 1997;1336:33–42.
26. Christen WG, Liu S, Glynn RJ, et al. A prospective study of dietary carotenoids, vitamins C and D, and risk of cataract in women. Arch Ophthalmol. 2008;126:102–9.

27. Lyle BJ, Mares-Perlman JA, Klein BE, et al. Antioxidant intake and risk of incident age-related nuclear cataract in the Beaver Dam Eye Study. Am J Epidemiol. 1999;149:801–9.
28. Brown L, Rimm EB, Seddon JM, et al. A prospective study of carotenoid intake and risk of cataract extraction in US men. Am J Clin Nutr. 1999;70:524–71.
29. Delcourt C, Carriere I, Delage M, et al. Plasma lutein and zeaxanthin and other carotenoids as modifiable risk factors for age-related maculopathy and cataract: the POLA study. Invest Ophthalmol Vis Sci. 2006;47:2329–35.
30. Gale CR, Hall NF, Phillips DIW, et al. Plasma antioxidant viatmins and carotenoids and age-related cataract. Ophthalmology. 2001;108:1992–8.
31. Vu HTV, Robman L, Hodge A, et al. Lutein and zeaxanthin and the risk of cataract: the Melbourne Visual Impairment Project. Invest Ophthalmol Vis Sci. 2006;47:3783–6.
32. Hayashi R, Hayashi S, Arai K, et al. Effects of antioxidant supplementation on mRNA expression of glucose-6-phosphate dehydrogenase, β-actin and 18S rRNA in the anterior capsule of the lens in cataract patients. Exp Eye Res. 2012;96:48–54.
33. Gomez-Zubeldia MA, Arbues JJ, Hinchado G, et al. Influence of estrogen replacement therapy on plasma lipid peroxidation. Menopause. 2001;8:274–80.
34. Mooradian AD. Antioxidant properties of steroids. J Steroid Biochem Mol Biol. 1993;45:509–11.
35. Ruiz-Larrea MB, Martin C, Martinez R, et al. Antioxidant activities of estrogens against aqueous and lipophilic radicals: differences between phenol and catechol estrogen. Chem Phys Lipids. 2000;105:179–88.
36. Borras C, Gambini J, Gomez-Cabrera MC, et al. 17β-Oestradiol up-regulates longevity-related, antioxidant enzyme expression via the ERK1 and ERK2(MAPK)/NFκB cascade. Aging Cell. 2005;4:113–8.
37. Prokai L, Prokai-Tatrai K, Perjesi P, et al. Quinol-based bioreversible metabolic cycle for estrogens in rat liver microsomes. Drug Metab Disp. 2003;31:701–4.

Chapter 18
Antioxidant Defense Network in the Lens and Benefits of Glutathione Prodrugs in Cataracts

Shakila Tobwala, Humeyra Karacal, and Nuran Ercal

Abbreviations

ATP	Adenosine triphosphate
BSO	L-Buthionine-(S,R)-sulfoximine
CAT	Catalase
G6PD	Glucose-6-phosphate dehydrogenase
GPx	Glutathione peroxidase
GR	Glutathione reductase
GRx	Glutaredoxin
GS	Glutamine synthetase
GSH	Glutathione
GST	Glutathione-S-transferase
LPO	Lipid peroxidation
MDA	Malondialdehyde
Msrs	Methionine sulfoxide reductases
NAC	N-acetylcysteine
NACA	N-acetylcysteineamide
PSSC	Protein-S-S-cysteine
PSSG	Protein-S-S-glutathione

S. Tobwala
Department of Chemistry, Missouri University of Science and Technology, Rolla, MO USA

H. Karacal
Department of Ophthalmology and Visual Sciences, Washington University,
St. Louis, MO, USA

N. Ercal (✉)
Department of Chemistry, Missouri University of Science and Technology,
236 Schrenk Hall, 400 W. 11th Street, Rolla, MO 65409-0010, USA
e-mail: nercal@mst.edu

© Springer Science+Business Media New York 2015
M.A. Babizhayev et al. (eds.), *Studies on the Cornea and Lens*,
Oxidative Stress in Applied Basic Research and Clinical Practice,
DOI 10.1007/978-1-4939-1935-2_18

PSSP   Protein-S-S-protein
PUFAs   Polyunsaturated fatty acids
ROS   Reactive oxygen species
SOD   Superoxide dismutase
TR   Thioredoxin reductase
Trx   Thioredoxin
TTase   Thioltransferase

18.1   Human Lens

18.1.1   Morphology

The human lens is located behind the iris and pupil in the anterior compartment of the eye. It is a clear, asymmetric oblate spheroid that does not have any nerves or vessels after fetal development [1]. The functions of the lens are to maintain its own clarity, refract light, and provide accommodation. It depends on the aqueous humor and vitreous to meet its metabolic requirements as well as to clear its waste [2]. The lens is suspended by the zonules of Zinn which attach to the non-pigmented epithelium of the pars plana and pars plicata of the ciliary body. It contributes 20 diopter (D) of the 60 D focusing power of the average adult eye. The equatorial diameter of the lens is 6.5 mm at birth and increases to 9–10 mm in diameter in later life, while its antero-posterior width is about 3 mm at birth and increases to 6 mm at 80 years of age [2]. Histologically, the lens has three components: capsule, epithelium, and lens substance [1].

Capsule The lens capsule is a transparent, elastic basement membrane composed of Type IV collagen, laminin, heparin sulfate proteoglycan, fibronectin, and entactin, which are laid down continuously throughout life by the epithelial cells. The lens is formed by lamellae (fibrils) stacked on top of each other and serves as the point of attachment for the zonular fibers supporting the lens. The capsule is made up of the basal membranes of the lens epithelium anteriorly and the basal membranes of the lens fiber cells posteriorly. The thickness of the capsule depends on the age and the region of the capsule. The thickest region is close to the equator on the anterior and posterior surfaces and measures 23 μ, whereas the posterior pole is the thinnest with 4 μ.

Lens Epithelium The lens epithelium is a single line of epithelial cells behind the lens capsule, extending to the equatorial lens bow. These cells are mitotic and metabolically active, synthesizing DNA, RNA, protein, and lipid. They supply the energy demands of the lens substance by generating adenosine triphosphate (ATP). The germinative zone is where the lens epithelial cells are formed. Newly formed cells migrate toward the transitional zone while they differentiate into fibers and then move to the bow region where terminal differentiation takes place [2]. Terminal differentiation entails an increase in cellular proteins in the fiber cell membranes and the loss of organelles, including nuclei, mitochondria, and ribosomes. This process is important for the transparency of the lens. However, these cells are now dependent on glycolysis for energy production [2].

Lens Substance The lens substance, which makes up the main mass, is composed of densely packed fibers with little extracellular space. An adult lens has a nucleus and a cortex, although these are histologically indistinct [1]. The nucleus is further divided into embryonic, fetal, infantile, and adult nuclei. As the lens fibers are continuously laid down, the oldest fibers make up the embryonic and fetal nuclei. As such, the outermost fibers are the most recently formed and make up the cortex of the lens [2]. The region between the hardened embryonic and fetal nuclear core and the soft cortex is referred to as the epinucleus, while the perinuclear region makes up the area between the deep cortex and adult nucleus.

The Y shaped features which are visible on slit lamp bio-microscopy are formed by the overlap of the ends of secondary fibers in each growth shell. They are found at both anterior ("erect" Y) and posterior poles ("inverted" Y) [1]. Multiple optical zones, as well as Y sutures are visible by slit lamp bio-microscopy. These zones of demarcation are due to the continuous layering of epithelial cells with different optical densities throughout life.

18.1.2 Physiology

After the tunica vasculosa lentis involutes, the metabolic needs of the lens are met by the aqueous and the vitreous. The lens capsule is permeable to water, ions, and some proteins like albumin. These can permeate the tight junctions between the epithelial cells. There are different pumps, transporters, and channels that allow movement from the extracellular space in and out of cells. The junctions between epithelial-epithelial, epithelial-fiber, and fiber-fiber allow exchange. Rapid endocytosis has been described between epithelial cells and elongating fibers [3]. The most important aspect of lens physiology is the mechanism controlling the electrolyte and water balance, which is essential for the transparency of the lens.

Transport of Ions The lens has higher levels of potassium ions (K^+) and amino acids than the surrounding aqueous and vitreous. The lens, on the other hand, contains lower levels of sodium ions (Na^+), chloride ions (Cl^-), and water than its surroundings. The cation balance across the lens is maintained by its permeable cell membrane and by the sodium pumps within the cell membrane of the lens epithelium and each lens fiber. The sodium pump, which pumps sodium ions out while taking potassium ions in, is regulated by the Na^+/ K^+-ATPase. Inhibition of this would elevate water content in the lens.

The pump-leak system of the lens combines active transport and membrane permeability. Potassium and other molecules, like amino acids, are actively transported into the lens through the anterior epithelium. These then diffuse out of the posterior side of the lens with the concentration gradient. Most of the Na^+/ K^+-ATPase is found in the lens epithelium and the superficial cortical fibers [4]. Chloride (Cl^-) is thought to be at equilibrium across the cell membrane, so its distribution is controlled by the resting voltage. The movement of Cl^- ion is via carrier mechanism but chloride channels have been found in the fiber plasma membrane [3]. In a normal lens, the intraocular concentration of calcium (Ca^{2+}) is below 0.1 % of that found in the ocular fluids. However, when the extracellular Ca^{2+} concentration is high, Ca^{2+} moves down the gradient to enter the cells through the cation channels and by receptor-mediated endocytosis. The low intracellular Ca^{2+} concentration is maintained by calmodulin-regulated Ca^{2+}-ATPase and a Na^+/Ca^{2+} exchange mechanism. Also, Ca^{2+} is sequestered at binding sites in the plasma membrane and cytosol. The endoplasmic reticulum, found in the epithelial cells and at the surface of cortical fibers, store, Ca^{2+} by using a calmodulin-independent Ca^{2+}-ATPase and another Ca^{2+} channel, thereby removing the Ca^{2+} from the cytoplasm. The pH gradient of the lens increases from the central nucleus to the peripheral layers. The intracellular pH of the lens is 7.0 [5].

Amino Acid and Sugar Transport Most amino acids are transported into a cell from the aqueous, an active transport that is mediated by the epithelial cell layer. The lens can also convert keto acids into amino acids. Amino acids are pumped into the lens though the anterior capsule and passively leak through the posterior capsule, so that a steady state concentration is maintained [6].

Optical isomers of sugars, D-forms (except L-arabinose), are selectively transported. D-glucose is transported by insulin-independent facilitated transport, with most of the glucose entering from the aqueous. Transporters are found in the epithelial cells, and the fibers. L-glucose passively diffuses and is found only in epithelial cells and outer cortical regions [7].

Electrophysiology The membrane potential of the human lens decreases from -50 mV at age 20 to -20 mV at age 80. Epithelial cells have three K^+ selective channels. Contrary to intracellular voltage, the extracellular voltage changes from the center toward the surface of the lens [3]. The Cl^- ion also plays a role in the control of transmembrane voltages [8], which create currents around the lens. The K^+ ions create an outward current around the equator, whereas Na^+ ions create an inward current both at the anterior and posterior poles [9].

18.1.3 Chemical and Molecular Composition of the Lens

Membranes The chemical composition of the lens fiber plasma membranes are saturated fatty acids, high cholesterol/phospholipid, and a high concentration of sphingomyelin. Lipids make up only 1 % of the total lens mass, but they constitute 55 % of the plasma membrane's dry weight. As the lens ages, the protein:lipid and cholesterol:phospholipid ratios increase due to loss of phospholipids, mainly in the nucleus [10].

Lens Proteins The lens has the highest protein ratio in the body at 33 % of its wet weight. There are two kinds of lens proteins, depending on their water solubility: crystallins are the water-soluble fractions of the young lens and account for 80 % of lens proteins. They play an important part in the transparency of the lens and refractive properties. Apart from crystallins, there are enzymes and regulatory proteins that are found primarily in the epithelium and in immature fiber cells where there is high metabolic activity [10].

Crystallins Crystallins are subdivided into α- and β,γ-crystallins. α-crystallins, which represent one third of the lens proteins by mass, are the largest of the crystallins, with a molecular mass of 600–800 kDa. There are two α-crystalline subunits, αA and αB. α-crystallin is a member of a small heat-shock protein family. α-crystallin complexes bind to denatured proteins and thus prevent protein aggregates; otherwise, the lens transparency would be lost.

β,γ crystallins are divided into two groups, based on their isoelectric points and molecular mass. The β-crystallins make up 55 % (by weight) of the water-soluble proteins. They are made up of polypeptides which form dimers and higher order complexes in their native state. β-crystallins can be separated into βH (high molecular mass) and βL (low molecular mass) fractions by gel chromatography. γ-crystallins are the smallest of the crystallins, and make up 15 % of adult lens proteins. These have the lowest molecular mass as they do not associate with each other or other proteins. The basic structures of the β- and γ-crystallins have been maintained through the evolution of vertebrates. Taxon-specific crystallins are described in various other animals [10].

Membrane Structural Proteins and Cytoskeletal Proteins The water-insoluble fractions of lens proteins are either soluble or insoluble in 8 M urea. The urea-soluble lens proteins, that provide the structural framework of cytoskeleton, are comprised of microtubules (microfilaments similar to those found in other cell types), and two unusual intermediate filaments. One is vimentin, which is not found in epithelial cells, and the other is beaded filaments (composed of phakinin and filensin proteins) which are specific to the lens.

The urea-insoluble fraction of the human lens is found in the plasma membranes of the fiber cells in the lens. Major intrinsic protein (MIP, also known as aquaporin O) makes up 50 % of membrane proteins. MIP is expressed only in lens fiber cells and regulates water transport [11].

18.2   Cataracts

18.2.1   Clinical Manifestations

Clinical manifestation of cataracts include: decreased visual acuity, increased glare, decreased contrast sensitivity, myopic shift, color shift, monocular diplopia or polyopia, and loss of visual field [12, 13].

Cataracts primarily cause concentric loss of visual field. Patients could realize decreased vision themselves, or visual problems might be recognized during an examination, especially if only one eye is involved or if the degree of cataracts in both eyes is asymmetric. The decrease in visual acuity depends on the type of cataract as well [13]. Decreased contrast sensitivity is also a manifestation, which may not be readily recognized by the patient, yet it provides a better estimate of the visual resolution of an eye. However, contrast sensitivity is not a very specific measure of visual disability caused by a cataract.

Increased glare is one of the most common and early complaints of patients with cataracts because of increased sensitivity to glare, which is caused by the scattering of light. Glare is more obvious when viewing oncoming car headlights in the dark or in the rain. This becomes more prominent with a posterior subcapsular type of cataract [13].

Nuclear cataract formation causes a myopic shift in the refractive index of the lens. This may lead to what is called "second sight"; as a person develops a nuclear cataract, the myopic shift enables the person to read up close [13]. Nuclear opacities in the lens absorb the blue end of the spectrum and this leads to poor hue discrimination [13]. Furthermore, double or multiple images of the same object is caused by the lower refractive index formed by the cortical spokes and water clefts in the lens, as compared to the surrounding lens [12].

18.2.2   Clinical Classification

Senile cataracts are classified as nuclear, cortical, and posterior subcapsular; combined forms of these can also occur.

Nuclear Cataracts Most people past middle age have nuclear sclerosis and yellowing of the lens. Progressive sclerosis and yellowing lead to a nuclear cataract, causing a central opacity (Fig. 18.1). In very advanced cases, the lens nucleus becomes opaque and brown and is called a brunescent cataract. This may interfere with the retinal photopic function [14].

Cortical Cataracts Cortical spokes (wedge shaped opacities) form in the periphery of the lens with the pointed end of these spokes oriented toward the center (Fig. 18.1). Changes in the ionic composition of the lens cortex with hydration of the lens fibers lead to the formation of cortical spokes [14].

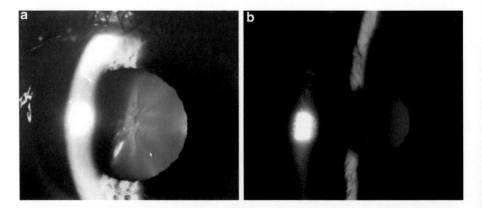

Fig. 18.1 Slit lamp photo of (**a**) age-related cortical cataract; (**b**) age-related nuclear cataract

Posterior Subcapsular Cataracts (PSC) PSCs, which are located in the posterior cortical layer may later turn into a granular and plaque-like opacity. PSCs are associated with posterior migration and enlargement of lens epithelial cells. They can occur due to aging as well as trauma, systemic or topical steroids, inflammation in the eye (uveitis), and exposure to ionizing radiation [14].

18.2.3 Etiology and Risk Factors

There are limited ways of repair and regeneration for a lens, and, thus, it may lose its transparency due to many different factors.

Aging Aging is a very common cause of cataract formation. The pathogenesis is multifactorial. The nucleus of a lens becomes compressed and hardened by the addition of cortical fibers. During aging, crystalline lens proteins are changed by chemical modification and they aggregate into high molecular weight proteins. The aggregation of lens proteins reduces transparency and causes a yellowish pigmentation. Increased concentration of sodium and calcium together with decreased potassium and GSH are other age-related changes [14].

Physical and Environmental Factors Physical factors like trauma and electric shock can lead to cataract formation. A lens may react to penetrating and non-penetrating chemical injuries and intralenticular foreign bodies in different ways. If the capsule of the lens is intact, a cataract forms; if the capsule is ruptured, water enters into the lens and a mature cataract forms [15]. If the lens is penetrated by a foreign object, the capsule may heal and a localized capsular opacity may form. Unless a seal forms, a severe inflammatory response may occur in the eye. A lens can also be damaged by lightening or accident. Voltages as low as 220 V may cause cataract formation [15].

Exposure to metals and cigarette smoke may also lead to cataract formation. Copper, for example, may form deposits in the eye from eye drops, from a copper-containing foreign body, or due to Wilson's disease (autosomal recessive deficiency of ceruloplasmin). These patients develop a typical sunflower cataract [16]. Iron forms deposits in the lens epithelium and iris leading to cataract and iris heterochromia [15], whereas gold forms deposits on the lens capsule and epithelium causing an anterior capsular cataract [15]. Cigarette smoke affects the eye mainly through ischemic or oxidative mechanisms. Both cataract development and age-related macular degeneration are directly accelerated by smoking [17].

The lens is very sensitive to radiation, including, ionizing, nonionizing, ultraviolet, and even microwave radiations. Although there is a long period of latency before the clinical appearance of a cataract, younger patients are more susceptible as they have more actively growing lens cells [14]. Ionization of water and the formation of free radicals, as well as hydration of electrons, damage cell membranes causing transcriptional errors and altering protein synthesis. The lens equatorial fibers that have high mitotic rates are affected and posterior subcapsular opacities form. Nonionizing radiation, i.e., infrared radiation, causes cataracts in glassblowers and furnace workers. A localized temperature rise of the iris pigment epithelium causes exfoliation of the anterior lens capsule [15]. UV radiations are also known to generate free radicals, which generally causes nuclear cataract. Interestingly, microwave radiation has been shown to cause cataracts in laboratory animals.

Medicinal Drugs Use of a number of medicinal drugs have been associated with cataract formation, including corticosteroids, phenothiazines, anticholinesterases, mitotic drugs, busulfan, amiodarone, and chloroquine. Systemic, topical, subconjunctival, and nasal steroid sprays can all cause cataracts, predominantly posterior subcapsular cataracts [14]. Chlorpromazine, a phenothiazine can generate free radicals leading to cataract formation. Initially, yellow-brown granules form under the anterior capsule and develop into stellate opacities [15]. Anticholinesterases, which were used for the treatment of chronic angle closure glaucoma, were shown to cause anterior subcapsular vacuoles. Busulfan, which is used in the treatment of leukemia may lead to posterior subcapsular cataracts, whereas amiodarone, which is used for cardiac arrhythmias, is known to cause visually insignificant anterior subcapsular opacities as well as corneal deposits [18]. Chloroquine, an antimalarial drug, causes white, flake-like posterior subcapsular cataracts [15].

Ocular Disorders Ocular disorders are also a risk factor for developing cataracts. Retinitis pigmentosa (RP), which is an inherited retinal dystrophy caused by the loss of photoreceptors, is characterized by typical posterior subcapsular cataracts [19]. Patients with gyrate atrophy, an autosomal recessive slowly progressive chorioretinal dystrophy, present with nyctalopia and limited peripheral visual fields. They also develop cystoid macular edema and cataracts [20]. Degenerative myopia is a major risk factor for posterior subcapsular, cortical, and nuclear cataracts [21]. Ciliary body tumors may also cause cataracts. These tumors may not be recognized behind a cataractous lens, which may lead to a delayed diagnosis. Ocular ischemia, another risk factor for cataract development, is characterized by vision loss, orbital pain, and changes of the visual field. Both anterior and posterior segment signs may be present. Anterior segment signs include iris neovascularization and secondary neovascular glaucoma, iridocyclitis, asymmetric cataracts, iris atrophy, and sluggish reaction to light [22]. Cataracts and glaucoma are the main complications of viral anterior uveitis due to the herpes simplex virus/varicella zoster virus (HSV/VZV), cytomegalovirus (CMV), and rubella [23]. Both intraocular inflammation and steroid therapy for inflammation can cause cataracts to form.

About a third of congenital cataracts is hereditary and is not associated with any systemic or metabolic disorder. Some of these may be associated with trisomy 13, 18, 21 and cri du chat syndrome [15]. The incidence of infantile cataracts is about 0.4 % of newborns [15]. These are important because, if unrecognized, they may lead to amblyopia, strabismus, and nystagmus. The causes of infantile cataracts may be maternal infections like rubella, systemic diseases, hereditary disorders, and local ocular disease.

Nutritional deficiencies have been demonstrated to cause cataracts in animal models; however, this has not been confirmed in humans [14]. One of the major risk factors for cataract formation is any type of intraocular surgery. Although there are numerous systemic disorders that are known to cause cataracts, they are beyond the scope of this chapter and, therefore, are not discussed here.

18.3   Oxidative Stress and Cataract Formation

18.3.1   Oxidative Stress

Oxidative stress, the loss of balance between the antioxidant defense and oxidant production in cells, has been implicated in the etiology and pathogenesis of cataracts and other eye disorders. Reactive oxygen species (ROS), a byproduct of cellular respiration and metabolism, are known to play a dual role as both deleterious and beneficial species. They are important physiologically in cellular signaling pathways and as defense against infectious agents and induction of mitogenic responses [24–28]. Their production is tightly regulated by various enzymes; however, overproduction of ROS, either through photooxidation or metabolic processes, can damage cell structures, including DNA, lipids, and proteins [25, 26, 29]. Balance between the dual roles of ROS is crucial and is achieved by redox regulation.

ROS can be generated endogenously or exogenously from the environment. The endogenous sources include mitochondria, peroxisomes, lipooxygenases, NADPH oxidase, and cytochrome P450. Leakage of electrons from complexes I and III of the electron transport chain in mitochondria is the major source of ROS in cells [30]. In addition to producing ROS, mitochondria are also a target for ROS which, in turn, reduces mitochondrial efficiency and leads to the generation of more ROS, a vicious self-destructive cycle. Exogenous sources include ultraviolet light, ionizing radiation, smoke, chemotherapeutics, inflammatory cytokines, environmental toxins, and hyperoxic environments [31, 32]. However, due to its anatomical locale, the lens is constantly exposed to light and radiation and is more vulnerable to these exogenous sources of ROS.

The superoxide radical ($O_2^{\cdot-}$), hydroxyl radical ($OH^\cdot$), singlet oxygen ($^1O_2$), and hydrogen peroxide (H_2O_2) are the most commonly formed ROS in a cell. Other oxygen-derived radicals like peroxyl ($ROO^\cdot$) radical and protonated superoxide ($HO_2^\cdot$) are also formed in living systems. Although, there are numerous types of ROS, the most reactive and damaging species are those that are free radicals due to the presence of unpaired valence electrons. The $O_2^{\cdot-}$ is the primary ROS which, in turn, generates secondary ROS species by dismutation to H_2O_2 and eventually $OH^\cdot$. H_2O_2 is a mild oxidizing agent which is relatively stable. However, the $OH^\cdot$ formed by Haber-Weiss reaction is very reactive with a short in vivo half-life of approximately 10^{-9} s [33]. When $OH^\cdot$ is produced, it reacts with macromolecules in its proximity resulting in lipid peroxidation as well as oxidation of proteins and nucleic acids.

Oxidation is the hallmark of many types of cataracts and plays a fundamental pathogenic role in their formation. Clinical and experimental studies have provided substantial evidence that ROS and the resulting oxidative damage are the key contributing factors to the development of a variety of cataracts [34–40]. Significant physiological and biochemical changes indicating oxidative damage in cataractous lenses have been identified by comparisons of the lipid compositions, protein and DNA modifications, and concentrations of antioxidants in cataractous and clear lenses [40–42]. Photooxidation [43, 44], xenobiotics [45–48], and impaired enzymatic and nonenzymatic defenses in the eye lens are some of the factors that contribute to oxidative stress [49–53].

In addition, hyperbaric oxygen therapy and oxidative stress produced similar biochemical changes and promoted cataract formation [54, 55], which substantiates the hypothesis that oxidative stress is involved in the pathogenesis of cataracts. A direct link between excessive oxygen exposure and nuclear cataracts has also been documented in some studies [55–58]. Elevated levels of $O_2^{\cdot-}$, H_2O_2, $OH^{\cdot}$, nitric oxide (NO), and glutathione disulfide (GSSG), peroxides, depletion of GSH, and oxidation of methionine and cysteine moieties in cataractous lenses, all indicate the involvement of ROS and RNS in cataract formation [59–65].

Under normal physiological conditions, the metabolically active cells of the lens have very high levels of antioxidants, like GSH and a series of defense mechanisms that protect it from the harmful effects of oxidations. These systems provide ample protection for the tissues against oxidation and various chemical modifications. However, under unusual stress, these defenses may not be sufficient to deter oxidative damage and visual impairment.

18.3.2 Oxidative Damage

Free radicals are generated by self-propagating chain reactions that involve initiation, propagation, and termination of free radicals. However, uncontrolled generation of free radicals results in oxidation of macromolecules and, eventually, in the disruption of function and integrity of the cell. The lens is subjected to oxidative stress throughout life, especially because of its anatomical locale, and is one of the leading causes of cataracts, especially maturity-onset cataracts [66–68]. Oxidation of proteins, lipids, and DNA has been implicated in cataractogenesis [69–73]. Furthermore, damaged macromolecules in a mature lens accumulate in the inner region as only the peripheral tissue is metabolically active [67, 68, 74–76]. This limited ability of fiber cells to repair themselves makes the lens susceptible to damage that can lead to the accumulation of protein, lipid, and DNA modifications in nucleated lens cells [67].

Protein Modifications A lens has an unusually high concentration of protein, about 35 % of its wet weight, although it may be as high as 50 % in the nuclear region. A majority of these proteins are crystallins, which are structural proteins with a high level of thiol groups in a reduced state [74]. α-crystallins help maintain the proper folded state of β-crystallins and γ-crystallins, which is crucial for maintenance of lens transparency by acting as a molecular chaperone and binding to β-crystallins and γ-crystallins [77, 78]. Under physiological conditions, the lens utilizes its various antioxidant defenses to effectively protect itself and maintain a reduced state of thiols, which is essential for retaining clarity of the lens. However, under oxidative stress, excess ROS can react with several amino acid residues resulting either in modified less active enzymes or in denatured nonfunctioning proteins [79, 80]. Sulfur-containing residues are the most susceptible ones to ROS attack. In addition, depletion of GSH leads to S-thiolation of lens proteins by oxidized nonprotein thiols to form protein-S-S-glutathione (PSSG), which is one of the initial events leading to conformational changes, including unfolding of crystallins and other modifications [81, 82]. These events are then followed by oxidation of methionines and cysteines and, subsequently, to the formation of protein-S-S-cysteine (PSSC) and protein–protein disulfide (PSSP). Accumulation of the resulting high molecular weight insoluble aggregates results in cataracts [83, 84]. A progressive loss of protein sulfhydryl groups, together with increased oxidation of methionine residues, is observed with worsening cataracts until >90 % of the cysteine and about 50 % of the methionine residues are oxidized in the advanced stages [84]. The amounts of PSSG and PSSC increase with age and are inversely proportional to the free GSH pool [85]. Elevated levels of PSSG and PSSC have been reported in hyperbaric oxygen-induced nuclear cataracts in old guinea pigs [86]. Oxidative stress induces formation of PSSG or PSSC, irrespective of the type of stress [87, 88]. A positive correlation between the level of protein thiolation and nuclear opacity has also been reported [89].

Oxidative damage to protein may lead to structural and functional changes, including conformational changes leading to inhibition of enzymatic and binding activities, fragmentation, denaturation, aggregation, altered gene regulation and expression, and modulation of cell signaling [90]. Crystallins and other proteins in lens fiber do not regenerate during the lifetime of the lens and, therefore, formation of inter- and intra-molecular conjugates, resulting from oxidative damage, will result in aggregation and accumulation of these damaged proteins and lead to cataract formation.

Lipid Peroxidation Oxidative stress, resulting from overproduction of ROS together with impaired enzymatic and nonenzymatic defenses of the lens, is believed to initiate lipid peroxidation (LPO), a causative factor in the mechanism of cataractogenesis. LPO is a widely accepted mechanism of cellular injury and lipid peroxides are used as markers of oxidative damage in cells and tissues. The LPO chain reaction is initiated by the formation of a lipid radical through hydrogen abstraction from a methylene carbon on a polyunsaturated fatty acid (PUFA) by an OH·. The resulting carbon-centered lipid radical rearranges and readily reacts with molecular oxygen in an aerobic environment to form a ROO·. These radicals are unstable and rearrange through cyclization to form endoperoxides and, eventually, a complex series of compounds, including a reactive carbonyl compound, of which malondialdehyde (MDA) is the most abundant [91, 92]. ROO· can also abstract hydrogen from the side chains of neighboring PUFAs, and form hydroperoxides. Hydroperoxides are toxic and may react with transition metals to form stable aldehydes like MDA that damage the cell membranes. This results in the propagation of the chain reaction which subsequently damages cytosolic, as well as membrane proteins [47, 51, 93–95]. LPO has been reported to cause changes in membrane function, including increased permeability [96–99], inactivation of Na^+ / K^+-ATPase and Ca^{2+}-ATPase [100, 101], and increased membrane rigidity. Furthermore, they may damage the lens by cross-linking with proteins and nucleic acids and also by disturbing the cellular thiol-redox status by depleting the total GSH content, which may lead to formation of protein aggregates with low solubility in the lens matter [102, 103]. Accumulation of these LPO products has been observed in the aqueous humor (AH) samples as well as in the lens membranes of some patients with cataracts [51, 104, 105].

DNA Damage ROS are reported to be mutagenic and can oxidatively damage DNA [106–108] by causing strand breaks and modifications of bases, including the oxidation of guanine residues into 8-hydroxy 2-deoxyguanosine (8-OHdG). Extensive DNA damage has been reported in the epithelial cells of cataractous lens [71, 106]. Mutations in the epithelial cells may lead to loss of transparency as they differentiate into fiber cells. The 8-OHdG is the most extensively studied DNA lesion and is a sensitive biomarker of oxidative DNA damage which can aid in the early diagnosis and treatment of patients with eye disorders [109, 110]. The extent of DNA damage and rate of repair depend on the nature of DNA-damaging agents [107].

18.4 Antioxidant Defense Network in the Lens

Because lenses are constantly exposed to light, they are extremely vulnerable to oxidative damage. However, they have several mechanisms that provide protection against oxidative stress and help maintain their redox state. Several layers of defense are employed to help prevent damage and to repair, recover, or degrade damaged molecules. These include nonenzymatic (e.g., GSH, ascorbic acid, vitamin E, and carotenoids) as well as enzymatic (e.g., superoxide dismutase, glutathione peroxidase, and catalase) systems. However, both of these defense networks are compromised with age and in cataracts [50, 83, 111–115]. In addition, lens cells contain enzymes that can repair damaged lens proteins.

18.4.1 Non Enzymatic Defense

Glutathione (γ-glutamylcysteinylglycine; GSH) and ascorbate serve as the primary nonenzymatic defenses against oxidative stress in the lens. In addition, vitamin E and carotenoids also provide some protection against oxidative stress.

Glutathione GSH is a tripeptide containing amino acids, glutamate, cysteine, and glycine (Fig. 18.2). An unusual γ-linkage between glutamate and cysteine prevents GSH from being readily hydrolyzed resulting in considerable stability of GSH within the cell. GSH is synthesized by the action of the two enzymatic processes. First, glutamate and cysteine are linked together through a gamma peptide bond by γ-glutamylcysteine synthetase (γ-GCS) to form γ-glutamylcysteine. In the second reaction, γ-glutamylcysteine reacts with glycine to form GSH by the enzyme glutathione synthetase (GS). GSH synthesis is primarily dependent on GCS activity, cysteine availability, and GSH feedback mechanisms.

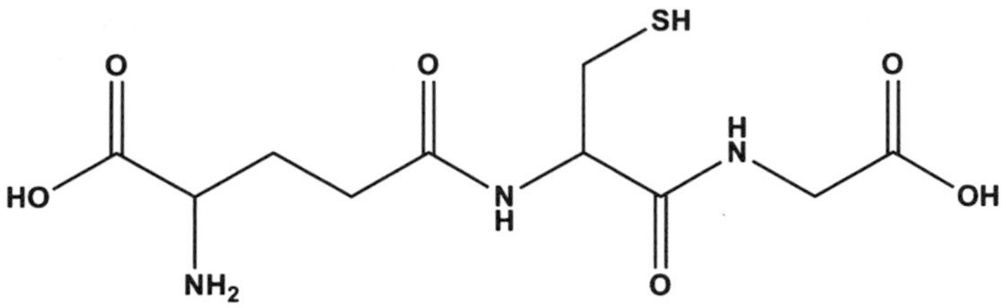

Fig. 18.2 Structure of glutathione (GSH)

GSH is the major thiol antioxidant (2–4 mM) and redox buffer in the lens [89]. It is the first line of defense against exogenous and endogenous ROS. Maintenance of redox homeostasis is crucial for lens transparency and the lens possesses both biosynthetic and regenerating systems for GSH. There is a concentration gradient of GSH with highest levels in peripheral lens fiber cells, where GSH is synthesized and reduced [63] from where it reaches the lens nucleus by diffusion from the surface cells [116]. GSH in the nucleus maintains the redox state of critical protein sulfhydryls required for DNA repair and expression. Oxidation of GSH results in formation of GSSG and decreases the ratio of GSH/GSSG, an important parameter of oxidative stress.

GSH maintains cellular redox homeostasis and protects against oxidative stress by [117]: (1) acting as a cofactor for key antioxidant enzymes like glutathione peroxidase (GPx), glutathione transferase, (2) participating in amino acid transport through the plasma membrane, (3) scavenging OH˙ radical and singlet oxygen directly, (4) detoxifying H_2O_2 and lipid peroxides by the catalytic action of glutathione peroxidase, and (5) regenerating Vitamins C and E.

GSH maintains the critical thiol groups in proteins involved in normal lens epithelial functions, such as Na^+/K^+-ATPase, certain cytoskeletal proteins, proteins associated with normal membrane permeability, and proteins involved in signaling such as receptors, protein kinases, and some transcription factors in a reduced state. Functions of these critical proteins can be altered by formation of mixed disulfides under oxidative stress when the GSSG content increases.

Decreased levels of GSH have been reported in a majority of the experimental cataracts [59, 63, 65]. Furthermore, depletion of lens GSH with age has been reported in rats [82] as well as in humans [113, 118]. A decrease in GSH is attributed to various factors, including inactivation of GR by disulfide bond formation and protein unfolding, decreased de novo synthesis, its use in detoxification, increased breakdown, or the failure to regenerate GSH from the oxidized form of GSSG [119, 120].

Depletion of GSH leads to oxidation of protein sulfhydryl groups, resulting in the formation of intra- and inter-molecular cross-links. An increase in disulfides has been reported in aging lenses due to a diminished GSH pool [121]. Mixed protein–thiol and protein–protein disulfide bonds have also been reported in lens proteins [81, 82, 122]. Damage to lens epithelial and fiber cells in early postnatal mice, due to GSH depletion, has been observed [123, 124], but was prevented by pretreatment with GSH prodrugs like glutathione ester [124]. Lou et al. reported a negative correlation between the level of GSH and nuclear opacity [89]. Formation of disulfide bonds, due to depletion of GSH, is believed to be an initiating event in the pathogenesis of cataract formation following oxidative stress [82].

The lens nucleus is highly susceptible to oxidative damage with age. Increased proportions of oxidized GSH in the nucleus are observed in older lenses as well as in nuclear cataracts [84, 86]. However, GSH levels in the outer cortex of the lens remain high, indicating local oxidation and reduced capacity of repair in the nucleus with age. With age, the lens enlarges and its cytoplasm stiffens, resulting in a lower rate of diffusion along with an increase in the distance that reduced GSH must diffuse in order to reach the nucleus [125].

Involvement of oxidative stress in cortical cataracts is not clear. Disruption of fiber cell membranes, followed by disintegration of its cytoplasmic content, occurs in cortical cataracts. It is not clear whether oxidative damage causes disruption of the fiber cells or if it occurs as a result of the loss of fiber cell membrane integrity. Disruption of the plasma membrane would allow GSH and other antioxidants to diffuse out of the fiber cells, resulting in the oxidation of the remaining cytoplasmic content.

Ascorbic Acid (Vitamin C) An aqueous-phase antioxidant, ascorbate, plays a major role in protecting against oxidative stress, especially in a lens where its concentration (1 mM) [126] is 30–35 times that of the plasma [127]. In addition to its antioxidant effects, its high concentration in the aqueous humor in diurnal animals protects the lens against the cataractogenic effect of ultraviolet (UV) radiation in sunlight by absorbing the radiation [128]. Ascorbate is a water-soluble reductant that not only absorbs UV light, but reacts with hydrogen peroxide, oxygen, and free radicals of oxygen [127]. A high level of ascorbic acid decreases the amount of oxygen in the vitreous by reacting with it and forming dehydroascorbate, which spontaneously hydrolyzes, thereby preventing the regeneration of ascorbate by cellular metabolism. Lower ascorbate levels and decreased oxygen consumption have been reported in patients with advanced vitreous liquefaction or previous vitrectomy due to the increased mixing of the vitreous, which exposes more of the ascorbate to oxygen. Vitamin C supplements have been reported to delay or prevent cataracts in several studies [129–132].

Vitamin E Vitamin E, one of the major chain-breaking, lipid-soluble antioxidants of cellular membranes, has four common forms: α-tocopherol, β-tocopherol, γ-tocopherol, and δ-tocopherol. Among these, α-tocopherol is the most effective scavenger of free radicals. The concentration of Vitamin E in the lens, unlike ascorbate, is very similar to that in plasma [133]. However, due to its lipid soluble nature, it is primarily concentrated in the lens fiber membrane, increasing its concentration by several orders of magnitude [134]. Vitamin E is a component of the total ROO$^{\bullet}$ radical trapping antioxidant system [135]. It protects against cataracts by (a) reacting directly with ROO$^{\bullet}$ and $O_2^{\bullet-}$ radicals and singlet oxygen, preventing photoperoxidation of lens lipids, and stabilizing lens cell membranes [136, 137] and (b) enhancing GSH recycling, thereby maintaining reduced GSH levels in the lens and aqueous humor.

Carotenoids Like Vitamin E, carotenoids are natural lipid-soluble antioxidants. Lutein, zeaxanthin, and riboflavin, the major carotenoids in the lens that are related to lens health, act as antioxidants by quenching singlet oxygen and triplet sensitizers, interacting with free radicals, and preventing lipid peroxidation. Low oxygen tension is reported to enhance the antioxidant activity of carotenoid's, which act synergistically with other antioxidants, like Vitamin C, to protect against oxidative damage to the lens.

18.4.2   Enzymatic Defenses

Several key antioxidant enzymes help maintain the redox status in the lens. These include superoxide dismutase (SOD), catalase, glutathione peroxidase (GPx), and glutathione reductase (GR). SOD represents the only enzyme family with activity against $O_2^{\cdot-}$ radicals. Eukaryotes possess two SOD isoforms: 80-kDa tetrameric Mn-SOD in mitochondria and the 32–34 kDa dimeric Cu/Zn-SOD in the cytosol [138]. Mitochondrial SOD is extremely important in combating ROS, as $O_2^{\cdot-}$ is formed in relatively high concentrations in the mitochondria due to the leakage of electrons from the respiratory chain. SOD converts $O_2^{\cdot-}$ radicals into a less reactive species, H_2O_2, which is detoxified by two enzymes: catalase and GPx. Catalases are heme-containing enzymes which decompose H_2O_2 to water and oxygen and specifically use H_2O_2 as a substrate. On the other hand, GPx (a selenoenzyme) can decompose H_2O_2 as well as other organic peroxides and requires GSH. The proposed catalytic mechanism [139] involves oxidation of the active site selenolate (Se^-) to selenenic acid (SeOH), which is transformed to a selenenylsulfide adduct with glutathione (Se-SG) upon addition of one molecule of GSH. Active selenolate is regenerated by the addition of a second molecule of GSH, resulting in the formation of GSSG. In addition, H_2O_2 can be degraded nonenzymatically by reduced GSH. The resulting GSSG is converted back to its reduced form by GR using NADPH as a cofactor [83]. In addition to the enzymes that help maintain the normal redox status of the lens, glucose-6-phosphate dehydrogenase (G6PD) is a major enzyme involved in protecting against ROS. G6PD catalyzes the first reaction of the pentose phosphate pathway, which is the conversion of glucose-6-phosphate to 6-phosphoglucono-δ lactone by reducing $NADP^+$ to NADPH. A second molecule of NADPH is generated in the next step of the pathway, catalyzed by another enzyme, 6-phosphogluconate dehydrogenase (6PGD). By these reactions, the cellular level of NADPH that is required for biosynthetic processes and protection against oxidative damage in the lens, are sufficiently maintained. Under conditions of oxidative stress, the pentose phosphate pathway increases to meet the higher demands of NADPH needed by the GSH protective system. However, a decreased efficiency of the pentose phosphate pathway can make the lens more vulnerable to oxidative injury.

Redundancy in the antioxidant defense enzymes provides additional protection against harmful peroxides and, therefore, animals lacking either catalase or GPx-1 appear normal [140, 141]. Interestingly, sensitivity to oxidative stress varies in these knockout animal models, with lenses from GPx-1 null mice being a little more sensitive to oxidative stress than the catalase knockout mice. Furthermore, lenses from SOD-1 (CuZn-SOD)-null mice appear normal, but they are more prone to develop photochemical cataracts [142].

18.4.3 Repair Enzymes

Repair enzymes play a key role in minimizing oxidative stress, along with antioxidant enzymes, by constantly dethiolating the protein–thiol mixed disulfides (protein thiolation) or protein–protein disulfides that result from oxidative insults. This helps restore the functions and activities of lens proteins. Oxidative stress can cause several protein modifications, including oxidation of methionines, oxidation of cysteines, and cross-linking of proteins through amino acids other than cysteine, which results in high molecular weight insoluble aggregates that are characteristic of cataracts [84, 143, 144]. The following three enzymes play a major role in the repair of proteins with oxidative damage:

GSH-Dependent Thioltransferase (TTase) or Glutaredoxin (GRx) This is a member of the thiol-disulfide oxidoreductase enzyme family with two isozymes: cytosolic TTase-1 (GRx1) and mitochondrial TTase-2 (Grx2) [145, 146]. It catalyzes the dethiolation of protein–thiol mixed disulfides [147] by using GSH as a cofactor. It is essential for maintaining the thiol-disulfide homeostasis in the lens by repairing the thiolated proteins [148], regulating GSH-dependent protein activity [149], and by preventing lens proteins from cross-linking.

Since thiolation is an initial way of protecting proteins from irreversible damage, a delicate balance between thiolation and dethiolation of proteins is crucial; TTase-1 is a key enzyme involved in the regulation of the ratio of intracellular SH/-S-S. TTase is a major antioxidant defense enzyme that is distributed in most ocular tissues [150]; however, it is two to threefold times higher in the epithelial layer. The iris has been shown to have the highest activity, followed by conjunctiva, corneal epithelial cells, and corneal endothelial cells [89]. Moderate activity has been demonstrated in the ciliary body, lens epithelial cells, and the retina, but none has been detected in the vitreous body. A striking feature of this enzyme is that it is resistant to oxidative stress under conditions where other thiol sensitive defense enzymes fail [151]. A higher concentration of this enzyme in the anterior segment of the eye (which is more vulnerable to oxidative stress) indicates the crucial protective role of TTase against oxidative stress-induced damage. In addition to dethiolation, these enzymes also act as dehydroascorbate reductase and reduce dehydroascorbate [152] in the presence of GSH. These enzymes prevent cataract formation by reducing dehydroascorbate (DHA), which is known to be toxic to the lens and can induce cataract formation [153]. Overexpression of thioltransferase in human lens epithelial cells (HLE B3) protected against oxidative stress induced by low levels of H_2O_2 [154]. In addition, Grx2 prevented disruption of the mitochondrial transmembrane potential during oxidative stress in lens epithelial cells [155].

NADPH-dependent thioredoxin/thioredoxin reductase system (Trx/TR system)
This system consists of two antioxidant oxidoreductases, namely: thioredoxin (Trx) and thioredoxin reducatse (TrxR). Trxs are small (10–12 kDa), ubiquitous proteins that catalyze the dissociation of disulfide bridges between proteins [156] and the reduction of protein thiols that have been oxidized to sulfenic acid [157]. The redox activity of their conserved active-site motif (Cys-Gly-Pro-Cys-) is crucial for their function as an oxidoreductase [158]. Oxidized thioredoxin is, in turn, reduced by selenoenzyme TrxR that uses NADPH as a cofactor [159]. TrxR isozymes are homodimers with one FAD per subunit that reduces the active site disulfide in an oxidized Trx [158, 160].

Trxs play numerous roles, including supplying reducing equivalents to enzymes like ribonucleotide reductase, methionine sulfoxide reductase, etc., selectively activating transcription factors, altering gene transcription, and functioning as cell growth factors, and also as inhibitors of apoptosis [161–164].

Oxidative stress induces the expression of Trx [165, 166], which is believed to protect against oxidative stress by (a) modulating signal transduction properties of ROS [167, 168], (b) scavenging radicals [169], (c) reducing intracellular disulfides, (d) selectively modulating the DNA-binding of transcription factors like AP-1 [170, 171] and NF-kB [172], (e) preventing apoptosis by binding to apoptosis signal-regulating kinase 1 (ASK-1) [173], (f) donating electrons to periredoxins, which are involved in the reduction of peroxides [174, 175], (g) reactivating partially oxidized and inactivated recombinant TTase [148], and (h) selectively inducing several antioxidant gene expressions [176].

TRx, a vital repair protein in the lens, is present in the epithelial, cortical, and nuclear regions of human lenses. However, the epithelial layer possesses three times more TRx 1 protein [159] than the other two regions since this layer is the initial target of oxidative insult. TrxR is present in the epithelial, as well as in the fiber cells of human lenses [177]. Upregulation of TRx 1 under oxidative stress was reported in an Emory mouse lens exposed to photooxidation [178] and in HLE B3 cells exposed to H_2O_2 [159]. This was believed to play a significant role in protecting the lens against oxidative insults [159]. Decline in the expression of the human lens thioredoxin gene and its protein in the human lens, with advanced age, indicates an association between altered Trx1 and senile cataractogenesis [177].

Methionine sulfoxide reductases (Msrs) Msrs are a family of thioredoxin-dependent oxidoreductases that catalyze reduction of oxidized methionine to methionine [179]. Two classes of Msrs, MsrA and MsrB, have been reported, which repair S- and R-epimers of methionine sulfoxide (MSO), respectively [180].

Oxidation of methionine to methionine sulfoxides has been reported in cataractous lenses [62, 64] and has been implicated in the etiopathogenesis of cataracts. Repair of oxidized methionine has been shown to protect against oxidative stress since oxidation of methionine may lead to significant changes in protein structure and functions [180–182]. Msr activity has been reported in lenses [179, 183] and both classes of Msrs protect lenses against oxidative stress-induced damages [183, 184]. Marchetti et al. provided evidence that MsrA is essential for cell viability and silencing it resulted in the loss of mitochondrial membrane potential and increased ROS levels in human lens cells [185]. Furthermore, Kantorow et al. reported high levels of MsrA transcript and protein throughout the human lens and showed that MsrA was essential for lens cell viability and protected the lens even in the absence of oxidative stress [183]. Overexpression of MsrA in lens epithelial cells was reported to protect against H_2O_2 mediated oxidative stress, whereas silencing of the MsrA gene using siRNA increased sensitivity to H_2O_2 mediated oxidative stress in lens cells. In addition, MsrB which contributes about 40 % of Msr activity in the lens played an important role in protecting against tert-butylhydroperoxide (tBHP)-induced oxidative stress [184].

Homeostasis of thiol/disulfide in cells is synergistically regulated by the TTase and TRx systems [158]. There is cross talk between the Trx/TrxR and TTase systems, as evidenced from the finding that Trx can induce TTase1 and TTase2 mRNA expressions. TTase dethiolates protein–thiol mixed disulfides, whereas TRx repairs enzymes and proteins that are oxidized to sulfenic or sulfunic acids. These enzymes act to restore cellular functions and energy production and, thereby, protect cells under oxidative stress. However, their activity is reported to decrease with age, which is one of the risk factors for cataracts.

Thus, a lens has several layers of primary protective mechanisms that help protect it against oxidative stress. However, when oxidative stress becomes overwhelming, a diminished pool of reduced GSH may cause a decrease in the ratio of the GSH/ GSSG, together with an impaired repair system. A decrease in total protein thiols, together with increases in protein–thiol mixed disulfides (PSSG and PSSC) and protein–protein disulfides result in the formation of high molecular weight protein aggregates, which are water insoluble and eventually cause light scattering and cataract formation.

18.5   Treatment Options

18.5.1   Surgery

The most common indication for cataract surgery is the patient's desire for improved visual function. Surgery is not based on a specific level of visual acuity. The patient and the physician determine whether the reduced visual function interferes with the adult's daily living (ADL) parameters [186].

Preoperative evaluation includes assessment of the general health of the patient, pertinent ocular history, social history, and postoperative care. The examination starts with an external exam and is followed by slit lamp examination of the conjunctiva, cornea, anterior chamber, iris, and lens. Fundus examination must be performed to assess any posterior segment pathology. After a decision is made that the cataract surgery is indicated to improve visual function, preoperative measurements with biometry, corneal topography should be done to determine the intraocular lens implant power. Topical anesthesia mostly used during cataract surgery but, depending on patient factors, local or general anesthesia can be preferred.

Extracapsular cataract extraction by nucleus expression involves removal of the lens nucleus and cortex through an opening in the anterior capsule, leaving the capsular bag in place. Phaco-emulsification is an extracapsular technique that uses an ultrasonically driven tip to emulsify the lens nucleus and aspirate the lens.

An incision is made in the sclera or clear cornea with a blade and special visco-elastic materials are used to maintain the intraocular space during the surgery. A circular opening is created in the anterior capsule and a phaco tip is inserted through the scleral or corneal incision toward the lens. There is continuous irrigation through the phaco tip so that ocular volume is maintained. The lens cortex is emulsified by the phaco handpiece, which oscillates at a frequency of 27,000–60,000 Hz. The melt is then aspirated through the handpiece. A femtosecond laser (1,053 nm wavelength) has been developed for creating corneal incisions, performing anterior capsulotomy, and fragmenting the nucleus; however, it is currently unable to completely emulsify the nuclear material. After the cortical and epinuclear materials are removed with a different handpiece, an intraocular lens implant is inserted within the capsular bag.

Today's intraocular lenses (IOLS) have evolved considerably since the initial implant in 1949. Harold Ridley, an English ophthalmologist, observed that the polymethylmethacrylate (PMMA) fragments from an airplane cockpit windshield embedded in the eyes of pilots were tolerated in the anterior segment of the eyes [186]. This led to the idea that, when cataractous lens is removed, refractive power of the eye can be restored using an intraocular lens.

IOLS have undergone significant changes in geometry (loops of different length, stiffness, vault and angle configuration; optics of different diameter, holes, style, and shapes) [186]. Then, with the advent of small incision cataract surgery, came the foldable IOLS. The majority of current IOLS is foldable and is either silicone or acrylic.

Monofocal IOLS have the drawback of a fixed focal point. Newer multifocal IOLS now offer refractive correction for near, intermediate, and distance vision at the expense of decreased contrast sensitivity, glare, and halos. Most IOLs have ultraviolet absorbing chromophores to protect the retina from radiation, while some IOLS have blue-blocker to prevent long-term damage to the retina caused by high frequency blue light.

The refractive power of IOLs should be specifically measured for each eye. IOL calculation is prone to some degree of error, but Light Adjustable Lens (Calhoun Vision) has currently been developed to circumvent this problem. The optic of the three piece foldable silicone lens is composed of an unpolymerized silicone matrix polymer, a polymerized macromer, a photo-initiator, and a UV-absorber. A digital light delivery device induces polymerization of the unpolymerized silicone. Irradiating the centre of the lens decreases the central curvature, with a resulting hyperopization of the IOL's refraction, while irradiating the periphery of the IOL increases its central curvature, with a resulting myopization. After a few weeks, a final lock-in of the IOL power is achieved [187].

Cataract Surgery Outcomes Modern cataract surgery has a very satisfying success rates as indicated by one large study where 95 % of adults were satisfied with the results of their surgery. Some tools assess how visual function is affected, including the VF-14 questionnaire and the Activities of Daily Vision Scale (ADVS), which is a measure of vision specific functional status [186]. Studies have shown significant improvement in the quality of life parameters such as driving, mental health, life satisfaction, and community and home activities [186].

Complications of Cataract Surgery Complication rates of modern cataract surgery are low, less than 5 %. Problems may occur during the surgery or in the immediate or late postoperative period. Intraoperative complications are wound related like Descemet's membrane detachment, thermal burns of the wound, and tears or breaks in the capsulotomy which may lead to extension of a tear toward the posterior capsule with loss of lens material to the posterior segment. Others are iris prolapse, hemorrhage in the retrobulbar space, anterior segment or suprachoroidal hemorrhage, and retained lens material. Some of these can be salvaged without loss of vision, while others like suprachoroidal hemorrhage may have grim consequences. Postoperative complications are wound dehiscence, wound leakage, corneal edema and bullous keratopathy, hemorrhage, uveitis, increased intraocular pressure, induced astigmatism, IOL decentration or dislocation, wrong IOL power, endophthalmitis, cystoid macular edema, macular infarction, retinal detachment, posterior capsule opacification, and epithelial downgrowth [188]. Although, cataract surgery is generally very successful, it should be kept in mind that it is not a risk-free procedure.

18.5.2   Alternative Treatment Options: GSH Prodrugs

Cataracts are routinely treated by surgical replacement of lens, a safe and effective approach. Although this surgery is currently the accepted treatment, it requires a significant and growing share of medical resources, especially through Medicare. Over three million procedures are performed in the United States each year at an estimated total cost of about $9 billion. In the face of the rapid increase in the prevalence of cataracts expected over the coming decades, an effective, nonsurgical treatment to prevent and treat cataracts would represent a cost-effective therapeutic alternative. Investigators at the Tufts University School of Nutrition estimate that delaying the onset of cataracts by 10 years would reduce the number of cataract extractions by half. Since oxidative stress is implicated in the pathogenesis of a cataract, a logical approach to drug development would be the prevention of oxidative damage to the lens. The lens, because of its anatomical location is subjected to significant photooxidative stress, requiring elevated levels of cellular antioxidants to neutralize photochemically generated ROS. In the lens, GSH is the primary cellular antioxidant and plays a critical role in protecting the lens from the toxic effects of ROS. It is found in lens epithelial cells at levels almost twice that of other cells in the human body. A young lens has a high concentration of enzymes, such as ascorbate and GSH, but the ability of the cell to synthesize GSH declines with age. This results in decreased GSH levels that leave the lens more vulnerable to oxidative

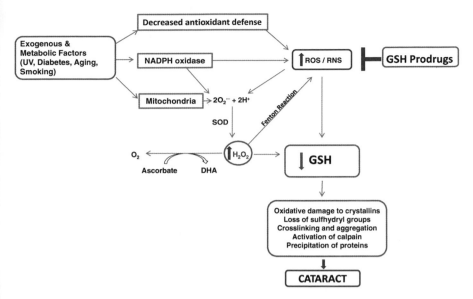

Fig. 18.3 Postulated beneficial role of GSH prodrugs in cataracts

damage and, eventually, cataract formation. Under conditions of chronic imbalance between antioxidants and ROS, oxidative damage results in structural perturbation of the crystallin proteins that comprise 90 % of the lens tissue and are responsible for maintaining its transparency. Since crystallins are long-lived stable proteins with minimal turnover, accumulation of oxidative damage leads to protein cross-linking, aggregation, insolubility, and fragmentation, resulting in the formation of cataracts. The important role of ROS and resulting oxidative damage in the development of various types of cataracts has been established by a number of studies.

The loss of antioxidants, such as GSH, that are critical to protecting eye tissue against oxidative stress, cannot be replaced directly. Instead, compounds that can easily pass into cells and result in increases in intracellular GSH levels are required. Therefore, a logical approach to enhancing antioxidant protection would be the use of pharmacologic doses of GSH prodrugs that can be delivered to the sites of damage by ROS. Use of GSH prodrugs represents a method for preventing or delaying the onset of cataracts in high-risk populations and slowing or reversing disease progression in patients who have these diseases (Fig. 18.3).

However, progress in this area has been modest. Although studies have indicated that GSH prodrugs like *N*-acetylcysteine (NAC) may ameliorate the risk of cataracts [189], NAC is not highly bioavailable and does not readily penetrate into cells, thereby requiring fairly high doses that can increase the side effect profile. Several other types of eye drops have been studied for counteracting oxidative stress, but most of these agents have significant drawbacks for clinical use, such as short half-life, adverse side effects, thermolability, or poor bioavailability.

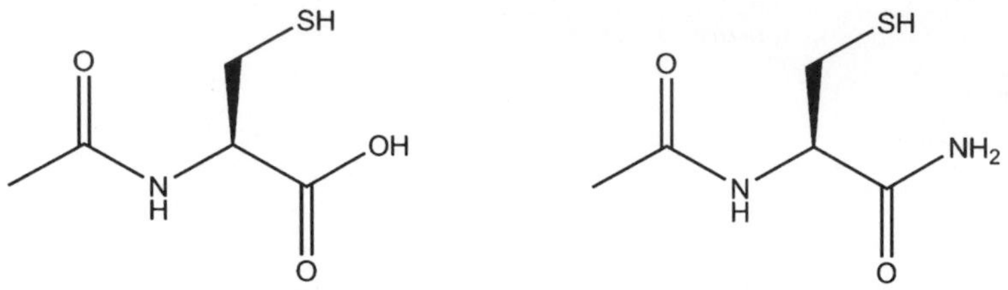

N-acetylcysteine   N-acetylcysteineamide

Fig. 18.4 The structures of *N*-acetylcysteine and *N*-acetylcysteineamide

A potential candidate, that possesses far better characteristics for development as an ophthalmologic agent to address oxidative stress damage, is the low molecular weight thiol antioxidant, *N*-acetylcysteineamide (NACA). NACA's characteristics as a drug were improved over NAC by neutralizing the carboxylic group of NAC, which makes the NACA molecule more lipophilic and, therefore, enhances its ability to penetrate cellular membranes (Fig. 18.4).

Evidence of greater efficient membrane penetration by NACA was demonstrated in a study that assessed NACA for its antioxidant and protective effects in a model using human red blood cells [190]. In addition, our own studies have demonstrated that NACA supplies significantly more thiols in plasma and liver tissues than NAC does. The enhanced ability to penetrate cells allows NACA to be administered at a lower dose than NAC, giving the drug a greater therapeutic index and lowering the risk of side effects that traditionally have been associated with higher doses of NAC [191]. NACA is an excellent source of sulfhydryl (SH) groups that can be converted by the cells into metabolites capable of stimulating GSH synthesis. The molecule can also promote intracellular detoxification and act directly as a free radical scavenger. NACA acts as a carrier of NAC and its antioxidant and free radical scavenging abilities are equal to or better than those of NAC [192]. While these features are promising, development of NACA has historically been avoided because of its lack of stability. Recently, a new chemical formulation process adequately addressed this problem. NACA now exists in a stable drug form that can be readily formulated into eye drops.

The effects of NACA on cataract development in our lab were evaluated in Wistar rat pups [193]. Cataract formation was induced in these animals with an intraperitoneal injection of a GSH synthesis inhibitor, L-buthionine-(S,R)-sulfoximine (BSO). The pups were sacrificed on postpartum day 15, after examination under a slit-lamp microscope. Their lenses were analyzed for selective oxidative stress parameters, including glutathione (reduced and oxidized), protein carbonyls, CAT, GPx, GR, and MDA. The lenses of pups in both the control and the NACA-only groups were clear, whereas all pups within the BSO-only group developed well-defined cataracts. It was found that supplemental NACA injections during BSO treatment prevented cataract formation in most of the rat pups in the NACA + BSO group. Only 20 % of these pups developed cataracts, and the rest retained clear lenses (Fig. 18.5).

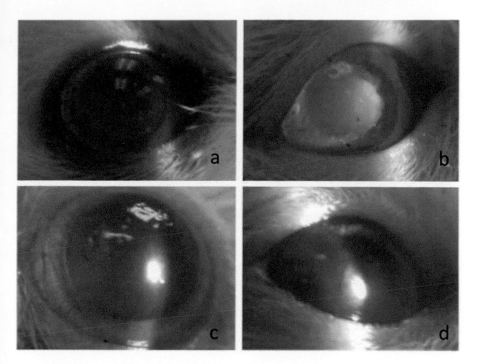

Fig. 18.5 Images of cataract formation, utilizing a slit-lamp microscope at 10× original magnification and a digital camera in macro mode are shown. These pictures were taken when the rat pups were 15 days of age, 1 day before sacrifice. A representative picture of the lenses observed for each group is shown. (**a**) Control lens (**b**) BSO-only lens (**c**) NACA-only lens (**d**) NACA + BSO lens

Further, GSH levels were significantly decreased in the BSO-only treated group, but rats that received NACA injections during BSO treatment had these levels of GSH replenished. GSH levels were restored in rats to 90 % of that of the control groups during BSO exposure.

Supporting these data, it was observed that NACA caused an increase in GR activity in BSO-treated rats. Increases in GSH levels and the GSH/GSSG ratio may be attributed to the increased activity of GR by preserving the integrity of cell membranes and by stabilizing the sulfhydryl groups of proteins. Further, NACA itself may act as a sulfhydryl group donor for GSH synthesis thereby decreasing the loss of protein sulfhydryl groups, as well as opacification of the lens. NACA was also able to reduce MDA levels by supplying an adequate amount of GSH as a substrate for glutathione peroxidase to effectively decompose lipid peroxides in the rats.

Oxidative damage to proteins, as reflected by protein carbonylation, was significant in BSO-treated animals and this was reversed by NACA. NACA was also able to restore both CAT and GPx to those levels seen within the control group. The possible mechanism for the restored CAT activity in BSO-exposed rats, when treated with NACA, may be the scavenging of free radicals by NACA. However, further investigation is needed to confirm this theory. Increased GPx activity is probably due to higher levels of GSH, which GPx uses as a substrate for its action.

In summary, these results strongly suggest that GSH prodrugs have the ability to protect against, or to delay, the onset of cataract formation by reducing oxidative damage. NACA could confer a protective effect by providing a substrate for the generation of GSH and the ability to maintain antioxidant levels within the lens and, possibly, through disulfide-exchange mechanisms.

The development of a useful GSH prodrug eye drop would significantly improve patient health and the clinical care of cataracts. Slowing the onset of cataract formation, or arresting the progression of the disease in its early stages, would mean that, in many cases, surgery could be avoided completely. Finally, a successful outcome to this research would provide additional momentum for research on antioxidant-based approaches to treating degenerative eye conditions. For example, in the case of cataract care, treatment broadly targets immune or inflammatory pathways, and emphasis would be shifted toward pharmacology rather than surgical intervention, which is better for patients.

Beneficial effects with GSH prodrugs support the hypothesis that depletion of GSH is an important step in the pathogenesis of cataracts and supplementation of GSH would be an effective noninvasive, cost-effective therapeutic option for both prevention and treatment of cataracts. Developing new therapeutics to both prevent cataract formation in high-risk populations and nonsurgically treat early stage cataracts will present great opportunities as well as challenges for researchers in the field.

References

1. Yanoff M, Duker JS. Ophthalmology. London: Mosby; 1999.
2. Rosenfeld SI BM, Bobrow JC, et al. Lens and cataracts In: Basic and clinical science course. American Academy of Ophthalmology, San Francisco; 2007e. p. 5–9
3. Kuszak JRBH. Embryology and anatomy of the lens. In: Albert DMJF, editor. Principles and practice of ophthalmology. Philadelphia: WB Saunders; 1994. p. 82–96.
4. Rosenfeld SI BM, Bobrow JC, et al. Lens and cataracts In: Basic and clinical science course. American Academy of Ophthalmology, San Francisco; 2007a. p. 19–23
5. Seland JH. The lens capsule and zonulae. Acta Ophthalmol Suppl. 1992;205:7–12.
6. Brown NP, Bron AJ. Lens structure. In: Brown NP, Bron AJ, Brown NP, editors. Lens disorders: a clinical manual of cataract diagnosis. Oxford: Butterworth-Heinemann; 1996. p. 32–47.
7. Olivero DK, Furcht LT. Type IV collagen, laminin, and fibronectin promote the adhesion and migration of rabbit lens epithelial cells in vitro. Invest Ophthalmol Vis Sci. 1993;34(10): 2825–34.
8. Lo WK, Harding CV. Tight junctions in the lens epithelia of human and frog: freeze-fracture and protein tracer studies. Invest Ophthalmol Vis Sci. 1983;24(4):396–402.
9. Patterson JW. Characterization of the equatorial current of the lens. Ophthalmic Res. 1988;20(3):139–42.
10. Cibis GWBH, Johns K, et al. Fundamentals and principles of ophthalmology. In: Basic and Clinical Science Course. American Academy of Ophthalmology, San Francisco; 2007. p. 323–30.
11. Rosenfeld SI BM, Bobrow JC, et al. Lens and cataracts In: Basic and clinical science course. American Academy of Ophthalmology, San Francisco; 2007b. p. 11–16

12. J C. The morphology and visual effects of lens opacities. In: Myron Y, JayS D editors. Ophthalmology. London: Mosby International; 1999. p. 1–2

13. Rosenfeld SI BM, Bobrow JC, et al. Lens and cataracts In: Basic and clinical science course. American Academy of Ophthalmology, San Francisco; 2007c. p. 75–88

14. Rosenfeld SI BM, Bobrow JC, et al. Lens and cataracts In: Basic and clinical science course. American Academy of Ophthalmology, San Francisco; 2007d. p. 45–69

15. Dk C. Cataract formation mechanism. In: Duker MYJS, editor. Ophthalmology. London: Mosby International; 1999. p. 1–8.

16. Becker M, Rohrschneider K. Ocular manifestations of Wilson disease. Ophthalmologe. 1997;94(11):865–70.

17. Solberg Y, Rosner M, Belkin M. The association between cigarette smoking and ocular diseases. Surv Ophthalmol. 1998;42(6):535–47.

18. Flach AJ, Dolan BJ, Sudduth B, Weddell J. Amiodarone-induced lens opacities. Arch Ophthalmol. 1983;101(10):1554–6.

19. Hamel C. Retinitis pigmentosa. Orphanet J Rare Dis. 2006;1:40. doi:10.1186/1750-1172-1-40.

20. Grover S, FG. Choroidal dystrophies. In: Myron Y,JayS D, editors. Ophthalmology. London: Mosby International; 1999. p. 1–6

21. Saw SM, Gazzard G, Shih-Yen EC, Chua WH. Myopia and associated pathological complications. Ophthalmic Physiol Opt. 2005;25(5):381–91. doi:10.1111/j.1475-1313.2005.00298.x.

22. Terelak-Borys B, Skonieczna K, Grabska-Liberek I. Ocular ischemic syndrome: a systematic review. Med Sci Monit. 2012;18(8):RA138–44.

23. Jap A, Chee SP. Viral anterior uveitis. Curr Opin Ophthalmol. 2011;22(6):483–8. doi:10.1097/ICU.0b013e32834be021.

24. Fleury C, Mignotte B, Vayssiere JL. Mitochondrial reactive oxygen species in cell death signaling. Biochimie. 2002;84(2–3):131–41.

25. Forman HJ, Torres M. Reactive oxygen species and cell signaling: respiratory burst in macrophage signaling. Am J Respir Crit Care Med. 2002;166(12 Pt 2):S4–8. doi:10.1164/rccm.2206007.

26. Nordberg J, Arner ES. Reactive oxygen species, antioxidants, and the mammalian thioredoxin system. Free Radic Biol Med. 2001;31(11):1287–312.

27. Rhee SG. Redox signaling: hydrogen peroxide as intracellular messenger. Exp Mol Med. 1999;31(2):53–9. doi:10.1038/emm.1999.9.

28. Visconti R, Grieco D. New insights on oxidative stress in cancer. Curr Opin Drug Discov Devel. 2009;12(2):240–5.

29. Valko M, Rhodes CJ, Moncol J, Izakovic M, Mazur M. Free radicals, metals and antioxidants in oxidative stress-induced cancer. Chem Biol Interact. 2006;160(1):1–40. doi:10.1016/j.cbi.2005.12.009.

30. Turrens JF. Mitochondrial formation of reactive oxygen species. J Physiol. 2003;552(Pt 2):335–44. doi:10.1113/jphysiol.2003.049478.

31. Church DF, Pryor WA. Free-radical chemistry of cigarette smoke and its toxicological implications. Environ Health Perspect. 1985;64:111–26.

32. Riley PA. Free radicals in biology: oxidative stress and the effects of ionizing radiation. Int J Radiat Biol. 1994;65(1):27–33.

33. Pastor N, Weinstein H, Jamison E, Brenowitz M. A detailed interpretation of OH radical footprints in a TBP-DNA complex reveals the role of dynamics in the mechanism of sequence-specific binding. J Mol Biol. 2000;304(1):55–68. doi:10.1006/jmbi.2000.4173.

34. Beebe DC. Maintaining transparency: a review of the developmental physiology and pathophysiology of two avascular tissues. Semin Cell Dev Biol. 2008;19(2):125–33. doi:10.1016/j.semcdb.2007.08.014.

35. Beebe DC, Holekamp NM, Shui YB. Oxidative damage and the prevention of age-related cataracts. Ophthalmic Res. 2010;44(3):155–65. doi:10.1159/000316481.

36. Elanchezhian R, Palsamy P, Madson CJ, Lynch DW, Shinohara T. Age-related cataracts: homocysteine coupled endoplasmic reticulum stress and suppression of Nrf2-dependent antioxidant protection. Chem Biol Interact. 2012;200(1):1–10. doi:10.1016/j.cbi.2012.08.017.

37. Spector A. Review: oxidative stress and disease. J Ocul Pharmacol Ther. 2000;16(2): 193–201.

38. Srivastava SK, Ansari NH, Bhatnagar A. Sugar induced cataractogenesis: a paradigm of oxidative tissue pathology? Lens Eye Toxic Res. 1990;7(2):161–71.

39. Varma SD, Kovtun S, Hegde KR. Role of ultraviolet irradiation and oxidative stress in cataract formation-medical prevention by nutritional antioxidants and metabolic agonists. Eye Contact Lens. 2011;37(4):233–45. doi:10.1097/ICL.0b013e31821ec4f2.

40. Zhang Y, Ouyang S, Zhang L, Tang X, Song Z, Liu P. Oxygen-induced changes in mitochondrial DNA and DNA repair enzymes in aging rat lens. Mech Ageing Dev. 2010;131 (11–12):666–73. doi:10.1016/j.mad.2010.09.003.

41. Berthoud VM, Beyer EC. Oxidative stress, lens gap junctions, and cataracts. Antioxid Redox Signal. 2009;11(2):339–53. doi:10.1089/ars.2008.2119.

42. Fan X, Zhang J, Theves M, Strauch C, Nemet I, Liu X, Qian J, Giblin FJ, Monnier VM. Mechanism of lysine oxidation in human lens crystallins during aging and in diabetes. J Biol Chem. 2009;284(50):34618–27. doi:10.1074/jbc.M109.032094.

43. Egorov S, Babizhaev MA, Krasnovskii Jr AA, Shvedova AA. Photosensitized generation of singlet molecular oxygen by endogenous substances of the eye lens. Biofizika. 1987;32(1): 169–71.

44. Goosey JD, Zigler Jr JS, Kinoshita JH. Cross-linking of lens crystallins in a photodynamic system: a process mediated by singlet oxygen. Science. 1980;208(4449):1278–80.

45. Bhuyan DK, Bhuyan KC. Oxy radicals in the eye tissues of rabbits after diquat in vivo. Free Radic Res Commun. 1991;12–13(Pt 2):621–7.

46. Bhuyan KC, Bhuyan DK, Chiu W, Malik S, Fridovich I. Desferal-Mn(III) in the therapy of diquat-induced cataract in rabbit. Arch Biochem Biophys. 1991;288(2):525–32.

47. Bhuyan KC, Bhuyan DK, Podos SM. Free radical enhancer xenobiotic is an inducer of cataract in rabbit. Free Radic Res Commun. 1991;12–13(Pt 2):609–20.

48. Bhuyan KC, Bhuyan DK, Santos O, Podos SM. Antioxidant and anticataractogenic effects of topical captopril in diquat-induced cataract in rabbits. Free Radic Biol Med. 1992;12(4): 251–61.

49. Babizhaev MA, Deev AI, Vladimirov Iu A, Deeva IB. Decomposition of H_2O_2 by human cataractous lenses. Biull Eksp Biol Med. 1986;102(8):158–60.

50. Bhuyan KC, Bhuyan DK. Superoxide dismutase of the eye: relative functions of superoxide dismutase and catalase in protecting the ocular lens from oxidative damage. Biochim Biophys Acta. 1978;542(1):28–38.

51. Bhuyan KC, Master RW, Coles RS, Bhuyan DK. Molecular mechanisms of cataractogenesis: IV. Evidence of phospholipid malondialdehyde adduct in human senile cataract. Mech Ageing Dev. 1986;34(3):289–96.

52. Fecondo JV, Augusteyn RC. Superoxide dismutase, catalase and glutathione peroxidase in the human cataractous lens. Exp Eye Res. 1983;36(1):15–23.

53. Rao GN, Sadasivudu B, Cotlier E. Studies on glutathione S-transferase, glutathione peroxidase and glutathione reductase in human normal and cataractous lenses. Ophthalmic Res. 1983;15(4):173–9.

54. Marsili S, Salganik RI, Albright CD, Freel CD, Johnsen S, Peiffer RL, Costello MJ. Cataract formation in a strain of rats selected for high oxidative stress. Exp Eye Res. 2004;79(5): 595–612. doi:10.1016/j.exer.2004.06.008.

55. Palmquist BM, Philipson B, Barr PO. Nuclear cataract and myopia during hyperbaric oxygen therapy. Br J Ophthalmol. 1984;68(2):113–7.

56. Holekamp NM, Shui YB, Beebe DC. Vitrectomy surgery increases oxygen exposure to the lens: a possible mechanism for nuclear cataract formation. Am J Ophthalmol. 2005; 139(2):302–10. doi:10.1016/j.ajo.2004.09.046.

57. Shui YB, Holekamp NM, Kramer BC, Crowley JR, Wilkins MA, Chu F, Malone PE, Mangers SJ, Hou JH, Siegfried CJ, Beebe DC. The gel state of the vitreous and ascorbate-dependent oxygen consumption: relationship to the etiology of nuclear cataracts. Arch Ophthalmol. 2009;127(4):475–82. doi:10.1001/archophthalmol.2008.621.

58. Zigler Jr JS, Goosey JD. Singlet oxygen as a possible factor in human senile nuclear cataract development. Curr Eye Res. 1984;3(1):59–65.

59. Calvin HI, Medvedovsky C, Worgul BV. Near-total glutathione depletion and age-specific cataracts induced by buthionine sulfoximine in mice. Science. 1986;233(4763):553–5.

60. David LL, Shearer TR. State of sulfhydryl in selenite cataract. Toxicol Appl Pharmacol. 1984;74(1):109–15.

61. Fu S, Dean R, Southan M, Truscott R. The hydroxyl radical in lens nuclear cataractogenesis. J Biol Chem. 1998;273(44):28603–9.

62. Garner MH, Spector A. Selective oxidation of cysteine and methionine in normal and senile cataractous lenses. Proc Natl Acad Sci U S A. 1980;77(3):1274–7.

63. Giblin FJ. Glutathione: a vital lens antioxidant. J Ocul Pharmacol Ther. 2000;16(2):121–35.

64. Truscott RJ, Augusteyn RC. Oxidative changes in human lens proteins during senile nuclear cataract formation. Biochim Biophys Acta. 1977;492(1):43–52.

65. Padgaonkar V, Giblin FJ, Reddy VN. Disulfide cross-linking of urea-insoluble proteins in rabbit lenses treated with hyperbaric oxygen. Exp Eye Res. 1989;49(5):887–99.

66. Babizhayev MA, Deyev AI, Linberg LF. Lipid peroxidation as a possible cause of cataract. Mech Ageing Dev. 1988;44(1):69–89.

67. Spector A. The search for a solution to senile cataracts. Proctor lecture. Invest Ophthalmol Vis Sci. 1984;25(2):130–46.

68. Ottonello S, Foroni C, Carta A, Petrucco S, Maraini G. Oxidative stress and age-related cataract. Ophthalmologica. 2000;214(1):78–85.

69. Babizhayev MA, Costa EB. Lipid peroxide and reactive oxygen species generating systems of the crystalline lens. Biochim Biophys Acta. 1994;1225(3):326–37.

70. Dische Z, Zil H. Studies on the oxidation of cysteine to cystine in lens proteins during cataract formation. Am J Ophthalmol. 1951;34(5:2):104–13.

71. Kleiman NJ, Spector A. DNA single strand breaks in human lens epithelial cells from patients with cataract. Curr Eye Res. 1993;12(5):423–31.

72. Spector A, Roy D. Disulfide-linked high molecular weight protein associated with human cataract. Proc Natl Acad Sci U S A. 1978;75(7):3244–8.

73. Zigler Jr JS, Huang QL, Du XY. Oxidative modification of lens crystallins by H_2O_2 and chelated iron. Free Radic Biol Med. 1989;7(5):499–505.

74. Harding JJ, Dilley KJ. Structural proteins of the mammalian lens: a review with emphasis on changes in development, aging and cataract. Exp Eye Res. 1976;22(1):1–73.

75. Rink H. Growth potential, repair capacity and protein synthesis in lens epithelial cells during aging in vitro. Monogr Dev Biol. 1984;17:94–107.

76. Wannemacher CF, Spector A. Protein synthesis in the core of calf lens. Exp Eye Res. 1968;7(4):623–5.

77. Bassnett S, Shi Y, Vrensen GF. Biological glass: structural determinants of eye lens transparency. Philos Trans R Soc Lond B Biol Sci. 2011;366(1568):1250–64. doi:10.1098/rstb.2010.0302.

78. Horwitz J. Alpha-crystallin can function as a molecular chaperone. Proc Natl Acad Sci U S A. 1992;89(21):10449–53.

79. Butterfield DA, Koppal T, Howard B, Subramaniam R, Hall N, Hensley K, Yatin S, Allen K, Aksenov M, Aksenova M, Carney J. Structural and functional changes in proteins induced by free radical-mediated oxidative stress and protective action of the antioxidants N-tert-butyl-alpha-phenylnitrone and vitamin E. Ann N Y Acad Sci. 1998;854:448–62.

80. Stadtman ER, Berlett BS. Reactive oxygen-mediated protein oxidation in aging and disease. Drug Metab Rev. 1998;30(2):225–43. doi:10.3109/03602539808996310.

81. Cui XL, Lou MF. The effect and recovery of long-term H_2O_2 exposure on lens morphology and biochemistry. Exp Eye Res. 1993;57(2):157–67. doi:10.1006/exer.1993.1111.

82. Lou MF, Dickerson Jr JE, Garadi R. The role of protein-thiol mixed disulfides in cataractogenesis. Exp Eye Res. 1990;50(6):819–26.
83. Reddy VN. Glutathione and its function in the lens: an overview. Exp Eye Res. 1990; 50(6):771–8.
84. Truscott RJ. Age-related nuclear cataract-oxidation is the key. Exp Eye Res. 2005;80(5): 709–25. doi:10.1016/j.exer.2004.12.007.
85. Dickerson Jr JE, Lou MF. Free cysteine levels in normal human lenses. Exp Eye Res. 1997;65(3):451–4.
86. Giblin FJ, Padgaonkar VA, Leverenz VR, Lin LR, Lou MF, Unakar NJ, Dang L, Dickerson Jr JE, Reddy VN. Nuclear light scattering, disulfide formation and membrane damage in lenses of older guinea pigs treated with hyperbaric oxygen. Exp Eye Res. 1995;60(3):219–35.
87. Xu GT, Zigler Jr JS, Lou MF. The possible mechanism of naphthalene cataract in rat and its prevention by an aldose reductase inhibitor (ALO1576). Exp Eye Res. 1992;54(1):63–72.
88. Zigman S, Paxhia T, McDaniel T, Lou MF, Yu NT. Effect of chronic near-ultraviolet radiation on the gray squirrel lens in vivo. Invest Ophthalmol Vis Sci. 1991;32(6):1723–32.
89. Lou MF. Redox regulation in the lens. Prog Retin Eye Res. 2003;22(5):657–82.
90. Hawkins CL, Davies MJ. Generation and propagation of radical reactions on proteins. Biochim Biophys Acta. 2001;1504(2–3):196–219.
91. Babizhayev MA. Current ocular drug delivery challenges for N-acetylcarnosine: novel patented routes and modes of delivery, design for enhancement of therapeutic activity and drug delivery relationships. Recent Pat Drug Deliv Formul. 2009;3(3):229–65.
92. Ohira A, Ueda T, Ohishi K, Hiramitsu T, Akeo K, Obara Y. Oxidative stress in ocular disease. Nippon Ganka Gakkai Zasshi. 2008;112(1):22–9.
93. Nishigori H, Lee JW, Yamauchi Y, Iwatsuru M. The alteration of lipid peroxide in glucocorticoid-induced cataract of developing chick embryos and the effect of ascorbic acid. Curr Eye Res. 1986;5(1):37–40.
94. Stark G. The effect of ionizing radiation on lipid membranes. Biochim Biophys Acta. 1991;1071(2):103–22.
95. Yagi K, Komura S, Ihara N, Abe H, Konishi H, Arichi S. Serum lipid peroxide levels in rats with inherited cataracts. J Appl Biochem. 1985;7(3):202–6.
96. Goosey JD, Tuan WM, Garcia CA. A lipid peroxidative mechanism for posterior subcapsular cataract formation in the rabbit: a possible model for cataract formation in tapetoretinal diseases. Invest Ophthalmol Vis Sci. 1984;25(5):608–12.
97. Zigler Jr JS, Bodaness RS, Gery I, Kinoshita JH. Effects of lipid peroxidation products on the rat lens in organ culture: a possible mechanism of cataract initiation in retinal degenerative disease. Arch Biochem Biophys. 1983;225(1):149–56.
98. Zigler Jr JS, Gery I, Kessler D, Kinoshita JH. Macrophage mediated damage to rat lenses in culture: a possible model for uveitis-associated cataract. Invest Ophthalmol Vis Sci. 1983;24(5):651–4.
99. Zigler Jr JS, Hess HH. Cataracts in the Royal College of Surgeons rat: evidence for initiation by lipid peroxidation products. Exp Eye Res. 1985;41(1):67–76.
100. Borchman D, Paterson CA, Delamere NA. Oxidative inhibition of Ca2 + -ATPase in the rabbit lens. Invest Ophthalmol Vis Sci. 1989;30(7):1633–7.
101. Paterson CA, Zeng J, Husseini Z, Borchman D, Delamere NA, Garland D, Jimenez-Asensio J. Calcium ATPase activity and membrane structure in clear and cataractous human lenses. Curr Eye Res. 1997;16(4):333–8.
102. Ferrer JV, Gasco E, Sastre J, Pallardo FV, Asensi M, Vina J. Age-related changes in glutathione synthesis in the eye lens. Biochem J. 1990;269(2):531–4.
103. Shin AH, Oh CJ, Park JW. Glycation-induced inactivation of antioxidant enzymes and modulation of cellular redox status in lens cells. Arch Pharm Res. 2006;29(7):577–81.
104. Babizhayev MA. Accumulation of lipid peroxidation products in human cataracts. Acta Ophthalmol. 1989;67(3):281–7.
105. Borchman D, Yappert MC, Rubini RQ, Paterson CA. Distribution of phospholipid-malondialdehyde-adduct in the human lens. Curr Eye Res. 1989;8(9):939–46.

106. Kleiman NJ, Wang RR, Spector A. Hydrogen peroxide-induced DNA damage in bovine lens epithelial cells. Mutat Res. 1990;240(1):35–45.
107. Kleiman NJ, Wang RR, Spector A. Ultraviolet light induced DNA damage and repair in bovine lens epithelial cells. Curr Eye Res. 1990;9(12):1185–93.
108. Spector A, Kleiman NJ, Huang RR, Wang RR. Repair of H_2O_2-induced DNA damage in bovine lens epithelial cell cultures. Exp Eye Res. 1989;49(4):685–98.
109. Dong QY, Cui Y, Chen L, Song J, Sun L. Urinary 8-hydroxydeoxyguanosine levels in diabetic retinopathy patients. Eur J Ophthalmol. 2008;18(1):94–8.
110. Loft S, Fischer-Nielsen A, Jeding IB, Vistisen K, Poulsen HE. 8-Hydroxydeoxyguanosine as a urinary biomarker of oxidative DNA damage. J Toxicol Environ Health. 1993;40(2–3):391–404. doi:10.1080/15287399309531806.
111. Bhuyan KC, Bhuyan DK. Regulation of hydrogen peroxide in eye humors. Effect of 3-amino-1H-1,2,4-triazole on catalase and glutathione peroxidase of rabbit eye. Biochim Biophys Acta. 1977;497(3):641–51.
112. Dwivedi RS, Pratap VB. Alteration in glutathione metabolism during cataract progression. Ophthalmic Res. 1987;19(1):41–4.
113. Harding JJ. Free and protein-bound glutathione in normal and cataractous human lenses. Biochem J. 1970;117(5):957–60.
114. Scharf J, Dovrat A, Gershon D. Defective superoxide-dismutase molecules accumulate with age in human lenses. Graefes Arch Clin Exp Ophthalmol. 1987;225(2):133–6.
115. Spector A, Garner WH. Hydrogen peroxide and human cataract. Exp Eye Res. 1981;33(6):673–81.
116. Sweeney MH, Truscott RJ. An impediment to glutathione diffusion in older normal human lenses: a possible precondition for nuclear cataract. Exp Eye Res. 1998;67(5):587–95. doi:10.1006/exer.1998.0549.
117. Masella R, Di Benedetto R, Vari R, Filesi C, Giovannini C. Novel mechanisms of natural antioxidant compounds in biological systems: involvement of glutathione and glutathione-related enzymes. J Nutr Biochem. 2005;16(10):577–86. doi:10.1016/j.jnutbio.2005.05.013.
118. Lou MF, Dickerson Jr JE. Protein-thiol mixed disulfides in human lens. Exp Eye Res. 1992;55(6):889–96.
119. Rathbun WB. Lenticular glutathione synthesis: rate-limiting factors in its regulation and decline. Curr Eye Res. 1984;3(1):101–8.
120. Yan H, Harding JJ, Xing K, Lou MF. Revival of glutathione reductase in human cataractous and clear lens extracts by thioredoxin and thioredoxin reductase, in conjunction with alpha-crystallin or thioltransferase. Curr Eye Res. 2007;32(5):455–63. doi:10.1080/02713680701257837.
121. Takemoto L. Increase in the intramolecular disulfide bonding of alpha-A crystallin during aging of the human lens. Exp Eye Res. 1996;63(5):585–90. doi:10.1006/exer.1996.0149.
122. Lou MF, McKellar R, Chyan O. Quantitation of lens protein mixed disulfides by ion-exchange chromatography. Exp Eye Res. 1986;42(6):607–16.
123. Laver NM, Robison Jr WG, Calvin HI, Fu SC. Early epithelial lesions in cataracts of GSH-depleted mouse pups. Exp Eye Res. 1993;57(4):493–8.
124. Martensson J, Steinherz R, Jain A, Meister A. Glutathione ester prevents buthionine sulfoximine-induced cataracts and lens epithelial cell damage. Proc Natl Acad Sci U S A. 1989;86(22):8727–31.
125. McGinty SJ, Truscott RJ. Presbyopia: the first stage of nuclear cataract? Ophthalmic Res. 2006;38(3):137–48. doi:10.1159/000090645.
126. Pirie A. A light-calalysed reaction in the aqueous humor of the eye. Nature. 1965;205:500–1.
127. Bunce GE, Kinoshita J, Horwitz J. Nutritional factors in cataract. Annu Rev Nutr. 1990;10:233–54. doi:10.1146/annurev.nu.10.070190.001313.
128. Reddy VN, Giblin FJ, Lin LR, Chakrapani B. The effect of aqueous humor ascorbate on ultraviolet-B-induced DNA damage in lens epithelium. Invest Ophthalmol Vis Sci. 1998;39(2):344–50.

129. Blondin J, Taylor A. Measures of leucine aminopeptidase can be used to anticipate UV-induced age-related damage to lens proteins: ascorbate can delay this damage. Mech Ageing Dev. 1987;41(1–2):39–46.

130. Devamanoharan PS, Henein M, Morris S, Ramachandran S, Richards RD, Varma SD. Prevention of selenite cataract by vitamin C. Exp Eye Res. 1991;52(5):563–8.

131. Varma SD, Kumar S, Richards RD. Light-induced damage to ocular lens cation pump: prevention by vitamin C. Proc Natl Acad Sci U S A. 1979;76(7):3504–6.

132. Varma SD, Srivastava VK, Richards RD. Photoperoxidation in lens and cataract formation: preventive role of superoxide dismutase, catalase and vitamin C. Ophthalmic Res. 1982;14(3):167–75.

133. Stephens RJ, Negi DS, Short SM, van Kuijk FJ, Dratz EA, Thomas DW. Vitamin E distribution in ocular tissues following long-term dietary depletion and supplementation as determined by microdissection and gas chromatography-mass spectrometry. Exp Eye Res. 1988;47(2):237–45.

134. Taylor A. Role of nutrients in delaying cataracts. Ann N Y Acad Sci. 1992;669:111–23; discussion 123–114.

135. Weber P, Bendich A, Machlin LJ. Vitamin E and human health: rationale for determining recommended intake levels. Nutrition. 1997;13(5):450–60.

136. Libondi T, Menzione M, Auricchio G. In vitro effect of alpha-tocopherol on lysophosphatidylcholine-induced lens damage. Exp Eye Res. 1985;40(5):661–6.

137. Varma SD, Beachy NA, Richards RD. Photoperoxidation of lens lipids: prevention by vitamin E. Photochem Photobiol. 1982;36(6):623–6.

138. McCord JM, Fridovich I. Superoxide dismutase. An enzymic function for erythrocuprein (hemocuprein). J Biol Chem. 1969;244(22):6049–55.

139. Epp O, Ladenstein R, Wendel A. The refined structure of the selenoenzyme glutathione peroxidase at 0.2-nm resolution. Eur J Biochem. 1983;133(1):51–69.

140. Ho YS, Xiong Y, Ma W, Spector A, Ho DS. Mice lacking catalase develop normally but show differential sensitivity to oxidant tissue injury. J Biol Chem. 2004;279(31):32804–12. doi:10.1074/jbc.M404800200.

141. Spector A, Kuszak JR, Ma W, Wang RR, Ho Y, Yang Y. The effect of photochemical stress upon the lenses of normal and glutathione peroxidase-1 knockout mice. Exp Eye Res. 1998;67(4):457–71.

142. Behndig A, Karlsson K, Reaume AG, Sentman ML, Marklund SL. In vitro photochemical cataract in mice lacking copper-zinc superoxide dismutase. Free Radic Biol Med. 2001;31(6):738–44.

143. Buckingham RH. The behaviour of reduced proteins from normal and cataractous lenses in highly dissociating media: cross-linked protein in cataractous lenses. Exp Eye Res. 1972;14(2):123–9.

144. Truscott RJ, Augusteyn RC. The state of sulphydryl groups in normal and cataractous human lenses. Exp Eye Res. 1977;25(2):139–48.

145. Gladyshev VN, Liu A, Novoselov SV, Krysan K, Sun QA, Kryukov VM, Kryukov GV, Lou MF. Identification and characterization of a new mammalian glutaredoxin (thioltransferase), Grx2. J Biol Chem. 2001;276(32):30374–80. doi:10.1074/jbc.M100020200.

146. Lundberg M, Johansson C, Chandra J, Enoksson M, Jacobsson G, Ljung J, Johansson M, Holmgren A. Cloning and expression of a novel human glutaredoxin (Grx2) with mitochondrial and nuclear isoforms. J Biol Chem. 2001;276(28):26269–75. doi:10.1074/jbc.M011605200.

147. Raghavachari N, Lou MF. Evidence for the presence of thioltransferase in the lens. Exp Eye Res. 1996;63(4):433–41. doi:10.1006/exer.1996.0133.

148. Qiao F, Xing K, Liu A, Ehlers N, Raghavachari N, Lou MF. Human lens thioltransferase: cloning, purification, and function. Invest Ophthalmol Vis Sci. 2001;42(3):743–51.

149. Fernandes AP, Holmgren A. Glutaredoxins: glutathione-dependent redox enzymes with functions far beyond a simple thioredoxin backup system. Antioxid Redox Signal. 2004;6(1):63–74. doi:10.1089/152308604771978354.

150. Wu F, Wang GM, Raghavachari N, Lou MF. Distribution of thioltransferase (glutaredoxin) in ocular tissues. Invest Ophthalmol Vis Sci. 1998;39(3):476–80.
151. Lou MF. Thiol regulation in the lens. J Ocul Pharmacol Ther. 2000;16(2):137–48.
152. Wells WW, Xu DP. Dehydroascorbate reduction. J Bioenerg Biomembr. 1994;26(4):369–77.
153. Sasaki H, Giblin FJ, Winkler BS, Chakrapani B, Leverenz V, Shu CC. A protective role for glutathione-dependent reduction of dehydroascorbic acid in lens epithelium. Invest Ophthalmol Vis Sci. 1995;36(9):1804–17.
154. Xing K, Lou MF. The possible physiological function of thioltransferase in cells. FASEB J. 2003;17(14):2088–90. doi:10.1096/fj.02-1164fje.
155. Fernando MR, Lechner JM, Lofgren S, Gladyshev VN, Lou MF. Mitochondrial thioltransferase (glutaredoxin 2) has GSH-dependent and thioredoxin reductase-dependent peroxidase activities in vitro and in lens epithelial cells. FASEB J. 2006;20(14):2645–7. doi:10.1096/fj.06-5919fje.
156. Fernando MR, Nanri H, Yoshitake S, Nagata-Kuno K, Minakami S. Thioredoxin regenerates proteins inactivated by oxidative stress in endothelial cells. Eur J Biochem. 1992;209(3):917–22.
157. Yoshitake S, Nanri H, Fernando MR, Minakami S. Possible differences in the regenerative roles played by thioltransferase and thioredoxin for oxidatively damaged proteins. J Biochem. 1994;116(1):42–6.
158. Holmgren A. Thioredoxin and glutaredoxin systems. J Biol Chem. 1989;264(24):13963–6.
159. Yegorova S, Liu A, Lou MF. Human lens thioredoxin: molecular cloning and functional characterization. Invest Ophthalmol Vis Sci. 2003;44(8):3263–71.
160. Holmgren A, Bjornstedt M. Thioredoxin and thioredoxin reductase. Methods Enzymol. 1995;252:199–208.
161. Arner ES, Holmgren A. Physiological functions of thioredoxin and thioredoxin reductase. Eur J Biochem. 2000;267(20):6102–9.
162. Baker A, Payne CM, Briehl MM, Powis G. Thioredoxin, a gene found overexpressed in human cancer, inhibits apoptosis in vitro and in vivo. Cancer Res. 1997;57(22):5162–7.
163. Gasdaska JR, Berggren M, Powis G. Cell growth stimulation by the redox protein thioredoxin occurs by a novel helper mechanism. Cell Growth Differ. 1995;6(12):1643–50.
164. Wakasugi N, Tagaya Y, Wakasugi H, Mitsui A, Maeda M, Yodoi J, Tursz T. Adult T-cell leukemia-derived factor/thioredoxin, produced by both human T-lymphotropic virus type I- and Epstein-Barr virus-transformed lymphocytes, acts as an autocrine growth factor and synergizes with interleukin 1 and interleukin 2. Proc Natl Acad Sci U S A. 1990;87(21):8282–6.
165. Nakamura H, Matsuda M, Furuke K, Kitaoka Y, Iwata S, Toda K, Inamoto T, Yamaoka Y, Ozawa K, Yodoi J. Adult T cell leukemia-derived factor/human thioredoxin protects endothelial F-2 cell injury caused by activated neutrophils or hydrogen peroxide. Immunol Lett. 1994;42(1–2):75–80.
166. Taniguchi Y, Taniguchi-Ueda Y, Mori K, Yodoi J. A novel promoter sequence is involved in the oxidative stress-induced expression of the adult T-cell leukemia-derived factor (ADF)/human thioredoxin (Trx) gene. Nucleic Acids Res. 1996;24(14):2746. -2752.
167. Dalton TP, Shertzer HG, Puga A. Regulation of gene expression by reactive oxygen. Annu Rev Pharmacol Toxicol. 1999;39:67–101. doi:10.1146/annurev.pharmtox.39.1.67.
168. Nakamura H, Nakamura K, Yodoi J. Redox regulation of cellular activation. Annu Rev Immunol. 1997;15:351–69. doi:10.1146/annurev.immunol.15.1.351.
169. Schallreuter KU, Wood JM. The role of thioredoxin reductase in the reduction of free radicals at the surface of the epidermis. Biochem Biophys Res Commun. 1986;136(2):630–7.
170. Arrigo AP. Gene expression and the thiol redox state. Free Radic Biol Med. 1999;27(9–10):936–44.
171. Galter D, Mihm S, Droge W. Distinct effects of glutathione disulphide on the nuclear transcription factor kappa B and the activator protein-1. Eur J Biochem. 1994;221(2):639–48.

172. Schenk H, Klein M, Erdbrugger W, Droge W, Schulze-Osthoff K. Distinct effects of thioredoxin and antioxidants on the activation of transcription factors NF-kappa B and AP-1. Proc Natl Acad Sci U S A. 1994;91(5):1672–6.

173. Saitoh M, Nishitoh H, Fujii M, Takeda K, Tobiume K, Sawada Y, Kawabata M, Miyazono K, Ichijo H. Mammalian thioredoxin is a direct inhibitor of apoptosis signal-regulating kinase (ASK) 1. EMBO J. 1998;17(9):2596–606. doi:10.1093/emboj/17.9.2596.

174. Chae HZ, Kang SW, Rhee SG. Isoforms of mammalian peroxiredoxin that reduce peroxides in presence of thioredoxin. Methods Enzymol. 1999;300:219–26.

175. Chae HZ, Kim HJ, Kang SW, Rhee SG. Characterization of three isoforms of mammalian peroxiredoxin that reduce peroxides in the presence of thioredoxin. Diabetes Res Clin Pract. 1999;45(2–3):101–12.

176. Yegorova S, Yegorov O, Lou MF. Thioredoxin induced antioxidant gene expressions in human lens epithelial cells. Exp Eye Res. 2006;83(4):783–92. doi:10.1016/j.exer.2006.03.018.

177. Bhuyan KC, Reddy PG, Bhuyan DK. Thioredoxin genes in lens: regulation by oxidative stress. Methods Enzymol. 2002;347:421–35.

178. Reddy PG, Bhuyan DK, Bhuyan KC. Lens-specific regulation of the thioredoxin-1 gene, but not thioredoxin-2, upon in vivo photochemical oxidative stress in the Emory mouse. Biochem Biophys Res Commun. 1999;265(2):345–9. doi:10.1006/bbrc.1999.1691.

179. Spector A, Scotto R, Weissbach H, Brot N. Lens methionine sulfoxide reductase. Biochem Biophys Res Commun. 1982;108(1):429–34.

180. Weissbach H, Etienne F, Hoshi T, Heinemann SH, Lowther WT, Matthews B, St John G, Nathan C, Brot N. Peptide methionine sulfoxide reductase: structure, mechanism of action, and biological function. Arch Biochem Biophys. 2002;397(2):172–8. doi:10.1006/abbi.2001.2664.

181. Brot N, Weissbach L, Werth J, Weissbach H. Enzymatic reduction of protein-bound methionine sulfoxide. Proc Natl Acad Sci U S A. 1981;78(4):2155–8.

182. Vogt W. Oxidation of methionyl residues in proteins: tools, targets, and reversal. Free Radic Biol Med. 1995;18(1):93–105.

183. Kantorow M, Hawse JR, Cowell TL, Benhamed S, Pizarro GO, Reddy VN, Hejtmancik JF. Methionine sulfoxide reductase A is important for lens cell viability and resistance to oxidative stress. Proc Natl Acad Sci U S A. 2004;101(26):9654–9. doi:10.1073/pnas.0403532101.

184. Marchetti MA, Pizarro GO, Sagher D, Deamicis C, Brot N, Hejtmancik JF, Weissbach H, Kantorow M. Methionine sulfoxide reductases B1, B2, and B3 are present in the human lens and confer oxidative stress resistance to lens cells. Invest Ophthalmol Vis Sci. 2005;46(6):2107–12. doi:10.1167/iovs.05-0018.

185. Marchetti MA, Lee W, Cowell TL, Wells TM, Weissbach H, Kantorow M. Silencing of the methionine sulfoxide reductase A gene results in loss of mitochondrial membrane potential and increased ROS production in human lens cells. Exp Eye Res. 2006;83(5):1281–6. doi:10.1016/j.exer.2006.07.005.

186. Rosenfeld SI BM, Bobrow JC, et al. Lens and cataracts In: Basic and clinical science course. American Academy of Ophthalmology, San Francisco; 2007g. p. 89–160

187. Dick HB. Light-adjustable IOL: refractive adjustability provides greater predictability in post-refractive surgery eyes. EUROTIMES, vol 17. European Society of Cataract and Refractive Surgery. 2012.

188. Rosenfeld SI BM, Bobrow JC, et al. Lens and cataracts In: Basic and clinical science course. American Academy of Ophthalmology, San Francisco; 2007f. p. 163–194

189. Zhang S, Chai FY, Yan H, Guo Y, Harding JJ. Effects of N-acetylcysteine and glutathione ethyl ester drops on streptozotocin-induced diabetic cataract in rats. Mol Vis. 2008;14:862–70.

190. Grinberg L, Fibach E, Amer J, Atlas D. N-acetylcysteine amide, a novel cell-permeating thiol, restores cellular glutathione and protects human red blood cells from oxidative stress. Free Radic Biol Med. 2005;38(1):136–45. doi:10.1016/j.freeradbiomed.2004.09.025.

191. Cotgreave IA. N-acetylcysteine: pharmacological considerations and experimental and clinical applications. Adv Pharmacol. 1997;38:205–27.
192. Ates B, Abraham L, Ercal N. Antioxidant and free radical scavenging properties of N-acetylcysteine amide (NACA) and comparison with N-acetylcysteine (NAC). Free Radic Res. 2008;42(4):372–7. doi:10.1080/10715760801998638.
193. Carey JW, Pinarci EY, Penugonda S, Karacal H, Ercal N. In vivo inhibition of l-buthionine-(S, R)-sulfoximine-induced cataracts by a novel antioxidant, N-acetylcysteine amide. Free Radic Biol Med. 2011;50(6):722–9. doi:10.1016/j.freeradbiomed.2010.12.017.

Part III
Clinical Practice

Chapter 19
Updates and Advances in Corneal Surgery

Alex W. Cohen

19.1 Introduction

Corneal transplantation is a very successful means for restoring vision to those afflicted by any number of diseases. While the cornea is a very thin structure, it has three anatomically distinct layers each of which can be affected by a disease process. Pathology confined to the surface epithelial layer can usually be treated medically or with minor surgical procedures and rarely results in corneal transplantation. Diseases of the thicker corneal stromal layer often result in opacification of the visual axis or loss of structural integrity and the development of irregular astigmatism. Dystrophies such as lattice corneal dystrophy and granular corneal dystrophy occur because of defects in keratocyte collagen production and result in opaque deposits in the central cornea. Keratoconus is a multifactorial progressive disease, resulting in corneal warpage and ectasia. The abnormal corneal contour can no longer refract light appropriately, leading to distorted and reduced vision. When these diseases become severe enough, they require either a full thickness corneal transplant or an anterior lamellar transplant to replace the defective tissue. Diseases of the corneal endothelium are perhaps the most common reason for corneal transplantation currently. The endothelial cells are terminally differentiated and function to maintain a clear compact cornea. When these cells are lost, the corneal stroma begins to swell and eventually opacify, resulting in reduced vision. Endothelial cells can be affected by trauma, surgery, and genetic disease. While cataract surgery has come a long way in the past 50 years, it does result in endothelial cell death and in some cases, corneal edema. Fuchs corneal dystrophy is a state in which the

A.W. Cohen, M.D., Ph.D. (✉)
Dean McGee Eye Institute, University of Oklahoma,
608 Stanton L. Young Blvd, Oklahoma City, OK, USA
e-mail: Alex-Cohen@dmei.org

© Springer Science+Business Media New York 2015
M.A. Babizhayev et al. (eds.), *Studies on the Cornea and Lens*,
Oxidative Stress in Applied Basic Research and Clinical Practice,
DOI 10.1007/978-1-4939-1935-2_19

endothelial cells die off more rapidly that in a normal eye. In addition, as the cells die, they leave behind small excrescences, called guttata, at the level of Descemet's membrane that can result in visual decline. Diseases confined to the corneal endothelium can be selectively targeted for replacement with the new corneal transplant techniques outlined below.

19.2 Body

The first successful full thickness corneal transplant was performed by Dr. Eduard Zirm in 1905 [1]. This occurred after centuries of theoretical discussion regarding the topic as well as years of failed attempts at transplantation in animal models. During the ensuing decades, several other key ophthalmologists, including Ramon Castroviejo, advanced and refined the technique [1, 2]. Despite these many advances however, the overall concept of the technique remained largely the same. That is, regardless of the layer of corneal pathology, the entire full thickness of the cornea was replaced (Fig. 19.1).

From external to internal, the cornea consists of a thin epithelial layer with its associated underlying Bowman's layer. Deeper to the epithelium lies the cornea stroma which makes up the majority of the corneal substance. This is composed of a disparate population of keratocytes interspersed between layers of collagen. The most internal structure is a single layer of endothelium, which rests upon its basement membrane, called Descemet's membrane. The average thickness of a human cornea is about 550 μm. Despite its small size, pathology can be limited to a single layer of the cornea. For example, Fuchs Dystrophy results in primary endothelial cell loss, which, in advanced states produces corneal edema and decreased vision. Despite the fact that the pathology is initially confined to one layer of the cornea, for many years the surgical therapy involved a full thickness penetrating keratoplasty (PKP). While largely successful, PKP can be fraught with both intraoperative and postoperative complications.

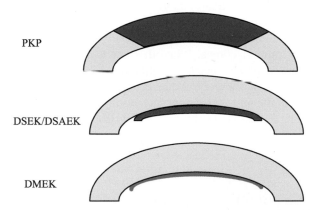

Fig. 19.1 Graphical representation of various surgical procedures. The blue area depicts the donor tissue. In a PKP the entire thickness of the central cornea is replace. In a DSEK/DSAEK the host Descemet's membrane is removed and replaced with donor Descemet's membrane and about 150 μm of donor stroma. In the DMEK the host endothelium is replaced with only donor endothelium

The eye is a pressurized system and a rapid decrease in this pressure, as is caused by the corneal incision in PKP, can result in intraocular or suprachoroidal hemorrhage and loss if intraocular contents. Postoperatively, the success of transplantation requires intensive topical steroid use. This can result in cataract formation and glaucomatous optic neuropathy. The localized immunosuppression also puts the recipient at risk for bacterial and fungal corneal infections which can be visually devastating.

The concept of lamellar corneal transplantation, or replacing just a layer of the cornea, is not new [2]. Early attempts at corneal transplantation included lamellar surgery in animals as well as in humans. In 1838, Richard Sharp Kissam attempted to transplant the anterior portion of a pig cornea onto a human cornea without success. For decades following, surgeons struggled with keratoplasty techniques but the anterior lamellar technique remained the predominant attempted form. These techniques fell out of favor with the success of the full thickness technique in the 1900s.

It was not until 1998 when Dutch ophthalmologist, Gerrit Melles, described a technique for replacing the inner layers of the cornea that a resurgence of interest in lamellar surgery came about [3]. Never before had successful attempts been made to selectively replace damaged endothelial tissue do largely to difficulties in surgical technique. Melles's technique involved creating a large (~9 mm) corneal scleral pocket that extended circumferentially into the clear cornea. The donor tissue was prepared in a similar fashion by hand dissecting the posterior corneal stroma to create a circular button. The donor tissue was placed into the eye and tucked into corneal pockets that had been created during the initial dissection. The technique relied upon air to aid in the dissection of the recipient bed as well as for holding the donor tissue in position. This technique was termed Posterior Lamellar Keratoplasty (PLK) [3].

Shortly after Melles's initial description of the technique, American ophthalmologist, Mary Terry described a revision of the technique, which he termed Deep Lamellar Endothelial Keratoplasty (DLEK) [4, 5]. In this technique, the eye is filled with a viscoelastic material rather than air and then the inner ~90 μm of posterior corneal tissue is dissected from the adjacent stroma and removed from the eye. The surgeon performs a similar dissection of donor tissue and then places it within the anterior chamber through a smaller (~5 mm) incision. The graft is oriented and then air is placed below it to hold it in place against that posterior host stroma. In addition, a circular corneal pocket is dissected into which the donor tissue is tucked. This surgery preserves the natural corneal curvature and spares the patient a large incision. However, it is technically very difficult and time consuming to perform. In addition, the postoperative best corrected visual acuity is generally less than that seen with a traditional PKP [6]. The reduction in vision has been attributed to scar tissue and haze that forms at the interface between the two hand dissected tissue planes [7]. Due to its difficulty, DLEK surgery was not widely adopted by cornea surgeons throughout the world.

In 2001 Melles again made history by describing an advancement in the above technique in which there was no tedious dissection of the host stromal tissue [2, 8]. Rather, Descemet's membrane is scrolled in the periphery and then removed from the eye. A donor lenticule consisting of endothelium, Descemet's membrane, and about 120 μm of stroma is then placed into that anterior chamber (Figs. 19.2 and 19.3). Similar to DLEK surgery, air is placed within the eye to facilitate attachment of the donor tissue to the posterior corneal surface (Fig. 19.4).

Fig. 19.2 DSAEK donor tissue after cutting with a trephine. The donor endothelium is facing up and the tissue has been inked in the periphery to aide in orientation

Fig. 19.3 DSAEK tissue being inserted into the eye via a 5 mm scleral incision. The tissue from Fig. 19.2 has been folded like a taco and is being placed into the eye with a pair of forceps

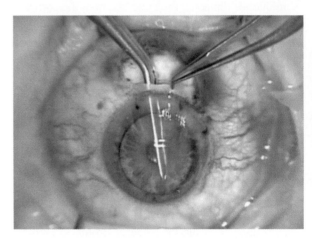

Fig. 19.4 DASEK tissue in correct position. After being inserted into the eye, the donor tissue is unfolded and floated up against the posterior corneal surface with an air bubble

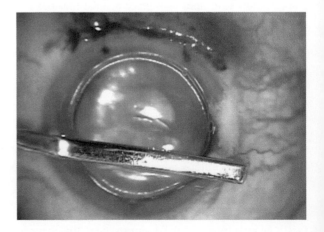

Eliminating the technically difficult dissection of the host corneal stroma results in a much easier surgical procedure as well as better visual results. In some instances, the results approach perfection and rival those obtained with PKP. This procedure was termed Descemet's stripping endothelial keratoplasty (DSEK).

Advances in donor tissue preparation also led to faster and easier surgery. In the beginning of DLEK and DSEK surgery, the surgeon was responsible for cutting and preparing the donor tissue. This led to the development of artificial anterior chambers and specialized blades to dissect free posterior corneal buttons for transplantation. Still, tissue preparation could be tedious and complicated by perforation and destruction of the donor tissue. For many years, eye banks have been procuring and processing tissue for full thickness transplantation. As lamellar surgery grew in popularity some eye banks began precutting the donor tissue for the surgeon's use. While a busy surgeon may perform 100 lamellar procedures in a year, a busy eye bank may prepare more than 1,000 precut corneas. This repetition and volume led to better and better tissue preparation, less tissue damage, and less tissue loss [9]. Now, the vast majority of surgeons use precut tissue in a procedure that has been termed Descemet's stripping automated endothelial keratoplasty (DSAEK) to indicate that an *automated* lamellar dissection was performed to create the tissue.

For the past several years, DSEK and DSAEK have been the mainstay of posterior lamellar keratoplasty in the United States. However, in 2006, Melles again described an advancement in the technique of endothelial keratoplasty that he termed Descemet's membrane endothelial keratoplasty (DMEK) [2, 10]. In this procedure, the patient's cornea is prepared in a similar fashion to DSAEK. Descemet's membrane is scrolled and removed from the posterior surface of the eye and the edges of the stroma are roughened. The difference however is in the tissue that is then inserted into the eye. Unlike DSAEK in which the endothelium is inserted along with 120 µm of donor stroma, in DMEK only the endothelium and Descemet's membrane are inserted into the eye. While this may sound simple, it is technically very challenging. Preparation of the donor tissue is very difficult given that the surgeon is dealing with a single cell layer. During preparation the tissue may be torn and destroyed. Again, some surgeons are turning to the eye banks for tissue preparation with good success. Once the tissue has been removed from the donor stroma is quickly scrolls tightly, making correct orientation difficult (Fig. 19.5). The tissue is injected into the patient's eye via a small incision (Fig. 19.6) and then an air bubble in placed on top of the tissue and used to unfold the graft. Once the graft has been unrolled, an air bubble is placed beneath the tissue to push it against the posterior aspect of the patient's cornea. Over time the donor cornea will stick in place as the new endothelial cells begin to function (Fig. 19.7).

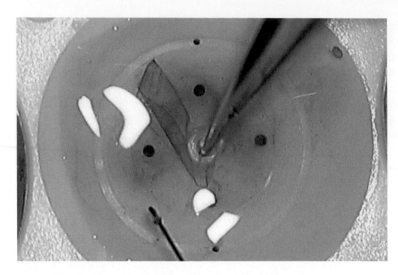

Fig. 19.5 DMEK tissue during processing. The tissue has been stained with trypan blue and peeled from the donor stroma. The tissue is so thin that it immediately scrolls onto itself

Fig. 19.6 DMEK tissue after insertion into the eye. It is tightly scrolled making orientation difficult

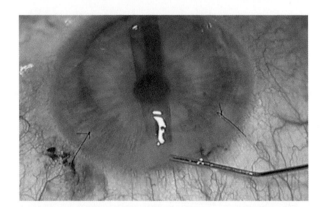

Fig. 19.7 DMEK graft 1 day after surgery. The eye is about 80 % full of air to facilitate graft adhesion. The edges of the graft are difficult to detect as the tissue is very thin

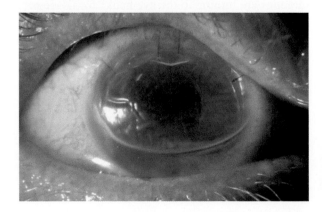

19.3   Conclusion

The past 15 years have seen a rapid advancement in the field of lamellar corneal transplantation. This has resulted in improved patient outcomes, reduced surgical times, and reduced transplant rejection rates. Dr. Melles is still working toward the next iteration of the endothelial transplant and others are working toward expanding autologous donor endothelial cells in culture for transplantation. While there is still a place for the full thickness transplant, the age of lamellar corneal surgery is here to stay.

References

1. Armitage WJ, Tullo AB, Larkin DF. The first successful full-thickness corneal transplant: a commentary on Eduard Zirm's landmark paper of 1906. Br J Ophthalmol. 2006;90:1222–3.
2. Crawford AZ, Patel DV, McGhee CN. A brief history of corneal transplantation: from ancient to modern. Oman J Ophthalmol. 2013;6:S12–7.
3. Melles GR, Eggink FA, Lander F, Pels E, Rietveld FJ, Beekhuis WH, Binder PS. A surgical technique for posterior lamellar keratoplasty. Cornea. 1998;17:618–26.
4. Terry MA, Ousley PJ. Deep lamellar endothelial keratoplasty in the first United States patients: early clinical results. Cornea. 2001;20:239–43.
5. Terry MA, Ousley PJ. Endothelial replacement without surface corneal incisions or sutures: topography of the deep lamellar endothelial keratoplasty procedure. Cornea. 2001;20:14–8.
6. Melles GR, Lander F, van Dooren BT, Pels E, Beekhuis WH. Preliminary clinical results of posterior lamellar keratoplasty through a sclerocorneal pocket incision. Ophthalmology. 2000;107:1850–6; discussion 7.
7. Terry MA, Ousley PJ. Replacing the endothelium without corneal surface incisions or sutures: the first United States clinical series using the deep lamellar endothelial keratoplasty procedure. Ophthalmology. 2003;110:755–64; discussion 64.
8. Melles GR, Kamminga N. [Techniques for posterior lamellar keratoplasty through a scleral incision]. Ophthalmologe. 2003;100:689–95.
9. Teba FA, Mohr A, Eckardt C, Wong D, Kusaka S, Joondeph BC, Feron EJ, Stalmans P, Van Overdam K, Melles GR. Trypan blue staining in vitreoretinal surgery. Ophthalmology. 2003;110:2409–12.
10. Melles GR, Ong TS, Ververs B, van der Wees J. Descemet membrane endothelial keratoplasty (DMEK). Cornea. 2006;25:987–90.

12.5   Conclusion

Chapter 20
Cataract and Diabetic Retinopathy

Iñigo Corcóstegui Crespo and Elío Díez-Feijóo Varela

Abbreviations

AGE	Advanced glycation end-products
DM	Diabetes mellitus
DR	Diabetic retinopathy
OCT	Optical coherence tomography.
VA	Visual acuity
VEGF	Vascular endothelial growth factor

20.1 Introduction

Diabetes mellitus (DM) is a metabolic disease characterized by the lack or inefficiency of insulin, which generates an increase in glycemia that, if left untreated over time, causes changes at multiple levels. According to the World Health Organization, there are more than 340 million people worldwide diagnosed with diabetes. Moreover, this figure is expected to rise due to the progressive ageing of the population, dietary changes, and the increasingly sedentary lifestyle found in developing countries [1].

Due to the metabolic changes that accompany DM, diabetics are more likely to develop cataracts and at an earlier age than their nondiabetic counterparts. In fact, diabetes is the third most significant risk factor for the opacification of the crystalline lens after genetic inheritance and age [2]. Cataracts are a major cause of

I.C. Crespo (✉) • E.D.-F. Varela
Surgical Clinical Institute of Ophthalmology (ICQO),
C/Virgen de Begoña, 34, 48006 Bilbao, Spain
e-mail: icorcos@gmail.com; eliodiez@hotmail.com

© Springer Science+Business Media New York 2015   371
M.A. Babizhayev et al. (eds.), *Studies on the Cornea and Lens*,
Oxidative Stress in Applied Basic Research and Clinical Practice,
DOI 10.1007/978-1-4939-1935-2_20

decreased vision in diabetic patients, and the incidence of cataract surgery in this population is two to five times higher than that of nondiabetic patients [3]. Diabetic patients with poor metabolic control and those who have suffered from the disease for many years also have a higher incidence of cataracts [4]. The most frequent types of DM-related cataracts are cortical cataracts and subcapsular cataracts [5].

Physiological or senile cataracts consist of a series of changes of the crystalline lens that include nuclear sclerosis from an increase in crystalline lens fibers and loss of water content, although crystalline lens changes that differ from senile sclerosis have also been observed in patients with DM. These changes include immature lens fibers, lamellae separation and nuclei present in deeper layers [6], as well as an increase in protein aggregation due to the alteration of intracellular calcium-activated cysteine proteinases (calpains) [7]. Several theories have been proposed to explain the cause for crystalline lens opacification in diabetic patients; however, the precise mechanism remains unknown but is likely to be multifactorial in origin. Among the theories put forward, the recent oxidation theory must be considered. The crystalline lens, through the actions of glutathione reductase (GSH), counteracts, in part, the effects of oxidation [8]. In diabetes, the increase in reactive oxygen species (ROS) has been proposed to generate oxidative stress that is not totally compensated by GSH; this stress generates disulfide cross-linking between proteins, which leads to their aggregation and alters the optical properties of the crystalline lens [8]. Additionally, in DM patients, the introduction of advanced glycation end-products (AGE) as a result of the abnormal metabolism of lipids, carbohydrates, and proteins could play a decisive role in the oxidative stress of the lens [9].

After the cataract is diagnosed, the inflammatory stimulus resulting from surgery involves the release of cytokines and proangiogenic factors that can lead to worsening of diabetic retinopathy (DR) and the ocular involvement associated with DM. In particular, the presence of significant macular edema, rubeosis iridis, and retinal neovascularization has been described after cataract surgery in DM patients [10]. Furthermore, poor metabolic control of DM or active DR increases the risk of an abnormal inflammatory response after cataract surgery, which may influence the occurrence of postsurgical complications such as postoperative uveitis, anterior chamber fibrin, or vitreous hemorrhage [11]. Glycated hemoglobin (HbA1c) is a serum parameter that provides objective information on the degree of metabolic control of DM and indirectly reports the risk of postsurgical complications related to this disease.

Therefore, it is advisable to achieve improved metabolic control and perform the necessary treatments for the stabilization of DR before cataract surgery. When such factors are controlled, the risk of postoperative complications related to DM is significantly reduced, thereby improving the prognosis of patients following surgery and decreasing the surgical risk.

The recent introduction of drugs capable of blocking vascular endothelial growth factor (VEGF) in clinical practice has resulted in a significant improvement in the control of DR [12]. Moreover, the application of such drugs together with laser photocoagulation of the retina prior to cataract surgery generally helps to avoid postsurgical complications resulting from DM [13–15].

Developments in surgical techniques have been a decisive factor in the improvement of cataract surgery results in DM patients. Specifically, the development of extra-capsular extraction techniques for the crystalline lens using phacoemulsification has reduced surgical time, inflammatory response and the complications resulting from surgery. In addition, this technique permits a smaller corneal incision, is less traumatic than the former intra-capsular extraction of the lens and achieves better postoperative results [16–18].

20.2 Indications for Surgery and Prognosis

The indication to perform cataract surgery in diabetic patients can be determined according to the following two situations:

(a) Opacity of the lens that significantly decreases the visual acuity (VA) of the patient.
(b) Opacity of the lens that prevents the examination or treatment of the retina in diabetic patients.

20.3 Preoperative Evaluation and Medical History

Before conducting cataract surgery on a diabetic patient, it is important to perform an initial assessment of the possible ocular and systemic repercussions of DM. This information is useful to determine if the patient requires complementary treatment prior to surgery and to assess the patient's prognosis.

The preoperative evaluation of a diabetic patient with cataracts must include the following data: (1) DM type and duration; (2) glycemic control level, preferably the glycated hemoglobin (HbA1c) level; [18] (3) other cardiovascular risk factors (blood pressure, cholesterol, possible cardiac diseases, and smoking history); (4) ophthalmologic history including previous diagnoses and treatments (e.g., previous history of retinal laser photocoagulation, possible neovascular glaucoma, history of previous surgeries).

The initial eye examination of a diabetic patient with cataracts is similar to that of a nondiabetic patient, although particular attention should be paid to the ocular structures affected by this disease. Therefore the examination of the anterior segment of the eye must evaluate the following components: (1) Corneal endothelium; (2) Iris and iris vessels, including the presence of neovascularization in the iris and pupillary border and the approximate measurement of the dilated pupillary diameter; and (3) Gonioscopy (with and without pupillary dilation) [19].

Once the exploration of the anterior segment is completed and with the pupil dilated, the examination of the posterior pole is performed. First, the presence of DR, its stage and its level of activity must be examined. The DR present is then classified as simple (absence of neovascularization) (Fig. 20.1) or proliferative

Fig. 20.1 Retinography of a patient with non-proliferative diabetic retinopathy and diffuse macular edema (*dotted line*). Hard exudates (*asterisks*) and micro-aneurysms (*arrows*) are present in the paramacular area

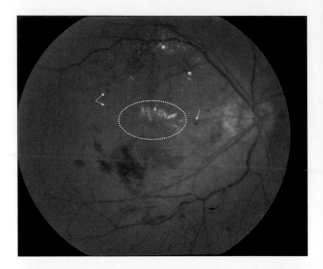

Fig. 20.2 Retinography of a patient with proliferative diabetic retinopathy. There is a macular hemorrhage (*asterisk*) and neovascular tissue (*arrow*) on the optic nerve

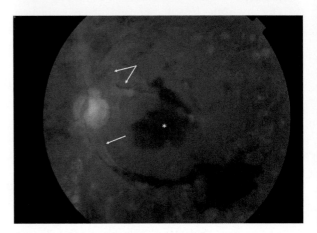

(presence of neovascularization or hemorrhage) (Fig. 20.2). Furthermore, simple DR can be classified as mild, moderate, or severe according to the number of retinal microaneurysms, the number of hemorrhages and the presence of venous anomalies [20].

For any patient with diagnosed DR, fluorescein angiography should be considered to assess the presence of retinal neovascularization, peripheral ischemia, or macular edema. Because the detection of diabetic macular edema with conventional ophthalmoscopy in cataract patients may be complicated, optical coherence tomography (OCT) provides a rapid noninvasive test that can more accurately identify possible macular neuroepithelial degeneration and the presence of macular edema (Figs. 20.3 and 20.4). OCT also enables the detection of other macular alterations in DM patients, such as vitreomacular traction syndrome or the epiretinal membrane. However, any of these alterations could represent a contraindication for cataract surgery and may influence surgical technique choice.

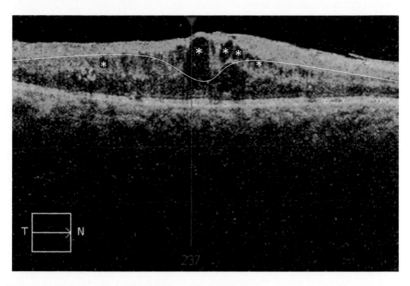

Fig. 20.3 Optical coherence tomography axial macular scan of a patient with diabetic macular edema showing intraretinal cysts (*asterisk*) and increased macular thickness. *White line* shows normal macular thickness and shape

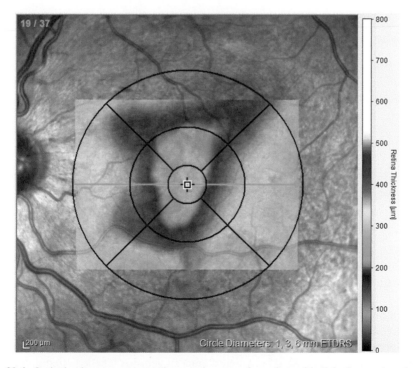

Fig. 20.4 Optical coherence tomography macular map of a patient with diabetic macular edema. Each of the *colors* that appear in the macular map represent a certain retinal thickness

20.4 Preoperative Treatment

Poor prognoses after cataract interventions on diabetic patients are due to postsurgical eye complications from this disease [15, 21]. To avoid such complications, it is important, whenever possible, to treat the ocular alterations (mainly retinal) associated with DM. The main therapeutic tools available include laser photocoagulation and the intravitreal administration of anti-VEGF drugs.

The prognosis of patients with active DR undergoing cataract treatment improves significantly with previous laser photocoagulation treatment [22]. An analysis of previous studies comparing the long-term visual outcomes between treated and untreated patients indicates that it is advisable to conduct appropriate laser treatment of DR before performing cataract surgery [4, 11].

Potent drugs that are currently available can help to stop the progression of DR when administered to the vitreous humor of the patient. Such drugs (Avastín®, Eylea®, Lucentis®, and triamcinolone acetonide) can also be combined with laser treatment to obtain better control of DR before cataract surgery [23].

20.5 Surgical Procedure and Additional Treatments

Once cataract treatment in diabetic patients has been indicated, one of the following three situations will be found:

20.5.1 Patient with Cataracts Without Significant Diabetic Retinopathy

Patients with a healthy retina that require only cataract treatment should receive conventional cataract surgery similar to nondiabetic patients.

20.5.2 Cataracts and Diabetic Macular Edema

Macular edema is the accumulation of fluid in the macular neuroepithelium and is the main cause of loss of vision in patients under 50 years of age in developed countries (Figs. 20.1, 20.2, and 20.4). Fluorescein angiography can be used to detect macular fluid extravasation. Although the opacity resulting from cataracts may hinder the detection of macular thickening during conventional funduscopic examination, the OCT approach permits an accurate measure of such thickening. Various studies have shown a potential relationship between cataract surgery in diabetic patients and postsurgical macular edema, for which some authors have recommended performing cataract surgery only when the condition is advanced and not operating on patients with an initial cataract to avoid the risk of macular edema [14, 16]. However, other authors have suggested carrying out the surgery before the macular edema is detected given that functional recovery after surgery is less successful when it is present [12, 20]. The ideal approach would be to definitively treat the macular edema before cataract surgery; however, in many cases, this condition is recurrent, and treatment may take months or years using combined laser and intravitreal anti-VEGF injection therapies [24].

Currently, with the existence of different drugs that show efficacy in the treatment of macular edema, its presence does not modify the indications for cataract surgery, and this condition can be treated simultaneously via intravitreal anti-VEGF injection and cataract surgery [25].

20.5.3 Cataracts and Vitreoretinal Alterations That Require Surgical Treatment

In cases where cataracts are diagnosed together with other alterations that require treatment, the typical approach is to combine both procedures. The most common surgical technique in these cases is phacoemulsification with a standard lens implant. However, we must consider that any retinal surgery (vitrectomy) performed in a patient over 50 years of age will cause lens opacification. As a result, when patients of this age group require vitrectomy, lens extraction is carried out during the same surgical procedure even though no significant cataract is present at that moment; this approach avoids a second operation a few months later when the cataract would develop. The cataract extraction surgical technique in these patients is the same as in patients without DM (phacoemulsification with standard lens implant), with the exception of patients with anterior tractional retinal detachment or a high degree of severity of the retinopathy [26].

The main vitreoretinal alterations that may accompany cataracts in diabetic patients and require vitrectomy include vitreomacular traction syndrome, epiretinal membrane, vitreous hemorrhage, and tractional retinal detachment.

Vitreomacular traction syndrome is a disease caused by the abnormal adherence between the vitreous and the retina. Specifically, the macula is pulled by the vitreous layer adhered to it, which leads to structural alterations that are frequent in diabetic patients [27].

An epiretinal membrane displays an abnormal growth of cells over the macular surface, which forms a tissue layer that can create traction over the macula, thereby altering its structure and producing vision loss [28].

Vitreous hemorrhage is a frequent complication of proliferative DR, which generally indicates active disease.

Tractional retinal detachment is typically a manifestation of advanced proliferative DR and consists of the separation of the retinal neuroepithelium and the retinal pigmented epithelium due to traction from the fibrous tissue.

These four situations previously described are treated using vitrectomy, and when these conditions coexist with a cataract, the combined procedure of vitrectomy and cataract surgery is performed [29, 30].

20.6 Surgical Technique

As discussed previously, the surgical treatment of cataracts in diabetic patients is, with the exception of the abovementioned cases, similar to that of non-diabetic patients [31]. In this section, we briefly describe the standard cataract surgical technique.

The cataract procedure is performed in most situations using only topical anesthesia. After the instillation of a few drops of anesthetic, the conjunctival sac, the ocular surface, and the periocular region are cleaned with povidone-iodine. Then, the eyelashes are isolated from the surgical field using sterile plastic material.

The first step of the operation consists of a corneal incision of approximately 1.8–2.4 mm, through which the entire procedure will be carried out. Secondly, an opening of the anterior capsule of the lens is formed, through which saline is injected for hyperhydration and rotation of the lens (hydrodissection and hydrodelamination). Then, the phacoemulsifier is introduced through the incision and the anterior capsule of the lens. This device emits ultrasound energy to produce liquefaction of the lens, which is aspirated. Once the lens capsule is empty, an intraocular lens is introduced into the crystalline sac.

As mentioned previously, the inflammatory reaction resulting from cataract surgery may favor the progression of DR, and it is therefore important to perform this operation with the least possible trauma [32]. Thus, it is essential to shorten surgical time and minimize the energy of phacoemulsification especially in medium-hard cataracts. Cataract surgery in a hypermature lens will involve a considerable increase in the surgical aggression delivered to the eye, with an increase in postoperative inflammation and a higher risk for progression of DR or other complications.

Another peculiarity of cataract surgery in DM patients is their higher incidence of retraction of the lens capsule after the operation (Fig. 20.5) [33]. The proliferation of remaining epithelial cells from the lens and the inflammatory environment are two factors that may play an important role in increasing the opacity and post-surgical contraction of the crystalline sac common in such patients [34, 35].

The specific aspects that must be taken into account when performing cataract surgery in a diabetic patient are the following [36]:

6.1 Surgical trauma should be avoided. As described earlier, surgical trauma may favor macular edema; it can accelerate the course of DR and favor contraction and opacification of the lens capsule. Additionally, the corneal endothelium receives greater damage in diabetic patients and demonstrates a poor recovery of postoperative edema following traumatic surgery [37].

Fig. 20.5 Anterior segment
image of a patient with
opacification and retraction
of the anterior lens capsule
(anterior capsule phimosis).
The *dotted line* shows the
original perimeter of the
capsulorhexis performed
during surgery. This
retraction (*arrows*) may be
difficult for future retinal
examinations or treatments

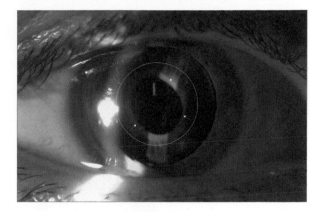

Fig. 20.6 Anterior segment
image of a patient with
persistent epithelial defect
(*arrows*) after a combined
surgical procedure,
phacoemulsification, and
vitrectomy

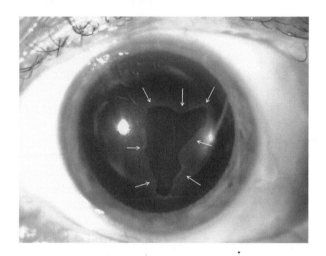

6.2 Corneal epithelium protection. The ocular surface is altered in diabetic patients,
which leads to poor healing of the surgical wound and an increase in the
incidence of postoperative epithelial erosions [38]. In some patients, these erosions may become complicated and develop into true persistent epithelial
defects with a difficult resolution (Fig. 20.6). To ensure protection of the corneal epithelium, hydration of the surface of the eye during surgery, avoiding
traumatic maneuvers and applying topical ointments in cases of intraoperative
de-epithelization are important. This is especially important for the combined
treatment of cataracts and vitrectomy [39].

6.3 Iris management. There is a higher incidence of poor dilation or intraoperative
miosis in diabetic patients [40]. In advanced retinopathies, the combination of
rubeosis and poor dilation may produce intraoperative hemorrhages of the
anterior chamber (Fig. 20.7).

Fig. 20.7 Anterior segment image of a diabetic patient with rubeosis iridis (*arrows*). These neovessels are weak and may bleed during surgery

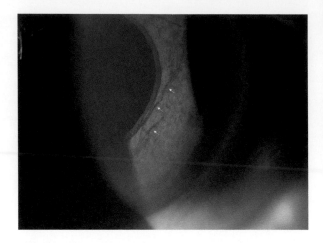

6.4 A larger capsulorhexis than in nondiabetic patients should be performed, so that any potential contraction and opacification of the capsule does not obstruct future retinal examination.

6.5 Wide optical intraocular lens optics (greater than 6.5 mm) should be used to facilitate future retinal examination.

6.6 The use of silicone intraocular lenses should be avoided. These lenses are avoided so that if it is necessary to introduce silicon oil into the vitreous cavity secondary to DR, complications from these types of lenses will be avoided. Additionally, it is not currently advisable to use multifocal lenses because these can obstruct retinal examination and decrease the visual quality of patients with macular lesions.

6.7 The combined use of intravitreal Triamcinolone or anti-VEGF drugs should be evaluated in cases of persistent macular edema or traumatic surgery due to hypermature lenses.

6.8 Overexposure to the surgical microscopes light should be avoided, as patients with DR can be more sensitive to phototoxicity produced by the microscope light [41].

6.9 In cases of vitreous hemorrhage, the use of dyes in the lens capsule should be contemplated to achieve better performance during capsulorhexis.

6.10 In cases receiving the combined procedure with vitrectomy or when gas or air is left as an intraocular tamponade, the injection of a miotic agent in the anterior chamber of the eye should be considered after surgery to decrease the risk of pupillary entrapment during the postoperative period [42].

6.11 In cases receiving the combined procedure with vitrectomy, posterior capsulotomy should be performed to complete the surgery to decrease the risk of postoperative opacity and capsular contraction.

20.7 Conclusions

Advances in the development of cataract surgical techniques, new intravitreally administered drugs and laser treatments have achieved a significant improvement in the prognosis of cataract surgery in diabetic patients. A multidisciplinary approach involving the metabolic control of the disease and the control of cardiovascular risk factors is of great importance to achieve a good long-term prognosis.

References

1. World Health Organization. Data and statistics. Global burden of diabetes 1995–2025: prevalence. Numerical estimates and projections. www.who.int/mediacentre/factsheets/fs312/es/. Accessed 25 Sept 2013.
2. Cotter SA, Varma R, Ying-Lai M, Azen SP, Klein R, Los Angeles Latino Eye Study Group. Causes of low vision and blindness in adult Latinos: the Los Angeles Latino Eye Study. Ophthalmology. 2006;113:1574–8.
3. Klein BE, Klein R, Moss SE. Incidence of cataract surgery in de Wisconsin epidemiologic study of diabetic study of diabetic retinopathy. Am J Ophthalmol. 1995;119:295–300.
4. Chew EY, Benson WE, Remaley NA, Lindley AA, Burton TC, Csaky K, Williams GA, Ferris FL. Results after lens extraction in patients with diabetic retinopathy: early treatment diabetic retinopathy study report number 25. Arch Ophthalmol. 1999;117:1600–6.
5. Saxena S, Mitchell P, Rochtchina E. Five-year incidence of cataract in older persons with diabetes and pre-diabetes. Ophthalmic Epidemiol. 2004;11:271–7.
6. Hegde KR, Varma SD. Morphogenetic and apoptotic changes in diabetic cataract: prevention by pyruvate. Moll Cell Biochem. 2004;262:233–7.
7. Tang D, Borchma D, Yappert MC, Vrensen GF, Rasi V. Influence of age diabetes, and cataract on calcium, lipid-calcium, and protein-calcium relationships in human lenses. Invest Ophthalmol Vis Sci. 2003;44:2059–66.
8. Ganea E, Harding JJ. Glutathione-related enzymes and the eye. Curr Eye Res. 2006;31:1–11.
9. Ahmed N. Advanced glycation endproducts-role in pathology of diabetic complications. Diabetes Res Clin Pract. 2005;67:3–21.
10. Jaffe GJ, Burton TC, Kuhn E, Prescott A, Hartz A. Progression of nonproliferative diabetic retinopathy and visual outcome after extracapsular cataract extraction and intraocular lens implantation. Am J Ophthalmol. 1992;114:448–56.
11. Hykin PG, Gregson RMC, Stevens JD, Hamilton PAM. Extracapsular cataract extraction in proliferative diabetic retinopathy. Ophthalmology. 1993;100:394–9.
12. Stewart MW. Anti-vascular endothelial growth factor drug treatment of diabetic macular edema: the evolution continues. Curr Diabetes Rev. 2012;8:237–46.
13. Dowler JG, Hykin PG, Hamilton AM. Phacoemulsification versus extracapsular cataract extraction in patients with diabetes. Ophthalmology. 2000;107:457–62.
14. Benson WE, Brown GC, Tasman W, McNamara JA, Vander JF. Extracapsular cataract extraction with placement of a posterior chamber lens in patients with diabetic retinopathy. Ophthalmology. 1993;100:730–8.
15. Schatz H, Atienza D, McDonald R, Johnson RN. Severe diabetic retinopathy after cataract surgery. Am J Ophthalmol. 1994;117:314–21.
16. Krupsky S, Zalish M, Oliver M, Pollack A. Anterior segment complications in diabetic patients following extracapsular cataract extraction and posterior chamber intraocular lens implantation. Ophthalmic Surg. 1991;22:526–30.

17. Pollack A, Leiba H, Bukelman A, Oliver M. Cystoid macular oedema following cataract extraction in patients with diabetes. Br J Ophthalmol. 1992;76:221–2.
18. Somaiya MD, Burns JD, Mintz R, Warren RE, Uchida T, Godley BF. Factors affecting visual outcomes after small incision phacoemulsification in diabetic patients. J Cataract Refract Surg. 2002;28:1364–71.
19. Blinder KJ, Friedman SM, Mames RN. Diabetic iris neovascularization. Am J Ophthalmol. 1995;120:393–5.
20. Wilkinson CP, Ferris FL, Klein RE, et al. Proposed internation clinical diabetic retinopathy and diabetic macular edema disease severity scales. Ophthalmology. 2003;110:1677–82.
21. Dowler JGF, Sehmi KS, Hykin PG, Hamilton AM. The natural history of macular edema after cataract surgery in diabetes. Ophthalmology. 1999;106:663–8.
22. Photocoagulation therapy for diabetic eye disease. Early Treatment Diabetic Retinopathy Study Research Group. JAMA. 1985;254(21):3086.
23. Ho AC, Scott IU, Kim SJ, Brown GC, Brown MM, Ip MS, Recchia FM. Anti-vascular endothelial growth factor pharmacotherapy for diabetic macular edema: a report by the American Academy of Ophthalmology. Ophthalmology. 2012;119:2179–88.
24. Arevalo JF, Lasave AF, Wu L, Diaz-Llopis M, Gallego-Pinazo R, Alezzandrini AA, Berrocal MH. Intravitreal bevacizumab plus grid laser photocoagulation or intravitreal bevacizumab or grid laser photocoagulation for diffuse diabetic macular edema: results of the Pan-American Collaborative Retina Study Group at 24 months. Pan-American Collaborative Retina Study Group (PACORES). Retina. 2013;33:403–13.
25. Akinci A, Muftuoglu O, Altınsoy A, Ozkılıc E. Phacoemulsification with intravitreal bevacizumab and triamcinolone acetonide injection in diabetic patients with clinically significant macular edema and cataract. Retina. 2011;31:755–8.
26. Nieto l. Cirugía de la catarata en enfermedades vitrorretinianas. In: Corcóstegui B, Adán A, García-Arumí J, Mateo C, Nieto l, editors. Cirugía vitreorretiniana: indicaciones y técnicas. Madrid: Tecnimedia; 1999. p. 91–103.
27. Sonmez K, Capone Jr A, Trese MT, Williams GA. Vitreomacular traction syndrome: impact of anatomical configuration on anatomical and visual outcomes. Retina. 2008;28:1207–14.
28. Ophir A, Martinez MR. Epiretinal membranes and incomplete posterior vitreous detachment in diabetic macular edema, detected by spectral-domain optical coherence tomography. Invest Ophthalmol Vis Sci. 2011;52:6414–20.
29. Chaudhry NA, Cohen KA, Flynn Jr HW, Murray TG. Combined pars plana vitrectomy and lens management in complex vitreoretinal disease. Semin Ophthalmol. 2003;18:132–41.
30. Lahey JM, Francis RR, Kearney JJ, Cheung M. Combining phacoemulsification and vitrectomy in patients with proliferative diabetic retinopathy. Curr Opin Ophthalmol. 2004;15:192–6.
31. Benson WE. Cataract surgery and diabetic retinopathy. Curr Opin Ophthalmol. 1992;3: 396–400.
32. Mittra RA, Barrillo JL, Dev S, Mieler WF, Koening SB. Retinopathy progression after phacoemulsification in patients with diabetes mellitus. Arch Ophthalmol. 2000;118:912–7.
33. Ebihara J, Kato S, Oshika Y, Yoshizaki M, Sugita G. Posterior capsule opacification after cataract surgery in patients with diabetes mellitus. J Cataract Refract Surg. 2006;32:1184–7.
34. Nishi O, Nishi K. Intraocular lens encapsulation by shrinkage of the capsulorhexis opening. J Cataract Refract Surg. 1993;19:544–54.
35. Nishi O, Nishi K, Fujiwara T, Shirasawa E, Ohmoto Y. Effects of the cytokines on the proliferation of and collagen synthesis by human cataract lens epithelial cells. Br J Ophthalmol. 1996;80:63–8.
36. Nieto I, Atienza R. Cataract surgery in diabetic patients. In: Lorent R, Mendicute J, editors. LXXXIV Ponencia oficial de la sociedad Española de Oftalmología. Madrid: SEO; 2008. p. 1192–202.
37. Morikubo S, Takamura Y, Kubo E, Tsuzuki S, Akagi Y. Corneal changes alter small-incision cataract surgery in patients with diabetes mellitas. Arch Ophthalmol. 2004;122:966–9.
38. Tsubota K, Chiba K, Shimazaki J. Corneal epithelium in diabetic patients. Cornea. 1991; 10:156–60.

39. Hiraoka M, Amano S, Oshika T, Katto S, Hori S. Factors contributing to corneal complications after vitrectomy in diabetic patients. Jpn J Ophthalmol. 2001;45:492–5.
40. Mirza SA, Alexandridou A, Marshall T, Stavrou P. Surgically induced miosis during phako-emulsification in patients with diabetes mellitas. Eye. 2003;17:194–9.
41. Cetinkaya A, Yilmaz G, Akova YA. Photic retinopathy after cataract surgery in diabetic patients. Retina. 2006;26:1021–8.
42. Demetriades AM, Gottsch JD, Thomsen R, Azab A, Stark WJ, Campochiaro PA, de Juan E, Haller Jr JA. Combined phacoemulsification, intraocular lens implantation and vitrectomy for eyes with coexisting cataract and vitreoretinal pathology. Am J Ophthalmol. 2003; 135:291–6.

Chapter 21
Traumatic Cataract: A Review

Mehul Shah and Shreya Shah

Abbreviations

BETTS   Birmingham Trauma Terminology systems
IOL   Intraocular lens
IOP   Intraocular pressure
OGI   Open globe injury
OTS   Ocular trauma score

21.1　Introduction

Trauma is probably the most under-recognized major health problem today. Trauma is a cause of monocular blindness in the developed world, although few studies have addressed the problem of trauma in rural areas [1]. The aetiology of ocular injury is likely to differ from that in urban areas and is worthy of investigation [2–4]. Any strategy for prevention requires knowledge of the cause of injury, which may enable more appropriate targeting of resources toward preventing such injuries. Both eye trauma victims and society bear a large, potentially preventable burden [3].

For prevention, knowledge of cause of injury is required for appropriate targeting of resources towards prevention of such injury. So, it will automatically reduce the economic burden of the society.

Ocular trauma can cause cataracts [1]. The methods used to evaluate the visual outcome in eyes managed for traumatic cataracts and senile cataracts are similar [5], but the damage to other ocular tissues due to trauma may compromise the visual

M. Shah (✉) • S. Shah
Drashti Netralaya, Nr. GIDC, Chakalia Road, Dahod 389151, Gujarat, India
e-mail: omtrust@rediffmail.com

© Springer Science+Business Media New York 2015   385
M.A. Babizhayev et al. (eds.), *Studies on the Cornea and Lens*,
Oxidative Stress in Applied Basic Research and Clinical Practice,
DOI 10.1007/978-1-4939-1935-2_21

gain in eyes operated on for traumatic cataracts. Hence, the success rates may differ between eyes with these two types of cataract.

With the introduction of the Birmingham Eye Trauma Terminology System (BETTS), the documentation of ocular trauma has been standardized [5]. Consequently, it would be interesting to study the visual outcomes following traumatic cataract surgery and the determinants predicting the outcome, especially in relation to the BETTS. There are approximately 1.6 million blind persons from injuries in the world, an additional 2.3 million people have bilateral low vision from this cause, and almost 19 million have unilateral blindness or low vision [6].

Work accidents are still frequent events despite efforts made for primary prevention, and the eye is a high-risk organ for work accidents. Although it represents only 0.27 % of the total body area and 4 % of the facial area, ocular trauma is very frequent [7]. Visual outcomes of traumatic cataracts have been reported in some cases [8, 9]. However, most studies involved small samples or were case studies.

In the present study, we examined the visual outcomes following cataract surgery in eyes sustaining injuries, and the predictors of satisfactory visual outcomes following the management of traumatic cataracts. Our study was conducted in a city located at the borders of three states in India: Gujarat, Madhya Pradesh, and Rajasthan [4]. Qualified ophthalmologists at our institute provide low-cost eye services mainly to the poor belonging to the tribal population of 4.2 million in this area.

Our primary objective is to determine visual outcome after treatment of traumatic cataract. Our secondary objectives are to determine the epidemiology of traumatic cataract, to compare the visual outcome in closed and open globe injury, and to study the effect of interval between time of injury and time of intervention on initial visual outcome, the effect of the morphology of traumatic cataract on final visual outcome, and to compare the effect of primary posterior capsulotomy and vitrectomy as primary step of surgery.

21.2   Patients and Methods

We obtained approval from the hospital administrators and research committee to conduct this study, and we received the participants' written consent.

This was a prospective study designed in 2002. All traumatic cataracts in either eye diagnosed and managed between January 2003 and December 2009 were enrolled in our study, and those consenting to participate and not having other serious body injuries were included.

For each patient enrolled in our study, we obtained a detailed history, including details of the injury and information on eye treatment and surgery performed to manage past ocular trauma. Data for both the initial and follow-up reports were collected using the online BETTS format of the International Society of Ocular Trauma. Details of the surgery were also collected using a specified pretested online form.

Fig. 21.1 Total cataract: when no clear lens matter was visible between the capsule and the nucleus, the cataract was defined as a total cataract

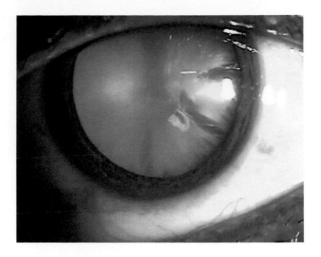

The cases of traumatic cataract were grouped as those with open- and those with closed globe injuries. The open globe injuries were further categorized into those with lacerations and those with rupture. Lacerations of the eyeball were subcategorized into eyes with perforating injuries, penetrating injuries, or injuries involving an intraocular foreign body. The closed globe group was subdivided into lamellar laceration and contusion.

Other demographic details collected included patient entry, residence, activity at the time of injury, object of injury, and previous examinations and treatments. After enrollment, all patients were examined using a standard method. Visual acuity was checked using the Snellen's chart, and the anterior segment was examined using a slit lamp.

Based on lenticular opacity, the cataracts were classified as total [10] (Fig. 21.1), membranous, in which both capsules fused with scant or no cortical material (Fig. 21.2), white soft (Fig. 21.3), and rosette types (Fig. 21.4). When an ophthalmologist did not observe clear lens matter between the capsule and nucleus, the cataract was defined as total. When the capsule and organized matter were fused and formed a membrane of varying density, it was defined as a membranous cataract. When loose cortical material was found in the anterior chamber together with a ruptured lens capsule, the cataract was defined as white soft. A lens with a rosette pattern of opacity was classified as a rosette type cataract.

For a partially opaque lens, the posterior segment examination was carried out with an indirect ophthalmoscope and a +20 D lens. When the optical medium was not clear, a B-scan was performed to evaluate the posterior segment.

The surgical technique was selected according to morphology [10] and the condition of tissues other than the lens. Phacoemulsification was used to operate on cataracts with hard, large nuclei. With a lens that had either a white soft or rosette type of cataract, unimanual or bimanual aspiration was used. Membranectomy and anterior vitrectomy, either via an anterior or pars plana route, were performed when the cataract was membranous.

Fig. 21.2 Membranous cataract: when the capsule and organized matter were fused and formed a membrane of varying density, the cataract was defined as a membranous cataract

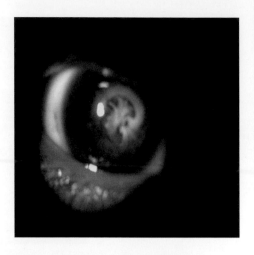

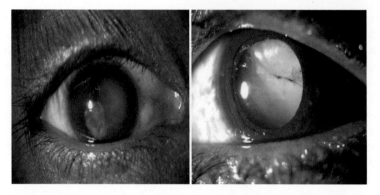

Fig. 21.3 White soft fluffy cataract: when loose cortical material was found in the anterior chamber together with a ruptured lens capsule, the cataract was defined as a white soft cataract

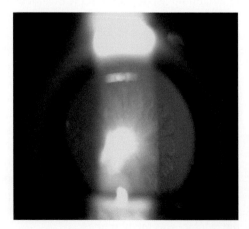

Fig. 21.4 Rosette cataract: lens with a rosette pattern of opacity was classified as a rosette type cataract

In all patients undergoing corneal wound repair, the traumatic cataract was managed in a second procedure. Recurrent inflammation was more prominent in patients who had undergone previous surgery for trauma [8, 9]. In such cases, the ocular medium will turn hazy due to condensation of the anterior vitreous unless a vitrectomy is performed. Hence, we performed a capsulectomy and vitrectomy via an anterior/pars plana route in adults.

In children younger than 2 years of age, both lensectomy and vitrectomy via a pars plana route were performed, and the same surgical procedures were used to manage the traumatic cataract. Lens implantation as part of the primary procedure was avoided in all children younger than 2 years of age.

All patients with injuries and without an infection were treated with topical and systemic corticosteroids and cycloplegics. The duration of medical treatment depended on the degree of inflammation in the anterior and posterior segments of the operated eye. The operated patients were re-examined after 24 h, 3 days, and 1, 2, and 6 weeks to enable refractive correction. Follow-up was scheduled for the third day, weekly for 6 weeks, monthly for 3 months, and every 3 months for 1 year.

At every follow-up examination, visual acuity was tested using the Snellen's chart. The anterior segment was examined with a slit lamp, and the posterior segment was examined with an indirect ophthalmoscope. Eyes with vision better than 20/60 at the glasses appointment (6 weeks) were defined as having a satisfactory grade of vision.

During the examination, data were entered online using a specified pretested format designed by the International Society of Ocular Trauma (initial and follow-up forms), which was exported to a Microsoft Excel spread sheet. The data were audited periodically to ensure completeness. We used the Statistical Package for Social Studies (SPSS 15) to analyze the data. The univariate parametric method was used to calculate frequency, percentage, proportion, and 95 % confidence interval (95 % CI). We used binominal regression analysis to determine the predictors of postoperative satisfactory vision (>20/60). The dependent variable was vision >20/60 noted at the follow-up 6 weeks after cataract surgery. The independent variables were age, gender, residence, time interval between injury and cataract surgery, primary posterior capsulectomy and vitrectomy procedure, and type of ocular injury.

21.3 Results

Our cohort consisted of 687 patients with traumatic cataracts including 496 (72.2 %) eyes with open globe ocular injuries and 191 (27.8 %) eyes with closed globe injuries (Fig. 21.5). The mean patient age was 27.1 ± 18.54 years (range, 1–80, Table 21.1). The patients were 492 (71.6 %) males and 195 (28.4 %) females, Table 21.2.

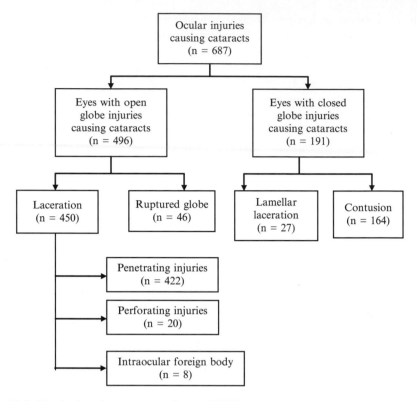

Fig. 21.5 Distribution of patients according to BETTS

Table 21.1 Age distribution

Age	Number (*n*)	Percentage (%)
0 to 10	155	22.6
11 to 20	184	26.8
21 to 30	90	13.1
31 to 40	79	11.5
41 to 50	95	13.8
51 to 60	55	8.0
61 to 70	24	3.5
71 to 80	4	0.5
Total	687	100

Table 21.2 Gender distribution

Gender	Number	Percentage
Female	195	28.4
Male	492	71.6
Total	687	100

Table 21.3 Entry of the patients

	Number	Percentage
Camp	181	26.3
Door to door	2	0.3
Other	7	1.0
Referral	6	0.9
School	16	2.3
Self	75	69.1
Total	687	100

$p=0.03$

Table 21.4 Socioeconomic status

	Number	Percentage
Poor	540	78.6
Rich	144	21.0
Very poor	3	0.4
Total	687	100

$p=0.008$

Table 21.5 Reporting interval according to residence

Days	Rural	Urban	Total
0–1	157	15	172
2–4	75	3	78
5–30	198	3	201
More	225	11	236
Total	655	32	687

We analyzed several demographic factors, including origin of patient referral (Table 21.3), socioeconomic status (79 % were from a lower socioeconomic class, Table 21.4), and residence (95 % were from a rural area, Table 21.5), none of which had a significant relationship ($p=0.3$) with final visual acuity. The object causing the injury (Table 21.7) and the activity at the time of the injury (Table 21.6) were also not significantly associated with satisfactory final visual acuity. Wooden sticks were the most common agent of injury (55.9 %, Table 21.7). A comparison of pre- and postoperative visual acuity showed that treatment significantly improved visual acuity (Table 21.8; Pearson's χ^2 test, $p<0.001$; ANOVA, $p=0.001$). An intraocular lens was implanted in 564 (82.1 %) cases (Table 21.9). The number of surgeries required varied significantly with morphology ($p=0.000$) (Table 21.10).

All traumatic cataracts were classified according to morphology (Tables 21.11 and 21.12) and were surgically treated using morphology as a guideline (Table 21.12).

Final visual outcome was found to vary according to morphology (Table 21.11) and surgical technique (Table 21.17). White soft cataracts were found to have a better prognosis and achieved significantly higher rates of positive outcome compared with other morphologies ($p=0.014$) (Table 21.11).

Table 21.6 Activity during injury

	N	Percent
Fall	11	1.6
Fighting	4	0.6
Firecrackers	5	0.7
House work	187	27.2
Job work	137	19.9
Other	114	16.6
Plain walk	14	2.0
Play	183	26.6
Travel top	29	4.2
Vehicular injury	3	0.4
Total	687	100.0

$p=0.03$

Table 21.7 Object of injury

Object	Number (n)	Percent (%)
Ball	6	9
Cattle horn	16	2.37
Finger	7	1.0
Firework	10	1.5
Glass	5	0.7
Iron wire	46	6.7
Other	58	8.4
Sharp	8	1.2
Stone	93	13.5
Unknown	54	7.9
Wooden stick	384	55.9
Total	687	100.0

Aspiration was performed using one or two ports in 48.6 % of the patients in the open globe group, and was significantly associated with improved visual acuity ($p<0.001$, Table 21.12).

We were able to do lens implants in 82 % of cases; details are shown in Table 21.9.

In comparing open globe and closed globe groups (Table 21.13), we found significant differences in variables other than final visual outcome, including age, gender, origin of patient referral, object of injury, early reporting, urban vs. rural residence, cataract morphology, surgical technique, number of surgeries, and lens implantation.

At 6 weeks postoperatively, the visual acuity in the operated eye was >20/60 in 298 (58 %) eyes in the open globe group, and 75 (39.1 %) eyes in the closed globe group ($p<0.001$, ANOVA, χ^2); this difference was significant (OR=1.61, 95 % CI 0.85–3.02). Overall, 373 (54.3 %) eyes regained a final visual acuity >20/60 (Table 21.14).

Table 21.8 Comparative visual acuity before and after treatment

Pre op vision	Posttreatment vision												
	Uncooperative		<1/60		1/60 to 3/60		20/200 to 20/120		20/80 to 20/60		20/40 to 20/20		Total E
Uncooperative	7	1.01	0	0	0	0	0	0	1	0.14	2	0.29	10	1.45
<1/60	10	1.45	164	23.87	53	7.71	54	7.86	110	16.01	191	27.80	582	84.71
1/60 to 3/60	0	0	4	0.58	3	0.43	8	1.16	21	3.05	10	1.45	46	6.69
6/60 to 6/36	0	0	0	0	0	0	1	0.14	7	1.01	12	1.74	20	2.91
6/24 to 6/18	0	0	3	0.43	0	0	1	0.14	6	0.87	7	1.01	17	2.47
6/12 to 6/6	0	0	0	0	0	0	0	0	0	0	1	0.14	1	0.14
Total	17	2.47	171	24.89	55	8.00	64	9.31	145	21.10	223	32.45	675	98.5

Legend: <1/60, 2/60 to 3/60, 6/60 to 6/24, 6/18 to 6/12, Unicooperative

Relationship between pre- and posttreatment visual acuity

χ^2 test, $p = 0.000$

Table 21.9 Lens implant in relation to morphology of traumatic cataract

	No implant		Implant		Total	
Morphology	N	%	N	%	N	%
Membranous	17	2.5	67	9.8	84	12.2
Rosette	2	0.3	6	0.9	8	1.2
White soft	58	8.4	354	51.5	412	60.0
Total cataract	46	6.7	137	19.9	183	26.6
Total	123	17.9	564	82.1	687	100

Table 21.10 Number of surgeries

Number	0–1	2–4	5–30	More	Total
1	128	58	181	216	584
2	39	19	18	19	94
3	5	1	2	1	9
Total	172	78	201	236	687

Table 21.11 Final visual outcome in relation to morphology of cataract

Final visual outcome	Morphology									
	Membranous		Rosette		White soft		Total cataract		Total	
	N	%	N	%	N	%	N	%	N	%
Uncooperative	3	0.4	0	0	11	1.6	5	0.7	19	2.7
<1/60	24	3.5	2	0.3	89	12.9	58	8.4	173	25.1
1/60 to 3/60	12	1.7	0	0	27	3.9	19	2.7	58	8.4
20/200 to 20/120	7	1.0	1	0.1	33	4.8	25	3.6	66	9.6
20/80 to 20/60	21	3.0	1	0.1	91	13.2	34	4.9	147	21.4
20/40 to 20/20	17	2.5	4	0.5	161	23.4	42	6.1	224	32.6
Total	83	12.1	7	1.0	412	59.9	183	26.6	687	100

χ^2 test, $p=0.014$

Table 21.12 Surgical techniques used according to morphology of cataract

	Morphology									
	Membranous		Rosette		White soft		Total cataract		Total	
Surgical technique	N	%	N	%	N	%	N	%	N	%
Aspiration	7	1	5	0.7	316	45.9	19	2.7	347	50.5
Lensectomy + vitrectomy	46	6.7	0	0	60	8.7	20	2.9	126	18.3
Phaco/SICS[a]	3	0.4	2	0.3	17	2.4	108	15.7	130	18.9
Delivery + vitrectomy	28	4	0	0	20	2.9	36	5.2	84	12.2
Total	83	12.1	7	1	412	59.9	183	26.6	687	100

[a]Small incision cataract surgery

Table 21.13 Comparative results amongst open and closed globe injuries

Parameter	Open globe		Closed globe		Total		
	No	%	No	%	No	%	p Value
Socioeconomic status							
Poor	386	56.2	154	22.4	540	78.6	0.008
Rich	108	15.7	36	5.2	144	21.0	
Very poor	2	0.3	1	0.15	3	0.4	
Total	496	72.2	191	27.8	687	100	
Entry							
Camp	114	16.6	67	9.6	181	26.3	0.03
Door-to-door	1	0.1	1	0.1	2	0.3	
Other	6	0.9	1	0.1	7	1.0	
Referral	15	2.2	2	0.3	6	0.9	
School	2	0.3	1	0.1	16	2.3	
Self	358	52.1	119	17.3	475	69.1	
Total	496	72.2	191	27.8	687	100	
Previous surgical treatment							
No	485	70.6	183	26.6	668	97.2	0.126
Yes	11	1.6	8	1.2	19	2.7	
Total	496	72.2	191	27.8	687	100	
Age distribution							
0 to 10	134	19.5	22	3.2	155	22.6	0.000
11 to 20	145	21.1	39	5.7	184	26.8	
21 to 30	70	10.1	20	2.9	90	13.1	
31 to 40	54	7.7	25	3.6	79	11.5	
41 to 50	52	7.5	43	6.3	95	13.8	
51 to 60	25	3.6	30	4.3	55	8.0	
61 to 70	15	2.1	9	1.3	24	3.5	
71 to 80	1	0.1	3	0.4	4	0.5	
Total	496	72.2	191	27.8	687	100	
Gender							
Female	152	22.1	43	6.3	195	28.4	0.020
Male	344	50.1	148	21.5	492	71.6	
Total	496	72.8	191	27.8	687	100	
Object of injury							
Ball	0	0	6	0.9	6	0.9	0.031
Cattle horn	6	0.9	6	0.9	12	1.60	
Cattle tail	1	0.1	3	0.4	4	0.6	
Fire work	6	0.9	4	0.5	10	1.5	
Other	34	4.9	40	5.8	74	10.8	
Sharp object	49	7.1	10	1.5	59	8.6	
Stone	55	8.0	38	5.5	93	13.5	
Wooden stick	312	45.4	20	2.9	54	7.9	
Unknown	34	4.9	72	10.5	384	55.9	
Total	496	72.8	191	27.8	687	100	

(continued)

Table 21.13 (continued)

Parameter	Open globe		Closed globe		Total		
	No	%	No	%	No	%	p Value
Object of injury compared to wooden stick							
Wooden stick object	312	45.4	72	10.5	384	55.9	0.000
Other object	184	26.8	119	17.3	303	44.1	
Total	496	72.8	191	27.8	687	100	
Reporting							
0 to 1	132	19.2	40	5.8	172	25.0	0.000
2 to 4	69	10.0	9	1.3	78	11.4	
5 to 30	163	23.7	38	5.5	201	29.3	
More	132	19.2	104	15.1	236	34.4	
Total	496	78.2	191	27.8	687	100	
Habitat							
Rural	479	69.7	176	25.6	655	95.3	0.014
Urban	17	2.5	15	2.1	32	4.7	
Total	496	78.2	191	27.8	687	100	
Morphology							
Membranous	63	9.1	20	2.9	83	12.1	0.000
Rosette	6	0.8	3	0.4	9	1.3	
Soft fluffy	329	47.9	83	12.1	412	60.0	
Total cataract	98	14.3	85	12.3	183	26.6	
Total	496	72.8	191	27.8	687	100	
Surgical technique							
Aspiration	292	42.5	55	8.0	347	50.5	0.000
Lensectomy and vitrectomy	90	13.1	36	5.2	126	18.3	
Delivery and vitrectomy	114	16.6	100	14.6	214	31.1	
Total	496	72.2	191	27.8	687	100	
Primary posterior capsulectomy vitrectomy							
Not performed	364	53.0	141	20.5	505	73.5	0.495
Performed	132	19.2	50	7.3	182	26.5	
Total	496	72.2	191	27.8	687	100	
Number of surgeries							
1.00	411	59.8	173	25.1	584	85	0.010
2.00	76	11.1	18	2.6	94	13.7	
3.00	9	1.3	0	0.0	9	1.3	
Total	496	72.2	191	27.8	687	100	
Lens implant							
No implant	65	9.5	58	8.4	123	17.9	0.000
Implant	431	62.7	133	19.4	564	82.0	
Total	496	72.2	191	27.8	687	100	

(continued)

Table 21.13   (continued)

Parameter	Open globe		Closed globe		Total		
	No	%	No	%	No	%	p Value
Final visual outcome							
<1/60	110	22.2	64	33.5	174	25.3	0.000
1/60 to 3/60	38	7.7	18	9.4	56	8.1	
20/200 to 20/80	38	7.7	27	14.1	65	9.4	
20/60 to 20/40	110	22.1	38	19.8	148	21.5	
20/40 to 20/20	188	37.9	37	19.3	225	32.7	
Uncooperative	12	2.4	6	3.1	18	2.6	
Total	496	100	191	100	687	100	

Table 21.14   Final visual outcome

Visual acuity	Open globe		Close globe		Total	
	NO	%	NO	%	NO	%
<1/60	110	22.2	64	33.5	174	25.3
1/60 to 3/60	38	7.7	18	9.4	56	8.1
20/200 to 20/80	38	7.7	27	14.1	65	9.4
20/60 to 20/40	110	22.1	38	19.8	148	21.5
20/32 to 20/20	188	37.9	37	19.3	225	32.7
Uncooperative	12	2.4	6	3.1	18	2.6
Total	496	100	191	100	687	100

Table 21.15   Primary posterior capsulectomy and vitrectomy and initial best corrected visual acuity

Procedure	<1/60		1/60 to 3/60		20/200 to 20/80		20/60 to 20/40		20/32 to 20/20		Total	
	N	%	N	%	N	%	N	%	N	%	N	%
Not performed	76	13.7	26	4.8	71	12.8	101	18.2	88	15.8	362	65.2
Performed	34	6.1	14	2.5	32	5.8	11	1.9	11	1.9	102	18.4
Total	110	19.8	40	7.2	103	18.5	122	20.2	99	17.8	464	83.6

Postoperative vision was compared according to the type of injury. Primary posterior capsulectomy and anterior vitrectomy, commonly performed for eyes with significant inflammation, resulted in a significant improvement in final visual acuity (Table 21.15; $p < 0.001$). We performed IOL relocation in five cases and secondary membranectomy in four cases.

When we have studied time interval between injury and initial intervention we found best results achieved if initial intervention done between 2 and 30 days (Tables 21.16 and 21.17). We have also studied the probable main reasons for no improvement in vision (Table 21.18).

Table 21.16 Final visual outcome according to reporting time

Vision	0–1	2–4	5–30	More	Total
Uncooperative	3	2	4	8	17
<1/60	51	13	49	59	172
1/60 to 3/60	12	7	9	30	56
20/200 to 20/120	22	7	12	26	67
20/80 to 20/60	34	16	48	50	148
20/40 to 20/20	50	33	79	63	225
Total	172	78	201	236	687

$p=0.02$

Table 21.17 Distribution of reporting time and percentage of patients with a final visual acuity >20/80

	Reporting interval (%)				
	0–1	2–4	5–30	More	Total
Vision >20/80	48.8	63.6	63	48	54.4

Table 21.18 Probable reasons for failure of improvement in vision

Reason for failure	Number	Percentage
No reason	202	36.4
Corneal scar	27	5.0
Extensive posterior segment damage	14	2.5
Glaucoma	2	0.4
Infection	3	0.5
Inflammation	28	5.0
Lens position	2	0.4
Lost follow-up	30	6.2
Other	7	1.3
Total	315	

We grouped our patients into four groups according to the interval between injury and the start of intervention: 25 % patients within 24 h, 11.4 % between 2 and 4 days, 29 % within 30 days, and 34.4 % after 30 days (Table 21.16). Patients residing in rural areas reported significantly later ($p=0.011$). A significant difference was found for socioeconomic status ($p=0.008$). Self-reporting patients reported early, while patients who came from an outreach program reported late ($p<0.001$). The final visual outcome differed significantly among the four groups (Table 21.17). Of the patients who reported between 2 and 30 days, 63 % recovered vision >20/80, while 48 % of the patients who reported within 24 h and after 30 days recovered vision >20/80 (Table 21.17). More surgery was performed in cases who reported early (Table 21.5). Patient who had open globe injury reported significantly earlier ($p<0.001$, Table 21.13).

We have studied the effect of age at time of intervention and found it to make a significant difference on final visual outcome ($p=0.000$, Table 21.19).

Lens implantation was done in 82 % of cases, and no significant difference was found among the groups ($p=0.357$). There was no significant difference in the incidence of infection among the groups ($p=0.931$).The final visual outcome in cases of injury with a wooden stick was significantly different from the outcomes following injuries caused by other objects (Table 21.20).

The overall incidence of infection was 0.4 % (Table 21.21). Injury with a wooden stick had a zero incidence of infection, but this was not significantly different from the infection incidence associated with injuries caused by other objects (Table 21.21). The final visual outcome was significantly affected by infection (Table 21.22).

21.4 Discussion

Visual gain following surgery for traumatic cataracts is a complex process. Electrophysiological [11] and radio-imaging [12, 13] investigations are important tools for assessing co-morbidities associated with an opaque lens.

Using a large database, we attempted to systematically classify the morphology of traumatic cataract and to select surgical techniques accordingly.

Various studies have touched on this topic. One study found 52.3 % total cataracts, whereas our results revealed 26.6 % total cataracts [6] Vajpayee reported an opening in the posterior capsule with types 1 and 2 openings with penetrating injury [6], whereas we found another membranous type of cataract (12.1 %) suggestive of late reporting, as membranous transformation of the lens with fusion of the anterior and posterior capsules may occur over time.

We were able to perform intraocular lens implants in 82.1 % of cases. Krishnamachary reported implants in 65.5 % cases, Churchill et al. reported implants in 46.8 % cases, and Fyodrove reported Sputnik implant in all cases [6, 7].

Fyodrove reported surgical techniques according to pathology and degree of lens absorption, but did not systematically classify similar factors taken into account in our study [6, 14].

These differences in findings may reflect the fact that of the 687 patients in our study, 30 % came from outreach activities and did not approach treatment facilities on their own. Therefore, the time period between ocular injury and reporting ranged widely, and morphology was also influenced by this interval.

We propose a specific morphological classification for traumatic cataract, which may provide guidelines for management that incorporate available technology and improve care for these patients.

In our study of patients with open globe and closed globe injuries leading to traumatic cataract, a satisfactory grade of vision following surgical management was significantly more common in those with open globe injuries (Table 21.14).

We are not aware of any other reported study that has compared final visual outcomes between these two groups, using the BETT classification system.

Table 21.19 Effect age group on final visual outcome

Age	Uncooperative		<1/6C		1/60 to 3/60		20/200 to 20/120		20/80 to 20/60		20/40 to 20/20		Total	
	n	%	n	%	n	%	n	%	N	%	n	%	n	%
0 to 10	9	1.31	46	6.69	15	2.18	12	1.74	37	5.38	34	4.94	153	22.27
11 to 20	2	0.29	30	4.36	14	2.03	17	2.47	33	4.80	87	12.66	183	26.63
21 to 30	2	0.29	19	2.76	8	1.16	6	0.87	12	1.74	42	6.11	89	12.95
31 to 40	1	0.14	23	3.35	5	0.73	7	1.01	16	2.32	25	3.63	77	11.2
41 to 50	3	0.43	23	3.35	6	0.87	12	1.74	28	4.07	22	3.20	94	13.68
51 to 60	0	0.0	21	3.00	7	1.01	6	0.87	8	1.16	11	1.60	53	7.71
61 to 70	0	0.0	8	1.16	0	0	2	0.29	10	1.45	2	0.29	22	3.20
71 to 80	0	0.0	1	0.14	0	0	2	0.29	1	0.14	0	0	4	0.58
Total	17	2.47	173	25.18	55	8.00	64	9.31	146	21.25	223	32.45	675	98.5

χ^2 test, $p = 0.000$

Table 21.20 Visual outcome comparison between wooden sticks (ws) and other objects

Vision	WS	Other	Total
Uncooperative	9	8	17
<1/60	82	89	171
1/60 to 3/60	32	24	56
20/200 to 20/120	29	35	64
20/80 to 20/60	83	62	145
20/40 to 20/20	144	79	223
Total	379	297	676

χ^2 test, $p = 0.002$

Table 21.21 Infection comparison between wooden sticks (ws) and other objects

	WS	Other	Total
Infection absent	384	300	684
Infection present	0	3	3
Total	384	303	687

χ^2 test, $p = 0.08$

Table 21.22 Visual outcome when infection is present

	Infection		Total
	Absent	Present	
Uncooperative	18	0	18
<1/60	168	3	171
1/60 to 3/60	60	0	60
20/200 to 20/120	66	0	66
20/80 to 20/60	146	0	146
20/40 to 20/20	223	0	223
Total	684	3	687

χ^2 test, $p = 0.02$

Brar et al. found that postoperative complications following ocular injuries were the main factor responsible for poor outcomes, with 20/40 or better vision seen in 38.8 % of eyes with closed globe injuries and in 86.4 % of eyes with open globe injuries [15]. This difference in success rates could be attributed to differences in the type of ocular trauma, the presence of other ocular tissue damage, or variation in surgical procedures. In contrast, in a case series of 60 eyes with traumatic cataracts, Wos et al. found no significant difference in visual outcome between those developing cataracts after perforating injuries and after non-perforating injuries [16].

Wos et al. noted that a large proportion of the population with traumatic cataracts in their series was male [16]. Baclouti et al. did not find a gender difference in traumatic cataracts in their study in Tunisia [17]. Although we had a large proportion of males in our cohort, the difference between the numbers of males and females was not statistically significant. Many working women in tribal areas may be at increased risk for ocular injuries and traumatic cataracts; this may explain the gender distribution noted in our study. We found a significant ($p = 0.020$) difference in open globe injuries by gender, with males predominantly affected.

Behbehani [18] reported 20/40 final visual outcome in 40 % cases of open globe injuries. Cillino [19] also reported final visual acuity of 20/40 in 48.3 % of all injury cases. Smith [20] similarly reported 47.8 % of cases reaching 20/40 vision, combining open- and closed globe injuries.

Our cohort of patients with traumatic cataracts was much younger than those in other studies [14]. Thus, appropriate intervention to avoid visual disability in our cohort would be more cost effective, because the disability-adjusted life years saved by successful intervention would be much higher.

Using a large database, we attempted to systematically classify the morphology of traumatic cataract and to select surgical techniques accordingly. We used a practical grading of cataracts to enable ophthalmologists to determine the best mode of managing them. This grading differs from the standard grading used for senile cataracts [8]. Various studies have touched on this topic. Krishnamachary et al. found 52.3 % of cataracts to be total, whereas our results revealed 26.6 % total cataracts [6]. Vajpayee reported type-1 and -2 openings in the posterior capsule with penetrating injury [14], whereas we found a membranous type of cataract in 12.1 % of cases. This is suggestive of late reporting, as membranous transformation of the lens with fusion of the anterior and posterior capsules may occur over time.

Of the patients who reported between 2 and 30 days after their injury, 63 % recovered vision >20/80, while 48 % of the patients who reported within 24 h or after 30 days recovered vision >20/80. Patients who underwent intervention within 24 h did not do as well as those who reported between 2 and 30 days in terms of the final visual outcome. This might be attributed to open globe injuries being reported early, undergoing primary repair, and developing inflammation, while late injuries were predominantly closed globe injuries and did not require more surgery [21]. Many studies have reported that early intervention is an important factor in better visual outcome [22].

Gupta et al. and Zhang et al. reported that the incidence of infection is higher if primary closure is late [23, 24], although the incidence of infection did not vary significantly in our study. Overall, the rate of infection in open globe injury in our study was 0.4 %, while reported values range from 2 to 17 % [25, 26].

Behbehani et al. and Woś reported that the time interval between injury and intervention did not make a difference in terms of the final visual outcome [16, 18]. Jonas and Yang reported that the final visual outcome was better with early intervention with an intraocular foreign body [27–29].

We are not aware of a study that examined the time interval between injury and treatment start for traumatic cataracts.

Many studies have reported that an object causing an open globe injury also causes infection [25, 26, 29–39]. In the present study, the total incidence of infection was 0.4 %, which is much lower than the incidence of endophthalmitis reported by Viestenz et al. (2–17 %) or by Cebulla (4–8 %) [40].

The final visual outcome following an open globe injury caused by a stick was significantly better than that following injuries caused by other objects ($p = 0.002$).

21.5   Controversies

A number of issues regarding the management of traumatic cataract remain unresolved. The high risk of amblyopia and intraocular inflammation as well as strong vitreoretinal adhesions in the pediatric age group requires management based on different principles. Prospective, controlled clinical studies of OGI are not possible. This article reviews pertinent data regarding these management issues and controversies, and provides recommendations for treatment based on the available published data and the authors' personal experience.

Blindness due to injury is a social and economical burden on society and the individual [6].

21.6   Controversies in Traumatic Cataract

21.6.1   Controversies Regarding Epidemiology and Incidence

The incidence varies in different regions of world, based mainly on retrospective studies or eye injury registries. Variability of incidence also arises due to different demographic conditions, age, sex, environment, and socio-economical conditions [4]. Within the pediatric age group, the incidence of traumatic cataract also varies [41, 42].

21.6.2   Controversies Over Ocular Trauma Classification

This is a proposed classification as an improved BETTS classification (Fig. 21.6).

Landmark for Zone III open globe injury. The International Ocular trauma classification group has defined zone III injuries as those extending 5 mm beyond the limbus. The length of 5 mm was determined arbitrarily, based on the justification that injuries in that zone may not extend into the pars plana. Therefore, any injury not involving zone III or the pars plana region is predicted to have a favorable prognosis.

It is imperative that surgeons realize that traumatic cataract is not a senile cataract. The injury is rarely limited to the lens alone, but may also be associated with the zonules, posterior capsule, and posterior segment. A patient with traumatic cataract should be warned about the potential visual outcome and the high risk of intraoperative problems.

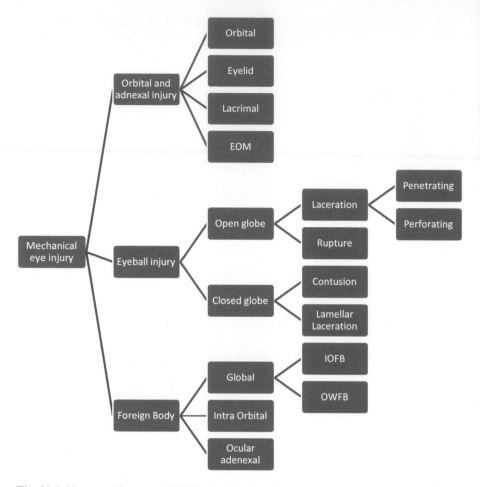

Fig. 21.6 Newer and improved BETTS classification of mechanical ocular trauma

21.6.3   Controversies About the Morphology of Traumatic Cataract

There is no standard morphological classification. Attempts at grading have been made, but these are arbitrary [42]. The morphology of traumatic cataract depends mainly upon the type of injury and the time interval between the injury and intervention.

One study proposed that all cases of cataract could be assigned to these groups [10].

21.6.4 Controversies in the Management of and Surgical Approaches to Traumatic Cataract

In adults in whom amblyopia is not an issue, the choice of surgery is governed by the surgeon's own preferences and to a certain extent by the status of the cataractous lens. If the anterior capsule is significantly disrupted and there is free floating lens matter in the anterior chamber, the surgeon may be justified in primary cataract extraction with or without intraocular lens (IOL) implantation (Fig. 21.3).

Eyes with a lens vitreous admixture should be considered for combined cataract extraction with limited anterior vitrectomy. Care should be taken to judiciously use a vitrector and not an aspirator while removing vitreous admixture in the ruptured lens matter (Fig. 21.3). Any traction on the vitreous can result in inadvertent retinal breaks [40]. When there is additional injury to the posterior segment, early pars plana lensectomy and vitrectomy by a posterior segment specialist is warranted [40]. In eyes with an intact anterior capsule and total traumatic cataract, second sitting cataract extraction with IOL implantation should be the best and safest approach for optimal visual outcome (Fig. 21.5) [43, 44].

Wherever possible, a multistep procedure after control of inflammation, with adequate corneal clarity and an appropriate IOL power calculation, should be adopted [44, 45]. A surgical approach of anterior vs. posterior via the pars plana route is debatable.

The following types of surgical approach may be used, according to morphology:

- Unimanual or bimanual aspiration.
  Lensectomy/membranectomy using the limbal or pars plana approach and insertion of the lens in the sulcus.
- Phacoemulcification or small-incision cataract surgery when the nucleus is harder [10].

21.6.5 Controversies Related to Pediatric Traumatic Cataract

Incidence: Incidence varies among reports, most of which were retrospective and with small databases. Studies of traumatic cataracts in children reported incidences of 25 % in southern India and 12 and 46 % in western India [4, 41, 42].

Challenges in children with traumatic cataract include amblyopia, a tendency for inflammation, synechiae, and after cataract [7].

Children account for approximately one-third of cases of serious eye injuries [46]. Despite this, the classification and scoring system in pediatric trauma is based on those developed for adults [46]. The controversy over the position of zones II and III is even more pronounced in pediatric trauma. During the first 5 years the length of the pars plana changes rapidly from ~1.8 mm in neonates to 3 mm by 1 year of age and reaches 5 mm by 5 years of age [47, 48]. Therefore, pediatric trauma assessment for research purposes is prone with inaccuracy, depending on how the injury is classified.

Children and younger patients exhibit stronger adherence between the posterior capsule and the anterior vitreous centrally; furthermore, the central vitreous is anatomically connected to the peripheral retina at the vitreous base [43]. Any traction on the anterior vitreous face is transmitted to the retina, and the younger the patient, the greater the risk. Additionally, children are at risk of amblyopia. In younger children, a surgeon may resort to a single-step procedure and perform lens extraction, IOL implantation, and, if required, anterior vitrectomy for an optimal outcome [15, 49–52]. Primary IOL implantation not only prevents amblyopia but also synechiae formation that can close the bag by the time secondary IOL implantation is due to be carried out [18].

Although the management of trauma in children has many similarities to that in adults, striking differences also exist. Foremost is that adults have reached visual maturity whereas amblyopia is a major contributor to poor outcomes in children, especially those under 5 years old. The worst outcomes reported were that less than half of the children with an OGI achieved good vision and amblyopia was a major confounding factor [16–18]. When treating children with OGI, to achieve a clear visual axis following globe repair this must be accompanied by accurate refraction and aggressive patching therapy to improve the visual outcomes [53]. Another difference is that a simple examination can prove difficult in a child, especially during the early stages after trauma, resulting in a requirement for general anesthesia or sedation for proper assessment. The involvement of a pediatric ophthalmologist and access to pediatric facilities is therefore essential for ultimate management.

Optical rehabilitation is important since an IOL may not be implanted in all cases [54]. Orthoptic treatment following optical correction, including patching and monitoring of visual regain, is vital because amblyopia is an important factor [54].

In conclusion, children with an OGI cannot simply be treated as small adults. Patient age and the effects of amblyopia should be included as additional negative prognostic factors.

21.6.6   Controversies Related to the Timing of Intervention

Controversy exists over the timing of cataract extraction and IOL implantation. Arguments can support both primary and secondary cataract extractions; thus a number of crucial facts should be considered before making a decision. Cataract extraction together with primary wound repair may have distinct advantages, such as controlling inflammation, but also has the possibility of raised intraocular pressure due to soft lens matter in the anterior chamber [10]. Secondary advantages include the direct visualization of the posterior segment and optic nerve [11]. Similarly, in pediatric patients the removal of media opacity may be crucial to prevent vision-deprivation amblyopia. In patients with lens vitreous admixture, this is a potent stimulator for further proliferative vitreoretinopathy and can also result in traction on the retina, hence primary extraction of the lens and vitreous is imperative in such patients [55, 56]. Minor advantages of primary lens removal are the patient's convenience and possibly cost effectiveness [11].

Proponents of second sitting cataract extraction recommend good control of intraocular inflammation, good media clarity, and a stable wound before planning for traumatic cataract extraction [11, 57]. If there is adequate control of inflammation, IOL implantation at second stage cataract extraction may be associated with a better outcome [11, 56]. An IOL power calculation is appropriate if IOL implantation is planned for a second sitting [44, 55].

Current data suggest that improved visual outcome results from intervention at 2–30 days [56].

A number of facts must be known before embarking on primary or delayed cataract extraction with IOL implantation. These include the age of the patient, the expertise of the surgeon and assisting staff, the infrastructure available, and the status of the cataractous lens and the lens vitreous admixture, which will act as a guide to the surgeon in planning the surgery.

Under no circumstances should the ophthalmologist subject a patient to half-completed or compromised surgery because of a lack of expertise or proper infra-structure. If the facility is not equipped to provide the surgeon after-hours with the full range of equipment, instruments, and material together with a full and knowl-edgeable staff, it is better not to contemplate primary lens removal. In most hospitals across the world when an OGI is seen during after hours, the state-of-art facilities are not normally available to attempt cataract extraction and IOL implantation.

21.6.7 Controversies Regarding Predictive Methods

Clinical conditions that might predict prognosis:

Relative Afferent Pupillary Defect (RAPD). There is sufficient evidence to suggest that RAPD may produce false positives with regard to damage to the optic nerve or retina in the presence of severe hyphema or subretinal vitreous hemorrhage, which may disappear after resorption or removal of the hemorrhage [11–13, 20]. Therefore, RAPD alone is a poor prognostic factor, may not be appropriate, and the possibility of reversal exists [58]. RAPD therefore should be weighted equally with other pre-operative variables following ocular trauma.

Type of injury. This may influence visual outcome. Reports have suggested open globe injury has a better outcome, whereas the opposite was true for penetrative injury [57, 59, 60]. Influences open globe of BETT (Birmingham Eye Trauma Terminology) as a predictive factor.

Predictive model. The Ocular Trauma Score (OTS) was developed to provide more accurate information on visual prognosis. However, it is not clear whether children were included in the databases of the over 2,500 serious ocular injuries from which the method was formulated [61, 62]. Reports from India validated OTS using 787 cases of traumatic cataract [63, 64].

Two important factors in calculating the OTS, initial visual acuity and RAPD, are very difficult to obtain for children after trauma, especially those in the younger age group, rendering the OTS inaccurate or even unusable. Two groups recently assessed the value of the OTS in pediatric patients aged from 2 years, coming to opposite conclusions, thus adding to the controversy [65, 66]. A new pediatric Ocular Trauma Score (POTS) to refine prognostic accuracy in children in whom initial vision is not accurate was published recently [67]. As in many other studies of pediatric trauma, it lacks the statistical power of the OTS due to its relatively small sample size, and its predictive power remains untested. OTS may be a valid predictor of visual outcome in children following surgery in a study of 354 traumatic cataracts in children [66].

Regression tree analysis has also been used, but has not been validated [68].

21.6.8 Controversies About Primary Posterior Capsulotomy

Determination of the correct IOL power prior to surgery may be difficult, if not impossible, for a range of reasons. Often the other eye serves as a guide, and inflammatory debris can settle on the IOL surface, cleansing of which may require postoperative YAG laser treatment or even surgery. In addition, the edge of the IOL will interfere with the surgeon's visualization of the peripheral retina should subsequent proliferative vitreo retinopathy development necessitate vitrectomy [55]. There has been considerable controversy over the use of primary posterior capsulotomy [56, 57]

21.6.9 Controversy Surrounding Preoperative Visual Acuity of No Light Perception and Poor Visual Prognosis

Visual acuity can be profoundly impaired, even to the extent of no light perception (NLP), in the presence of significant media opacity (e.g., corneal edema, hyphema, cataract, dense vitreous hemorrhage), retinal detachment, associated subretinal or subhyaloid hemorrhage, hemorrhagic choroidals, and even psychological factors (e.g., hysteria). Assessment of light perception is a subjective measurement and not a full-proof test in the presence of severe media opacity secondary to dense vitreous hemorrhage, traumatic cataract, dense hyphema, or corneal edema [69]. Even with the bright light of an indirect ophthalmoscope, the assessment of light perception can give a false impression of NLP [59]. Ultrasonography is useful for assessment of the posterior segment in eyes with media opacity and to differentiate between retinal detachment and vitreous hemorrhage [59]. However, using this methodology it is sometimes difficult to differentiate a detached retina from blood clots in the vitreous cavity or membranes [59]. Before deciding on enucleation in patients with NLP, reversible causes of vision loss should be excluded, including psychological factors [59, 69]. Even in situations where enucleation appears inevitable, the ophthalmologist should still discuss the possible options with the patient before making a final decision. Primary enucleation for severely traumatized eyes with NLP is controversial due to the risk of sympathetic ophthalmia. Sympathetic ophthalmia with the potential for bilateral blindness is a relative indication for enucleation of an injured eye [9]. Therefore, primary surgical repair should not necessarily be abandoned because of the risk of sympathetic ophthalmia in eyes with NLP. Currently, most surgeons recommend a globe salvaging procedure for eyes with severe trauma with NLP vision at initial presentation.

Controversy remains regarding the age in children at which IOL insertion and rehabilitation of aphakia can be undertaken.

21.6.10 Controversies About Infection Following Open Globe Injury (OGI)

Many studies have reported infection and endophthalmitis following OGIs [21–24, 70]. Intraocular infection is also common with a retained intraocular foreign body [57–59].

Absence of infection following wooden stick injuries due to the antimicrobial and antifungal properties of certain plants has been reported [60–66]

21.7 Conclusion

Despite the advances in state-of-the-art surgery and in the understanding of ocular trauma, a range of unresolved, controversial issues remain in the management and treatment of OGIs. Over the last two decades all have been addressed and real progress in many aspects of definitive management of traumatic cataracts has been made.

The timing of intervention in traumatic cataract appears to be a never-ending debate. "Sooner the better" was the traditional view; however, an alternative view is that a better outcome results from intervention at 3–30 days. Although a controlled, prospective clinical trial would be ideal, no two ocular trauma cases are alike, and confounding factors can affect the final outcome.

Morphological assessment might be used to guide management.

The age at intervention and laterality play important roles as predictors of visual outcome.

The accuracy of predictive models varies between adults and pediatric traumatic cataract cases.

Controlling for the significant differences that occur among individual injuries is difficult. This makes highly problematic independent assessment of potential risk factors or treatment variances for visual and anatomical outcome. Current management is based on the surgeon's experience and will continue to be based on a retrospective review of accumulated data and the personal preferences of the treating ophthalmologist open globe injury. These types of management problems are dealt with on a case-by-case basis, and even the most experienced ophthalmologist open globe injury will at some time find themselves in a dilemma regarding strategic planning of ocular trauma management. This report has attempted to provide a comprehensive overview of most of the controversies pertaining to the management of OPEN GLOBE INJURYs, and has presented our preferred guidelines. This should aid ophthalmologist open globe injury in patient counseling and the decision-making process regarding the management of OPEN GLOBE INJURYs involving the anterior or posterior segment.

In summary, the vast majority of ophthalmologists who encounter a traumatic cataract have sufficient experience in lens extraction and IOL implantation in the non-traumatic setting. What every ophthalmologist must accept is that an injured lens requires many individualized, conscious decisions regarding what to do, when to do it, and how to achieve the best possible outcome.

21.8 Literature Review

In this review, we identified pertinent articles on traumatic cataract from a combination of sources including electronic database searches and manual searches through the literature. A computerized search of the PubMed database (National Library of Medicine) in Endnote was performed up to July 2012. The term traumatic cataract was used for a broad and sensitive search. Subsequently all the abstracts were carefully scanned and were divided into subcategories covering topics including type of injuries, *morphology*, *presenting visual acuity*, *surgical techniques*, and *ocular trauma scores*. Non-English articles were included when deemed necessary. Copies of the entire articles were obtained. Reference lists of identified sources were used to glean more articles on the same topic. Additional books cited from these references were also used.

21.8.1 Conclusion

1. Morphology of traumatic cataracts and hence surgical technique may influence final visual outcome
2. We obtained good visual outcomes after managing traumatic cataracts. According to results satisfactory visual acuity following cataract surgery was more likely with open globe injuries than with closed globe injuries
3. Following primary repair of any ocular injury, the final visual outcome in terms of traumatic cataracts may be better if treatment is delayed for a week or so.

Disclosure   No financial support received from any company or institution.
This study is not presented at any conference or meeting.
Authors do not have any financial interest in any aspect of this study.

References

1. Khatry SK, Lewis AE, Schein OD, Thapa MD, Pradhan EK, Katz J, et al. The epidemiology of ocular trauma in rural Nepal. Br J Ophthalmol. 2004;88:456–60.
2. Abraham DI, Vitale SI, West SI, Isseme I. Epidemiology of eye injuries in rural Tanzania. Ophthalmic Epidemiol. 1999;6:85–94.
3. Alfaro DV, Jablon EP, Fontal MR, Villalba SJ, Morris RE, Grossman M, Roig-Melo E, et al. Fishing-related ocular trauma. Am J Ophthalmol. 2005;139:488–92.
4. Shah M, Shah S, Khandekar R. Ocular injuries and visual status before and after their management in the tribal areas of Western India—a historical cohort study. Grafes Arch Clin Exp Ophthalmol. 2008;246:191–7.
5. Kuhn F, Morris R, Witherspoon CD, Mester V. The Birmingham Eye Trauma Terminology system (BETT). J Fr Ophtalmol. 2004;27:206–10.
6. Krishnamachary M, Rathi V, Gupta S. Management of traumatic cataract in children. J Cataract Refract Surg. 1997;23 Suppl 1:681–7.
7. Churchill AJ, Noble BA, Etchells DE, George NJ. Factors affecting visual outcome in children following uniocular traumatic cataract. Eye (Lond). 1995;9(Pt 3):285–91.
8. Thylefors B, Chylack Jr LT, Konyama K, Sasaki K, Sperduto R, Taylor HR, et al. A simplified cataract grading system. Ophthalmic Epidemiol. 2002;9:83–95.
9. Weinand F, Plag M, Pavlovic S. Primary implantation of posterior chamber lenses after traumatic cataract penetration. Ophthalmology. 2003;100:843–6.
10. Shah MA, Shah SM, Shah S, et al. Morphology of traumatic cataract: does it play a role in final visual outcome? BMJ Open. 2011;1:e000060.
11. Corbett MC, Shilling JS, Holder GE. The assessment of clinical investigations: the Greenwich Grading System and its application to electro diagnostic testing in ophthalmology. Eye. 1995;9:59–64.
12. Segev Y, Goldstein M, Lazar M, Reider-Groswasser I. CT appearance of a traumatic cataract. AJNR Am J Neuroradiol. 1995;16:1174–5.
13. McWhae JA, Crichton AC, Rinke M. Ultrasound biomicroscopy for the assessment of zonules after ocular trauma. Ophthalmology. 2003;110:1340–3.
14. Vajpayee RB, Sharma N, Dada T, Gupta V, Kumar A, Dada VK. Management of posterior capsule tears. Surv Ophthalmol. 2001;45:473–88.
15. Brar GS, Ram J, Pandav SS, Reddy GS, Singh U, Gupta A. Postoperative complications and visual results in uniocular pediatric traumatic cataract. Ophthalmic Surg Lasers. 2001;32:233–8.
16. Wos M, Mirkiewicz-Sieradzka B. Traumatic cataract—treatment results. Klin Oczna. 2004; 106:31–4.
17. Baklouti K, Mhiri N, Mghaieth F, El Matri L. Traumatic cataract: clinical and therapeutic aspects. Bull Soc Belge Ophtalmol. 2005;298:13–7.
18. Behbehani AM, Lotfy N, Ezzdean H, Albader S, Kamel M, Abul N. Open eye injuries in the pediatric population in Kuwait. Med Princ Pract. 2002;11:183–9.
19. Cillino S, Casuccio A, Di Pace F, Pillitteri F, Cillino GA. Five-year retrospective study of the epidemiological characteristics and visual outcomes of patients hospitalized for ocular trauma in a Mediterranean area. BMC Ophthalmol. 2008;22:6–8.
20. Smith AR, O'Hagan SB, Gole GA. Epidemiology of open- and closed-globe trauma presenting to Cairns Base Hospital, Queensland. Clin Experiment Ophthalmol. 2006;34:252–9.

21. Cebulla CM, Flynn Jr HW. Endophthalmitis after open globe injuries. Am J Ophthalmol. 2009;147:567–8.
22. Andreoli CM, Andreoli MT, Kloek CE, et al. Low rate of endophthalmitis in a large series of open globe injuries. Am J Ophthalmol. 2009;147:601–8.
23. Gupta A, Srinivasan R, Gulnar D, Sankar K, Mahalakshmi T. Risk factors for post-traumatic endophthalmitis in patients with positive intraocular cultures. Eur J Ophthalmol. 2007; 17:642–7.
24. Zhang Y, Zhang MN, Jiang CH, Yao Y, Zhang K. Endophthalmitis following open globe injury. Br J Ophthalmol. 2010;94:111–4.
25. Chhabra S, Kunimoto DY, Kazi L, et al. Endophthalmitis after open globe injury: microbiologic spectrum and susceptibilities of isolates. Am J Ophthalmol. 2006;142:852–4.
26. Gupta A, Srinivasan R, Kaliaperumal S, Saha I. Post-traumatic fungal endophthalmitis— a prospective study. Eye (Lond). 2008;22:13–7.
27. Jonas JB, Knorr HL, Budde WM. Prognostic factors in ocular injuries caused by intraocular or retrobulbar foreign bodies. Ophthalmology. 2000;107:823–8.
28. Jonas JB, Budde WM. Early versus late removal of retained intraocular foreign bodies. Retina. 1999;19:193–7.
29. Yang CS, Lu CK, Lee FL, Hsu WM, Lee YF, Lee SM. Treatment and outcome of traumatic endophthalmitis in open globe injury with retained intraocular foreign body. Ophthalmologica. 2010;224:79–85.
30. Unejo SA, Ahmed M, Alam M. Endophthalmitis in paediatric penetrating ocular injuries in Hyderabad. J Pak Med Assoc. 2010;60:532–5.
31. Al-Mezaine HS, Osman EA, Kangave D, Abu El-Asrar AM. Risk factors for culture-positive endophthalmitis after repair of open globe injuries. Eur J Ophthalmol. 2010;20:201–8.
32. Wade PD, Khan SS, Khan MD. Endophthalmitis: magnitude, treatment and visual outcome in northwest frontier province of Pakistan. Ann Afr Med. 2009;8:19–24.
33. Viestenz A, Schrader W, Behrens-Baumann W. Traumatic Endophthalmitis Prevention Trial (TEPT)]. Klin Monbl Augenheilkd. 2008;225:941–6.
34. Wykoff CC, Flynn Jr HW, Miller D, Scott IU, Alfonso EC. Exogenous fungal endophthalmitis: microbiology and clinical outcomes. Ophthalmology. 2008;115:1501–7. 07 e1-2.
35. Al-Omran AM, Abboud EB, Abu El-Asrar AM. Microbiologic spectrum and visual outcome of posttraumatic endophthalmitis. Retina. 2007;27:236–42.
36. Essex RW, Yi Q, Charles PG, Allen PJ. Post-traumatic endophthalmitis. Ophthalmology. 2004;111:2015–22.
37. Lieb DF, Scott IU, Flynn Jr HW, Miller D, Feuer WJ. Open globe injuries with positive intraocular cultures: factors influencing final visual acuity outcomes. Ophthalmology. 2003; 110:1560–6.
38. Sabaci G, Bayer A, Mutlu FM, Karagul S, Yildirim E. Endophthalmitis after deadly-weapon-related open-globe injuries: risk factors, value of prophylactic antibiotics, and visual outcomes. Am J Ophthalmol. 2002;133:62–9.
39. Duch-Samper AM, Chaques-Alepuz V, Menezo JL, Hurtado-Sarrio M. Endophthalmitis following open-globe injuries. Curr Opin Ophthalmol. 1998;9:59–65.
40. Kuhn F. Traumatic cataract: what, when, how. Graefes Arch Clin Exp Ophthalmol. 2010;248:1221–3.
41. Eckstein M, Vijayalakshmi P, Killedar M, et al. Aetiology of childhood cataract in south India. Br J Ophthalmol. 1996;80:628–32.
42. Johar SR, Savalia NK, Vasavada AR, et al. Epidemiology based etiological study of pediatric cataract in western India. Indian J Med Sci. 2004;58:115–21.
43. Rumelt S, Rehany U. The influence of surgery and intraocular lens implantation timing on visual outcome in traumatic cataract. Graefes Arch Clin Exp Ophthalmol. 2010;248:1293–7.
44. Cohen KL. Inaccuracy of intraocular lens power calculation after traumatic corneal laceration and cataract. J Cataract Refract Surg. 2001;27:1519–22.
45. Mohammad pour M, Jafarinasab MR, Javadi MA. Outcomes of acute postoperative inflammation after cataract surgery. Eur J Ophthalmol. 2007; 17:20-8.

46. Maltzman BA, Pruzon H, Mund ML. A survey of ocular trauma. Surv Ophthalmol. 1976; 21:285–90.
47. Lemley CA, Han DP. An age-based method for planning sclerotomy placement during pediatric vitrectomy: a 12-year experience. Trans Am Ophthalmol Soc. 2007;105:86–9.
48. Hairston RJ, Maguire AM, Vitale S, et al. Morphometric analysis of pars plana development in humans. Retina. 1997;17:135–8.
49. Baykara M, Dogru M, Ozçetin H, et al. Primary repair and intraocular lens implantation after perforating eye injury. J Cataract Refract Surg. 2002;28:1832–5.
50. Synder A, Kobielska D, Omulecki W. Intraocular lens implantation in traumatic cataract. Klin Oczna. 1999;101:343–6.
51. Morgan KS. Cataract surgery and intraocular lens implantation in children. Curr Opin Ophthalmol. 1993;4:54–60.
52. Koenig SB, Ruttum MS, Lewandowski MF, et al. Pseudophakia for traumatic cataracts in children. Ophthalmology. 1993;100:1218–24.
53. Mireskandari K, Bunting H. Paediatric open globe injuries: a seventeen year experience. Invest Ophthalmol Vis Sci. 2011;52(suppl):1572.
54. Shah MA, Shah SM, Shah AH, Pandya JS. Visual outcome of cataract in pediatric age group: does etiology have a role? Eur J Ophthalmo 2014;24:76–83.
55. Cardillo J, Stout J, LaBree L, et al. Posttraumatic proliferative vitreoretinopathy. The epidemiologic profile, onset, risk factors, and visual outcome. Ophthalmology. 1997;104: 1166–73.
56. Shah MA, Shah SM, Shah SB, et al. Effect of interval between time of injury and timing of intervention on final visual outcome in cases of traumatic cataract. Eur J Ophthalmol. 2011;21:760–5.
57. Shah MA, Shah SM, Shah SB, et al. Comparative study of final visual outcome between open- and closed-globe injuries following surgical treatment of traumatic cataract. Graefes Arch Clin Exp Ophthalmol. 2011;249:1775–81.
58. Rabinowitz R, Yagev R, Shoham A, et al. Comparison between clinical and ultrasound findings in patients with vitreous hemorrhage. Eye. 2004;18:253–6.
59. Striph GG, Halperin LS, Stevens JL, et al. Afferent pupilary defect caused by hyphema [letters]. Am J Ophthalmol. 1988;106:352–3.
60. Shah M, Shah S, Shah SB, et al. Visual recovery and predictors of visual prognosis after managing traumatic cataracts in 555 patients. Indian J Ophthalmol. 2011;59(3):217–22.
61. American Society of Ocular Trauma. Ocular trauma score (OTS). http://www.asotonline.org/ots.html. Accessed 10 Dec 2008.
62. Kuhn F, Maisiak R, Mann L, et al. The Ocular Trauma Score (OTS). Ophthalmol Clin North Am. 2002;15:163–5.
63. Shah MA, Shah SM, Shah SB, et al. OcularTrauma Score: a useful predictor of visual outcome at six weeks in patients with traumatic cataract. Ophthalmology. 2012;119(7):1336–41.
64. Shah MA, Shah SM, Shah SB, et al. Ocular Trauma Score as a predictor of final visual outcomes in traumatic cataract cases in pediatric patients. J Cataract Refract Surg. 2012; 38(6):959–65.
65. Unver YB, Acar N, Kapran Z, et al. Visual predictive value of the ocular trauma score in children. Br J Ophthalmol. 2008;92:1122–4.
66. Uysal Y, Mutlu FM, Sobaci G. Ocular Trauma Score in childhood open-globe injuries. J Trauma. 2008;65:1284–6.
67. Acar U, Tok OY, Acar DE, Burcu A, et al. A new ocular trauma score in pediatric penetrating eye injuries. Eye. 2011;25:370–4.
68. Schmidt G, Broman A, Grant M. Vision survival after open globe injury predicted by classification and regression tree analysis. Ophthalmology. 2007;115:202–9.
69. Morris R, Kuhn F, Witherspoon CD. Management of the opaque media eye with no light perception. Vitreoretinal surgery of the injured eye. Philadelphia: Lippincott-Raven; 1999. p. 113–24.
70. Narang S, Gupta V, Simalandhi P, et al. Paediatric open globe injuries. Visual outcome and risk factors for endophthalmitis. Indian J Ophthalmol. 2004;52:29–34.

Chapter 22
Cataract in Pediatric Age Group: A Review

Mehul Shah and Shreya Shah

Abbreviations

BETTS   Birmingham Trauma Terminology Systems
IOL   Intraocular lens
IOP   Intraocular pressure
OGI   Open globe injury
OTS   Ocular trauma score

22.1 Introduction

Childhood cataracts are responsible for 5–20 % of blindness in children worldwide and for an even higher percentage of childhood visual impairment in developing countries [1–5].

The overall incidence of clinically significant cataracts (unilateral or bilateral) in childhood is unknown, but has been estimated to be as high as 0.4 % [6, 7]. The prevalence of childhood cataract varies from 1.2 to 6.0 cases per 10,000 infants. Pediatric cataracts are responsible for more than one million cases of childhood blindness in Asia. In developing countries, such as India, 7.4–15.3 % of childhood blindness is due to cataracts [8, 9]. Internationally, the incidence is unknown. Although the World Health Organization and other health organizations have made outstanding progress in vaccination and disease prevention, the rate of congenital cataracts remains much higher in underdeveloped countries.

M. Shah (✉) • S. Shah
Drashti Netralaya, Nr. GIDC, Chakalia Road, Dahod 389151, Gujarat, India
e-mail: omtrust@rediffmail.com

© Springer Science+Business Media New York 2015
M.A. Babizhayev et al. (eds.), *Studies on the Cornea and Lens*,
Oxidative Stress in Applied Basic Research and Clinical Practice,
DOI 10.1007/978-1-4939-1935-2_22

The visual results of cataract surgery in children have generally [10, 11] been poorer than in adults [1–3, 6, 11, 12]. This difference is due, in part, to the various types of amblyopia that develop in children with cataracts, the association of nystagmus with early onset cataracts, and the presence of other ocular abnormalities that adversely affect vision in eyes with developmental lens opacities. Since the introduction of the aspiration technique for cataract removal by Scheie in 1960, surgical procedures for the removal of the lens in childhood have improved [13–15]. and earlier surgery for congenital cataracts has been encouraged [16, 17].

Congenital cataracts are one of the most common causes of treatable blindness in children, particularly in developing countries [1]. A recent report indicated that infants with bilateral congenital cataract who underwent early surgery (within 1 month of birth) and received appropriate optical rehabilitation could obtain visual acuity of better than 0.4 and could even achieve stereopsis [2]. However, because of typically relatively late detection and diagnosis, the nonavailability of facilities for infant anesthesia, and poor compliance with long-term follow-up, the visual prognosis for infants with congenital cataract in developing countries differs markedly from that in industrialized countries. Visual loss is primarily attributable to amblyopia, most importantly, to "stimulus-form deprivation amblyopia," with the additional factor of ocular rivalry in unilateral disease. Thus, improved understanding of the critical periods of visual development has resulted in surgical intervention for dense cataracts being deemed necessary within the first 3 months of life, possibly as early as the first 6 weeks in unilateral disease. Clinical factors believed to be important to visual outcome in children include age at diagnosis and surgery, type of refractive correction, type of cataract surgery, compliance with occlusion regimen, aetiology of the cataract, presence of non-ophthalmic disorders, development of capsular opacity or secondary membrane, and serious ocular postoperative complications.

22.2   Materials and Methods

Study approval taken from hospital ethical committee, written consent was obtained from parents/guardians for participation in study. A Prospective hospital-based study was conducted in a tertiary care eye hospital in western India over 20 year's period from January 1992 to April 2012. All demographic details were documented including socioeconomical details according to Kuppuswamy grading [18]

All pediatric patients (0 to 18) having cataract were enrolled in the study from January 1992 to April 2012. Patient's primary details, history, and all details were documented in pretested online format. An ocular trauma detail was documented in online world eye injury registry form. Vision checkup of each eye was assessed according to the American academy of pediatric vision checkup protocol. Anterior segment examination was done using the slit lamp biomicroscope. The pupils were dilated according to dilatation protocol.

The ocular pressure was measured using Perkin's hand-held tonometer. If this was not possible, the pressure was measured under general anesthesia. This procedure was omitted for eyes having open globe injuries. The posterior segment

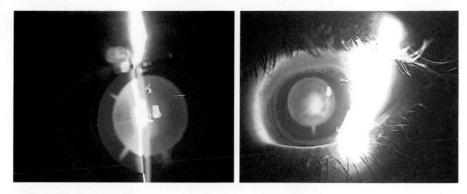

Fig. 22.1 (**a**, **b**) A Lamellar (Zonular) cataract

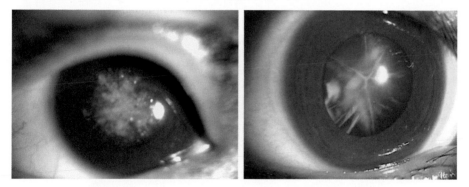

Fig. 22.2 (**a**, **b**) Floriform cataract

of the eye was evaluated with the help of an indirect ophthalmoscope and a +20 D Lens and an ultrasound "B" scan if media was not clear.

Surgical technique was decided according to the aetiology, morphology of cataract, and position of lens (Figs. 22.1, 22.2, 22.3, 22.4, and 22.5). Surgery was either done by anterior or pars plana route. Anterior route surgeries were done either by phacoemulsifier or manual suction. Membranectomy and lensectomies were performed using pneumatic cutter. Intraocular lens was not implanted in patients of less than 1.5 years. Children less than this age group were operated by lensectomy/membranectomy and secondary implant was done after that. Patient was rehabilitated by glasses or contact lenses in-between. For IOL power calculation we have used a deducted power according to guidelines [19, 20].

22.2.1 Specific for Traumatic Cases

For each patient enrolled in our study, we obtained a detailed history including the details of the injury and information on eye treatment and surgery performed to manage past ocular trauma. Data for both the initial and follow-up reports were

Fig. 22.3 Blue dot and
sutural cataract

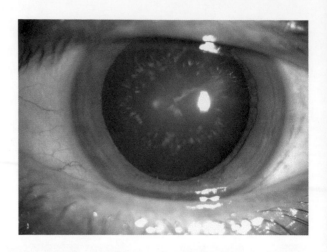

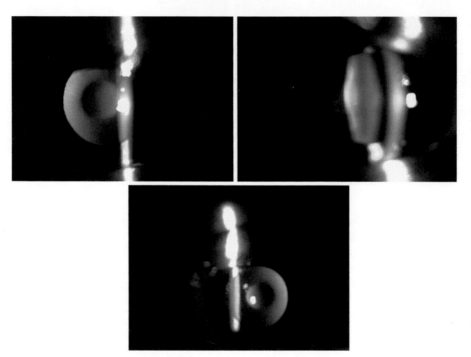

Fig. 22.4 (**a–c**) Lens with showing anterior and posterior lenticonus

collected using the online BETTS format of the International Society of Ocular
Trauma. Details of the surgery were also collected using a pretested online form.
Ocular trauma score was calculated [9].

The cases of traumatic cataract were grouped as those involving open globe versus those involving closed globe injuries. The open globe injuries were further

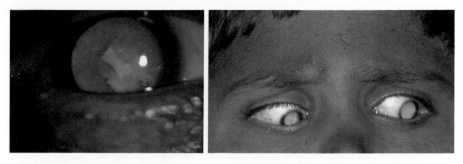

Fig. 22.5 (**a**, **b**) Neglected cataract

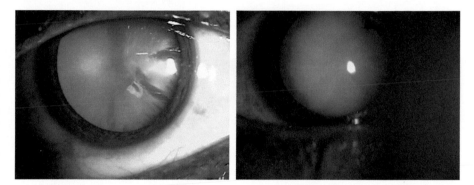

Fig. 22.6 (**a**, **b**) When no clear lens matter was visible between the capsule and the nucleus, the cataract was defined as a total cataract

categorized into those with lacerations versus rupture. Lacerations of the eyeball were subcategorized into eyes with perforating injuries, penetrating injuries, or injuries involving an intraocular foreign body. The closed globe group was subdivided into lamellar laceration and contusion.

Based on lenticular opacity, the cataracts were classified as total when an examiner did not observe clear lens matter between the capsule and nucleus (Fig. 22.6), membranous when the capsule and organized matter were fused and formed a membrane of varying density (Fig. 22.7), white soft cataract with a ruptured capsule when loose cortical material was found in the anterior chamber together with a ruptured lens capsule (Fig. 22.8), and rosette type cataract for a lens with a rosette pattern of opacity (Fig. 22.9). We could classify all of the cataract cases seen with this classification. Morphology was influenced mainly by the type, force, and object of injury and the time interval between the injury and examination [21].

In case of globe rupture type open globe injury, wound repair was done as first stage and cataract was operated at second sitting. All steps of surgical techniques were documented in pretested online format. All traumatic cataract patients without infection were treated with systemic corticosteroids. In all patients with inflammation

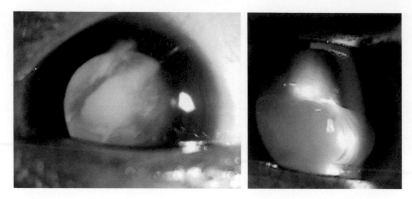

Fig. 22.7 (**a, b**) When loose cortical material was found in the anterior chamber together with a ruptured lens capsule, the cataract was defined as a white soft cataract

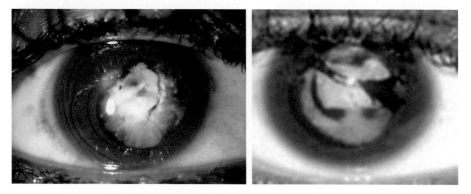

Fig. 22.8 (**a, b**) When the capsule and organized matter were fused and formed a membrane of varying density, the cataract was defined as a membranous cataract

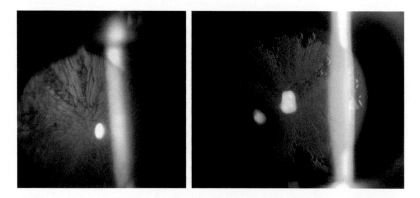

Fig. 22.9 (**a, b**) A lens with a rosette pattern of opacity was classified as a rosette type cataract

and membranous cataracts, primary posterior capsulotomy and anterior vitrectomy was performed [21].

The surgical technique was selected according to morphology and the condition of tissues other than the lens. Phacoemulsification was used to operate on cataracts with hard, large nuclei. With a lens that had either a white soft or rosette type cataract, unimanual or bimanual aspiration was used. Membranectomy and anterior vitrectomy, via either an anterior or a pars plana route, were performed when the cataract was membranous.

In all patients undergoing corneal wound repair, the traumatic cataract was managed in a second procedure. Recurrent inflammation was more prominent in patients who had undergone previous surgery for trauma [9, 10]. In such cases, when the ocular medium was hazy due to inflammation of the anterior vitreous, we performed a capsulectomy and vitrectomy via an anterior/pars plana route in adults.

In children younger than 1.5 years of age, both a lensectomy and vitrectomy via a pars plana route were performed, leaving the rim of the anterior capsule available for secondary lens implantation, and the same surgical procedures were used to manage the traumatic cataracts. Lens implantation as part of the primary procedure was avoided in all children younger than 1.5 years of age; these children were rehabilitated with optical correction, and secondary implantation was done after that.

All patients with injuries and without an infection were treated with topical and systemic corticosteroids and cycloplegics. The duration of medical treatment depended on the degree of inflammation in the anterior and posterior segments of the operated eye.

All children received supportive amblyopia therapy from a qualified pediatric orthoptist, and a pediatric ophthalmologist treated strabismus.

Postoperative follow-up was done according to pretested online format including vision, anterior and posterior segment findings and intraocular pressure, at proper follow-up schedule.

Glasses were prescribed as media was clear and final prescription was at 6 weeks postoperative. Patients undergone orthotic evaluation and amblyopic patients were treated with appropriate patching. Aphakic patients were rehabilitated either by glasses or by contact lenses.

22.2.1.1   Follow-Up

The operated patients were re-examined after 24 h, 3 days, and 1, 2, and 6 weeks to enable refractive correction. Follow-up was scheduled for day 3, weekly for 6 weeks, monthly for 3 months, and every 3 months for 1 year.

At every follow-up examination, visual acuity was tested according to age using the AAO guidelines. The anterior segment was examined with a slit lamp and the posterior segment with an indirect ophthalmoscope. Eyes with vision better than 20/60 at the glasses appointment (6 weeks) were defined as having a satisfactory grade of vision.

Patients developing after cataract underwent membranectomy. For children operated below the age of 1.5 years, secondary lens implantation was done after the age of 1.5 years. Patients were evaluated for stereopsis and contrast sensitivity using titmus vision tester or titmus fly test.

During the examination, data were entered online using a pretested format designed by the International Society of Ocular Trauma (initial and follow-up forms), which was exported to a Microsoft Excel spreadsheet. The data were audited periodically to ensure completion. We used the Statistical Package for the Social Studies (SPSS 17) to analyze the data. We used descriptive statistics and cross tabulation to compare the cause and effect of different variables. The dependent variable was vision >20/60 noted at the follow-up 6 weeks after cataract surgery. The independent variables were age, gender, residence, time interval between the injury and cataract surgery, primary posterior capsulectomy and vitrectomy procedure, and type of ocular injury.

We compared all of the variables for both the open and closed globe groups using various statistical tests.

22.2.2   Results

The enrolled patient group consisted of 1,305 eyes of 1,045 patients with cataract in pediatric age group, comprising 858 (65.7 %) males and 447 (34.3 %) females (Table 22.1). The mean patient age was 9.5 ± 4.7 years ranging from 0 to 18, mean 9.4. Among these eyes, 610 (46.7 %) were traumatic, and 695 (53.3 %) were congenital or developmental cataract (Table 22.2). One thousand one hundred seventeen (85.6 %) (Table 22.3) presented with diminished vision and 188 (14.4 %) presented with leukocoria.

Amongst nontraumatic group eyes further subdivided in congenital 293 (22.5 %), developmental 373 (28.6 %), and secondary cataract 29 (2.2 %) (Table 22.2). According to cross tabulation and statistical analysis, the demographic factors analyzed, including socioeconomic status (74.5 % were of lower socioeconomic status) and residence (92 % were from rural areas), had no significant relationship to the final visual acuity.

With regard to patient entry, 9.2 % of the patients had received primary treatment prior to reaching our center, and this was not associated with a significant difference

Table 22.1 Age and sex distribution

	Sex	
Age (years)	Female	Male
0 to 2	37	56
3 to 5	67	119
6 to 10	156	293
11 to 18	187	390
Total	447	858

Table 22.2 Aetiology
of cataract

Type of cataract	Frequency	Percent
Complicated	29	2.2
Congenital	293	22.5
Developmental	373	28.6
Traumatic	610	46.7
Total	1,305	100.0

Table 22.3 Comparison of visual outcome according to aetiology

	Type of cataract			
Visual acuity	Complicated	Congenital	Developmental	Traumatic
Uncooperative	0	9	6	5
<1/60	20	66	66	201
1/60 to 3/60	3	61	65	69
20/200 to 20/120	2	17	50	46
20/80 to 20/60	3	128	92	94
20/40 to 20/30	0	5	30	53
20/20 to 20/16	1	5	62	140
Total	29	291	371	608

$p=0.000$

in the final visual outcome ($p=0.2$). Of the total patients enrolled, 26.4 % entered via an outreach department, and 71 % were self-referred

Among the injuries, 30 % were reported within the first 24 h, 30 % were reported within 3 days, and 33.9 % were reported within 1 month. A wooden stick was the most common object causing eye injury (51.4 %). Neither the injury-causing object ($p=0.3$) nor the activity at the time of injury ($p=0.3$) was significantly associated with the final visual acuity.

A comparison between pre- and postoperative visual acuities showed that treatment significantly improved the visual acuity (Table 22.4; $p=0.000$, Pearson's χ^2 test; $p=0.001$, ANOVA). An intraocular lens was implanted in 1,205 cases (92.3 %) and was significantly associated with improved visual acuity ($p=0.000$).

When we compared visual outcome between traumatic and nontraumatic group, traumatic group is doing well (Table 22.5, $p=0.000$).

Final visual acuity following cataract surgery was >20/200 in 396 eyes (56.8 %) and ≥20/40 in 108 eyes (14.8 %) in nontraumatic group. In the traumatic group, the visual acuity was >20/200 in 334 eyes (54.6 %) and ≥20/40 in 108 eyes (14.8 %). The difference between the groups was significant ($p=0.000$). In total, 296 (22.6 %) eyes had a final visual acuity ≥20/40 and 728 (57.8 %) eyes had a final visual acuity >20/200 (Table 22.4). The follow-up period ranged from 1 to 3,084 days, with a mean of 117.4 days.

Table 22.4 Comparison of visual outcome according to preoperative visual acuity

Visual acuity (post op)	Visual acuity (preoperative)						
	Uncooperative	<1/60	1/60 to 3/60	20/200 to 20/120	20/80 to 20/60	20/40 to 20/20	20/30 to 20/20
Uncooperative	7	13	0	0	0	0	0
<1/60	7	321	11	11	3	1	0
1/60 to 3/60	1	128	54	11	5	0	0
20/200 to 20/120	0	74	28	10	3	0	0
20/80 to 20/60	7	132	37	99	41	0	2
20/40 to 20/30	1	62	15	3	7	1	0
20/20 to 20/16	5	140	24	23	12	3	3
Total	28	870	169	157	71	5	5

$p=0.000$

Table 22.5 Comparison of visual outcome amongst traumatic and nontraumatic group

Visual acuity	Cataract	
	Nontraumatic	Traumatic
Uncooperative	15	5
<1/60	153	201
1/60 to 3/60	130	69
20/200 to 20/120	70	46
20/80 to 20/60	224	94
20/40 to 20/30	36	53
20/20 to 20/16	69	140
Total	697	608

$p=0.000$

Table 22.6 Comparison of visual outcome according to presence of sensory nystagmus

Visual acuity	Sensory_nystagmus	
	No	Yes
Uncooperative	13	7
<1/60	302	52
1/60 to 3/60	150	49
20/200 to 20/120	96	20
20/80 to 20/60	230	88
20/40 to 20/30	88	1
20/20 to 20/16	204	5
Total	1,083	222

$p=0.000$

We found sensory nystagmus present in 221 (17 %) eyes which caused significant difference in final visual outcome (Table 22.6, $p=0.00$).

We have studied unilateral versus bilateral (Table 22.7, $p=0.000$) and found better results with unilateral cases. In nontraumatic group bilateral cases have better results (Table 22.7, $p=0.000$).

Table 22.7 Comparative study of visual outcome according to laterality over all category of cataract in pediatric age group

Visual acuity	Laterality	
	Bilateral	Unilateral
Uncooperative	8	12
<1/60	76	278
1/60 to 3/60	98	101
20/200 to 20/120	59	57
20/80 to 20/60	187	131
20/40 to 20/30	29	60
20/20 to 20/16	59	150
Total	516	789

$p=0.000$

Table 22.8 Visual outcome according to age, laterality, aetiology

No	Category	p Value	Conclusion
1	Total	0.000	UL better
2	Above five	0.000	UL better
3	Under five	0.001	BL better
4	Traumatic	0.020	UL better
5	Nontraumatic	0.000	BL better
6	Nontraumatic ≤5	0.000	BL better
7	Traumatic ≤5	No bilateral	NA
8	Nontraumatic >5	0.000	BL better
9	Traumatic >5	0.061	NA

We have studied cases with amblyopia in unilateral as well as bilateral cataracts, and significant difference was found in bilateral cases (Table 22.8, $p=0.000$). In nontraumatic group we had 184 eyes with unilateral cataract with deprivational amblyopia vision improvement was not significant ($p=0.510$).

We have studied and compared visual outcome taking laterality into account according to age (<5/= and >5), and according to category (traumatic, nontraumatic) (Table 22.8).

We have concluded that visual outcome is better in unilateral cases if we consider all subgroups, better with bilateral cases in group ≥5 ($p=0.000$) with unilateral cases in group >5 ($p=0.000$). Visual outcome is better in traumatic group in unilateral ($p=0.02$) and nontraumatic cases bilateral ($p=0.000$).

We have also compared visual outcome for both groups ≤5 and >5 (Table 22.8), found that in case of nontraumatic cases visual outcome is better in bilateral conditions ($p=0.000$) and in case of traumatic no significant difference is found ($p=0.061$) (Table 22.8).

When we studied traumatic group our cohort consisted of 354 patients with traumatic cataracts, including 287 (82.6 %) eyes with open globe ocular injuries and 67 (17.3 %) eyes with closed globe injuries (Fig. 22.10). The patients included 252 (71.1 %) males and 102 (28.8 %) females. The mean patient age was 10.6±4.3 years (range 0–17) (Table 22.9).

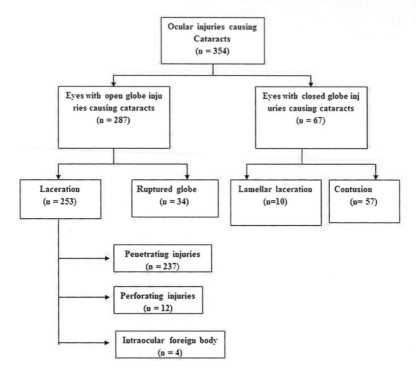

Fig. 22.10 BETTS Distribution of traumatic cataract

Table 22.9 Age and sex distribution of traumatic cases

	Sex		
Age group	Female	Male	Total
0 to 1	2	2	4
2 to 4	12	33	45
5 to 10	44	88	132
11 to 16	44	129	173
Total	102	252	354

We analyzed several demographic factors, including patient entry ($p=0.000$), cases self-reported as having done well (Table 22.10), socioeconomic status (79 % were from lower socioeconomic classes), and residence (95 % were from rural areas). None was significantly related to the final visual acuity according to cross tabulation and statistical tests.

The object causing the injury (Table 22.12) and the activity at the time of the injury ($p=0.3$; Table 22.11) were also not significantly associated with a visual acuity at 6 weeks. A stick was the most common agent of injury (56.1 %; Table 22.12). Better outcome was achieved above 5 years (Table 22.13).

A comparison of the pre- and postoperative visual acuity showed that treatment significantly improved visual acuity (Pearson's χ^2 test, $p=0.000$; ANOVA, $p=0.001$;

Table 22.10 Patient entry and visual outcome

Vision	Entry Self	ORD	Total
Uncooperative	6	2	8
No PL	10	0	10
HMPL	36	16	52
1/200 to 19/200	39	28	67
20/200 to 20/50	66	29	95
≥20/40	94	28	122
Total	250	104	354

$p=0.000$

ORD out reach department

Table 22.11 Activity at the time of the injury

Object	Number (n)	Percentage (%)
Fall	6	1.7
Making a fire	10	2.8
Housework	58	16.4
Employment	20	5.6
Other	45	12.7
Walking	4	1.1
Playing	195	55.1
Travelling	12	3.4
Unknown	4	1.1
Total	354	100.0

Table 22.12 Object causing the injury

Object	Number (n)	Percentage (%)
Ball	5	1.4
Cattle horn	6	1.7
Cattle tail	1	.3
Finger	3	.8
Fire	10	2.8
Glass	4	1.1
Thorn	12	3.4
Other	31	8.8
Sharp object	31	8.8
Stone	38	10.7
Unknown	31	8.8
Stick	182	51.4
Total	354	100.0

Table 22.14). Morphologically, lens ruptured with soft material in anterior chamber has done significantly better (Tables 22.15 and 22.16). Aspiration was significantly associated with improved visual acuity ($p=0.000$) and was performed using one or two ports in 48.6 % of the patients in the open globe group.

Table 22.13 Comparative study of age and visual outcome in traumatic cataract in children

Vision	Age group (years)				Total
	0 to 2	3 to 5	6 to 10	11 to 18	
Uncooperative	0	6	2	0	8
No PL	1	2	4	3	10
HMPL	0	11	23	19	53
1/200 to 19/200	0	7	30	29	66
20/200 to 20/50	2	14	41	37	94
≥20/40	1	5	32	85	123
Total	4	45	132	173	354

$p=0.007$
UC uncorrected vision

Table 22.14 Pre-treatment and post-treatment vision following surgical treatment for traumatic cataract in children

	Vision						Total
	UC	No PL	PLHM	1/200 to 19/200	20/200 to 20/50	≥20/40	
UC	7	0	1	0	0	0	8
No PL	0	8	2	0	0	0	10
PLHM	0	4	48	1	0	0	53
1/200 to 19/200	0	1	48	16	0	1	66
20/200 to 20/50	1	3	64	22	4	0	94
≥20/40	2	0	83	32	6	0	123
Total	10	16	246	71	10	1	354

$p=0.000$
UC uncooperative

Table 22.15 Morphology and visual outcome following surgical treatment for traumatic cataract in children

Vision	Morphology				Total
	Membrane	Rosette	Lens ruptured	Total	
Uncooperative	1	0	6	1	8
No PL	2	0	4	4	10
HMPL	14	0	27	11	53
1/200 to 19/200	17	0	38	11	66
20/200 to 20/50	21	2	62	9	93
≥20/40	12	4	97	11	124
Total	67	6	234	47	354

$p=0.02$

Primary posterior capsulotomy and anterior vitrectomy, commonly performed for eyes with significant inflammation, caused no significant improvement in the visual acuity at 6 weeks ($p=0.23$).

Table 22.16 Distribution according to morphology of cataract

Morphology	Number (*n*)	Percent (%)
After cataract	25	1.9
Ant lenticonus	2	.2
Ant subcapsular	2	.2
Aphakia	38	3.0
Blue dot, floriform	10	.9
Complicated	18	1.4
Cortical + blue dot	2	.2
Cortical + PSC	2	.2
Decentred IOL	10	.8
Dislocated	1	.1
Ectopia lentis	14	1
Lamellar	233	17.8
Membranous	159	12.2
Microspherophakia	3	.2
Nuclear	8	.6
PHPV	1	.1
Postlenticonus	1	.1
Postsubcapsular	47	3.6
Rosette	15	1.1
Soft fluffy with ruptured capsule	295	22.6
Sutural	2	.2
TOTAL	419	32.1
Total	1,305	100.0

Table 22.17 Type of injury and visual outcome following surgical treatment for traumatic cataract in children

	Category		
Vision	Open	Closed	Total
UC	3	5	8
No PL	3	7	10
PLHM	10	43	53
1/200 to 19/200	15	51	66
20/200 to 20/50	21	73	94
≥20/40	16	109	125
Total	67	287	354

p = 0.05

UC uncorrected vision

We also compared these variables in the open globe and closed globe subgroups (Tables 22.17 and 22.18) and found significant differences in different subcategories.

Six weeks postoperatively, the visual acuity in the operated eye was >20/200 in 181 (63 %) and ≥20/40 109 (38 %) eyes in open globe group and >20/200 in 36 (53 %) and ≥20/40 16 (22.4 %) eyes in closed globe group (*p* = 0.143), and the difference

Table 22.18 Type of injury and visual outcome following surgical treatment for traumatic cataract in children

Vision	Type of injury						Total
	Lamellar laceration	Contusion	Globe rupture	Penetrating	Perforating	IOFB	
UC	2	1	0	5	0	0	8
No PL	1	2	3	2	1	1	10
PLHM	2	9	8	32	4	0	55
1/200 to 19/200	2	15	5	43	2	0	67
20/200 to 20/50	1	18	16	54	3	1	93
≥20/40	2	12	2	101	2	2	121
Total	10	57	34	237	12	4	354

$p=0.000$
UC uncooperative, *IOFB* intraocular foreign body

Table 22.19 Time interval between injury and intervention

Vision	Days before reporting				Total
	0 to 2	3 to 7	8 to 30	>30	
Uncooperative	4	0	2	2	8
No PL	5	2	0	3	10
HMPL	19	6	7	21	53
1/200 to 19/200	16	5	12	33	66
20/200 to 20/50	29	16	18	31	94
≥20/40	37	17	39	30	121
Total	110	46	78	120	354

$p=0.122$

between the groups was not significant in children. Overall, 125 (35.3 %) eyes gained visual acuity at 6 weeks ≥20/40 and >20/200 in 214 (61.3 %) cases (Tables 22.14, 22.17, and 22.18). Penetrating subgroup of open globe improved significantly better than other subcategories (Tables 22.14, 22.17, and 22.18, $p=0.004$).

We implanted an intraocular lens in 82 % of the cases, 90 % of which was polymethyl methacrylate and 10 % acrylic, 5 IOL subluxated or dislocated, polymethyl methacrylate lenses used for secondary implant and 30 % of the cases underwent more than one operative procedure. Of the children, 30 % reported within the first 24 hours of the injury. Time interval between injury and intervention did not make significant difference in outcome at 6 weeks (Table 22.19, $p=0.172$).

Table 22.20 is displaying the list of complications encountered after surgery.

We have studied the visual outcome of children under 5 years-old, as this is an age of amblyopia.

The enrolled patient group consisted of 128 eyes of 128 patients with cataract in pediatric age group (0 to 5), comprising 80 (62.5 %) males and 48 (37.5 %) females (Table 22.21). The mean patient age was 2.5 ± 0.73 years ranging from 0 to 5, mean 2.5. Among these eyes, 85 (66.4 %) were traumatic, and 43 (33.6 %) were congenital

Table 22.20 Complications following surgical treatment

	Number	Percent
After cataract	6	.5
Aphakia	17	1.3
Corneal opacity	12	.9
Endophthelmitis	3	.3
Hyphema	7	.5
Inflammation	16	1.2
IOL dislocation	7	.5
Iridodialysis	2	.2
Optic atrophy	1	.1
Optic capture	8	.7
Phithisis	3	.2
RD	7	.6
Secondary glaucoma	6	.6
Total	95	7.6

Table 22.21 Age and sex distribution under five category

	SEX		Total
Age groups	F	M	F
0 to 1	6	12	18
2 to 3	17	13	30
4 to 5	25	55	80
Total	48	80	128

or developmental cataract. One hundred six (82.8 %) presented with diminished vision and 22 (17.2 %) presented with leukocoria.

Amongst nontraumatic group, eyes were further subdivided into congenital 30 (23.4 %), developmental 9 (7 %), and complicated cataract 4 (3.1 %). According to cross tabulation and statistical analysis, the demographic factors analyzed, including socioeconomic status (74.5 % were of lower socioeconomic status) and residence (92 % were from rural areas), had no significant relationship to the final visual acuity.

Regarding patient entry, 9.2 % of the patients had received primary treatment prior to reaching our center; this was not associated with a significant difference in the final visual outcome ($p=0.2$). Of the total patients enrolled, 26.4 % entered via an outreach department, and 71 % were self-referred.

A comparison between pre- and postoperative visual acuities showed that treatment significantly improved the visual acuity overall and subcategories (Table 22.22; $p=0.005$, Pearson's χ^2 test; $p=0.001$, ANOVA). An intraocular lens was implanted in 114 cases (89.1 %), and was significantly associated with improved visual acuity ($p=0.000$).

When we compared visual outcome between the traumatic and nontraumatic groups, the nontraumatic group did significantly better (Table 22.27; $p<0.001$). Final visual acuity following cataract surgery was >6/60 in 22 eyes (51.1 %) and ≥6/12 in 1 eye (2.3 %) in the nontraumatic group. In the traumatic group, the visual

Table 22.22 Comparison of visual outcome according to preoperative visual acuity under five category

Post op vision	Preoperative vision					
	Uncooperative	<1/60	1/60 to 3/60	6/60 to 6/36	6/24 to 6/18	Total
Uncooperative	2	6	0	0	0	8
<1/60	0	39	2	2	2	45
1/60 to 3/60	0	5	4	2	2	13
6/60 to 6/36	0	8	0	0	0	8
6/24 to 6/18	1	23	0	17	5	46
6/12 to 6/9	0	3	0	0	0	3
6/6 to 6/5	0	3	0	2	0	5
Total	3	87	6	23	9	128

$p < 0.001$

Table 22.23 Comparison of visual outcome according to age of intervention

	Age category			
	0 to 1	2 to 3	4 to 5	Total
Uncooperative	1	4	3	8
<1/60	2	8	35	45
1/60 to 3/60	1	2	10	13
6/60 to 6/36	0	1	7	8
6/24 to 6/18	12	15	19	46
6/12 to 6/9	0	0	3	3
6/6 to 6/5	2	0	3	5
Total	18	30	80	128

$p = 0.008$

acuity was >6/60 in 40 eyes (47.1 %) and ≥6/12 in 7 eyes (8.2 %). The difference between the groups was significant ($p < 0.001$). In total, 62 (48.4 %) eyes had a final visual acuity ≥6/60 and 8 (6.3 %) eyes had a final visual acuity >6/12 (Table 22.22).

We have compared visual outcome according to age of intervention and found significant difference; better results were achieved in age range between 0 to 1 (Tables 22.23 and 22.27, $p = 0.000$).

Insertion of an IOL resulted in no significant improvement in visual outcome (Tables 22.24 and 22.27; $p < 0.001$).

We compared final visual outcome between the anterior and pars plana routes. An anterior approach resulted in a better outcome (Tables 22.25 and 22.27; $p < 0.001$).

We also studied visual outcome when a primary posterior capsulotomy and vitrectomy were performed. Primary posterior capsulotomy and vitrectomy were performed in 53 (41.4 %) cases and no significant difference was found (Tables 22.26 and 22.27; $p < 0.001$

We compared results by morphology of the cataract in nontraumatic group. Good results were achieved with lamellar cataracts ($p < 0.001$).

Table 22.24 Comparison of visual outcome according to presence of IOL

Postoperative vision	IOL		Total
	N	Y	
Uncooperative	2	6	8
<1/60	6	39	45
1/60 to 3/60	0	13	13
6/60 to 6/36	0	8	8
6/24 to 6/18	5	41	46
6/12 to 6/9	0	3	3
6/6 to 6/5	1	4	5
Total	14	114	128

$p=0.511$

Table 22.25 Comparison of visual outcome according to surgical approach

Postoperative vision	SURGICAL_A		Total
	A	P	
Uncooperative	1	7	8
<1/60	18	27	45
1/60 to 3/60	8	5	13
6/60 to 6/36	7	1	8
6/24 to 6/18	31	15	46
6/12 to 6/9	3	0	3
6/6 to 6/5	5	0	5
Total	73	55	128

$p=0.001$

Table 22.26 Comparison of visual outcome according to presence of primary posterior capsulotomy

Postoperative vision	PPC		Total
	No	Yes	
Uncooperative	4	4	8
<1/60	32	13	45
1/60 to 3/60	3	10	13
6/60 to 6/36	3	5	8
6/24 to 6/18	27	19	46
6/12 to 6/9	2	1	3
6/6 to 6/5	4	1	5
Total	75	53	128

$p=0.054$

Table 22.27 Comparison of traumatic and nontraumatic group under five category

Variable		Nontraumatic		Traumatic		Total		p Value
		N	%	N	%	N	%	
Habitate	Rural	30	23.4	58	45.3	88	68.8	0.12
	Urban	23	17.9	27	21.1	50	39.1	
Socioeconomical category	Rich	5	3.9	12	9.4	17	13.3	0.11
	Poor	38	29.7	73	57.0	111	86.7	
Age	0 to 1	11	8.6	7	5.5	18	14.1	0.01
	2 to 3	12	9.4	18	14.1	30	23.4	
	4 to 5	20	15.6	60	46.9	80	62.5	
	Total	43	33.6	85	66.4	128	100	
Pre op vision	Uncooperative	2	1.6	1	0.8	3	2.3	0.003
	<1/60	21	16.4	66	51.6	87	67.9	
	1/60 to 3/60	4	3.1	2	1.6	6	4.7	
	6/60 to 6/36	9	7.0	14	10.9	23	18.0	
	6/24 to 6/18	7	5.5	2	1.6	9	7.0	
	6/12 to 6/9	0	0	0	0	0	0	
	6/6 to 6/5	0	0	0	0	0	0	
	Total	43	33.6	85	66.4	128	100	
Post op vision	Uncooperative	5	3.9	3	2.3	8	6.3	0.005
	<1/60	8	6.3	37	28.9	45	35.2	
	1/60 to 3/60	8	6.3	5	3.9	13	10.2	
	6/60 to 6/36	1	0.8	7	5.5	8	6.3	
	6/24 to 6/18	20	15.6	26	20.3	46	35.9	
	6/12 to 6/6	1	0.8	2	1.6	3	2.3	
	6/9 to 6/6	0	0	5	3.9	5	3.9	
	Total	43	33.6	85	66.4	128	100	
Sensory nystagmus	No	34	26.6	84	65.6	118	92.2	0.000
	Yes	9	7.0	1	0.8	10	7.8	
	Total	43	33.6	85	66.4	128	100	
Amblyopia	No	41	32	76	59.4	117	91.4	0.000
	Yes	2	1.6	9	7.0	11	8.6	
	Total	43	33.6	85	66.4	128	100	
PPC	No	23	18.0	52	40.6	75	58.6	0.259
	Yes	20	15.6	33	25.8	43	33.6	
	Total	43	33.6	85	66.4	128	100	
IOL	No	5	3.9	9	7.0	14	10.9	0.538
	Yes	38	26.7	76	59.4	114	89.1	
	Total	43	33.6	85	66.4	128	100	
Surgical approach	Anterior	23	18.0	50	39.1	73	57.0	
	Posterior	20	15.6	35	27.3	55	42.9	0.349
	Total	43	33.6	85	66.4	128	100	

Table 22.28 Type of strabismus

Type	Number (n)	Percent (%)
Esotropia	33	50.8
Vertical	4	7.2
Exotropia	27	42.0
Total	64	100.0

Table 22.29 Amblyopia according to age

	Amblyopia		
Age	Yes	No	Total
≤1	9	41	50
1 to 3	16	66	82
4 to 5	11	122	133
6 to 10	19	443	462
11 to 18	18	560	578
Total	73	1,232	1,305

$p=0.000$

We found strabismus in 64 (4.9 %) cases, of which 33 (50.8 %) was esotropia, 27 (42 %) was exotropia, and 4 (7.2 %) was vertical strabismus with esotropia or exotropia (Table 22.28). Out of all 36 % cases had special forms of strabismus like A pattern, V pattern, or Dissociated vertical deviation. In cases with strabismus 37 (53.6 %) had sensory nystagmus, 56 (81.2 %) had amblyopia (Table 22.28).

We compared strabismus according to age of intervention. Significantly association with strabismus was found with younger age group (Table 22.29; $p<0.001$).

When compared with category strabismus is less commonly associated with traumatic cataract ($p<0.001$).

We examined unilateral versus bilateral cataracts ($p<0.001$) and found significant association with strabismus in bilateral cases. Sensory nystagmus was present in 250 (18.8 %) eyes and was significantly associated with strabismus ($p<0.001$). An intraocular lens was implanted in 1,187 cases (89.1 %) and was significantly associated with strabismus ($p=0.000$).

We have studied amblyopia according to age of intervention which was found significantly differing (Table 22.29; $p=0.000$) and patching has made significant difference in final visual outcome (Table 22.30).

22.2.3 Discussion

We have focused discussion for the points which has controversies and compared findings of other studies too with conclusions.

Table 22.30 Comparative study of pre and post patch vision

Post patch vision	Pre patch vision				Total
	<1/60	1/60 to 3/60	6/60 to 6/36	6/24 to 6/18	<1/60
<1/60	7	0	0	1	8
1/60 to 3/60	2	21	0	0	23
6/60 to 6/36	0	4	4	0	8
6/24 to 6/18	1	0	2	27	30
6/12 to 6/9	0	0	3	1	4
Total	10	25	9	29	73

$p = 0.000$

22.2.3.1 Epidemiology and Demography

Our study agrees with Gogate et al. and Ram et al. that demographic factors did not have significant difference on visual outcome [22–25].

22.2.3.2 Aetiology

In terms of aetiology, pediatric cataracts occur due to genetic diseases, metabolic diseases, maternal infections, and trauma, and can also be idiopathic. The aetiology of cataracts can be established in up to half of the children with bilateral cataracts, but in only a small proportion of children with unilateral cataracts.

Aetiology causes significant difference in visual outcome (Tables 22.3 and 22.5, $p = 0.000$).

When we compared traumatic versus nontraumatic group we found age of intervention and laterality also play important role.

22.2.3.3 Morphology

Any opacification of the lens and its capsule in children is defined as a pediatric cataract. Pediatric cataracts can be unilateral or bilateral. They can be subdivided based on morphology, as well as aetiology. Morphologically, the most common type of pediatric cataract is the zonular cataract, characterized by opacification of a discrete region of the lens. This type includes nuclear, lamellar, sutural, and capsular cataracts [21].

Polar cataracts are opacities of the subcapsular cortex in the polar regions of the lens. Almost all (90 %) anterior polar cataracts are unilateral; bilateral anterior polar cataracts are commonly asymmetric and typically do not progress over time. Posterior polar cataracts are often small, but even a small posterior polar cataract can impair vision. A distinctive type of posterior polar cataract is the posterior lentiglobus or lenticonus, in which a protrusion of the posterior capsule is present.

Membranous cataracts form when the lens, cortex, and nucleus are partially or completely reabsorbed, leaving a small amount of opacified lens material between the anterior and the posterior lens capsules.

Persistent hyperplastic primary vitreous (PHPV) is usually a unilateral ocular condition associated with a retrolenticular fibro vascular membrane. Although the lenses in most eyes with PHPV are initially clear, they often become opacified over time. Even when the lens remains clear, the retrolenticular membrane is usually sufficiently opaque to affect vision.

Nevertheless, we feel that some useful observations can be made on the basis of this review of patients. There seem to be two general categories of patients with congenital and developmental cataracts. One is characterized by extensive lens opacity and an early, obvious reduction in vision. These patients, who come for cataract surgery in the first year of life, often have smaller-than-normal corneal diameters, poorly dilating pupils, and a vulnerability to delayed postoperative open-angle glaucoma. The other category includes patients with partial, often lamellar lens opacities, corneas of normal size, and a remarkably good visual prognosis. Lamellar cataract did significantly better when compared with other morphologies in the nontraumatic group, similar to other studies (Tables 22.15 and 22.16). Of the patients in the first category, 222 (17 %) developed nystagmus at 2–4 months of age, which was accompanied by a reduction in visual acuity despite a good anatomical result from surgery (Table 22.6). Early surgery, within the time frame indicated in [10], did not appear to stop the development of nystagmus, although some investigators have suggested that this may be so, based on anecdotal experiences with small numbers of patients [10, 17]. It is possible that even earlier surgery than we have done, undertaken in the first few weeks of life, would have a more favorable influence on the development of nystagmus. On the other hand, the nystagmus may be a manifestation of a more general ophthalmic disorder that would not be influenced by the timing of surgery. Further evaluation of this question will require a randomized controlled study, because patient selection could influence the prevalence of nystagmus in any small series of patients, especially if all patients with congenital cataracts, regardless of type, were subjected to early surgery

Morphology of cataract play important role for visual outcome, as surgical techniques may be decided by morphology and have significant difference in final visual outcome. Penetrating injury with ruptured capsule and soft cataract has better outcome (Tables 22.15 and 22.16, $p=0.02$, [21]).

22.2.3.4 Age of Intervention and Laterality

Mean age in other studies was 7.1 [22, 23] which is less than our study. Age of intervention made significant difference for visual outcome (Tables 22.8 and 22.13, $p=0.000$); other investigators also have similar findings [23].

When compared traumatic and nontraumatic groups, the group with traumatic cataract is doing significantly well as compared to the other group, which may be attributed to fully developed visual system in children above 5 years of age and other local or systemic anomalies in cases of congenital or developmental cataracts.

When we tried to compare unilateral and bilateral we found that unilateral cataracts are doing well in contrast to other studies (Table 22.7) [10, 22, 23], may be attributed to large number of traumatic cases.

A prospective study of the outcome of surgery for cataracts in pediatric age group has several limitations. Although we believe that all patients included in the study had congenital, developmental, or traumatic lens opacities, not all patients were seen and followed by us from the time of birth. In particular, some patients with lamellar cataracts were not seen by us until they were several years old.

Regarding timing of intervention our study suggests that visual outcome is affected by age of intervention, aetiology, and laterality. In case of unilateral cataract sooner the better. Findings are similar in case of nontraumatic group [10, 24, 25].

As study was over 20 long years surgical techniques evolved so surgeries performed in our series of patients were not identical in all cases. The surgery performed in our series of patients was not identical in all cases. For instance, the posterior capsule was handled differently at different times during the study period.

Additionally, the timing of surgery was not dictated by an established protocol, but was determined by the age at the time of referral and by the visual status of individual patients. Finally, some observations that in our series would have been useful for analysis were missing from our records because of lost follow-up.

In present study we have tried to study role of laterality along with other variables. Our results establishing importance of age group (≤ 5 and >5), traumatic and nontraumatic (Table 22.8) and amblyopia (Table 22.29) which is different than other studies [10, 24–26]. These are studies with very small sample size with retrospective design.

Deprivational amblyopia due to asymmetry of cataracts from the outset is very difficult to reverse, similar to the situation in patients with monocular congenital cataracts. An early start of treatment would seem to be the only hope of success in these asymmetrical cases [10].

Our study suggesting various outcomes according to age (>5 and ≤ 5) similar to the study by Robb [10]. Our study reflects, if we consider traumatic cases, better outcome in older age group, maybe because of less intense inflammation. In case of nontraumatic group, sooner intervention has better outcome in type-1 cataract and later intervention in case of type-2 cataract (Table 22.8). We are not aware about any study which has specifically studied this aspect except small numbers [10, 27–29].

The visual prognosis in this group of patients, whose surgery is usually performed after 5 years of age, at a time when increasing visual needs begin to exceed the limits imposed by 507 Bilateral Congenital Cataracts the lens opacities, is excellent [27].

Conclusion: Laterality has a very important role to play in addition to other factors like age group and aetiology. Age of intervention affects visual outcome significantly and our conclusions out of this study are—Aetiologically if cataract is of traumatic variety late intervention (age 11 to 18) has better outcome ($p = 0.000$). In nontraumatic group, if cataract morphology is type-1 early age intervention has better outcome ($p = 0.000$). In nontraumatic group, if cataract morphology is type-2 with partial opacities late age interventions has better outcome ($p = 0.000$).

22.2.3.5   Traumatic Consideration

Visual gain following surgery for traumatic cataracts is a complex problem because it is not only the lens that determines the visual outcome, comorbidities play very important role. Our study concluded traumatic cataracts has better outcome in age group above five [30, 31].

Our study examined patients with open and closed globe injuries who developed traumatic cataracts. A satisfactory grade of vision following the management of traumatic cataracts was significantly more frequent in the eyes with open globe injuries [30, 31] (Tables 22.17 and 22.18).

Various authors have reported different results in children with traumatic cataracts: Shah et al. [32] reported 20/60 or better in 56 % of their cases; Kumar [33] reported 20/60 or better in 50 %; Staffieri [34], 6/40 or better in 35 %; Bekibele [35] 20/60 or better in 35.6 %; Gradin [36] 20/60 or better in 64.7 % Brar [37] 0.2 or better in 62 %; Cheema [38] 20/80 in more than 68 % of their cases; Karim [39] 0.2 or better in 62 %; Krishnamachary [40] 20/60 or better in 74 %; Knight-Nanan [41] 20/60 or better in 64 %; Bienfait [42] 0.7 in 27 %; and Anwar [43] reported 20/40 or better in 73 % of their cases.

Using a polymethyl methacrylate (PMMA) lens, Verma [44] reported a visual outcome similar to that found in our study. Eckstein [45] and Zou [46] reported that primary intraocular lens implantation is important for a better visual outcome, similar to our results. Also similar to our results, Vajpayee [47] and Gupta [48] reported primary insertion of an intraocular lens with posterior capsule rupture.

Shah [49] reported that a better visual outcome was achieved when intervention was done between 5 and 30 days in adults with traumatic cataracts. As in our study, Rumelt [50] found no significant difference between primary and secondary implantation.

Staffieri [34] performed primary implantation in 62 % of cases versus 82 % in our study. Kumar [33] and Verma [44] advocated primary posterior capsulotomy and vitrectomy for a better outcome; our results concurred.

We are not aware of any study that has compared the visual acuity at 6 weeks in children between two groups classified using the Birmingham Eye Trauma Terminology Systems (BETTS). Shah et al. [29, 51, 52] reported a comparison between open and closed globe injuries in general population. We are also not aware of another large series of successfully treated traumatic cataracts in children.

Satisfactory visual outcome can be achieved in children with traumatic cataracts, no significant difference were found amongst open and closed globe injuries in pediatric age group.

22.2.3.6   Analyses of Under 5

The enrolled patient group consisted of 128 eyes of 128 patients with cataract in pediatric age group (0 to 5), comprising 80 (62.5 %) males and 48 (37.5 %) females (Table 22.21). The mean patient age was 2.5 ± 0.73 years ranging from 0 to 5, mean

2.5. Among these eyes, 85 (66.4 %) were traumatic, and 43 (33.6 %) were congenital or developmental cataract. One hundred six (82.8 %) presented with diminished vision and 22 (17.2 %) presented with leukocoria.

Mean age in other studies was 3.6 ± 1.5 [28] which is more than our study. Age of intervention made significant difference for visual outcome (Tables 22.8 and 22.13, $p = 0.000$); other investigators also have similar findings [29].

Compared with or without primary posterior capsulotomy and vitrectomy we did not find significant difference over all (Table 22.26). Randomized controlled trial may be done to establish importance of this tool [14, 53–55].

When compared anterior and posterior approach anterior approaches it was found that patients with anterior approach were doing well (Table 22.25) but posterior approach was selected for more complex cases. A randomized controlled trial may be done to eliminate confounding conditions [56–59].

The incidence of traumatic cataracts in children was higher than that reported previously [26, 60–62]. In a comparison of the traumatic and nontraumatic groups, the group with nontraumatic cataracts did significantly better under age of 5 years in unilateral cases, likely attributable to comorbidities.

Patients having cataract surgery in the first year of life, often have smaller than-normal corneal diameters, poorly dilating pupils, and a vulnerability to delayed postoperative open-angle glaucoma. The other category includes patients with partial, often lamellar lens opacities, corneas of normal size, and a remarkably good visual prognosis. For traumatic cataract soft white with rupture did significantly well.

Visual outcome in cases of pediatric cataract is multifactorial. Amongst these factors aetiology particularly trauma plays important role, traumatic cataract has poorer prognosis than nontraumatic group, particularly if age <5.

22.2.3.7 Complications

Detection of posterior capsule rupture in presence of traumatic cataract is possible with various investigations. Surgical approach may change according to presence or absence of capsular tear.

There is a list of complications in Table 22.20 reporting complications in 95 (7.6 %) cases such as aphakia, complications related lens mal position, inflammation, and corneal opacities. The majority of them are attributed to traumatic causes.

Deprivational amblyopia is an important limitation to visual outcome. Some concern has been raised about the possibility of a higher complication rate following cataract surgery performed in the first 2 months of life. While it is well-known that deprivational amblyopia is difficult to surmount in patients with unilateral congenital cataracts (Table 22.29 [10];) it is not a major factor in patients with bilateral partial or lamellar opacities, as long as their lens opacities are symmetrical.

Development of open-angle glaucoma is a known complication of early cataract surgery; surprisingly, there was no such case in our study. The cause of delayed open-angle glaucoma following congenital cataract surgery is not well-understood [10, 63]. It has been suggested that the trabecular meshwork is less porous because of early postoperative inflammation. Another possibility is that there is an underlying developmental abnormality of the anterior chamber angle, predisposing to the glaucoma and related to the developmentally small corneas and poorly dilating pupils that are seen in association with more severe congenital cataracts. The 27 % incidence of delayed open-angle glaucoma in patients who have had surgery in the first 15 months is probably an underestimate, because we do not yet have postoperative pressure measurements in all patients and the glaucoma develops late. Certainly, ocular pressures should be monitored regularly after early congenital cataract surgery. The incidence of glaucoma following later surgery for congenital lamellar opacities was zero in our series [10].

22.2.3.8   Amblyopia and Strabismus

Treatment of strabismic amblyopia following bilateral congenital cataract surgery is useful, although the ocular misalignment is sometimes hard to identify, and the amblyopia may be profound by the time it is recognized [10]. Deprivational amblyopia due to asymmetry of cataracts from the outset is very difficult to reverse, similar to the situation in patients with monocular congenital cataracts. An early start of treatment would seem to be the only hope of success in these asymmetric cases (Table 22.29 [10, 62, 64];)

Spanou, Alexopoulos et al. reported strabismus in 23 % and Magli, Iovine et al. [62, 64] Reported 34 % incidence of strabismus in contrast to only 4.9 % in our study [62, 64]. may be because of larger number of traumatic cataract.

With regard to unilateral and bilateral cases, we found that more incidence of strabismus in bilateral cataracts as contradicting other reports [62].

When studied according to sensory nystagmus incidence of strabismus is higher in presence of sensory nystagmus.

Conclusions: Incidence of strabismus in case of pediatric cataract is only 5 %, more common in nontraumatic group, bilateral cases also in younger age group. Orthoptic treatment made significant difference in visual outcome.

Disclosure

- No financial support received from any company or institution
- This study is not presented at any conference or meeting
- Authors do not have any financial interest in any aspect of this study

References

1. Long V, Chen S. Surgical interventions for bilateral congenital cataract. Cochrane Database Syst Rev. 2006; CD003171.
2. Zimmermann-Paiz MA, Quiroga-Reyes CR. Pediatric cataract in a developing country: retrospective review of 328 cases. Arq Bras Oftalmol. 2011;3:163–5.

3. Gogate P, Kalua K. Blindness in childhood in developing countries: time for a reassessment? PLoS Med. 2009;12:e1000177.
4. Wilson ME, Pandey SK. Paediatric cataract blindness in the developing world: surgical techniques and intraocular lenses in the new millennium. Br J Ophthalmol. 2003;87:14–9.
5. de Carvalho KM, Minguini N. Characteristics of a pediatric low-vision population. J Pediatr Ophthalmol Strabismus. 1998;35:162–5.
6. Owens WC, Hughes WF. Results of surgical treatment of congenital cataract. Arch Ophthalmol. 1948;39:339–50.
7. Msukwa G, Njuguna M. Cataract in children attending schools for the blind and resource centers in eastern Africa. Ophthalmology. 2009;116:1009–12.
8. Dandona R, Dandona L. Review of findings of the Andhra Pradesh Eye Disease Study: policy implications for eye-care services. Indian J Ophthalmol. 2001;49:215–34.
9. Vasavada AR, Vasavada SA, Bobrova N. Outcomes of pediatric cataract surgery in anterior persistent fetal vasculature. J Cataract Refract Surg. 2011;38:849–57.
10. Robb RM, Petersen RA. Outcome of treatment for bilateral congenital cataracts. Trans Am Ophthalmol Soc. 1992;19:183–200.
11. Francois J. Late results of congenital cataract surgery. Ophthalmology. 1979;86:1586–98.
12. Hiles DA, Wallar PH. Visual results following infantile cataract surgery. Int Ophthalmol Clin. 1977;17:265–82.
13. Scheie HG. Aspiration of congenital or soft cataracts: a new technique. Am J Ophthalmol. 1960;50:1048–56.
14. Taylor D. Choice of surgical technique in the management of congenital cataract. Trans Ophthalmol Soc UK. 1981;101:114–7.
15. Parks MM. Posterior lens capsulectomy during primary cataract surgery in children. Ophthalmology. 1983;90:344–5.
16. Rogers GL, Tishler CL, Tsou BH. Visual acuities in infants with congenital cataracts operated on prior to 6 months of age. Arch Ophthalmol. 1981;99:999–1003.
17. Gelbart SS, Hoyt CS, Jastrebski G. Long-term visual results in bilateral congenital cataracts. Am J Ophthalmol. 1982;93:615–21.
18. Kumar N, Shekhar C, Kumar P, Kundu AS. Kuppuswamy's socioeconomic status scale-updating for 2007. Indian J Pediatr. 2007;12:1131–2.
19. Trivedi RH, Wilson ME. Accuracy of the Holladay 2 intraocular lens formula for pediatric eyes in the absence of preoperative refraction. J Cataract Refract Surg. 2011;37:1239–43.
20. McClatchey SK. Choosing IOL power in pediatric cataract surgery. Int Ophthalmol Clin. 2011;50:115–23.
21. Shah MA, Shah S ,Shah SB, et al. Morphology of traumatic cataract: does it play a role in final visual outcome? BMJ Open 2011;1:e000060. doi:10.1136/ Bmjopen-2011-000060
22. Congdon NG, Ruiz S. Determinants of pediatric cataract program outcomes and follow-up in a large series in Mexico. J Cataract Refract Surg. 2007;33:1775–80.
23. Yang ML, Hou CH. Clinical characteristics and surgical outcomes of pediatric cataract in Taiwan. Graefes Arch Clin Exp Ophthalmol. 2006;11:1485–90.
24. Gogate P, Khandekar R, Shrishrimal M, Dole K, Taras S, Kulkarni S, Ranade S, Deshpande M. Delayed presentation of cataracts in children: are they worth operating upon? Ophthalmic Epidemiol. 2010;17:25–33.
25. Ram J, Sukhija J, Thapa BR, Arya VK. Comparison of hospital versus rural eye camp based pediatric cataract surgery Middle east Africa. J Ophthalmol. 2012;19:141–6.
26. Khandekar R, Sudhan A. Pediatric cataract and surgery outcomes in Central India: a hospital based study. Indian J Med Sci. 2007;61:15–20.
27. Hochstrasser P, Gloor B. Surgical results of uni- and bilateral congenital and traumatic cataract in infancy to adolescence. Klin Monbl Augenheilkd. 1994;204:274–8.
28. Keech RV, Tongue AC. Complications after surgery for congenital and infantile cataracts. Am J Ophthalmol. 1989;108:136–41.

29. You C, Wu X. Visual impairment and delay in presentation for surgery in Chinese pediatric patients with cataract. Ophthalmology. 2011;118:17–23.
30. Shah MA, Shah SM, Shah SB, Patel CG, Patel UA, Appleware A, Gupta A. Comparative study of final visual outcome between open- and closed-globe injuries following surgical treatment of traumatic cataract. Graefes Arch Clin Exp Ophthalmol. 2011;249(12):1775–81.
31. Shah MA, Shah SM, Shah AH, Pandya JS. Visual outcome of cataract in pediatric age group: does etiology have a role? Eur J Ophthalmo 2014; 24:76–83.
32. Shah M, Shah S, Khandekar R. Ocular injuries and visual status before and after their management in the tribal areas of Western India—a historical cohort study. Grafes Arch Clin Exp Ophthalmol. 2008;246:191–7.
33. Kumar S, Panda A, et al. Safety of primary intraocular lens insertion in unilateral childhood traumatic cataract. JNMA J Nepal Med Assoc. 2008;47:179–85.
34. Staffieri SE, Ruddle JB, et al. Rock, paper and scissors? Traumatic paediatric cataract in Victoria 1992-2006. Clin Experiment Ophthalmol. 2010;38:237–41.
35. Bekibele CO, Fasina O. Visual outcome of traumatic cataract surgery in Ibadan, Nigeria. Niger J Clin Pract. 2008;11:372–5.
36. Gradin D, Yorston D. Intraocular lens implantation for traumatic cataract in children in East Africa. J Cataract Refract Surg. 2001;27:2017–25.
37. Brar GS, Ram J, et al. Postoperative complications and visual results in uniocular pediatric traumatic cataract. Ophthalmic Surg Lasers. 2001;32:233–8.
38. Cheema RA, Lukaris AD. Visual recovery in unilateral traumatic pediatric cataracts treated with posterior chamber intraocular lens and anterior vitrectomy in Pakistan. Int Ophthalmol. 1999;23:85–9.
39. Karim A, Laghmari A, et al. Therapeutic and prognostic problems of traumatic cataracts. Apropos of 45 cases. J Fr Ophtalmol. 1998;21:112–7.
40. Krishnamachary M, Rathi V, Gupta S. Management of traumatic cataract in children. J Cataract Refract Surg. 1997;23:681–7.
41. Knight-Nanan D, O'Keefe M, et al. Outcome and complications of intraocular lenses in children with cataract. J Cataract Refract Surg. 1996;22:730–6.
42. Bienfait MF, Pameijer JH, et al. Intraocular lens implantation in children with unilateral traumatic cataract. Int Ophthalmol. 1990;14:271–6.
43. Anwar M, Bleik JH, et al. Posterior chamber lens implantation for primary repair of corneal lacerations and traumatic cataracts in children. J Pediatr Ophthalmol Strabismus. 1994; 31:157–61.
44. Verma, N., J. Ram, et al. Outcome of in-the-bag implanted square-edge polymethyl methacrylate intraocular lenses with and without primary posterior capsulotomy in pediatric traumatic cataract. Indian J Ophthalmol 59: 347-51.
45. Eckstein M, Vijayalakshmi P, et al. Use of intraocular lenses in children with traumatic cataract in south India. Br J Ophthalmol. 1998;82:911–5.
46. Zou Y, Yang W, et al. Primary posterior chamber intraocular lens implantation in traumatic cataract with posterior capsule breaks. Yan Ke Xue Bao. 1995;11:140–2.
47. Vajpayee RB, Angra SK, et al. Pre-existing posterior capsule breaks from perforating ocular injuries. J Cataract Refract Surg. 1994;20:291–4.
48. Gupta AK, Grover AK, et al. Traumatic cataract surgery with intraocular lens implantation in children. J Pediatr Ophthalmol Strabismus. 1992;29:73–8.
49. Shah MA, Shah SM, Shah SB. Effect of interval between time of injury and timing of intervention on final visual outcome in cases of traumatic cataract. Eur J Ophthalmol. 2011;21: 760–5.
50. Rumelt S, Rehany U. The influence of surgery and intraocular lens implantation timing on visual outcome in traumatic cataract. Graefes Arch Clin Exp Ophthalmol. 2010;248(9): 1293–7.
51. Kuhn F, Morris R, Witherspoon CD, Mester V. The Birmingham Eye Trauma Terminology system (BETT). J Fr Ophtalmol. 2004;27:206–10.

52. American Society of Ocular Trauma. Ocular trauma score (OTS). http://www.asotonline.org/ots.html. Accessed 10 Dec 2008.
53. Taylor D, Vaegan, Morris JA. Amblyopia in bilateral infantile and juvenile cataract: relationship to timing of treatment. Trans Ophthalmol Soc UK. 1979;99:170–5.
54. Thouvenin D. Management of infantile cataracts: surgical technics and choices in lens implantation. J Fr Ophtalmol. 2011;34(3):198–202.
55. Morgan KS, Karcioglu ZA. Secondary cataracts in infants after lensectomies. J Pediatr Ophthalmol Strabismus. 1987;24(1):45–8.
56. Basti S, Ravishankar U, et al. Results of a prospective evaluation of three methods of management of pediatric cataracts. Ophthalmology. 1996;103(5):713–20.
57. Grossman SA, Peyman GA. Long-term visual results after pars plicata lensectomy-vitrectomy for congenital cataracts. Br J Ophthalmol. 1988;72(8):601–6.
58. Kanski JJ, Crick MD. Lensectomy. Trans Ophthalmol Soc UK. 1977;97(1):52–7.
59. Gessner B, Wiese S, et al. Results of pars plana lensectomy for childhood cataract. Ophthalmologe. 2004;101(9):901–6.
60. Abraham DI, Vitale SI, West SI, Isseme I. Epidemiology of eye injuries in rural Tanzania. Ophthalmic Epidemiol. 1999;6:85–94.
61. Yang ML, Hou CH. Clinical characteristics and surgical outcomes of pediatric cataract in Taiwan. Graefes Arch Clin Exp Ophthalmol. 2006;244:1485–90.
62. Magli A, Iovine A. Strabismus in developmental cataract. Eur J Ophthalmol. 2006;18:540–3.
63. Acuna OM, And K, Yen G. Outcome and prognosis of pediatric patients with delayed diagnosis of open-globe injuries. J Pediatr Ophthalmol Strabismus. 2009;46:202–7.
64. Spanou N, Alexopoulos L. Strabismus in pediatric lens disorders. J Pediatr Ophthalmol Strabismus. 2011;48:163–6.

Index

© Springer Science+Business Media New York 2015
M.A. Babizhayev et al. (eds.), *Studies on the Cornea and Lens*,
Oxidative Stress in Applied Basic Research and Clinical Practice,
DOI 10.1007/978-1-4939-1935-2